The Principles
of
Engineering Materials

The Principles
of
Engineering Materials

Craig R. Barrett / William D. Nix

Department of Materials Science and Engineering
Stanford University

Alan S. Tetelman

Materials Department
University of California at Los Angeles

Prentice-Hall, Inc.   *Englewood Cliffs, New Jersey*

Library of Congress Cataloging in Publication Data
BARRETT, CRAIG R.
    Principles of engineering materials

    Includes bibliographies
    1. Materials. I. NIX, WILLIAM D., joint author.
    II. TETELMAN, A. S., joint author. III. Title
    TA403.B24   620.1'12   72-8908
ISBN 0–13–709394–2

© 1973 by PRENTICE-HALL, INC. Englewood Cliffs, New Jersey

10   9   8   7   6   5   4

Printed in the United States of America

PRENTICE-HALL INTERNATIONAL, INC., *London*
PRENTICE-HALL OF AUSTRALIA, PTY., LTD., *Sydney*
PRENTICE-HALL OF CANADA, LTD., *Toronto*
PRENTICE-HALL OF INDIA PRIVATE LIMITED, *New Delhi*
PRENTICE-HALL OF JAPAN, INC., *Tokyo*

dedicated to

O. Cutler Shepard

for his early leadership
in establishing the study of materials
at Stanford

Preface

Materials Science—the study of engineering materials—has become a notable addition to engineering education during the past decade. It has gained its position in the curriculum in part because of the increased level of sophistication required of engineers in a rapidly changing technological society. The properties and characteristics of materials figure prominently in almost every modern engineering design, providing problems as well as opportunities for new invention, and setting limits for many technological advances. The study of solids and the relationship between structure and physical properties is therefore an important component of modern engineering education.

It is our belief that an introductory course on engineering materials should emphasize principles rather than empirical facts; consequently, this book does not give the student extensive knowledge about the myriad engineering materials now in existence; it does provide a conceptual framework for understanding the behavior of engineering materials by emphasizing important relationships between internal structure and properties. It attempts to present a general picture of the nature of materials and the mechanisms that act upon, modify, and control their properties. The subject matter in this book is meant to provide prospective engineers with sufficient background and understanding for them to appreciate existing materials and to exploit new materials development effectively.

The Principles of Engineering Materials is designed for a second- or third-year undergraduate course and therefore draws upon the student's knowledge of introductory physics, chemistry, and elementary calculus. The book has three main parts. The first part treats the internal structure of materials, both perfect and imperfect. It also deals with kinetic problems of diffusion, phase transformations, and structure control. The second and third parts treat the mechanical and electronic properties of materials showing the relation to structure. As these last two sections are essentially independent of one another, an instructor could choose to emphasize either structural materials by covering the first and second parts of the book or electronic materials by combining the first and third parts.

Something must be said about the systems of units employed in this text. We have purposely avoided using a single set of units (that is, cgs, mks, S.I., etc.) throughout the text for a very important reason. It is a fact that the units commonly used in the diverse areas of materials science belong to no common group. This is true in spite of international agreements. At present we are faced with a choice of either using a common set of units thereby promoting communication barriers between students and workers in various engineering disciplines, or of employing totally different units when discussing different topics. We have chosen the latter approach, judging that familiarity with commonly accepted units is of more value than religious adherence to a single set of units. Conversion tables for the various unit systems used in the text are contained in the appendix.

The authors wish to acknowledge the assistance and able advice of many colleagues at Stanford. This book is an outgrowth of the introductory materials science course offered at Stanford, to which many people have contributed substantially. Special thanks are due Professors R. A. Huggins, O. D. Sherby, D. A. Stevenson, and T. H. Geballe and to the many students who offered comments and suggestions for improvement of the preliminary version of the text. The authors are also greatly indebted to the many individuals who helped in the preparation of this book. We extend special thanks to Phyllis Parks, Anne Morse, and Trilby Nattkemper who typed the many versions of the manuscript and who helped in many other ways. We are also indebted to Margaret McAbee of Prentice-Hall for her able assistance in the preparation of this text.

Finally we wish to thank our families and many personal friends who indirectly contributed to this book through their love and friendship during the last five years.

Contents

6

INTRODUCTION TO THE MECHANICAL PROPERTIES OF SOLIDS   193

7

PLASTIC DEFORMATION IN CRYSTALLINE SOLIDS   225

8

STRENGTHENING MECHANISMS   251

13
ELECTRICAL PROPERTIES OF JUNCTIONS   427

14
MAGNETIC PROPERTIES OF MATERIALS   457

15
OPTICAL PROPERTIES   506

The Principles
of
Engineering Materials

1
The Nature of Materials Science

1-1 INTRODUCTION

During the last generation we have witnessed and benefited from the development of numerous new technological systems. A large fraction of our electrical energy now comes from nuclear powered generating plants. High-speed jet aircraft carry hundreds of passengers at speeds over 700 miles per hour today and will carry them even faster within a few years. Spacecraft have already visited the moon and will soon be capable of placing human life on other planets. Closer to home, high-speed computers affect many facets of our lives, from the processing of our checking accounts at the bank to controlling our speed on crowded freeways. Advances in the electronics industry have led to the development of satellite communication, color television, and the use of lasers in complicated medical operations. Each of these technologies has been advanced by the development of materials with new and exotic properties. For example, the development of the transistor and other solid state electronic devices such as tunnel diodes, solar batteries, and integrated circuits has facilitated the rapid advances in the electronics industry. Without these developments, it would be impossible to meet the size, weight, and power requirements of such devices as communication satellites, computers, and guidance systems for missiles.

These developments emphasize the importance of materials as the primary building blocks for engineering developments. On the one hand, the properties of materials have dictated nearly every design and every useful application that the engineer could devise. But with the present sophistication of our engineering science, it is no longer simply a question of being satisfied to design with existing materials. We are now requiring new materials with new properties to fit our designs. This is true in all fields of engineering, whether it be the mechanical engineer trying to design high-strength, lightweight casings for rocket hulls, the electrical engineer trying to design a solid state electronic device that will operate at temperatures above a few hundred degrees Celsius, or the nuclear engineer concerned with the materials needed to contain, control, and utilize a nuclear reaction. This search for new materials with improved properties now occupies an important position in the engineering world.

Along with the search for new materials has come the realization that effective usage of materials can be realized only when the engineer fully understands the various properties of materials. The reason for this is that *practically all of the useful properties of materials are strongly dependent on their internal structure.* The rather broad term *internal structure* is defined as the arrangement of electrons and atoms within a material. We shall see shortly that *for a material of given chemical composition, the internal structure is not constant,* but can vary greatly, depending on (1) how the material was manufactured (exactly what processing conditions were involved) and (2) under what conditions (temperature, pressure, exposure to radiation, etc.) the material is placed into service. This seemingly simple observation has some rather profound implications. It means that by altering the internal structure of a material in a controlled manner it is possible to effectively control the properties of the material. And because a single material may be treated to have different internal structures and correspondingly different properties, one material may be used for many applications, each calling for different physical properties.

To illustrate the fact that variations in internal structure can lead to large changes in the physical properties of a given material, we will briefly consider two examples.

In the first example, let us examine the mechanical strength of a carbon steel (iron + 0.8 % carbon) as a function of the thermal history of the material. Prior to testing, three specimens are heated to 900°C and then cooled to room temperature at different rates. The variation in yield strength (stress needed to produce permanent deformation) with cooling rate is given in Fig. 1-1. Note that as the cooling rate increases, the yield strength increases drastically. That is, even though all three samples have the same chemical composition and are tested at the same temperature, the mechanical strength varies with the thermal history.

In order to understand why these changes in strength occur, it is necessary to examine the internal structure of the material in some detail. For example, if the steel specimens are polished to produce a very smooth surface and then placed in contact with a suitable chemical solution, the etching action of the solution reveals that each specimen has a different structure. This structure is generally too fine to be resolved with the naked eye, and it is necessary to use the optical microscope (10–1600×) or the electron microscope (2000–100,000×) to resolve it. Consequently, this structure is called the *microstructure*.

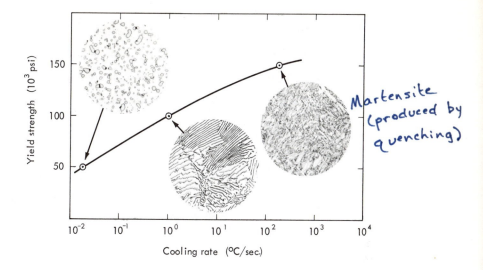

Martensite (produced by quenching)

FIG. 1-1   Variation of yield strength with cooling rate for an Fe + 0.8%C steel initially held
at 900°C. Photos courtesy U.S. Steel Corp. from *The Making, Shaping, and
Treating of Steel* (1971). Used with permission. And from *Physical Metallurgy for
Engineers* by Donald S. Clark and Wilbur R. Varney © 1952 by Litton Educational
Publishing, Inc., Reprinted by permission of Van Nostrand Reinhold Company.

Variations in structure occur over distances (or periods) of the order of microns
(1 micron $= 10^{-4}$ cm). Figure 1-1 illustrates the microstructure of the steel test
specimens cooled at the three different rates used to obtain the data shown in Fig.
1-1. It is apparent that a change in cooling rate produces a change in microstructure,
and that this variation in microstructure is responsible for the variation in mechanical strengths.

As a second example of the importance of internal structure on physical properties, let us consider the phenomenon of *superconductivity*. Superconductivity
occurs in certain metals and alloys at temperatures near absolute zero and is characterized by a loss of all electrical resistance. This means that an electric current
induced in a superconducting circuit will flow indefinitely without any resistance
loss. The superconducting properties of any material are very dependent on the
imposed current density and any applied magnetic fields. Superconductivity will
disappear if a sufficiently strong magnetic field or current density is imposed.
Figure 1-2 illustrates the relationship between current density and magnetic field,
showing the regions over which a material will exhibit superconductivity and normal conductivity (with resistance losses). For a given magnetic field, the maximum
current density that can be tolerated in the superconducting state is referred to as
the *critical current density*. Figure 1-3 shows that the critical current density is extremely sensitive to variations in internal structure. The data in this figure show
that the critical current density varies by a factor of as much as 1000 for a Nb-

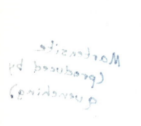

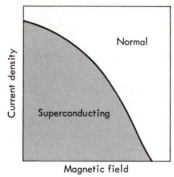

FIG. 1-2 Relation between current density, applied magnetic field, and the conductivity of a superconducting material.

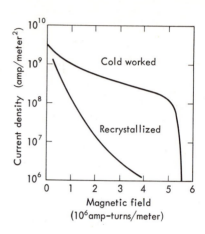

FIG. 1-3 Superconducting properties of Nb_3Zr for two different internal structures.

25% Zr alloy, depending on the exact thermal mechanical history of the wire. To briefly explain the nomenclature in this figure, the *recrystallized* sample has been heated to a temperature slightly below the melting point and then cooled to room temperature, whereas the *cold-worked* sample was given the same recrystallization treatment and then plastically deformed at room temperature. Apparently the plastic deformation has changed the internal structure of the Nb_3Zr alloy in such a fashion as to change the superconducting properties. In fact, examination of the microstructure with the aid of an electron microscope would reveal that recrystallized and cold-worked solids have vastly different internal structures. In Fig. 1-4 the cold-worked sample (in this case a Ni-Cr alloy that is not superconducting) contains a large number of defects (known as *dislocations*) in its crystalline structure, whereas the recrystallized sample has very few of these defects. On this basis we can conclude that the higher the density of these structural defects, the higher the critical current density. It is clear that since these structural defects can be induced in a material by permanent deformation, a simple method is available for changing the superconducting properties of this alloy.

1-2 ELEMENTS OF MATERIALS SCIENCE

The two examples described above offer a good starting point to attempt a definition of exactly what the field of materials science is and an explanation of how it relates to other engineering and scientific disciplines. In the broadest sense, materials science is concerned with the relations between the structure and properties of materials and the processes for altering the structure and, hence, the properties of solids. It brings together in a unified form the developments in physical metallurgy, ceramics, polymer science, and the physics and chemistry of solids in

 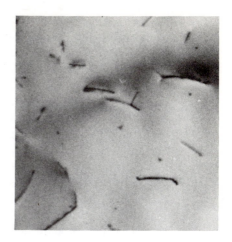

cold worked (20,000×)   recrystallized (20,000×)

FIG. 1-4   Transmission electron micrographs illustrating the cold-worked and recrystallized
structure of a Ni-Cr alloy. Each dark line represents a defect (dislocation) in the
crystalline structure. Photos courtesy W. C. Harrigan, Jr. Stanford University. Used
with permission.

order to understand and improve upon the engineering properties of solids. To
show exactly how materials science interacts with other fields of study, it is convenient to classify the various engineering and scientific disciplines in terms of the
dimensional range that is of most interest to the particular discipline (Fig. 1-5).
Consider the case of structural materials. Mechanical, civil, and aeronautical engineers are concerned with the stresses and strains acting on structural components
such as I-beams, wing spars, pressure vessels, and bridges. Typically, the minimum
size range of interest is $\approx 10^{-1}$ cm (e.g., the thickness of a thin pressure vessel wall),
while the maximum size can extend up to $\approx 10^5$ cm (the length of a long bridge).
Solid state physics, at the other extreme, deals with electron and atomic distributions (10^{-12} up to about 10^{-5} cm). Materials science attempts to bridge the gap
between such disciplines as solid state physics and structural engineering, since it

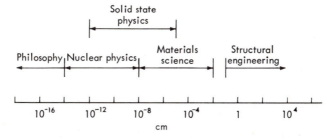

FIG. 1-5   Dimensional range of interest to scientific and engineering disciplines.

deals with structures whose periodicity varies from about 10^{-8} up to about 10^{-2} cm. Consequently, the fields of solid state physics and structural engineering are related, but the relation is indirect. Solid state physics provides part of the basis for understanding materials science (microstructure), and materials science in turn contributes significantly to the understanding of the properties of materials of interest in structural engineering.

We have just suggested that materials science provides a relationship between the findings of solid state physics and the problems of structural engineering. Similarly, materials science acts as a bridge between the concepts of particular interest in solid state physics and the properties of materials utilized in electronic devices by electrical engineers. For example, while the province of solid state physics is to predict the electronic properties that can be expected for a given arrangement of atoms (microstructure), it does not include the prediction of how a material should be processed to achieve this structural state. It remains, therefore, for materials science to act as a link between the predictions of solid state physics and the materials requirements of electrical engineering.

Clearly, then, materials science has two significant roles to play in the engineering world. First, and most obvious, is the practical role of emphasizing to engineers that a material is not homogeneous and structureless, but is an engineering structure in itself, composed of microscopic structural features, and that the properties of the structure will depend on how it is processed and under what conditions it is used. It is expected that a better understanding of materials science will lead to proper materials usage and less failures in service. Secondly, and perhaps more important, materials science fills the gap between atomic physics and chemistry, and the engineering properties of materials. Consequently, materials science, like its counterpart in living materials (biology), shows how the abstractions of the world of physics and chemistry can be related to the technological problems faced by the world of engineering. To emphasize these points more clearly, let us consider the interaction between materials science and engineering in some detail.

1-3 MATERIALS SCIENCE IN THE CONTEXT OF ENGINEERING

The relationship between materials science and engineering involves several levels of scientific and engineering sophistication. For the purpose of discussion we may group all such encounters into one of three domains: (1) intelligent use of materials, (2) relation between design and materials development, and (3) submicroscopic or molecular engineering. Since most complex engineering systems of today involve all three categories, it is worthwhile to illustrate these encounters with some examples.

Intelligent Use of Materials

Surely the utilization of materials in the construction of engineering systems is the most obvious relationship between materials science and engineering. In this respect, it is important that both the potentials and limitations of materials be factored into engineering endeavors. The following examples serve to illustrate how knowledge of the behavior of materials is important in engineering. These examples demonstrate the intelligent use of engineering materials.

The mechanical properties of materials depend sensitively on the temperature at which they are used. During World War II, for example, cracks were observed in 25% of the all-welded Liberty cargo ships that were constructed in the United States. Of the 4700 ships that were constructed, 233 were cracked so seriously that they were lost or considered to be unsafe. Figure 1-6 illustrates one of the fractures which occurred in the T-2 tanker *Schenectady*. Most of these failures occurred on very cold nights; the low temperature was considered to be an important factor in causing the brittle fracture. In fact, on testing the steel used in the construction of the Liberty ships, it was found that its fracture or crack resisting properties changed dramatically with temperature. Over a narrow range of only about 20°C, the steel changed from being a strong, tough, fracture resisting material to a brittle material that would break or crack quite easily. This example illustrates that the properties of materials may change appreciably with only small changes in environmental conditions.

A second example, which illustrates the need for awareness of the variation in properties of materials, involves the use of rubber and plastic in and about nuclear reactors. Rubber or plastic used for seals must be elastic and pliable if it is to be

FIG. 1-6   Photograph of the T-2 tanker, *Schenectady*, that failed at the pier. Photo courtesy John Wiley & Sons, Inc., from *Brittle Behavior of Engineering Structures* by Earl R. Parker (1957). Used with permission.

effective. These properties are due mainly to the fact that the molecules of the material are linked together into long chains. Neutron irradiation can cause the chains to be broken and can promote the formation of interchain bonds in such a way as to render the rubber hard and unpliable, making it useless as a seal material. In this case, the intelligent use of materials involves the realization that the structure, and hence the properties, of a rubbery substance can be greatly affected by a neutron flux.

The fact that the properties of materials depend on the service environment can also be useful to the engineer. With this in mind, the engineer may incorporate the change in properties into his design to produce a useful object. For example, a photoconductive material, whose electrical conductivity depends on the intensity and wavelength of the electromagnetic radiation incident on the surface, can be incorporated into an electrical circuit to act as a light detector. Or a thermistor, a semiconducting device whose electrical resistance varies greatly with temperature, can be used as a very accurate thermometer or as part of a temperature control system.

The above examples involve materials in engineering in a traditional way. Materials are used in a common sense of the word. The needed insight relates to the restrictions that are placed on the engineering function by the limitations of the material.

Relations Between Design and Materials Development

In many engineering systems the overall performance of the system is directly limited by the capabilities of the component materials. In these cases it is usually necessary to work near the limits of capabilities of the materials in order to meet engineering performance requirements. Consequently, in these special cases, engineering achievements are directly linked with improvements in the properties of materials. Because of this interrelation, engineering design and the development of new materials must be interconnected in such a way as to allow for the optimization of the performance of the overall system. The design and materials application may not be optimized separately when we are working close to the limit of capability of the material. To illustrate how this interrelation can be established, let us consider a specific example.

The magnetic materials used in the cores of transformers and inductors are known as *ferromagnetic materials*. The property of ferromagnetic materials that makes them useful is their ability to be strongly magnetized by the application of relatively weak magnetic fields. One of the desirable characteristics for ferromagnetic transformer cores is that the core material have as high a permeability (ratio of induced magnetization to applied field) as possible. The reasons for this are twofold. First, with a high permeability, lower applied fields are required to attain a given magnetization in the core. Furthermore, the increase in permeability generally results in a slight decrease in power losses associated with the alternating currents (alternating magnetic fields) that the core experiences.

Historically, iron was used for transformer cores until the early part of the twentieth century when it was found that iron-silicon alloys (iron with 3 to 4% silicon) had higher permeabilities and lower power losses than pure iron. The magnetization behavior of iron and iron–3% silicon is shown in Fig. 1-7. In addition, it was later observed that if iron-silicon was prepared in the form of thin sheets, with a very special thermal-mechanical history, the magnetic properties of the sheet varied with the direction of measurement (see Fig. 1-7). Specifically, the

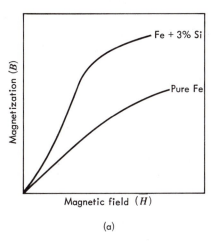
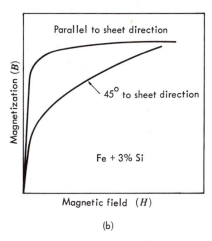

(a)   (b)

FIG. 1-7   Variation in magnetic permeability (B/H) of (a) pure Fe and Fe + 3% Si and (b) specially prepared Fe + 3% Si sheet.

permeability measured along the length of the sheet was different from that found at 45° to the long axis of the sheet. This variation in properties with orientation is common to many materials and is referred to as *anisotropy.*

The question one now must ask is, "Is it possible for a designer to make use of the fact that iron-silicon sheet, processed in a certain fashion, has anisotropic properties?" The answer is most emphatically, "Yes!" Utilizing the anisotropic iron-silicon sheet as a core material with the direction of the applied magnetic field parallel to the direction of easy magnetization for the sheet has allowed designers to create transformers with only one-third the power loss found in earlier transformers with iron cores. This example is important because it illustrates the necessity for the designer to realize that it is possible to process available materials in such a way that they will have unique properties. In this particular instance the properties depend on the direction in which they are measured, This concept is especially important when we are utilizing materials near the limit of their capabilities. For example, a 10% enhancement of a certain property in a given direction may allow for an appreciable increase in efficiency in the overall system. This will be true whether we are considering ferromagnetic materials for cores of transformers or composite materials such as fiberglass or plywood for use as structural members. This type of design flexibility is possible only because the properties of materials can be engineered through structural control.

Perhaps the most significant relationship between materials science and engineering is in the realm of materials engineering or the *development of materials with new and unique engineering properties.* Rather than tailoring the use of materials to their properties, we are now in a position to synthesize new materials, often of dissimilar elements, and obtain a much broader range of properties, thereby tailoring the material to suit our needs. In essence, the importance of materials engineering was demonstrated when we considered the variation of the yield strength of steel as a function of thermal history, or how cold work influences the superconducting properties of Nb-25% Zr wire. In both these instances, by suitably engineering or changing the internal structure, it is possible to bring about a significant change in properties. To give one further example of materials engineering, let us consider the case of fiber-reinforced composite materials. Here we will try to point out the principles involved and indicate the wide variety of properties that can be engineered into fiber composites.

It has long been known that improved mechanical properties can be achieved by reinforcing materials with fibers. For example, ancient cultures reinforced clay bricks with straw and took advantage of the composite properties of bamboo. And for the last several decades extensive use has been made of fiberglass-reinforced plastics. The present-day interest in fiber composites stems from the fact that materials are generally far stronger in fiber form than in bulk or massive form. For example, glass fibers are strong because they have fewer surface flaws due to the forming process. Vapor-deposited materials may be in the form of single crystals or whiskers which are relatively free from harmful imperfections. And thin metal wires are strong because the material has been extensively deformed. Some typical fiber strengths are given in Table 1-1.

While the high strengths of the fibers make them very attractive, because of their small size (generally <0.01 cm in diameter) and the fact that they are often brittle and crack easily, they are not, by themselves, useful structural materials. To produce a useful engineering material from these fibers, they must be bonded together by a glue or matrix material. Here the matrix must be able to deform sufficiently without fracturing to transfer the load to the fibers. Thus, in the fiber composite the fiber serves to support the bulk of the applied load, while the matrix serves to bind the fibers together to protect them from mechanical and chemical damage and to distribute the load to the individual fibers. The principal concept embodied in the design of fiber composites is identical to that of all other man-made

TABLE 1-1

Representative Tensile Strengths of Fibrous Materials

Material	Tensile strength (psi)	Material	Tensile strength (psi)
Graphite whiskers	up to 3.5×10^6	Steel wire	up to 0.45×10^6
Silica fibers	up to 3.5×10^6	Fiberglass	up to 0.2×10^6
Alumina whiskers	up to 2.2×10^6	Nylon	up to 0.008×10^6
Iron whiskers	up to 1.9×10^6	Spider thread	up to 0.027×10^6
Tungsten wire	up to 0.5×10^6	Hemp rope	up to 0.015×10^6

composite materials, namely, that composites require the joining of two or more different materials for the purpose of gaining advantageous characteristics from each, or of overcoming disadvantageous characteristics of each. The wide variety of properties obtainable from fiber composites is immediately evident when one considers the different properties of the various fibers and matrices and the added variables of fiber size, volume fraction, and distribution.

The attractive properties of fiber composites include improved strength-to-density ratios, improved high temperature strength, increased thermal shock resistance, and improved oxidation and corrosion resistance. To give an idea of the strength-to-density ratios attainable, some typical values are listed in Table 1-2 and are compared with that for a high strength steel.

TABLE 1-2

Room Temperature Tensile and Specific Strengths for Various Fiber Reinforced Materials

Matrix	Fiber	Tensile strength (10^3 psi)	Strength/density (10^3 in.)
Al	B	110	1690
Al	SiO_2	118	1340
Al	Al_2O_3	161	1425
Ni	B	384	1470
Ni	Al_2O_3	171	600
Ag	Al_2O_3	232	720
Epoxy	B	320	4320
Epoxy	Glass	250	3120
Steel	—	220	770

Thus, even though the fibers by themselves have no useful engineering application and the matrix may have low strength, if the two are incorporated into a fiber composite we can obtain the utmost of their properties. This concept of materials development suggests a simple way to use the strongest known materials in useful shapes. Certainly there are problems involved in synthesizing various fiber composites, such as the chemical compatability and stability of fiber and matrix and attaining the necessary interfacial fiber bond strength matrix, but, nevertheless, the technique has been established. All that remains now is for new composites to be developed and exploited with the understanding of materials behavior and the application of competent materials engineering.

Submicroscopic or Molecular Engineering

Thus far in this introductory chapter we have considered the ways in which materials are utilized in engineering and the ways in which materials developments can lead to new engineering applications. In every case the engineering application has involved a macroscopic component of a more complex technological system. We shall now consider the role of materials in a type of engineering which is sometimes called *submicroscopic* or *molecular* engineering. As the names imply, this branch of engineering is concerned with devices and instruments that are designed and built within the context of a molecule or single crystal. The building blocks for this type of engineering are not the traditional engineering system components such as wing spars, I-beams, electric windings, vacuum tubes, and resistors. Rather, they are such atomic level features as *impurity atoms*, *surfaces*, and *trapping centers*. To illustrate this new form of engineering and the importance of materials in it, let us consider the various ways a radio frequency controlled switch may be constructed.

For many years, the conventional method for constructing electronic devices involved the use of vacuum tubes and other electronic components such as resistors, capacitors, and inductors. Figure 1-8(a) illustrates the arrangement of the electronic components for a radio receiver. A receiver similar to this would be needed, for example, to actuate a mechanism to open a garage door by remote control. Notice that the size of this device, is about 15 in. × 12 in. × 6 in. With the advent of the *transistor* (a single-crystal device which replaces the vacuum tube function in electronic circuits), it became possible to construct electronic circuits at great savings in both space and cost. A transistorized circuit, which performs a function similar to the vacuum tube circuit just mentioned, is shown in Fig. 1-8(b). Notice that the use of transistors has allowed the receiver to be greatly reduced in size. It should be noted that the components shown here would be packaged in a more compact manner in a commercial device.

In the past few years, the concept of the transistor has been generalized in such a way that whole circuits are now constructed within small single crystals of silicon or germanium. This form of engineering is called *integrated solid state circuitry*. Devices of this type are constructed by diffusing impurity atoms into various regions of single crystals of silicon or germanium.

The solid state circuit shown in Fig. 1-8(c) performs the same function as the vacuum tube or transistorized circuits just described. When the solid state device is fully constructed, there are regions in the single crystal which exhibit the properties of some of the electronic components in the vacuum tube circuit and perform that same function. For example, a transistor can replace a *triode* in the vacuum tube circuit. In addition, the capacitors and resistors can also be built into the single crystal circuit. The conductors that connect the elements in the solid state circuit are made by vapor-depositing thin metallic strips along the surface of the device.

It should be noted that there is not always a one-to-one relation between the electronic components in a solid state circuit and a conventional vacuum tube circuit or transistor circuit. The electronic performance of solid state circuits is generally more precise and predictable; hence, solid state circuitry is often much simpler than conventional circuitry and can be designed with fewer components.

A solid state circuit, which performs the same function as the other circuits, is so small that it may easily be placed on the head of a small nail. The integrated circuit shown in Fig. 1-8(c) is also shown in Fig. 1-8(b) for size comparison. Ordinarily these solid state circuits are packaged in small cans. Even after such

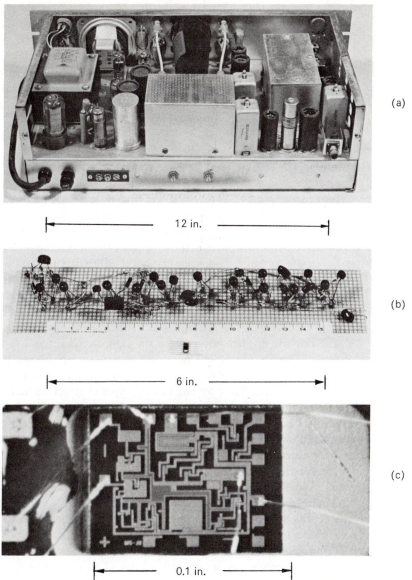

(a)

(b)

(c)

FIG. 1-8 Illustration of a radio receiver with conventional electronic components (a), and a radio frequency controlled switch with transistorized components (b), and with a solid state integrated circuit (c). The integrated circuit is also shown in (b) for size comparison. Photos courtesy R. A. Menezes, Stanford University and G. Perlegos and J. D. Meindl, Stanford University. Used with permission.

packaging, the solid state circuit is very much smaller than either of the other circuits. The integrated circuit shown here is used to turn on and off battery powered devices implanted in the human body. By being able to turn the implanted devices on and off by remote control, we can make more efficient use of the power available in the implanted battery.

This example demonstrates that microstructural features within a single crystal can be engineered in such a way as to produce a useful engineering device. The reduction in size and weight of such devices allows the miniaturization of engineering systems and is partially responsible for opening new vistas for engineering in space, medicine, and communications.

1-4 SCOPE OF THE TEXT

In the most general sense the field of materials science treats all classes of materials (solids, liquids, and gases), but in this text we shall limit our attention to the most common class of engineering materials—solids. The text does not attempt to emphasize the differences in the properties of various types of engineering materials (metals, organic materials, ceramics, composite materials, etc.), but instead presents a general picture of the nature of materials and the mechanisms that control their properties. We have attempted to present a conceptual framework that will lead to an understanding of the internal structure of solids as well as to the physical basis of the important engineering properties and their relationship to structure.

The book is divided into three main sections. Chapters 1–5 treat the problem of internal structure of both pure and impure materials. Specifically, Chapters 2 and 3 deal with atomic arrangements in materials and the presence of defects in these arrangements. Next we consider the types of structure that exist under a given set of conditions. First there are thermodynamic considerations, which are discussed in Chapter 4. The laws of *thermodynamics* enable us to predict the microstructure that would exist under a given set of thermodynamic conditions: specifically, temperature, pressure, and chemical composition, provided that equilibrium has been achieved. In this sense thermodynamics is an "idealistic" science because equilibrium in materials is never actually reached. *Kinetics*, the subject of Chapter 5, is more "realistic," on the other hand, as it permits us to deal directly with the rate at which equilibrium is approached and to predict the microstructure that is formed in a *finite* period of time.

The differing roles of thermodynamics and kinetics may be illustrated with the following example. According to the laws of thermodynamics, alumimun in the metallic state is unstable in air at room temperature; that is, aluminum should combine with oxygen to form Al_2O_3, as this substance is more stable than elemental aluminum when in contact with oxygen. However, we know that a piece of aluminum does not readily react to form Al_2O_3 even after a very long time. The reason for this is that once a thin layer of oxide is formed on the surface, further oxidation is very slow because diffusion of oxygen and aluminum through the oxide is extremely slow. Thus, even though Al_2O_3 is the equilibrium state, pure aluminum will exist almost indefinitely because the rate of approach to equilibrium is limited by the kinetics of the oxide formation. The study of kinetics is very important in materials science because it deals with the conditions that determine what reactions can, in fact, occur during processing or exposure to a given environment.

After considering the factors that are important in determining the nature of the microstructure, we turn our attention to the influence of microstructure and operating conditions on the mechanical properties of materials (Chapters 6–10). Typical questions to be answered are: Why does the mechanical strength of many materials, such as metals and polymers, depend sensitively on temperature? Why are glasses brittle while metals deform in a ductile fashion? What is the influence of a finely dispersed second phase on the mechanical properties of materials? Once we have established the structure–property relationships we reverse the process and ask how we should alter the chemical composition and process a material to obtain a given set of desired mechanical properties (Chapter 9). Here we combine the principles of thermodynamics and kinetics with a knowledge of structure–property relationships to synthesize a desired material. This process of materials synthesis is the ultimate step in the application of materials science to engineering.

The last section, comprising Chapters 11–15 is concerned with the electronic properties of materials. In Chapters 11 and 12 we consider some of the possible distributions of electrons in solids. Distinctions are drawn between the various types of electrical conduction processes which take place in solids and the resultant classification of materials as *conductors, semiconductors,* or *insulators.* In Chapters 13–15 we relate the electrical, magnetic, and optical properties of materials to their electronic structure and microstructure. To give a few examples, we shall treat such problems as electrical properties of junctions, the differences between hard and soft magnets, photoconductivity, and the scientific basis of the process known as *light amplification* by the *stimulated emission* of *radiation* (lasers).

This introductory text is not intended to develop a deep knowledge of current-day engineering practice. Rather, it is intended to provide a perspective and understanding which will be useful to students in acquiring such knowledge. Neither does the text cover the structure–property relations for all fields of professional engineers. However, it does provide a broad base for most of the engineering sciences and a foundation for developing and applying materials in the years to come.

2
The Structure of Perfect Solids

2-1 INTRODUCTION

It was pointed out in the previous chapter that an explanation of the properties of materials requires a detailed knowledge of the atomic structure. This rather broad term *atomic structure* includes the electronic structure of the atoms making up the material, the binding forces between the atoms, and the three-dimensional arrangement of the atoms. So important are these three factors that materials with identical chemical compositions but with different atomic arrangements can have vastly different properties. This fact was illustrated in Chapter 1, where it was shown that the strength of a carbon-steel was very dependent on the rate of cooling from an elevated temperature. In this instance the strength is determined by the exact atom arrangement, which in turn depends on the rate of cooling. If the exact atom arrangement can have such a profound effect on the observed properties, then a logical starting point in our study of materials is to consider the possible ways in which atoms can join together.

In this chapter we shall find that there are some relatively simple principles governing the agglomeration of atoms and that with these principles we can explain large-scale atom arrangements. The term *large scale* does not refer to the macroscopic size of the material under consideration, but rather to the number of atoms taking part in the synthesis of even the smallest object. To give some idea of the numbers involved, in a cubic centimeter of material there are something like 10^{22} atoms. Fortunately, the arrangement of atoms in most materials is rather simple, and the structure can be specified by describing the arrangement of a small group of atoms much in the same way that a wallpaper pattern can be subdivided into a small design which, when repeated over and over, creates the whole pattern.

An appropriate starting point for our discussion of the *arrangement* of atoms in materials is to consider briefly the *structure* of atoms. This is necessary because the forces that bind atoms together find their origins in the electronic structure of atoms. These interatomic forces result from the rearrangement of a few of the outermost electrons from each atom into a lower-energy configuration when another atom or group of atoms is nearby. This rearrangement lowers the energy of the group of atoms and provides the binding force between them. The following sections deal with the nature of the arrangements of electrons in materials and the characteristics of the interatomic forces that develop.

2-2 ELECTRONIC STRUCTURE OF THE ATOM

A complete description of the motion of electrons in atoms requires the use of quantum mechanics, or in the form we shall use it, wave mechanics. Wave mechanics differs from the more familiar classical mechanics in that the latter study assumes a complete knowledge of the position and motion of any object under consideration. For an event such as the motion of a golf ball this is a reasonable assumption. However, for the motion of small particles, such as electrons, it can be shown that the uncertainty involved in specifying both the position and motion of the particle is appreciable. This basic fact, usually referred to as the *Heisenberg uncertainty principle*, can be stated mathematically in the form

$$\Delta x \cdot \Delta p = \frac{h}{2\pi} \tag{2-1}$$

where Δx is the uncertainty in position, Δp is the uncertainty in momentum, and h is Planck's constant (6.63×10^{-27} erg·sec).

As an example of the uncertainty principle, consider the motion of an electron with a velocity corresponding to an energy of about 1 electron volt (eV). As will be discussed in Chapter 11, this is a typical value of energy for an electron in a solid. If the uncertainty of the energy is 10^{-3} eV, what is the uncertainty in position Δx? Expressing the kinetic energy of the electron as $\frac{1}{2}mv^2$, where m is the electronic mass and v is the velocity, we can calculate the uncertainty in velocity corresponding to the uncertainty in energy as $\Delta v = 3 \times 10^4$ cm/sec. Since $\Delta p = m\Delta v$, we can calculate Δp and use Eq. (2-1) to evaluate Δx. Following this procedure, we find $\Delta x \approx 4 \times 10^{-5}$ cm. Although this value of Δx seems fairly insignificant in terms of the size of objects we are normally concerned with, the diameter of an atom is only about 10^{-8} cm. Thus, with an accurate measure of the energy of the electron we find that the uncertainty in specifying its location is about 4000 atomic diameters! And if we specify the velocity any more accurately the uncertainty in position becomes even larger.

Clearly, if we cannot specify both the position and velocity any more accurately than this, we cannot expect to predict the electron's position accurately at any later time by employing the classical laws of motion. To overcome this problem we find it necessary to resort to a probability or statistical theory. In this theory we do not deal with the exact values of position or velocity of the electron, but rather with the probability that these values lie within a certain range. An example of this concept is given in Fig. 2-1, where we consider the position of a moving electron

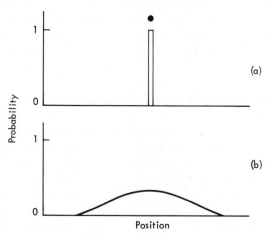

FIG. 2-1   The position of an electron according to (a) classical mechanics and (b) quantum mechanics

according to classical and quantum mechanics. Classical mechanics assumes the knowledge of the *exact* position of the electron, whereas wave mechanics indicates that there is finite probability of finding the electron over a considerable distance. Thus, wave mechanics cannot tell us exactly where the electron is, nor can it tell us exactly what path the electron follows. But it does give us a time average, similar to a long time exposure of the motion of the electron.

To formulate the motion of an electron in terms of wave mechanics we need to consider the wave nature of particles. DeBroglie first proposed that the properties of a particle can be described by a wave with a wavelength λ given by

$$\lambda = \frac{h}{p} \tag{2-2}$$

where p is the momentum of the particle. This postulate has been verified by numerous experiments, perhaps the most dramatic being the experiment of Davison and Germer showing that electrons can be diffracted in the same manner as electromagnetic waves. (See Fig. 2-39.) This wave nature of electrons led Schrödinger in 1926 to postulate that the motion of electrons in an atom can be described by a wave equation similar to the type used to describe waves in vibrating strings. For the one-dimensional case, the Schrödinger wave equation has the form

$$\frac{d^2\psi}{dx^2} + \frac{8\pi^2 m}{h^2}(E - V)\psi = 0 \tag{2-3}$$

where ψ is a quantity called the *wave function* of the electron, E is the total energy, and V is the potential energy of the electron. The wave function ψ is a function of x and has the unique property that $|\psi|^2\, dx$ gives the probability of finding the electron in the interval between x and $x + dx$. The plot in Fig. 2-1(b) is thus a plot of $|\psi|^2$ versus x.

Equation (2-3) can be easily extended to treat the three-dimensional motion of electrons in an atom. If this is done and we seek solutions for ψ we find the very interesting result that there are only specific values of E for which there is a solution. This means that there is not a *continuous* range of energy states available to the electron, but instead only a *discrete* set of possible energy states, Rather than demonstrate this concept by employing the complicated mathematics necessary to solve Eq. (2-3), we shall make use of a simple model to approximate the more detailed calculations for low atomic number elements. In this model, usually referred to as the *Bohr model*, we assume that the electrons are moving in circular orbits of radius r about the nucleus. In addition, we postulate that the angular momentum mvr of the electron be equal to an integral multiple of $h/2\pi$ or

$$mvr = \frac{nh}{2\pi} \tag{2-4}$$

where n is an integer. Equation (2-4) is nothing more than the requirement that the circumferential path length of the electron be equal to an integral number of its deBroglie wavelengths λ, where λ is given by Eq. (2-2). This is analogous to the vibrating string problem, where there is always an integral number of standing waves (the position of zero amplitude does not change with time) in a vibrating string that is clamped at both ends and vibrating in resonance. In Eq. (2-4) v and r are not independent variables, as we require for a stable orbit that the centripetal (electrostatic) force just balance the centrifugal force such that

$$\frac{Ze^2}{r^2} = \frac{mv^2}{r} \tag{2-5}$$

where Z is the atomic number (or the number of positive charges in the nucleus). The total energy of the electronic state E is then the sum of the kinetic energy $\frac{1}{2}mv^2$ and the potential energy $-Ze^2/r$. Combining Eqs. (2-4) and (2-5), we find

$$E = -\frac{2\pi^2 Z^2 e^4 m}{h^2 n^2} = -\frac{13.6 Z^2}{n^2}\,\text{eV} \tag{2-6}$$

where the integer n may take on any nonzero value. The negative energy of the electronic state means that the electron has less energy when it is in the atom than when it is free. The energy of a free electron is taken as zero. This means that the electron is bound to the atom and that it takes energy to remove the electron or ionize the atom.

The integer n is known as the principal quantum number and in the Bohr model uniquely defines the energy of the electron. The physical picture of the Bohr atom can be thought of as shells of electrons in circular orbits about the nucleus, the smallest orbit corresponding to the lowest energy state $n = 1$, and so on. Since there are only certain "allowed" energy levels, an electron can change its energy only by making a transition between these allowed energy levels. Accompanying the transition of an electron from one energy level to another must be either the absorption or emission of energy, depending on the direction of the transition. The electron must absorb energy to move to a higher energy state (less negative energy) and must emit energy when it moves to a lower energy state. For example, during a transition to a lower energy state an electron will emit a discrete quantum of energy (electromagnetic radiation) known as a *photon*, where the wavelength λ of the photon is related to the energy change

$$\Delta E = \frac{hc}{\lambda} = h\nu \tag{2-7}$$

where c is the speed of light and ν is the frequency. The discrete nature of these electronic transitions can serve to identify or differentiate between various elements. From Eqs. (2-6) and (2-7) we see that the possible energy changes ΔE for an electron depend only on Z, the atomic number. Thus, if we analyze the photons emitted by an excited atom and determine their wavelength and the corresponding values of ΔE, we can identify the element. This procedure is the basis for spectrographic analysis, a procedure in which electrons in an atom are excited or raised to high-energy states by exposure to the intense heat of an electric arc.

An important consequence of electronic transitions between the innermost electron shells of elements of atomic number greater than about 20 is that the photons emitted have a wavelength in the range of 0.1–10 A (constant λ for a given element) and are classified as X rays. X rays are important in the study of materials because the wavelength of X rays is comparable to the spacing between atoms in solids, and the resolution or smallest object separation distance to which any radiation can yield useful information is about equal to the wavelength of the radiation. Thus, the diffraction of X rays by materials can be used to determine the arrangements of atoms within those materials. This subject will be treated in more detail in Section 2-8.

EXAMPLE 2-1 Calculate the energy and wavelength of a photon emitted by a hydrogen atom when an electron moves from the $n=3$ state to the $n=1$ state.

Solution: From Eq. (2-6) we can express the difference in energy between the n_1 and n_2 energy levels as

$$\Delta E = \frac{13.6Z^2}{n_1^2} - \frac{13.6Z^2}{n_2^2}\,\text{eV}$$

or

$$\Delta E = \frac{13.6}{1} - \frac{13.6}{9} = 12.1\ \text{eV} = 1.94 \times 10^{-11}\ \text{erg}$$

and

$$\lambda = \frac{(6.63 \times 10^{-27}\ \text{erg}\cdot\text{sec})(3 \times 10^{10}\ \text{cm/sec})}{1.94 \times 10^{-11}\ \text{erg}} = 1025\ \text{A}$$

In reality, atoms are more complicated than the simple Bohr model. Electrons can have noncircular, or elliptical, orbits about the nucleus, and those electrons with small radii tend to screen or shield those further out from the full positive charge of the nucleus. The more precise calculations of the electronic structure using equations similar to Eq. (2-3) show that the state of motion of an electron about the nucleus is characterized by not one but by four quantum numbers, n, l, m_l, and m_s.

1. The principal quantum number n is the major determining factor for the energy; it can take on only integer values $n = 1, 2, 3, 4, \ldots$. The corresponding electronic shells are sometimes called the $K, L, M, N, \ldots$ shells.

2. l determines the ways in which the orbital angular momentum is quantized and can vary in integers from 0 to $n-1$. The value of l has a small effect on the energy level. To simplify the notation it is usual to specify an electron with $l=0$ as an s electron or as an electron in an s state. Energy states with $l = 1, 2, 3, \ldots$ are called p, d, f, $\ldots$ states. Electrons in an s state have zero angular momentum and hence move in all directions with equal probability; they essentially have a spherically symmetric orbit. Electrons in p, d, or f states have nonzero angular momentum, move preferentially in certain directions, and have a certain directionality in their orbit. Figure 2-2 illustrates the electron probability distribution for electrons in s and p states, showing the directionality of the p electrons. This observation will be very important when we consider the directionality of atomic bonds.

3. m_l, the magnetic quantum number, specifies the orientation of the angular

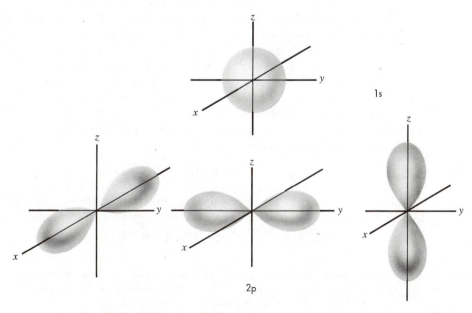

FIG. 2-2   Electron charge distribution for the 1s and 2p electrons

momentum in space; m_l can take on only integer values between $-l$ and $+l$.

4. The quantum number m_s arises because the electron itself has an angular momentum usually referred to as the *spin*. That is, the electron behaves as if it were not only orbiting the nucleus but also spinning on its own axis. The quantum number m_s describes the possible angular momentum components of the spin and has only two possible values: $+\frac{1}{2}$ or $-\frac{1}{2}$.

Each set of quantum numbers n, l, m_l, m_s specifies a particular energy state or quantum state for an electron. These quantum numbers uniquely define the state of motion of the electron. Now to specify the total electronic structure of the atom we need to know how many electrons are in each state and which of the many available quantum states are occupied. The answer to the first question can be found in a fundamental postulate of quantum mechanics, the *Pauli exclusion principle*. This principle states that *no two electrons in a given atom can have the same set of quantum numbers*. Thus, there can be at most only one electron in each quantum state. To determine which quantum states are occupied, we must resort to the general principle that dynamical systems tend toward the states where they have the lowest energy, or, in more specific terms, the electrons will occupy those quantum states of lowest energy. Now the energy of a quantum state depends primarily on the principal quantum number n, with the energy increasing with increasing n.* In addition, for a given value of n, or for a given electron shell, the energy also depends on the quantum number l, increasing with increasing l. Thus, the energy of the $n=2$, $l=0$ state is less than that of the $n=2$, $l=1$ state. To simplify the notation we will use the letter notation for the quantum number l such that the $n=2$, $l=0$ and $n=2$, $l=1$ states become, respectively, the $2s$ and $2p$ states. The electronic structure of an atom of atomic number Z can be then specified by putting one electron in each of the Z lowest energy quantum states.

Allowing for all the possible permutations of the quantum numbers, l, m_l, and m_s, we find that for each value of n, or for each electron shell, there are $2n^2$ quantum states. Thus, for a copper atom with 29 electrons, both quantum states in the $n=1$ shell would be filled, all eight quantum states in the $n=2$ shell would be filled, all 18 quantum states in the $n=3$ shell would be filled, and one of the 32 states in the $n=4$ shell would be filled. The shorthand notation for the electronic structure of the copper atom would be

$$1s^2\ 2s^2\ 2p^6\ 3s^2\ 3p^6\ 3d^{10}\ 4s$$

The integers refer to the principal quantum numbers; the s, p, and d refer to the orbital quantum numbers; and the superscripts refer to the number of electrons with the same principal and orbital quantum numbers.

---

*There are some exceptions to this rule. For example, the $4s$ state sometimes has a lower energy than the $3d$ state, the $5s$ has a lower energy than the $4d$, etc. A complete list of the relative energies of the different quantum states is given in Appendix 1.

2-3 ELECTRONIC STRUCTURE AND CHEMICAL PROPERTIES

A simple relationship exists between the chemical properties of atoms and their electronic structure. We can get some hint of this relationship by examining the electronic structure of the inert gases, Ne, Ar, and Kr:

Ne $1s^2\, 2s^2\, 2p^6$

Ar $1s^2\, 2s^2\, 2p^6\, 3s^2\, 3p^6$

Kr $1s^2\, 2s^2\, 2p^6\, 3s^2\, 3p^6\, 3d^{10}\, 4s^2\, 4p^6$

All of these elements have completely filled *s* and *p* states in the outermost shell. The resultant electron charge distribution is spherically symmetric; that is, even though each of the *p* electrons has some directionality in its orbit (Fig. 2-2) the resultant charge distribution from the six *p* electrons is spherically symmetric. The reluctance of these elements to undergo chemical reactions of any type indicates that when the outer *s* and *p* states are completely filled the electronic structure is stable. There is no tendency for the atom either to give away or accept new electrons.

We find in practice that all atoms tend toward the state in which their outer *s* and *p* states are filled. Elements with only one or two electrons in the outer shell, Li, Na, Mg, Cu, Ag, Au, etc., tend to give up these electrons during chemical reactions. These atoms have a relatively low *ionization energy*—the energy necessary to remove an electron from the atom. Elements needing only one or two electrons to complete their outer shell, O, F, Cl, Br, I, etc., tend to undergo chemical reactions in which they accept electrons. As might be expected, these elements have a high ionization energy and a reasonably large *electron affinity*—the energy given up when an initially free electron is added to their outer shell. Typical values of ionization energy and electron affinity are given in Table 2-1. Those elements that give up electrons to complete their outer shell are sometimes termed *electropositive*, and the elements that accept electrons are termed *electronegative*. Electropositive elements are usually metallic in nature, while electronegative elements are usually classified as nonmetallic. Elements with three, four, or five electrons in their outer *s* and *p* shells do not always behave as either metals or nonmetals. In some reactions they donate electrons, while in others they accept electrons. Typical elements in this category are C, Si, Ge, Sn, and Bi.

EXAMPLE 2-2

Using the Bohr model of the atom, calculate the first ionization energy of Na and compare this to the tabulated value given in Table 2-1.

Solution:

The outermost electron in the Na atom is a *3s* electron with an energy given by Eq. 2-6 as

$$E = -\frac{13.6Z^2}{n^2} = -\frac{13.6(11)^2}{(3)^2} = -183 \text{ eV}$$

Thus, it would require 183 eV to remove the one *3s* electron. This calculated value is much higher than the ionization energy of 5.2 eV listed in Table 2-1 because in our calculation we have neglected the fact that the ten inner electrons screen the $3s$ electron from the nucleus. The effective positive charge on the nucleus as seen by the $3s$ electron is then only about $+2e$.

TABLE 2-1

First Ionization Energy, *I*, and Electron Affinity, *A*, at 0°K

Atomic Number	Element	*I*	*A*
1	H	13.6 eV	0.71 eV
2	He	24.6	—
3	Li	5.4	0
4	Be	9.4	—
5	B	8.3	—
6	C	11.2	—
7	N	14.5	—
8	O	13.6	—
9	F	17.4	3.6
10	Ne	21.6	—
11	Na	5.2	0
17	Cl	13.0	3.7
19	K	4.3	0
35	Br	11.9	3.5
53	I	10.5	3.2
55	Cs	3.9	0

2-4 INTERATOMIC FORCES

The different types of bonds formed between atoms are usually grouped into two general categories, depending on the magnitude of the interatomic forces that develop. Strong bonds that are a result of large interatomic forces are set up in *ionic*, *covalent*, and *metallic* bonds. Weak bonds or small forces of attraction are found in *Van der Waals* and *hydrogen* bonded substances. In most materials the interatomic bonds are predominantly of one type although quite often we find mixtures of several types.

Ionic Bonding

Ionic bonds form between strongly electropositive elements (metals) and strongly electronegative elements (nonmetals). In this type of bond the electropositive atom gives up one or more electrons to the electronegative atom, producing two oppositely charged ions. The bonding forces are then a result of the electrostatic attraction between the ions. A typical example of a substance showing ionic bonding is LiF. The electronic structures of Li and F are shown in Fig. 2-3. In Li (the metal) the $2s$ electron is weakly bound to the atom (the two $1s$ electrons screen it from the nucleus) and has an ionization energy of only 5.4 eV. F, on the other hand, has a large electron affinity, each F atom giving up 3.6 eV when

Li $1s^2 2s^1$ F $1s^2 2s^2 2p^5$

FIG. 2-3   Electronic structure of Li and F

an electron is added to its outer shell. The pertinent ionization reactions are then

$$Li + 5.4\,eV \rightarrow Li^+ + e^-$$
$$F + e^- \rightarrow F^- + 3.6\,eV$$
$$Li + F + 1.8\,eV \rightarrow Li^+ + F^-$$

We can get some idea of the bond energy by considering the coulombic attraction between the two ions. If the ions have charges of plus and minus e, then the attractive force is e^2/r^2 where r is the separation distance. The decrease in potential energy, V, as they approach one another from infinity to some distance r_0, is

$$V = \int_\infty^{r_0} \frac{e^2}{r^2}\,dr = -\frac{e^2}{r_0} \approx -7.2\,eV \tag{2-8}$$

Writing this as

$$Li^+ + F^- \rightarrow LiF + 7.2\,eV$$

and combining with the ionization reactions above we have

$$Li + F \rightarrow LiF + 5.4\,eV$$

Thus the electrostatic attraction of the ions produces a decrease in energy that more than compensates for the energy required to ionize the atoms and a stable ionic bond is formed.

The ions will not approach closer than some distance r_0 (about 2A for LiF) because, as they get very near, their completely filled electron shells begin to overlap and a repulsive force is set up. This repulsive force increases much more rapidly with decreasing separation than the electrostatic attractive force, and consequently the repulsive interaction dominates the change in energy of the system at small atomic separations. This is illustrated in Fig. 2-4, where the attractive energy is proportional to $1/r$ [Eq. (2.8)] and the repulsive energy is proportional to $1/r^m$, where $m \approx 10$. At an interatomic separation r_0 the potential energy of the atom pair is a minimum, and correspondingly this distance is their equilibrium separation distance.

The decrease in potential energy as the atoms come together (V_0 in Fig. 2-4) is a measure of the bonding energy or the energy to dissociate the atom pair into separate ions. Some typical examples of binding energies for compounds with ionic bonding are given in Table 2-2.

It should be noted that the binding energies in Table 2-2 are expressed in

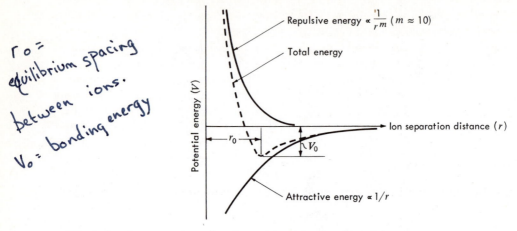

$r_0 =$ equilibrium spacing between ions.

$V_0 =$ bonding energy

FIG. 2-4 Energy vs. separation distance for two oppositely charged ions

TABLE 2-2

Examples of Substances with Different Types of Interatomic Bonds

Type of Bond	Substance	Bond Energy* kcal/mole	Melting Temperature (°C)	Characteristics
Ionic	CaCl	155	646	Low electrical conductivity; transparent; brittle; high melting temperature
	NaCl	183	801	
	LiF	240	870	
	CuF_2	617	1360	
	Al_2O_3	3618	3500	
Covalent	Ge	75	958	Low electrical conductivity; very hard; very high melting temperature
	GaAs	≈75	1238	
	Si	84	1420	
	SiC	283	2600	
	Diamond	170	3550	
Metallic	Na	26	97.5	High electrical and thermal conductivity; easily deformable; opaque
	Al	74	660	
	Cu	81	1083	
	Fe	97	1535	
	W	201	3370	
Van der Waals	Ne	0.59	−248.7	Weak binding; low melting and boiling points; very compressible
	Ar	1.8	−189.4	
	CH_4	2.4	−184	
	Kr	2.8	−157	
	Cl_2	7.4	−103	
Hydrogen	HF	7	−92	Higher melting points than Van der Waals bonding; tendency to form groups of many molecules
	H_2O	12	0	

*The bond energy is the energy necessary to dissociate the solid into separated atoms, ions, or molecules, as appropriate.

kcal/mole. Here is meant the energy needed to separate the atoms (or ions or molecules, as appropriate) from each other for 1 mole of substance (6.02×10^{23} molecules). In other places in this chapter and throughout the book, we will express the binding energy, usually in electron volts, as the energy needed to separate the atoms or ions in one molecule.

The electronic structure of the charged ions participating in ionic bonding is identical to that of the inert gases. This means that the outer electron shells are filled, and consequently the atoms have a spherically symmetric electron cloud distribution. Each atom can thus be represented as a spherically symmetric charge. And since the electrostatic attractive forces between spherically symmetric charges is independent of the orientation of the charges, *there is no preferred directional character in the ionic bond.* With the absence of any directionality in the ionic bond, the atomic arrangements in ionic solids are dictated by two factors: (1) The atoms must arrange themselves so as to preserve local charge neutrality; positive and negative ions must alternate in a symmetrical fashion. (2) The packing sequence must be consistent with the size of the ions. The ion size is important because positive ions tend to surround themselves with as many negative ions as possible, and vice versa. However, a large positive ion can surround itself with many small negative ions, whereas a small positive ion can have only a few large negative ions around it. Ionic solids with ions of considerably different size would therefore be expected to have an atomic arrangement different from those with ions of nearly the same size. This is indeed the case, and Fig. 2-5 shows the type of

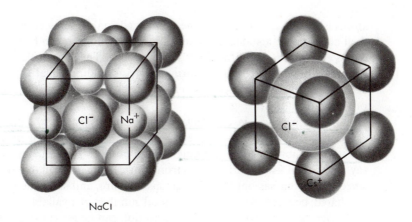

NaCl

FIG. 2-5   Atom packing of NaCl and CsCl

atomic packing found in CsCl (ions about the same size) and NaCl (ions considerably different in size). In both instances local charge neutrality is maintained, each ion having an equal number of oppositely charged nearest neighbors. In CsCl this number of nearest neighbors is eight while in NaCl, it is six. The topic of atomic arrangements in ionic solids will be discussed in more detail in Section 2-6.

Covalent Bonding

Whereas ionic bonding takes place between atoms that are strongly electropositive and electronegative (elements on opposite sides of the periodic table), covalent bonding takes place between atoms of elements near one another in the periodic table. In covalent bonding the atoms do not give up their electrons, but rather share their outer s and p electrons with other atoms, so that each atom has a complete outer electron shell (i.e., eight electrons). This can be illustrated with the example shown in Fig. 2-6. The electronic structure of each of the two fluorine

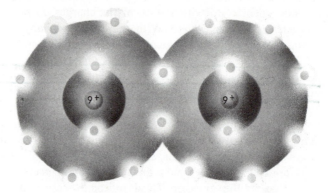

FIG. 2-6   Covalent bonding in the F_2 molecule

atoms is $1s^2 2s^2 2p^5$. Each fluorine atom needs only one $2p$ electron to fill its outer shell and achieve a stable configuration. This condition can be attained if the two atoms each share one of the other's $2p$ electrons. The shared electrons can be considered as belonging to both fluorine atoms, spending equal time in each atom's $2p$ orbital. This type of bond is very directional because of the directional nature of the $2p$ orbital (see Fig. 2-2).

Electrons involved in covalent bonding can be shared by only two atoms. Because of this, a pure element must have at least four electrons in the outer s and p states to exhibit covalent bonding. If the number of electrons is less than four, say three, then five additional electrons must be made available to each atom. But since each atom has only three electrons to share, and since each electron can only be shared between two atoms, the total number of available electrons per atom can never be more than six. This observation can be generalized into the following rule: *If N is the group number of an element in the periodic table, then in a covalently bonded material the element must have 8–N covalent bonds (or nearest neighbor atoms) to complete its outer electron shell.*

Limiting our attention to covalent bonding in pure elements for the moment, group VII elements (F, Br, I, etc.) have only one covalent bond per atom and hence form diatomic molecules. Group VI, V, and IV elements need two, three, and four covalent bonds per atom, respectively, and usually do not form molecules but rather large-scale, three-dimensional atom arrangements because of the directional characteristics of the bonds. Some typical examples are shown in Fig. 2-7.

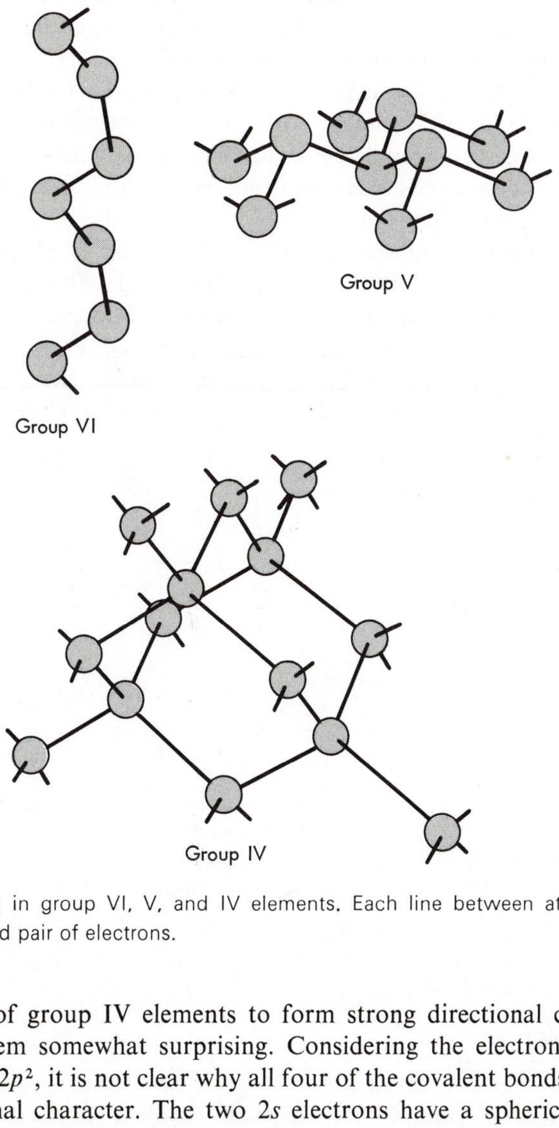

FIG. 2-7   Covalent bonding in group VI, V, and IV elements. Each line between atoms represents a shared pair of electrons.

The ability of group IV elements to form strong directional covalent bonds may at first seem somewhat surprising. Considering the electronic structure of carbon, $1s^2 2s^2 2p^2$, it is not clear why all four of the covalent bonds should have a strong directional character. The two $2s$ electrons have a spherically symmetric wave function and would not be expected to give rise to a strongly directional bond. Only the $2p$ electrons which have directional wave functions (nonzero angular momentum) should have directional bonding properties. Yet the structure of diamond (pure carbon), illustrated in Fig. 2-7, shows each carbon atom to have four strongly directional covalent bonds. The structure can be represented by the placement of a carbon atom in the center of the space formed by four tetrahedrally placed atoms. The tetrahedrally oriented bonds of each carbon atom therefore make spatial angles of 109.5°. The explanation to this problem lies in an effect called *hybridization*. Instead of the 2s and 2p wave functions, there is an equivalent set of wave functions which are solutions to Schrödinger's wave equation. These hybrid wave functions give a high electron density in the directions of the corners of a regular tetrahedron. Although the hybrid orbitals have a slightly higher energy than the normal 2s and 2p orbitals, this increase in energy is more than compensated by the energy decrease accompanying bonding.

Covalent bonds may also form between atoms of different types. The numerous carbon-hydrogen compounds are an excellent example. A carbon atom needs four electrons to complete its outer shell and can share electrons with four hydrogen atoms, each of which has a single 1s electron. The hydrogen atom needs only two 1s electrons for a stable electron configuration. Other possible carbon-hydrogen configurations involve the carbon atom sharing electrons with any combination of hydrogen atoms and other carbon atoms to give complete outer electron shells for all the atoms involved. Several examples are given in Fig. 2-8.

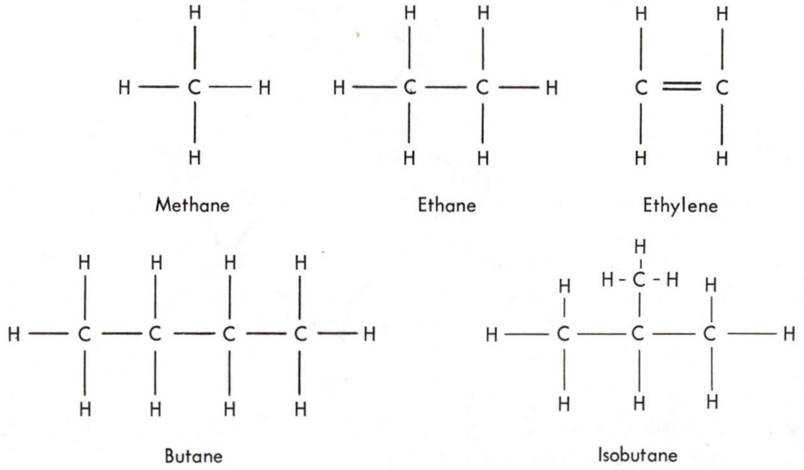

FIG. 2-8   Carbon-hydrogen covalently bonded molecules. Each dash represents two covalent bonds.

Another group of materials with covalent bonding comprises those compounds formed between elements found symmetrically spaced on either side of the group IV elements in the periodic table. Compounds of elements from groups III and V, such as GaAs, are typical examples. In GaAs the Ga atom contributes three electrons and the As atom contributes five electrons to the covalent bonds. The total of eight electrons are then shared by both types of atoms. Compounds formed from groups III and V (GaAs, GaP, InSb, etc.) are usually known as III-V compounds, and those from groups II and VI (MgTe, ZnS, etc.) are called II-VI compounds. Many of the III-V and II-VI compounds have the same type of tetrahedral covalent bonding found in the group IV elements.

Covalent bond energies are generally high, indicating strong bonds (see Table 2-2). However, many materials with covalent bonding have neither great strength nor high melting temperature, although both these qualities are indicative of large bond energies. This type of behavior usually occurs in molecular solids where the bonds between atoms in the molecule are covalent and strong but the attraction between adjacent molecules is very weak. For example, methane CH_4 has an *intramolecular* bond energy of 396 kcal/mole, whereas the *intermolecular* bond energy is only 2.4 kcal/mole. Because of this low intermolecular bond energy methane melts at only $-183°C$.

Many solids have components of both ionic and covalent bonding. It is usually very difficult to estimate the amount of either type of bonding, but it is generally true that compounds between atoms with nearly filled shells (NaCl, LiF, etc.) tend to be ionic and compounds between atoms with about half-filled shells (C, GaAs, etc.) tend to be covalent. The III-V compounds, for example, while predominantly covalent, may also have a sizeable ionic bond component, whereas II-VI compounds such as ZnS may have about equal amounts of ionic and covalent bonding.

Metallic Bonding

The metallic bond occurs between atoms with only a few electrons in their outer s and p orbitals. We know from certain of the properties of metals that unlike the ionic and covalent bonds, the bonding electrons in metals are essentially free. The high electrical and thermal conductivity of metals, for example, requires that some of the electrons be able to move through the solid very easily in response to either electrical or thermal gradients. This can be the case only if some of the electrons are relatively free to move from atom to atom.

The simplest model of the metallic bond is one in which the outer (valence) electrons are given up by the individual atoms, resulting in a geometrical array of positive ions surrounded by a free electron cloud. The bonding force is pictured as the electrostatic attraction between the electron cloud and the ions. From the wave mechanical standpoint, this means that when metal atoms are brought into close proximity the wave functions of the outer electrons spread out in such a manner that the electron has a finite probability of being near several different atoms. This lowers the energy of these electrons and provides the bonding force.

Since the bonding electrons are not localized between two atoms, as is the case in covalent bonding, metals are thought to have nondirectional bonds. This is indeed the case for atoms with completely filled inner shells (Mg, Cu, Zn; see Appendix 1). However, the atoms with incomplete inner shells (transition metals Fe, W, Ti, etc.) do have some directional bonding characteristics. This arises from covalent bonds formed by the inner shell orbitals. These covalent bonds, aside from being directional, increase the bond energy and generally result in a higher melting temperature (Table 2-2).

The metallic bond, in view of its free electron character, affords metals many unique properties. Metals are opaque because the free electrons can absorb the incident light photons and then reemit this energy as they fall back into lower energy states (Chapter 15). Metals can also be plastically deformed easily, as the absence of both primary directional bonds and requirements for local charge neutrality allow metal atoms to slide by one another fairly easily (see Chapter 7).

Van der Waals Bonding

Thus far we have considered bonding between atoms that have incomplete outer electron shells and for which the driving force for bonding is a rearrangement of the bonding electrons. But what of those atoms with filled outer shells, such as inert gases or covalently bonded molecules? How do these atoms bond together? To answer this question we must look in some detail at the electronic structure of these atoms and molecules.

Very weak bonds can develop between atoms with filled electron shells because the atoms attain a dipole character due to an asymmetry in their electron charge distribution. This is caused by the instantaneous location of a few more electrons on one side of the nucleus than on the other. Although the orientation of this dipole continually changes with time due to the motion of the electrons, there is a weak attractive force set up between the fluctuating dipoles in adjoining atoms. This type of fluctuating dipole bond is usually called a *Van der Waals bond* and is generally very weak with a binding energy of only a few kcal/mole. Van der Waals bonds are nondirectional due to the fluctuating character of the dipoles causing the bonding, and the low binding energies result in very low melting temperatures (Table 2-2).

Hydrogen Bonding

Under some conditions hydrogen atoms can be strongly attracted to two different atoms, thereby acting as a bond between them. Such bonding is called *hydrogen bonding*. When a hydrogen atom is bonded to a single electronegative atom, like Cl (see Fig. 2-9), the bond formed is a directional covalent bond. However, since the hydrogen atom contains only one electron it can form only one purely covalent bond. To explain how a hydrogen atom can be simultaneously bonded to two different electronegative atoms we must consider the forces to be largely ionic. Since a hydrogen ion is simply a bare proton, with no electrons about it, it is a vanishingly small cation. We may think of this cation as attracting, simultaneously, two large anions. A third anion could not participate in this bond because it would be repelled by the first two anions long before it comes into contact with the cation (proton). In this way the hydrogen atom itself acts as a bond by providing a proton bridge between strongly electronegative atoms.

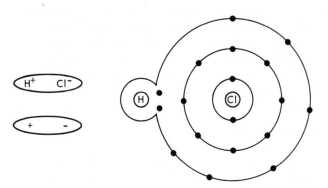

FIG. 2-9   Dipole character of the HCl molecule

2-5 POTENTIAL WELL CONCEPT

In Section 2-4 we discussed the concept of an equilibrium separation distance for atoms with ionic bonds. In Fig. 2-4 we considered only the interaction between two ions and showed that the ions would assume a position where the potential energy of the system is a minimum and there is no net force on the ions. This calculation can be generalized to treat solids with any type of interatomic bond. To do this we focus our attention on a single atom in a solid and calculate the potential energy of this atom as a function of the position of all other atoms. In theory we must consider all the other atoms because each atom interacts with every other atom. However, in practice the calculation is greatly simplified by the fact that the magnitudes of the first few nearest neighbor interactions are much larger than any of the others.

Carrying out the approximate summation, we obtain the potential energy of the atom as a function of the interatomic or nearest neighbor distance. For all types of bonding we find a relationship similar to that shown in Fig. 2-10. This is

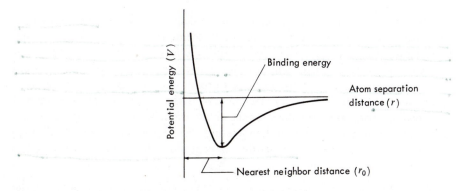

FIG. 2-10   Potential energy well for atoms in a solid

analogous to the case we treated earlier for two oppositely charged ions. Each atom sits at the bottom of a potential energy well. The exact shape of the energy well, which depends on the type of bonding, is important in determining several physical properties of the material. The most significant among these include the elastic properties of the material, the amplitude of atomic vibrations, and the coefficient of thermal expansion.

Elastic Modulus

Consider what happens to the equilibrium atom spacing when an external force is applied to a material. If the force tends to push the atoms together (compressive), then the atoms will assume a new interatomic spacing $r < r_0$ at which the sum of the atomic and applied forces acting on the atoms is zero. A similar situation with $r > r_0$ exists for tensile forces that tend to pull the atoms apart.

When the applied forces are small the displacement of the atoms, $r - r_0 = \Delta r$, is small and proportional to the force. We can show this by expressing the potential energy of the displaced atom $V(\Delta r)$ as a Taylor series

$$V(\Delta r) = V_{r_0} + \left(\frac{dV}{dr}\right)_{r_0} \Delta r + \frac{1}{2}\left(\frac{d^2V}{dr^2}\right)_{r_0} \Delta r^2 \tag{2-9}$$
$$+ \text{ higher order terms}$$

where V_{r_0} is the energy at r_0 and all the derivatives are evaluated at r_0. Noting that Δr is very small,* we can simplify Eq. (2-9) by neglecting all terms with higher order than those shown. In addition, as $dV/dr = 0$ at $r = r_0$, we can write

$$V(\Delta r) = V_{r_0} + \frac{1}{2}\left(\frac{d^2V}{dr^2}\right)_{r_0} \Delta r^2 \tag{2-10}$$

Now since the potential energy is related to the atomic force (negative of the applied force) by

$$\text{Force} = -\frac{dV(\Delta r)}{d(\Delta r)} = -\left(\frac{d^2V}{dr^2}\right)_{r_0} \Delta r \tag{2-11}$$

we see that the force is linearly related to the displacement Δr. The constant of proportionality is merely the curvature of the potential energy well at its minimum. If we express force and displacement in terms of stress and strain, the proportionality constant is called the *elastic modulus*. A narrow, steep potential energy well corresponds to a high modulus, whereas a broad, shallow energy well represents a low modulus. Note that the elastic modulus is the same for both tension and compression. Some typical values for the elastic modulus are given in Table 2-3. Generally the higher the melting temperature, the higher the modulus, although this is not always the case. The melting temperature is determined mainly by the

---

*We are interested only in the range $\Delta r < 0.01r_0$. For larger displacements, materials do not deform by simply *stretching* the atomic bond. This problem will be discussed more fully in Chapters 6, 7, and 10.

TABLE 2-3

Typical Room Temperature Values of the Elastic Modulus and Linear Thermal
Expansion Coefficient for Several Materials

Material	Elastic Modulus		Linear Thermal Expansion Coefficient, α
	(10^6 psi)	(10^{11} dynes/cm²)	(length/length·°C)
Diamond	114	77.5	1.2×10^{-6}
W_2C	90	61.2	$\approx 7.0 \times 10^{-6}$
W	56.5	38.4	4.4×10^{-6}
Al_2O_3	50	34.0	8.7×10^{-6}
MgO	40	27.2	$\approx 10.0 \times 10^{-6}$
Ni	30	20.4	13.0×10^{-6}
Si	29	19.7	7.6×10^{-6}
Ge	23	15.7	—
LiF	19	12.9	—
Cu	17	11.5	16.8×10^{-6}
SiO_2	10	6.8	8.0×10^{-6}
Mg	6.3	4.3	26.0×10^{-6}
NaCl	4.7	3.7	40.4×10^{-6}
Polystyrene	0.4	0.27	$\approx 79 \times 10^{-6}$
Nylon	0.4	0.27	$\approx 100 \times 10^{-6}$
Polytetra- fluorethylene	0.06	0.041	$\approx 100 \times 10^{-6}$
Polyethylene	0.02	0.014	$\approx 300 \times 10^{-6}$
Natural rubber	10^{-3}–10^{-2}	$\approx 7 \times 10^{-4}$ 7×10^{-3}	$\approx 650 \times 10^{-6}$

depth of the energy well, whereas the modulus is determined by the *curvature* at
the bottom of the well.

Thermal Properties

When a solid is at any temperature above 0°K the atoms have thermal energy
and vibrate about their equilibrium positions. The amplitude of this vibration is
dictated by the shape of the potential energy well. The atom vibrates back and forth
in the well at a height above the bottom equal to its thermal energy. This situation
is shown schematically for several different temperatures in Fig. 2-11. The higher
the temperature, the larger the amplitude of vibration. Because of the asymmetrical
nature of the energy well, the mean position of the atom changes with increasing
temperature; the atoms tend to move apart as the temperature is raised. For a
deep, narrow energy well the change in interatomic distance is small, but for a
broad, shallow well, being generally more asymmetric, the change can be relatively
large. This phenomenon is known as thermal expansion and is usually described
by the *thermal expansion coefficient*, which is a measure of the length increase of a
macroscopic sample for a given temperature increase. The units of the thermal
expansion coefficient are length/length·°C. Some typical examples are given in

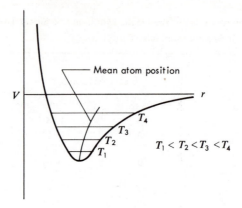

FIG. 2-11 Increase in the nearest neighbor distance with increase in temperature

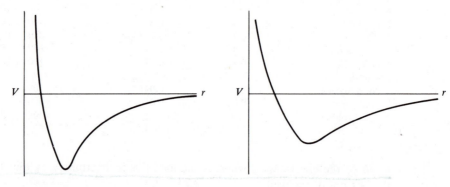

(a) High melting temperature, high elastic
      modulus, low thermal expansion coefficient

(b) Low melting temperature, low elastic
      modulus, high expansion coefficient

FIG. 2-12 Schematic illustration of relation of physical properties to potential energy well

Table 2-3. Figure 2-12 illustrates the different properties associated with potential energy wells of different shapes.

2-6 ATOM ARRANGEMENTS

All solid materials may be classified on the basis of their atom arrangements as being either *crystalline* or *amorphous*. Crystalline materials exhibit *long-range order* in the position and stacking sequence of atoms or molecules. Thus, if we know the precise atom arrangement at one position in a solid, we can move to any new position in the solid and predict exactly what the atom arrangement will be. Two-dimensional analogies to crystalline solids are numerous, three examples being the arrangement of bricks in the wall of a building, the repetitive pattern of wallpaper, and the arrangement of a set of parallelograms produced by the intersection of two coplanar sets of equally spaced parallel lines. Amorphous solids, on the other hand, have no long-range order. The atoms or molecules in these solids are not periodically located over large distances. Although there is no long-range order in amorphous solids, there is usually some *short-range order* or definite first nearest neighbor coordination. This short-range order satisfies the directional character of the covalent bonds found in most amorphous solids. Figure 2-13 shows the essential difference between crystalline (having both short- and long-range order) and amorphous (having only short-range order) materials.

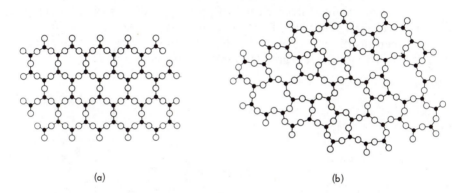

(a)   (b)

FIG. 2-13   Schematic illustration of (a) crystalline and (b) amorphous structure

Many amorphous materials have internal structures similar to liquids. In fact, the only obvious distinction between amorphous materials, such as glass, and liquids is the high viscosity (resistance to flow) of the amorphous solids.

A general rule is that solids with nondirectional bonding are always crystalline, whereas directional bonding may yield either crystalline or amorphous solids. This is because nondirectional bonding can be satisfied best when each atom or molecule has identical surroundings (long-range order). Any local change in the packing sequence will force some directionality to occur in the bonding. On the other hand, directional bonding may be satisfied when the atoms or molecules have either short-range or long-range order.

Crystalline Solids

The fact that crystalline solids have long-range order requires that each atom, group of atoms, or molecule, whichever is the basic unit of the material, must have identical surroundings in the solid. Thus, if we take a sodium and chlorine atom pair as the basic unit of NaCl, the surroundings of each atom pair, in terms of nearest neighbor ions, etc., must be identical.

A simple way to interpret the crystalline structure is to consider that it consists of a three-dimensional arrangement of points in space (lattice points), where each lattice point has identical surroundings. Associated with each of these lattice points is a single atom or identical group of atoms, depending on the solid under consideration. If it is a group of atoms, then each atom group must be identical with respect to number, type, and position of atoms and the way in which the atom group is positioned on the lattice point. Once we know the atom group associated with each lattice point, then we can consider the crystalline structure in terms of the array of lattice points and the atom group associated with any one lattice point. The great simplification here involves the fact that there are only 14 different ways to arrange the lattice points in space so that each point has identical surroundings. Thus, each of the many thousands of crystalline materials can be described by one of these arrangements of points plus the associated atom group. The size and the complexity of the atom group vary, but for most materials the group consists of only a few atoms. Figure 2-14 illustrates how the point lattice is combined with an atom group to generate a periodic atom array.

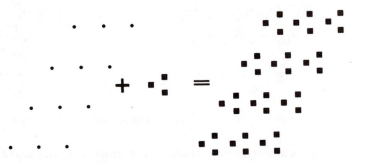

Point lattice + atom group = periodic atom array

FIG. 2-14 Schematic illustration of the generation of a periodic atom array using a point lattice and basic atom group

The 14 different point lattices are called the *Bravais lattices*. They are most easily described by a *unit cell* or small parallelepiped which reflects the symmetry of the entire Bravais lattice. The unit cell is defined by three convenient small *lattice translation vectors* (vectors connecting lattice points). An example of a unit cell of the most general shape, defined by the three lattice vectors **a**, **b**, and **c**, is shown in Fig. 2-15. The magnitudes of **a**, **b**, and **c** are usually referred to as the *lattice parameters*. When the unit cell is translated by all possible lattice translation vectors it will generate the Bravais lattice. Thus, all we need to define the Bravais lattice is the unit cell. Correspondingly, the problem of specifying the positions of the $\approx 10^{22}$ atoms in a cubic centimeter of typical crystalline material has been reduced to merely a specification of the appropriate unit cell and the atom group that is associated with each of the lattice points.

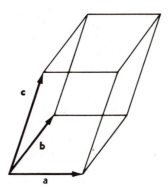

FIG. 2-15 Unit cell defined by the three lattice translation vectors **a**, **b**, **c**

The three lattice translation vectors that specify the unit cell define a coordinate system that is usually referred to as the *crystal system*. The 14 Bravais lattices can be defined by only seven different crystal systems. These are described in Table 2-4, and the unit cells of the Bravais lattices are shown in Fig. 2-16.

TABLE 2-4

The Characteristics of the Seven Different Crystal Systems

System	Axes* and Interaxial Angles	Examples
Triclinic	Three axes not at right angles, of any length $a \neq b \neq c$† $\alpha \neq \beta \neq \gamma \neq 90°$	$B(OH)_3$, $K_2S_2O_8$, Al_2SiO_5, $NaAlSi_3O_8$
Monoclinic	Three axes, one pair not at right angles, of any lengths $a \neq b \neq c$ $\alpha = \gamma = 90° \neq \beta$	$C_{18}H_{24}$, KNO_2, $K_2S_4O_6$, As_4S_4, $KClO_3$
Orthorhombic	Three axes at right angles, all unequal $a \neq b \neq c$ $\alpha = \beta = \gamma = 90°$	I, Ga, Fe_3C, FeS_2, $BaSO_4$
Tetragonal	Three axes at right angles, two equal $a = b \neq c$ $\alpha = \beta = \gamma = 90°$	In, TiO_2, $C_4H_{10}O_4$, KIO_4
Cubic	Three axes at right angles, all equal $a = b = c$ $\alpha = \beta = \gamma = 90°$	Cu, Ag, Ar, Si, Ni, NaCl, LiF
Hexagonal	Two axes of equal length at 120°, third axis at 90° to these $a = b \neq c$ $\alpha = \beta = 90°$ $\gamma = 120°$	Zn, Cd, Mg, NiAs
Rhombohedral	Three axes equally inclined, not at right angles, all equal $a = b = c$ $\alpha = \beta = \gamma \neq 90°$	Hg, Sb, Bi

*a, b, and c refer to the lattice parameters or dimensions of the unit cell.

†In this table $\neq$ means "not necessarily equal to, and generally different from."

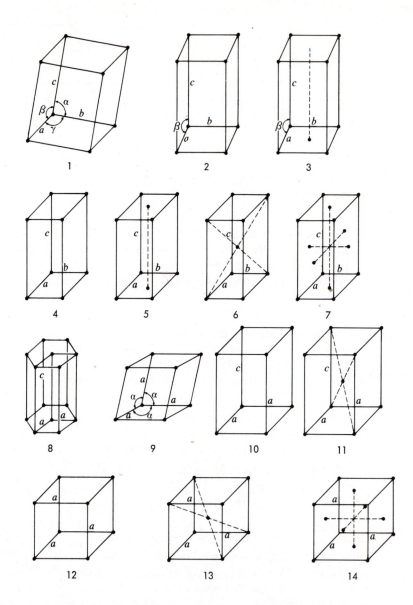

FIG. 2-16 The 14 point lattices illustrated by a unit cell of each: (1) triclinic, simple; (2) monoclinic, simple; (3) monoclinic, base centered; (4) orthorhombic, simple; (5) orthorhombic, base centered; (6) orthorhomic, body centered; (7) orthorhombic, face centered; (8) hexagonal; (9) rhombohedral; (10) tetragonal, simple; (11) tetragonal, body centered; (12) cubic, simple; (13) cubic, body centered; (14) cubic, face centered.

EXAMPLE 2-3   Calculate the number of lattice points associated with each unit cell for the body-centered-cubic (BCC) Bravais lattice.

Solution: Although there would seem to be nine lattice points associated with each unit cell, some of these lattice points are shared by more than one unit cell. This can be seen in Fig. 2-17, which shows eight adjacent unit cells in the body-centered-cubic lattice. Each corner lattice point, N_c, is shared by eight unit cells. Each interior lattice point, N_I, belongs to just one unit cell. And each lattice point on the face of a unit cell, N_F

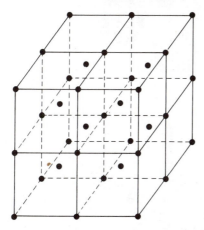

FIG. 2-17   Eight adjacent BCC unit cells

(there are none in this case), is shared by two unit cells. The total number of lattice points per unit cell, N, is then given by

$$N = \frac{N_c}{8} + N_I + \frac{N_F}{2}$$

For the body-centered-cubic lattice $N = \frac{8}{8} + 1 + \frac{0}{2} = 2$. Similarly, for any noncubic body-centered unit cell $N = 2$.

The *crystal structure*, or the description of atom positions, is given by specifying the appropriate Bravais lattice unit cell along with the atom positions about one lattice point. The coordinates of the atom positions are specified in terms of the three lattice translation vectors defining the unit cell, choosing one corner of the unit cell as the origin of the crystal system. An atom at the center of any unit cell thus has coordinates $\frac{1}{2}\frac{1}{2}\frac{1}{2}$. Consider the CsCl structure shown in Fig. 2-18 as an example. The Bravais lattice of this structure is simple cubic. There is one lattice point per unit cell and two atoms (one cesium and one chlorine) associated with each lattice point. If we choose to give the coordinates of the atoms about the lattice point at the origin of the unit cell (coordinate, 000), then the cesium atom is located at 000 and the chlorine atom at $\frac{1}{2}\frac{1}{2}\frac{1}{2}$. This is the complete description of the CsCl structure. At first glance the CsCl structure might seem to have a body-centered-cubic (BCC) unit cell, but if this were the case it would violate our requirement that each lattice point have the *same identical* atom or groups of atoms associated with it.

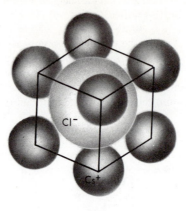

FIG. 2-18 Unit cell of CsCl

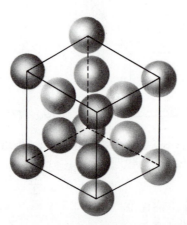

FIG. 2-19 The copper crystal structure (FCC Bravais lattice)

EXAMPLE 2-4

The copper crystal structure is shown in Fig. 2-19. If the lattice parameter (cube edge) is 3.61 A, what is the density of copper?

Solution:

The Bravais lattice is face-centered-cubic (FCC), and there is one atom associated with each lattice point. The number of atoms (or lattice points) associated with each unit cell is given by:

$$N = + \frac{N_c}{8} + N_I + \frac{N_F}{2} = \frac{8}{8} + \frac{0}{1} + \frac{6}{2} = 4$$

Therefore the density of copper is the weight of four copper atoms divided by the volume they occupy, or a^3. Thus,

$$\text{Density} = \frac{(4 \text{ atoms/unit cell})(63.5 \text{ g/mole})}{(6.02 \times 10^{23} \text{ atoms/mole})[(3.61)^3 \times 10^{-24} \text{ cm}^3/\text{unit cell}]} = 8.95 \text{ g/cm}^3$$

The simplest crystal structures are composed of pure elements with completely nondirectional interatomic bonds. These atoms pack together in such a way to minimize the open space in the structure and form what we call the *close-packed* crystal structures. The atom arrangement in these crystal structures is precisely the same as that found in the most economical (space saving) arrangement of spherical objects, say Ping-Pong balls or oranges. Figure 2-20(a) shows a single

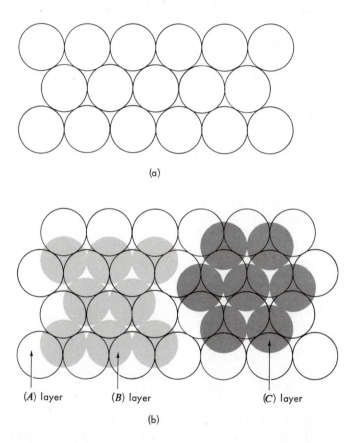

(a)

(A) layer   (B) layer   (C) layer

(b)

FIG. 2-20   (a) Close-packed layer of atoms and (b) stacking of two close-packed layers

close-packed layer of atoms. To build up the three-dimensional close-packed structure we must place layers of this type on top of one another in a systematic fashion. There are several distinct ways of performing this stacking operation, the two simplest stacking sequences resulting in the two basic close-packed crystal structures. Let us call this first layer of atoms an *A* type, where the *A* designates the positions of the atoms in the plane parallel to the sheet of the paper. The second layer of atoms is placed on top of the first with the atoms of the second layer fitting into one of the two sets of hollows provided by the first layer. We will designate this second layer as a *B* layer or a *C* layer, depending on which set of hollows it fits in [Fig. 2-20(b)]. At this point the structure is the same, regardless of which of the two sets of hollows the second layer sits in. The difference in stacking comes with the placing of the third layer. If we take the second layer as a *B* layer, there are two possible positions for this third layer of atoms. One position is the set of hollows directly over the first layer, i.e., another *A* layer. The second possibility is the set of hollows that is not over the first layer, i.e., a *C* layer.

These two basic stacking sequences of close-packed layers can be thought of as *ABABAB*... and *ABCABCABC*.... The *ABAB* packing generates the hexagonal close-packed (HCP) crystal structure shown in Fig. 2-21(a), while the *ABCABC* stacking produces the face-centered-cubic (FCC) crystal structure illustrated in Fig. 2-21(b). In the FCC structure the close-packed planes are perpendicular to the body diagonal. Both structures have equal packing densities and equal numbers of nearest neighbor atoms (12). The main difference in the structures is the second nearest neighbor distance; this influences bond energy in a small way. Many metals, inert gases, and symmetric molecules have one of these two crystal structures.

The transition metals, which have components of covalent bonding due to their incomplete inner *d* electron shells, do not have close-packed crystal structures. Instead, these metals usually have a BCC crystal structure, where the atom arrangement satisfies the directional character of the *d* shell bonds.

Throughout our discussion of crystal structures we will refer to the sites in between the atoms as *interstices*. For the close-packed structures two important kinds of interstitial sites are the *tetrahedral* sites having 4 nearest neighbor atoms (at the corners of a tetrahedron) and the *octahedral* sites having 6 nearest neighbor atoms (which form an 8-sided polyhedron).

EXAMPLE 2-5

Calculate the percent volume change that would occur if a material were to transform from FCC to BCC. Assume that the atomic diameter is fixed and that the atoms pack together as hard spheres.

Solution:

First, we express the lattice parameter of the unit cell in terms of the atom radius. For the FCC case, the atoms are close-packed (or touch) along the face diagonal. For the BCC case, the atoms are close-packed along the body diagonal. If the radius of the atom is r, then the lattice parameter a can be written

$$a_{FCC} = \frac{4r}{\sqrt{2}} \quad \text{and} \quad a_{BCC} = \frac{4r}{\sqrt{3}}$$

The volume associated with each atom can then be expressed as

$$\frac{\text{Volume of unit cell}}{\text{atoms per unit cell}}$$

or

$$V_{FCC} = \frac{(4r/\sqrt{2})^3}{4} \quad \text{and} \quad V_{BCC} = \frac{(4r/\sqrt{3})^3}{2}$$

The percent volume change in going from FCC to BCC is then

$$\frac{\Delta V}{V} = \frac{V_{BCC} - V_{FCC}}{V_{FCC}} = \frac{\frac{1}{2}(4r/\sqrt{3})^3 - \frac{1}{4}(4r/\sqrt{2})^3}{\frac{1}{4}(4r/\sqrt{2})^3} = 0.088 = 8.8\%$$

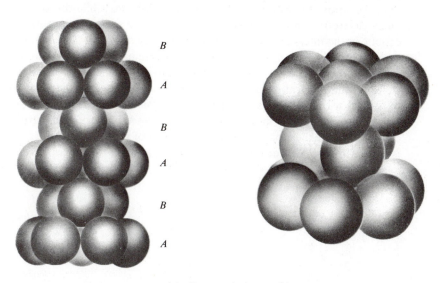

(a)   Hexagonal close packing

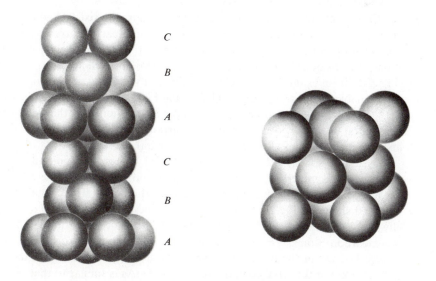

(b)   Cubic close packing

FIG. 2-21   Packing sequence and unit cell for (a) HCP and (b) FCC structures

In ionic solids the atom packing is determined by the ways positive and negative ions can be stacked together to minimize electrostatic repulsion and maximize the electrostatic attractive forces. These packing sequences depend very critically on the respective sizes of the cations and anions. Usually the cation is much smaller than the anion (the cation has lost its outer electrons while the anion has gained outer electrons), and the crystal structure can be built up by putting the small cations in the interstices (holes) between the larger anions. The number of anions surrounding the cation, the *coordination number* or number of nearest neighbors, depends on the radius of the ions in the following manner. As many anions as possible surround the cation, subject only to the conditions that the anions are always in contact with the cations and that the electron shells of the anions do not overlap. Figure 2-22 shows these conditions for the stability of ionic structures. One can thus predict the atom arrangement in terms of the radius ratio r_{cation}/r_{anion}. This is shown in Fig. 2-23 for the only important cases which arise.

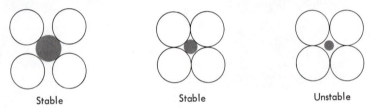

Stable   Stable   Unstable

FIG. 2-22   Stable and unstable ionic coordination configurations

An example of eightfold coordination in ionic solids has already been described, i.e., the case of CsCl shown in Fig. 2-18. Ionic solids with sixfold and fourfold coordination can be built up with the anions in a close-packed array (either FCC or HCP) with the cations in the available octahedral or tetrahedral sites. Figure 2-24 shows the positions of these sites in both the close-packed anion structures. NaCl (Fig. 2-5) is an example of sixfold coordination with the small sodium ions in all the available octahedral sites provided by the FCC arrangement of chlorine atoms. BeO (Fig. 2-25) is an example of fourfold coordination with the Be atoms in the tetrahedral sites provided by the FCC arrangement of oxygen atoms. In this instance only one-half of the available tetrahedral sites are filled.

Crystalline solids with directionally bonded atoms and molecules have only one general characteristic in common. The crystal structure is generally not close-packed due to the directional requirements of the bonds. The crystal structure of diamond shown in Fig. 2-26(a) is perhaps the best known example of a directionally bonded material. Each carbon atom is tetrahedrally coordinated with four other carbon atoms. The Bravais lattice is FCC, but the structure is not close-packed because of the presence of the four atoms in the interior of the unit cell. The structure of the III-V compounds such as GaAs is similar to that of diamond, with tetrahedral coordination between all the atoms. Each gallium atom is surrounded by four arsenic atoms, and vice versa [Fig. 2-26(b)].

The structure of *molecular crystals*, covalent-bound molecules with weak inter-

Ratio of cation radius to anion radius	Disposition of ions about central ion	Coordination number	
1–0.732	Corners of cube	8	
0.732–0.414	Corners of octahedron	6	
0.414–0.225	Corners of tetrahedron	4	
0.225–0.155	Corners of triangle	3	

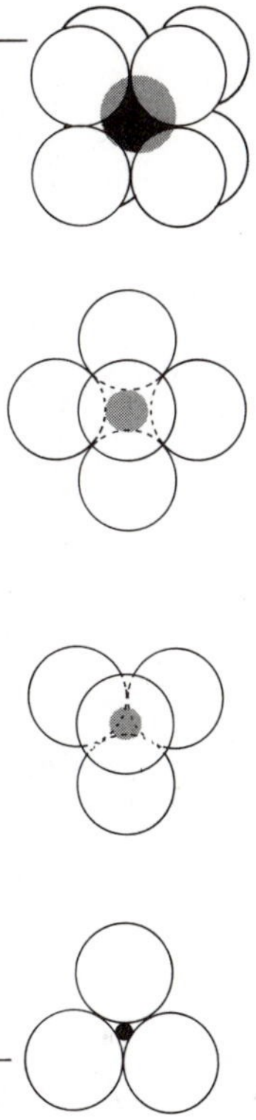

FIG. 2-23   Radius ratios for various atom arrangements in ionic bonding

molecular forces, is most important when considering the long-chained organic molecules commonly called *polymers*. Typical of these molecules are polyethylene and Teflon (Fig. 2-27). These molecules are generally very long and may contain thousands of atoms. For example, most polymers useful for plastics, rubbers, or fibers have molecular weights between 10,000 and 1,000,000. The weak dipole bonds between the molecules tend to align the molecules into a paralled array with a definite crystal structure, but this is only rarely achieved in practice. The great length of the molecules and the fact that they usually have a great many kinks and bends along their length usually make it very difficult for the molecules to align themselves. Also, if the molecules are asymmetric in cross section, they generally must rotate along an axis parallel to their length during the alignment procedure, making the process even more difficult. Crystallinity in polymers is usually limited therefore to the shorter-length symmetric molecules with relatively strong intermolecular forces (hydrogen bonds). Examples of crystalline polymers include polyethylene, nylon, and polypropylene. Figure 2-28 shows the crystalline arrangement of polyethylene, which has an orthorhombic unit cell.

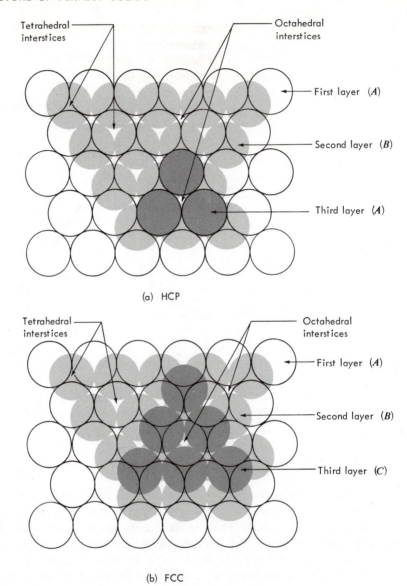

(a) HCP

(b) FCC

FIG. 2-24 Octahedral and tetrahedral sites in HCP and FCC packing of atoms

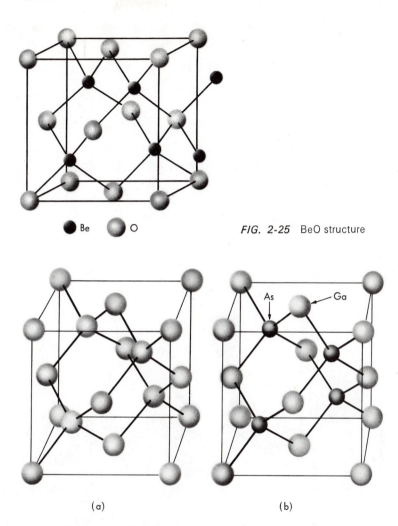

Be ● ○ O

FIG. 2-25   BeO structure

(a)   (b)

FIG. 2-26   Structure of (a) diamond and (b) GaAs

Amorphous Solids

All solids tend to exist in the crystalline state rather than the amorphous state because the crystalline structure always has a larger binding energy. However, in numerous instances amorphous solids are formed when liquids are cooled below the melting temperature. This occurs for two reasons: (1) the structure of the molecules is so complex that they cannot easily rearrange themselves to form a crystalline structure, and/or (2) the solid forms so rapidly that the atoms or molecules do not have time enough to rearrange themselves into a crystalline structure.

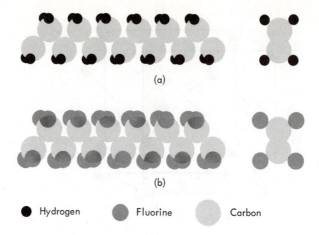

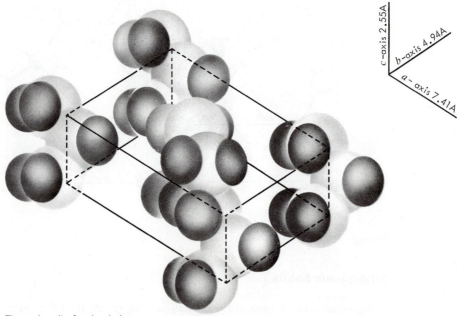

FIG. 2-27 Atom arrangements in polymer chains of (a) polyethylene and (b) polytetra-fluoroethylene (Teflon).

FIG. 2-28 The unit cell of polyethylene

During our discussion of amorphous solids, then, we must always remember that the exact atom arrangement will depend on exactly how the material is prepared.

Generally, amorphous solids have one of two distinct atomic arrangements: either a tangled mass of long-chained molecules or a three-dimensional network of atoms with no long-range order. Amorphous materials with long-chained molecules (S, Se, polymers) have a structure similar to that shown in Fig. 2-29. The

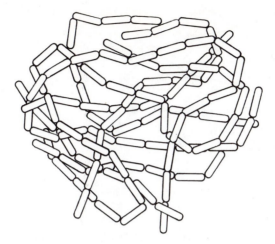

FIG. 2-29   A schematic representation of a long-chain polymer. Each segment represents one of the repeating units of the polymer chain.

arrangement of the molecules is fairly random, resulting in a loosely packed structure. Network amorphous solids are usually oxides, the most common being silica $(SiO)_2$. The structure of amorphous silica is composed of oxygen tetrahedra sharing corners (Fig. 2-30). This structure has short-range order but none of the long-range order found in crystalline silica. Thus, in both amorphous and crystalline silica, each silicon atom and each oxygen atom have essentially the same local surroundings, even though there is no long-range periodicity in the amorphous structure.

2-7 CRYSTALLOGRAPHIC NOTATIONS

In a discussion of the properties of crystalline solids it is often convenient to be able to specify a given direction or plane in a crystal. This can be done in a straightforward manner using the crystal axes **a, b, c** which define the unit cell. To specify a direction, say **OA** in Fig. 2-31, we merely resolve the components of **OA**—call them uvw—onto the three crystal axes, taking the unit distances along the three axes to be equal to the lattice parameters a, b, and c. For the case of **OA** the components uvw are 1, 1, and $\frac{1}{2}$. We then remove fractions or reduce to the smallest set of integers having the same ratio and place square brackets around the three numbers to indicate we are considering a specific direction in the crystal. A complete description of **OA** is then [221]. Negative components are introduced by putting a bar over the appropriate number. For example, the negative **OA** direction would be written [$\bar{2}\bar{2}\bar{1}$]. The set of directions which are *crystallographically equivalent*, such as cube edges, are written as $\langle uvw \rangle$ where $\langle \quad \rangle$ refers to a family of directions. Thus, for a cubic unit cell the [100], [$\bar{1}$00], [010], [0$\bar{1}$0], [001], and [00$\bar{1}$] directions are crystallographically equivalent and are members of the

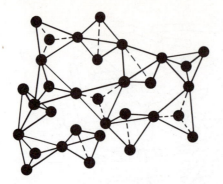

FIG. 2-30 Amorphous SiO_2 represented by a random arrangement of SiO_4 tetrahedra sharing corners. Only the oxygen atoms are shown (corners of tetrahedra). There is a silicon atom at the center of each tetrahedron.

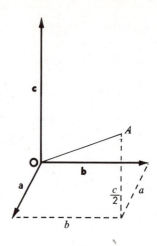

FIG. 2-31 Determination of a crystallographic direction

$\langle 100 \rangle$ family. Some examples of different directions in a cubic unit cell are shown in Fig. 2-32.

Planes in a crystal can be specified using a notation called *Miller indices*. To specify the plane *ABCD* in Fig. 2-33 we use the following procedure. Find the intercepts of the plane on the crystal axes, again taking the unit distance along the three axes to be the lattice parameter. For plane *ABCD* the intercepts are $\frac{1}{2}$, 1, and ∞. Next we take the reciprocals of the intercepts and clear fractions. We refer to the

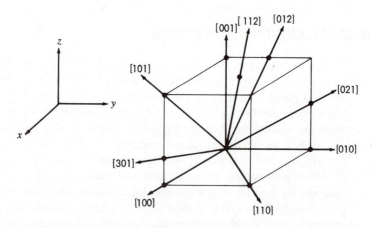

FIG. 2-32 Some common directions in a cubic unit cell. The dot indicates the point where the vector exits the unit cell.

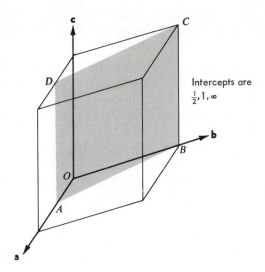

FIG. 2-33 Determination of the Miller indices of plane *ABCD*

set of the three numbers so obtained as (*hkl*). For *ABCD* this representation is (210). These three numbers (*hkl*) are the Miller indices of the plane and they completely describe its orientation.

EXAMPLE 2-6 Determine the Miller indices for the plane *ABC* in the orthorhombic unit cell in the accompanying diagram.

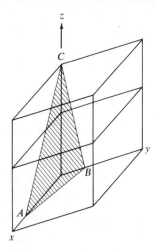

Solution: The intercepts of plane ABC along the crystal axes are $\frac{1}{2}$, $\frac{1}{3}$, and 2. Taking reciprocals, we have 2, 3, and $\frac{1}{2}$; after clearing fractions the Miller indices are (461).

If the plane happens to pass through the origin, giving intercepts of zero, we can overcome this problem by merely shifting the origin to an equivalent position in an adjacent unit cell. When a large number of adjacent unit cells are considered, the arbitrary position of the origin means that a whole set of equispaced parallel planes may be described by the same Miller indices. The interplanar spacing d_{hkl} is defined as the perpendicular distance between the (hkl) plane closest to the origin and a parallel plane that goes through the origin. This is shown in Fig. 2-34 for the (110) planes in a cubic crystal. Note that the spacing between planes of different order, i.e., (111) and (222) planes, is different.

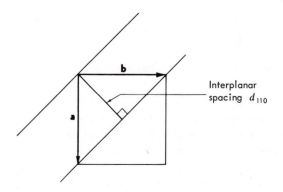

FIG. 2-34 View along **c** direction, showing the spacing between (110) planes.

EXAMPLE 2-7 Show that the spacing between (hkl) planes in the cubic system d_{hkl} is given by

$$d_{hkl} = \frac{a}{\sqrt{h^2 + k^2 + l^2}}$$

where a is the lattice parameter.

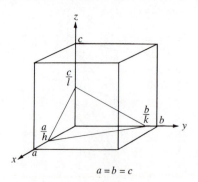

$a = b = c$

Solution: A perfectly general (hkl) plane is shown in the accompanying diagram. The interplanar spacing is the length of the line drawn from the origin perpendicular to the plane. If α_1, α_2, and α_3 are the angles between the plane normal and the x, y, and z axes, respectively, then we can write

$$d_{hkl} = \frac{a}{h} \cos \alpha_1 = \frac{a}{k} \cos \alpha_2 = \frac{a}{l} \cos \alpha_3$$

By squaring and adding these equations we obtain

$$\cos^2 \alpha_1 + \cos^2 \alpha_2 + \cos^2 \alpha_3 = d_{hkl}^2 \left(\frac{h^2 + k^2 + l^2}{a^2} \right)$$

For an orthogonal coordinate system

$$\cos^2 \alpha_1 + \cos^2 \alpha_2 + \cos^2 \alpha_3 = 1$$

and

$$d_{hkl} = \frac{a}{\sqrt{h^2 + k^2 + l^2}}$$

For other crystal systems the expression for d_{hkl} is slightly more complicated.

A group of planes that are crystallographically equivalent is called a *family of planes* and is represented by $\{hkl\}$. In cubic crystals, members of the $\{111\}$ family include (111), $(\bar{1}\bar{1}\bar{1})$, $(\bar{1}11)$, $(1\bar{1}\bar{1})$, $(1\bar{1}1)$, $(\bar{1}1\bar{1})$, $(11\bar{1})$, and $(\bar{1}\bar{1}1)$. Some examples of different planes are shown in Fig. 2-35.

2-8 CRYSTAL STRUCTURE ANALYSIS

The analysis of crystal structures is normally performed using X-ray diffraction techniques. As mentioned previously, we use X-rays (electromagnetic radiation with a wavelength in the range of 0.1–10 A) because the resolution or smallest object separation distance to which any radiation can yield useful information is about equal to the wavelength of the radiation, and the average distance between adjacent atoms is about 1 A. We do not try to look directly at atoms, for unlike visible radiation there is no convenient way to focus X-rays with lenses and magnify the image. Rather, we look at the interference effects when X-rays are scattered by the atoms comprising the crystal lattice. This is analogous to studying the structure of an optical diffraction grating by examining the interference pattern produced when we shine visible light on the grating. For the optical grating, the ruled lines act as scattering centers, whereas in the case of the crystal it is the atoms (or more correctly the electrons) that scatter the incoming radiation.

Before we discuss the application of X-ray diffraction to crystal structure analysis, we must consider the geometrical conditions that must be satisfied if we are to observe a diffraction peak or a maximum in diffracted intensity. Consider a monochromatic (single wavelength) beam of X-rays to be incident on a crystal as shown in Fig. 2-36. Let us imagine that we can replace the scattering centers (atoms) by a set of parallel planes which act as mirrors and reflect the X-rays. The spacing of these planes is d_{hkl}. For constructive interference of the scattered X-rays we require that all of the scattered beams be in phase after they leave the

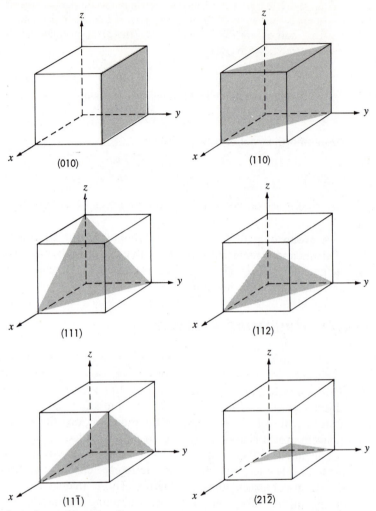

FIG. 2-35 Several planes in a cubic unit cell.

surface of the crystal. In terms of the beams labeled 1 and 2, this requires that the distance $AB + BC$ be equal to an integral number of wavelengths, λ. A similar criterion exists for X rays scattered from any of the more deeply lying planes. Thus, we can write that a condition for constructive interference (the presence of a diffraction peak) is

$$n\lambda = AB + BC$$

or

$$n\lambda = 2d_{hkl} \sin \theta \qquad (2\text{-}12)$$

where n is an integer. Equation (2-12) is known as *Bragg's law* and describes the angular position of the diffracted beam in terms of λ and d_{hkl}. For all cases of interest we can take $n = 1$ and write this expression as $\lambda = 2d_{hkl} \sin \theta$. We are able to

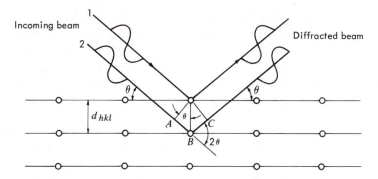

FIG. 2-36   Derivation of Bragg's law, assuming the planes of atoms behave as reflecting planes

do this because we can always interpret a diffraction peak for $n = 2, 3$, etc., as diffraction from the $(nh\ nk\ nl)$ planes, i.e., planes with one-nth the interplanar spacing of d_{hkl}.

There are numerous experimental X-ray diffraction techniques used to study the structure of crystalline solids. One of the most useful techniques for accurately identifying the crystal structure of a substance is the *powder method*, shown schematically in Fig. 2-37. In this method a monochromatic X-ray beam strikes the

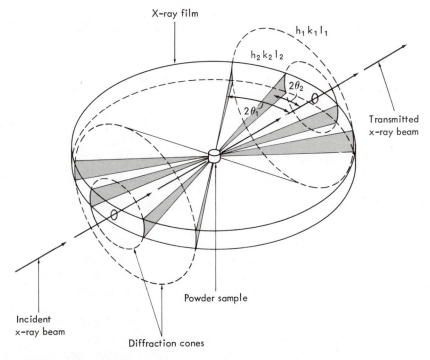

FIG. 2-37   Schematic illustration of the powder method. A strip of film surrounds the sample to record different diffraction peaks.

sample, which is in the form of a fine powder. Each particle in the powder sample can be thought of as a small crystal with a specific crystallographic orientation. All the particles in the sample are assumed to be randomly oriented with respect to one another. The random orientation of the individual crystals insures that the proper Bragg angle will be achieved for all possible crystallographic planes. This means that a large number of diffracted beams are produced. The diffracted beams exist as cones of radiation, where the semiangles of the cones are 2θ, or twice the Bragg angle for that particular crystallographic plane. The reason a diffracted beam forms a cone is that the planes in question will give rise to a diffracted beam for any orientation around the incident beam as long as the incident beam makes the appropriate Bragg angle with the planes. Thus, there is a rotational symmetry of the diffracted beam about the direction of the incident beam.

Those planes with the largest interplanar spacing have the smallest Bragg angle, θ. In fact, only planes with $d_{hkl} > \lambda/2$ will give rise to diffracted beams as θ cannot exceed 90°. Thus, there is a limit to the number of diffraction peaks observed, and this number depends on the wavelength of X-rays. The fact that the diffraction cones have a semiangle 2θ can be explained with reference to Fig. 2-36, where it is shown that the diffracted beam makes an angle of 2θ with respect to the incident beam.

EXAMPLE 2-8

Calculate the angular position of the first diffraction peak (lowest θ) for CsCl if the lattice parameter is 4.1 A and the wavelength of the X-rays is 1.54 A.

Solution:

The first diffraction peak occurs for the largest value of d_{hkl}. Since CsCl has a cubic unit cell, d_{hkl} is largest for the $\{100\}$ type reflection. The value of d_{100} is given by (see Example 2-7)

$$d_{100} = \frac{a}{\sqrt{h^2 + k^2 + l^2}} = \frac{4.1}{\sqrt{1}} = 4.1 \text{ A}$$

The corresponding value of θ is given by

$$\sin\theta = \frac{\lambda}{2d} = \frac{1.54}{8.2} = 0.188$$

$$\theta = 10.8°$$

There are two bits of information we can obtain from studying the diffraction pattern of a crystal. First, from the positions of the diffraction peaks we can determine the size and shape of the unit cell. And secondly, from the relative intensities of the diffraction peaks we can deduce the atom positions in the unit cell. This second proposition is by far the more complicated of the two, and we will limit our attention to the former, giving only a few examples of how to determine the crystal structure once the shape of the unit cell is known.

Using Bragg's law it is possible to determine the size and shape of the unit cell from the angular position of the diffraction peak. We are able to do this because the value of θ is directly related to the value of d_{hkl}. The interplanar spacings are, in turn, a unique function of the lattice parameter and interaxial angles. In general, each particular unit cell has a different set of d_{hkl}. For example, a cubic unit cell with lattice parameter a has values of d_{hkl} equal to the following:

(hkl)	(100)	(110)	(111)	(200)	(210)	. . .
d_{hkl}	a	$\dfrac{a}{\sqrt{2}}$	$\dfrac{a}{\sqrt{3}}$	$\dfrac{a}{\sqrt{4}}$	$\dfrac{a}{\sqrt{5}}$	. . .
$\dfrac{1}{d_{hkl}^2}$	$\dfrac{1}{a^2}$	$\dfrac{2}{a^2}$	$\dfrac{3}{a^2}$	$\dfrac{4}{a^2}$	$\dfrac{5}{a^2}$	. . .

The values of $1/d_{hkl}^2$ are in the simple ratio of 1, 2, 3, 4, 5, ... Rearranging Bragg's law to

$$\frac{\sin^2\theta_1}{\sin^2\theta_2} = \frac{d_2^2}{d_1^2} \tag{2-13}$$

where the subscripts refer to two different reflections, we see that the observed values of $\sin^2\theta$ must follow a similar relationship. Thus, by comparison of the $\sin^2\theta$ values corresponding to observed diffraction peaks it is possible to determine which planes give rise to a given diffraction peak. Then, merely by identifying the diffraction peaks with the appropriate Miller indices and knowing λ, we can calculate the lattice parameter a. The problem is somewhat more complex for noncubic unit cells, since in these cases d_{hkl} depends on more than a single variable; i.e., the interplanar spacing depends on the lattice parameters and interaxial angles.

The intensity of diffraction peaks depends on the phase relationship between radiation scattered by all the atoms in the unit cell. Quite often the intensity of a particular peak, whose presence is predicted by Bragg's law, will be zero. This is because Bragg's law deals not with atom positions but only with the size and shape of the unit cell. For example, consider the intensity of the (100) diffraction peak for a crystal that has a BCC unit cell. Figure 2-38 shows the phase relationship for

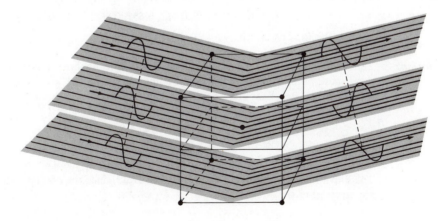

FIG. 2-38 Body-centered cubic unit cell showing phase relationship for reflected beams from (100) planes. Resultant intensity is zero.

the reflected beams. The X rays scattered from the atoms at the top and bottom faces of the unit cell [(100) planes] constructively interfere but are 180° out of phase with X rays scattered by the atom at the center of the unit cell. The resultant intensity is therefore zero, and no diffraction peak is observed. There are some simple rules, summarized in Table 2-5, governing the presence of

TABLE 2-5

Rules Governing the Presence of Diffraction Peaks

Bravais Lattice	Reflections Present	Reflections Absent
Simple	All	None
Body centered	$(h + k + l) =$ even	$(h + k + l) =$ odd
Face centered	h, k, l unmixed*	h, k, l mixed

*h, k, and l are either all odd or all even.

diffraction peaks in different Bravais lattices. These rules are strictly true only for unit cells where a single atom is associated with each lattice point. Unit cells with more than one atom per lattice point may have these atoms arranged in such positions that reflections normally present will be missing. For example, diamond has an FCC Bravais lattice and has two atoms associated with each lattice point. All the reflections present for diamond have unmixed indices, but reflections such as {200}, {222}, and {420} are also missing. The fact that all reflections present have unmixed indices indicates the Bravais lattice is FCC, while the extra missing reflections give some further information as to the exact atom arrangement. Thus, we can consider Table 2-5 as really only listing conditions that must be fulfilled for a diffracted beam to exist. The rules themselves do not necessarily predict the existence of the beam.

EXAMPLE 2-9

The element Q belongs to the cubic crystal system. From X-ray measurements it is known that the seven lowest order diffraction peaks occur at

$$\sin^2 \theta = 0.137, 0.275, 0.412, 0.551, 0.688, 0.826, 0.962$$

What is the Bravais lattice?

Solution:

The $\sin^2 \theta$ values are in the ratio 1, 2, 3, 4, 5, 6, 7. For the three cubic Bravais lattices the lowest order, $1/d_{hkl}^2$ and corresponding $\sin^2 \theta$ values are in the ratios of $(h^2 + k^2 + l^2)$ (see Example 2.7). From Table 2.5 we can evaluate the ratio of the allowed $(h^2 + k^2 + l^2)$ values as:

Simple cubic	1, 2, 3, 4, 5, 6, 8, 9, 10
BCC	2, 4, 6, 8, 10, 12, 14, 16, 18
FCC	3, 4, 8, 11, 12, 16, 19

The Bravais lattice of element Q is therefore BCC, since the observed $\sin^2 \theta$ values are identical to the allowed $(h^2 + k^2 + l^2)$ values.

Information can also be obtained about the structure of solids using diffraction techniques associated with high energy particles. The most common of these techniques are electron and neutron diffraction. The diffraction patterns obtained using these particles may be interpreted in terms of Bragg's law, where we take λ equal to the deBroglie wavelength [Eq. (2-2)]. Electron diffraction is a useful technique because the scattering efficiency of atoms for electrons is much greater than that for X-rays. This means that very thin layers of material (≈ 100 A) can be studied by electron diffraction. Figure 2-39 shows an electron diffraction pattern of a thin

FIG. 2-39 Transmission electron diffraction pattern from a stainless steel sample containing many small randomly oriented grains. The pattern is similar to that obtained from the powder pattern technique (Fig. 2-37), with film perpendicular to the incident beam. Photo courtesy A. J. West and R. J. Asaro, Stanford University. Used with permission.

nickel foil composed of a number of small crystals. Neutron diffraction is important because (1) the scattering of neutrons by light elements is relatively much stronger than the scattering of X rays, thus permitting the study of the position of light atoms, such as carbon and hydrogen, in polymers, and (2) there is an interaction of the magnetic moment associated with the spin of the neutron and the magnetic moments of the atoms. This latter point allows one not only to deduce the location of atoms but also to determine the direction of their magnetic moment. This point is very important in the study of the magnetic behavior of materials (Chapter 14).

PROBLEMS

2-1   In a sodium-vapor lamp visible light is produced with a wavelength equal to 5893 A. How many photons are emitted per second if the lamp has a rating of 1 W?

2-2   Calculate the wavelength and the energy of the photon emitted when an electron in a copper atom jumps from the $n=2$ state to the $n=1$ state.

2-3   Suppose a photon of wavelength 0.1 A is absorbed by an $n = 1$ electron in potassium and the electron is removed from the atom. Only part of the photon's energy was used to remove the electron. If the remainder of the energy is stored in kinetic energy, find the velocity of the electron.

2-4   Assume that the potential energy between atoms can be represented by the following relation

$$v(r) = -\frac{A}{r} + \frac{B}{r^{10}}$$

where A and B are constants. Write an expression for the force between two atoms. Make a sketch of the force–distance curve between two atoms, and illustrate the feature of the curve that corresponds to the elastic modulus of the solid in which the atoms reside. Derive an equation for the force that would be necessary to separate the two atoms (analogous to a theoretical fracture strength).

2-5   The elastic modulus of diamond is measured by applying a force parallel to the body diagonal of the unit cell and measuring the displacement. A similar measurement is made applying the force parallel to one of the cube axes. Explain why the two measurements give different values of the elastic modulus.

2-6   Calculate the volume change for a material which transforms from a simple cubic structure to a hexagonal close-packed structure. Assume the atoms to behave as hard spheres.

2-7   Explain why a simple ionic compound AB, which has ions of equal size, cannot adopt a crystal structure in which each atom has 12 unlike nearest neighbors.

2-8   A hypothetical compound AB has the NaCl structure. In this compound the A ions are larger than the B ions. Each A ion is in physical contact with 12 A ions and six B ions. Using the hard-sphere approximation, calculate the ratio of ionic radii r_A/r_B for this compound.

2-9   Write the coordinates of the largest interstitial hole in the FCC structure (i.e., where should we put an extra atom if we are looking for the most roomy spot?). How many of these sites are there per unit cell? Do the same for the BCC structure.

2-10 Suppose that the compounds BeO, MgO, and CaO could be constructed by placing the oxygen ions in an FCC array and by putting the cations into the octahedral sites. The ionic radii for the ions in question are

$$r_{Be}^{++} = 0.31 \text{ A} \qquad r_{Mg}^{++} = 0.65 \text{ A}$$
$$r_{Ca}^{++} = 0.99 \text{ A} \qquad r_{O}^{--} = 1.40 \text{ A}$$

(a) If the oxygen ions are arranged in a close-packed array (FCC), which of the cations (Be, Mg, or Ca) most nearly fits into the octahedral hole provided by the oxygen ions?

(b) Which of the cations causes most distortion in the oxygen lattice?

(c) Calculate the radius of the cation which would fit exactly into the octahedral hole.

2-11 The unit cell of $BaTiO_3$ is shown in the accompanying diagram. The crystal system is cubic.

(a) What is the Bravais lattice?

(b) How many atoms are there per unit cell?

(c) Give the type and number of nearest neighbors for each atom.

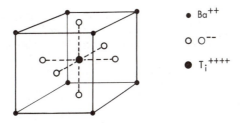

- Ba^{++}
- O^{--}
- Ti^{++++}

2-12 Draw the following planes in an orthorhombic unit cell: (001), (011), (112), (201), ($\bar{1}$01).

2-13 In a hexagonal close-packed structure, if the atoms are imagined to pack together like hard balls, what is the c/a ratio, where c and a are the dimensions of the unit cell?

2-14 How many $\langle 110 \rangle$ directions are contained in the (111) plane in the FCC structure?

2-15 What are the directions of closest atomic packing in the BCC, FCC, and HCP structures?

2-16 The FCC unit cell, defined by three orthogonal vectors **a**, **b**, and **c**, where $a = b = c$, has four lattice points per unit cell. Another unit cell sometimes used to describe the FCC structure is defined by the vectors $(\bar{a} + \bar{b})/2$, $(\bar{a} + \bar{c})/2$, and $(\bar{b} + \bar{c})/2$. Show that this unit cell has only one lattice point associated with it.

2-17 Show that the spacing between (111) planes in the FCC structure is the same as the spacing between (002) planes in the HCP structure. Imagine the atoms to behave as hard spheres.

2-18 BaO has the NaCl crystal structure with a lattice parameter of 5.50 A. Calculate the number of atoms per unit cell and the density of BaO.

2-19 During a lunar exploration an unknown crystalline substance is discovered whose external morphology indicates that it is a cubic material. A diffraction pattern of this material, using radiation of wavelength 1.54 A, provides the following data:

Diffraction Peak	Diffraction Angle, θ	Sin² θ
#1	13.70°	0.0561
#2	15.97°	0.0748
#3	22.85°	0.1495
#4	27.05°	0.2055
#5	28.30°	0.2245
#6	32.61°	0.2900

(a) What structure does this unknown material possess?

(b) What is its lattice parameter?

2-20 Gold has an FCC crystal structure with a lattice parameter of 4.07 A. Calculate the Bragg angle θ for the (112) diffraction peak for X rays with a wavelength of 1.0 A and for electrons with an energy of 100,000 eV.

BIBLIOGRAPHY

J. C. ANDERSON and K. D. LEAVER, *Materials Science*, Ch. 1–6. New York: Van Nostrand-Reinhold, 1969.

L. V. AZAROFF, *Introduction to Solids*, Ch. 3. New York: McGraw-Hill, 1960.

———, *Elements of X-ray Crystallography*, Ch. 1–4. New York: McGraw-Hill, 1968.

J. B. COHEN, *Diffraction Methods in Materials Science*, Ch. 1–2. New York: Macmillan, 1966.

B. D. CULLITY, *Elements of X-ray Diffraction*, Ch. 2–3. Reading, Mass.: Addison-Wesley, 1956.

A. T. DIBENDETTO, *The Structure and Properties of Materials*, Ch. 1, 3–5. New York: McGraw-Hill, 1967.

T. S. HUTCHISON and D. C. BAIRD, *The Physics of Engineering Solids*, Ch. 1–2. New York: John Wiley, 1968.

C. KITTEL, *Introduction to Solid State Physics*, Ch. 1. New York: John Wiley, 1956.

W. G. MOFFATT, G. W. PEARSALL, and J. WULFF, *The Structure and Properties of Materials*, Vol. I, Ch. 1–3. New York: John Wiley, 1964.

3
The Structure of
Imperfect Solids

3-1 INTRODUCTION

Real crystalline solids are perfect neither with regard to atomic configuration nor with respect to their electronic energy spectrum. Fortunately, however, even though crystals are not perfect, their departures from perfection are quite discrete and finite. For this reason we may visualize a real crystal as the superposition of discrete imperfections onto a perfect crystalline lattice. These imperfections may be conveniently classified with respect to their dimensionality. In this scheme we find that all defects can be classified into one of the following groups:

1. *Point defects*—zero-dimensional imperfections
2. *Line defects*—one-dimensional imperfections
3. *Interfacial defects*—two-dimensional or planar imperfections
4. *Bulk defects*—three-dimensional imperfections

Point defects are localized defective regions of the crystal that are confined to a volume which is of atomic dimensions. Foreign atoms, vacant lattice sites, and extra or missing electrons are examples of imperfections belonging to this class.

Line imperfections, as the name implies, are defective regions of the crystal that extend through the crystal along a line. This line is not necessarily straight; it may be curved and even close upon itself and form a loop. Although there are many types of line imperfections having rather different properties, all are referred to by the term *dislocations*.

Interfaces are two-dimensional imperfections that serve as boundaries between one orderly region of a crystal and another. The interface is commonly a transition region in which the *order* of the crystal is interrupted. In this description, *order* is a generalized term that may refer to the magnetic state, the electronic state, the chemical composition, or the configurational arrangement of atoms.

Bulk defects are macroscopic, or large-scale, defects that represent an inhomogeneity in the shape or structure of the solid. Typical of bulk defects are voids, cracks, and inclusions (undesired included particles of another material, such as an iron oxide inclusion in a steel).

Amorphous solids also have imperfections in their structures. However, because of the lack of long-range order in amorphous solids, only point, planar, and bulk imperfections are important in these materials. The presence of line defects, or dislocations, requires that the material have a crystalline structure.

The size range over which the various classes of imperfections exist is illustrated in Table 3-1. This size range is somewhat approximate and is intended only to convey a feeling for the relative dimensions involved.

TABLE 3-1

Relative Size Range of the Different Classes of Defects

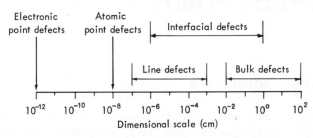

Imperfections in solids are not merely interesting intricacies of the solid state. In fact, most of the properties of solids are so intimately related to imperfection behavior that it would be grossly misleading to consider the hypothetical properties of a perfectly ordered arrangement of atoms with a perfectly orderly electronic energy spectrum. We will consider briefly the role of defects in the determination of a few of the important properties of solids.

The electrical behavior of semiconductors is controlled by crystal imperfections. The conductivity of silicon, for example, can be increased 10,000 times by the addition of minute amounts (0.01 %) of arsenic. Each atom of arsenic in this case represents a point defect in the silicon lattice. The fact that small amounts of impurity atoms can so significantly alter the electrical properties of semiconductor materials is responsible for the development of the transistor and more recently has opened up the entire field of solid-state circuits. Practically none of the semiconducting properties that lead to these engineering accomplishments would be found in a perfect crystal. They are properties peculiar to the defect in the crystalline solid.

The presence of dislocations in crystals provides a mechanism by which permanent change of shape or mechanical deformation can occur. If a crystalline solid free of dislocations could be produced, it would be completely brittle and practically useless as an engineering material. While the existence of dislocations in crystals insures ductility (ability to deform), the strength of crystalline solids is drastically reduced by their presence. It has been found, for example, that a crystal of copper can be permanently deformed by the application of stress as low as 10 psi. However, the strength of such a crystal can be increased by a factor of 1000 simply by severe deformation at room temperature. The increase in strength is related uniquely to the increase in the number of line imperfections that accompany such a treatment.

From these statements we see that dislocations play a central role in the determination of such important properties as strength and ductility. We shall see later that essentially all mechanical properties of crystalline solids are controlled by the behavior of line imperfections.

The ability of a ferromagnetic material (such as iron, nickel, or iron oxide) to be magnetized and demagnetized depends in large part on the presence of two-dimensional imperfections known as *Bloch walls*. These interfaces are boundaries between two regions of the crystal which have a different magnetic state. As magnetization occurs, these defects migrate and by their motion provide the material with a net magnetic moment. Without the existence of Bloch walls, all ferromagnetic materials would be permanent magnets. In fact, electromagnets would not exist were it not for these defects.

Finally, the presence of bulk defects such as cracks causes brittle materials like glass to break at very small applied stresses. For example, glass in the absence of any surface cracks has a fracture strength of $\approx 10^6$ psi. However, if as is generally the case there are small surface cracks present, which are only $\approx 10^{-4}$ cm deep, the fracture strength of the glass decreases to $\approx 10^4$ psi. And if the cracks are deeper than 10^{-4} cm the fracture strength decreases even more. This fact is familiar to everyone who has broken a glass tube by first filing a small notch (or crack) into the surface. Removal of cracks from the surface of glass either by etching in hydrofluoric acid or by flame polishing almost always raises the fracture strength.

In the following sections we shall briefly describe the various imperfections found in materials. Our picture of imperfections will be limited mainly to describing imperfections in the configurational arrangement of atoms, although we shall briefly consider the influence of these atomic imperfections on the electronic structure. Chapters 11–15 will consider imperfections in the electronic structure in considerably more detail.

3-2 POINT DEFECTS IN CRYSTALLINE SOLIDS

Point defects in pure crystalline materials are defects of atomic dimensions that usually result from the presence of an impurity atom, the absence of a matrix atom, or the presence of a matrix atom in the wrong place. These defects are shown in Fig. 3-1. An impurity atom that occupies a normal lattice site is called a *substitutional impurity atom*, and an impurity atom found in the interstice between matrix atoms is called an *interstitial impurity atom*. Whether a foreign atom will occupy a substitutional or interstitial site depends on the size of the atom relative to the size of the site. Small atoms are usually interstitial impurities, while larger atoms are usually substitutional impurities.

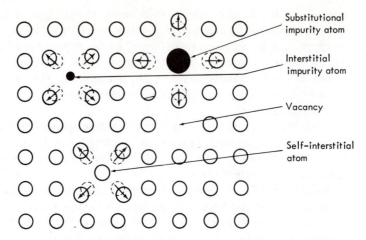

FIG. 3-1 Various point defects in crystalline materials

A vacant lattice site is referred to as a *vacancy*, and a matrix atom in an interstitial site is called a *self-interstitial*. The concentration of vacant lattice sites in pure materials is very small at low temperatures—about one vacancy every 10^8 atom sites—and increases with increasing temperature to about one vacancy every 10^3 sites at the melting temperature.* Vacancies are important because they control the rate of matrix or substitutional atom diffusion; that is, atoms are able to move around in a crystalline solid because of the movement of vacancies. This is shown schematically in Fig. 3-2. Self-interstitials generally do not occur naturally, but may be introduced into a structure by irradiation. For example, high-energy neu-

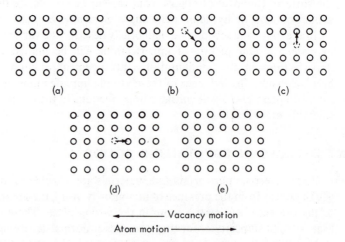

FIG. 3-2 Importance of vacancies in matrix atom motion (solid state diffusion)

---

*The calculation of the number of vacancies present at any given temperature is given in Chapter 4.

trons from atomic fission can knock atoms from their regular sites into interstitial sites, creating vacancy-interstitial pairs.

EXAMPLE 3-1

Assuming atoms to pack together as hard spheres, find the largest interstitial atom that can fit in a BCC arrangement of iron atoms. Take the radius of the iron atom as 1.24 A. Carry out the same calculation for an FCC arrangement of iron atoms. On the basis of these calculations, would you expect carbon with an atomic radius of 0.77 A to occur as an interstitial in BCC iron?

Solution:

The largest interstitial site in the BCC unit cell is of the type found centered at $\frac{1}{2}\frac{1}{4}0$. An atom at this site has four nearest neighbors, the atoms at 000, $\frac{1}{2}\frac{1}{2}\frac{1}{2}$, 100, and $\frac{1}{2}\frac{1}{2}\overline{1}$. To find the largest atom that can just fit into this site without disturbing the iron atoms we need to know the lattice parameter of BCC iron. In the BCC structure the close-packed crystallographic direction is $\langle 111 \rangle$, and thus the lattice parameter a_{BCC} is equal to

$$a_{BCC} = \frac{4(\text{radius of iron atom})}{\sqrt{3}} = \frac{4}{\sqrt{3}}(1.24) \text{ A} = 2.86 \text{ A}$$

Now for an interstitial atom at $\frac{1}{2}\frac{1}{4}0$ to be just touching the iron atom at 000, we have

(Radius of iron atom) + (radius of interstitial atom) = (distance from 000 to $\frac{1}{2}\frac{1}{4}0$)

$$(1.24\text{A}) + (r_{\text{interstitial}}^{BCC}) = a_{BCC}\sqrt{\tfrac{5}{16}} = 1.59 \text{ A}$$
$$r_{\text{interstitial}}^{BCC} = 0.35 \text{ A}$$

In an FCC lattice the largest interstitial site is of the type found at $\frac{1}{2}\frac{1}{2}\frac{1}{2}$ (or $\frac{1}{2}00$), where there are six nearest neighbors. To determine the lattice parameter of FCC iron we know the closed-packed direction is $\langle 110 \rangle$, consequently,

$$a_{FCC} = \frac{4(\text{radius of iron atom})}{\sqrt{2}} = 3.5 \text{ A}$$

And for the largest interstitial atom, we have

$$(\text{Radius of iron atom}) + (\text{radius of interstitial atom}) = \frac{a_{FCC}}{2}$$
$$r_{\text{interstitial}}^{FCC} = 1.75 - 1.24 = 0.51 \text{ A}$$

Since the radius of the carbon atom is larger than the interstitial sites in BCC iron, carbon cannot be dissolved interstitially without distorting the lattice. In practice, it is found that the atoms that dissolve interstitially in BCC iron (e.g., carbon and nitrogen) distort the arrangement of iron atoms, creating a stress field about the interstitial.

Point defects in ionic structures differ from those found in pure elements because of the charge neutrality requirement in ionic solids. For example, in a pure mono-valent ionic material a cation vacancy must have associated with it either a cation interstitial or an anion vacancy to maintain charge neutrality. Similar requirements hold for anion vacancies. The vacancy pair defect is usually called a *Schottky imperfection*, and the vacancy-interstitial pair is called a *Frenkel imperfection*. These two types of imperfections are shown in Fig. 3-3(a). Self-interstitials are much more common in ionic structures than in pure elements because many ionic com-

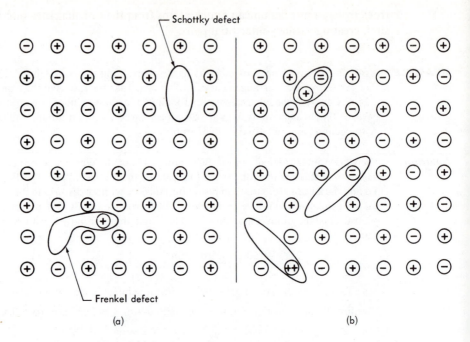

FIG. 3-3   Imperfections in ionic crystals: (a) imperfections in pure crystals, (b) imperfections caused by impurity atoms

pounds have relatively large interstitial sites available. That is, there are often interstitial sites in the unit cell that have nearly the same surroundings as normal atom sites. For example, in BeO (see Fig. 2-25) the Be atoms fill up only one-half the available tetrahedral sites, leaving four possible cation interstitial sites per unit cell. Thus, a Be atom could go from a regular lattice site to an almost equivalent interstitial site with little distortion of the lattice.

Foreign atoms in ionic crystals produce defects that also must maintain charge neutrality. For example, in NaCl a monovalent cation such as lithium may simply replace one of the sodium ions as a substitutional impurity. But a divalent cation, such as calcium, replacing a sodium ion must be accompanied by either a cation vacancy or an anion interstitial if charge neutrality is to be maintained. Correspondingly, monovalent impurity cations in a divalent structure, (e.g., Na in MgO) must be accompanied by an appropriate number of cation interstitials or anion vacancies. Some of these defects are shown schematically in Fig. 3-3(b).

Substitutional impurities in covalently bonded materials can create a unique imperfection in the electronic structure if the impurity atom is from a group in the periodic table other than the matrix atoms. For example, consider a group V element in a group IV matrix, such as As in Si. In the tetrahedrally covalent silicon structure (see Fig. 2-7) the arsenic atom shares four of its five outer electrons in covalent bonds. Those four electrons plus one electron from each of the four adjacent silicon atoms complete the outer electron shell of the arsenic atom. The remaining outer electron of the arsenic atom is *not* involved in a covalent bond and is therefore not involved in completing the outer electron shell of the arsenic atom. Consequently, this electron is only very weakly bound to the arsenic atom and may be considered to be an essentially free electron in the solid, similar to the bonding electrons in metals. As a free electron it can respond to electrical fields and move easily through the lattice. This is in contrast to the electrons involved in the covalent bonds, which are strongly bound electrons and cannot move under the influence of an external electric field. These essentially free electrons donated by the arsenic atoms cause a large increase in the electrical conductivity of silicon when minute amounts of arsenic are added.

The addition of group III elements to a group IV matrix results in a different type of imperfection in the electronic structure, but an imperfection that also greatly increases the electrical conductivity. In this case there is one electron too few to complete all the covalent bonds, and consequently there is a missing electron, or *hole*, in the covalent bond structure. As shown in Fig. 3-4, these holes in the

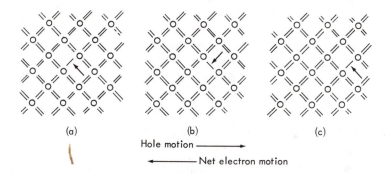

(a)   (b)   (c)

Hole motion ———→

←——— Net electron motion

FIG. 3-4   Electrical conduction via hole motion. The arrowheads indicate hole position.

electron structure can contribute to the electrical conductivity in the same fashion that atom vacancies allow a net movement of atoms. Under the influence of an electric field, electron holes move in one direction (jumping from bond to bond) and electrons move in the opposite direction. The resultant current flow is due entirely to the presence of the electron holes, which in turn are directly related to the presence of the impurity atoms. Although we have described electronic imperfections only in the group IV elements, these imperfections can be found in all covalently bonded elements and compounds (e.g., the III-V and II-VI compounds). In general, we call the impurity atoms that donate free electrons *donor atoms* and the impurities producing electron holes (or accepting an electron) *acceptor atoms*. In Chapter 12 we shall consider the role of donor and acceptor atoms in the electrical properties of covalent materials in much greater depth.

When foreign atoms are incorporated into a crystal structure, whether in substitutional or interstitial sites, we say that the resulting phase is a *solid solution* of the matrix material (solvent) and the foreign atoms (solute). The term *solid solution* is not restricted to low solute contents, for there are many systems that are solid solutions over a wide composition range. For example, silicon and germanium are completely soluble in one another in the solid state.

The two different types of solid solutions are shown in Fig. 3-5. In a *substitutional solid solution* the atoms distribute themselves on the normal atom sites in a random fashion such that the probability of finding a particular type of atom at a given

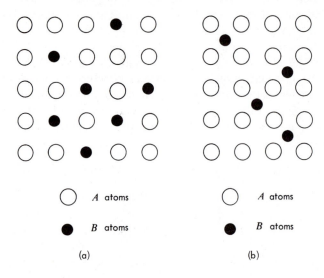

○ *A* atoms

● *B* atoms

(a)

○ *A* atoms

● *B* atoms

(b)

FIG. 3-5   Substitutional (a) and interstitial (b) solid solutions

site is equal to the atomic fraction of that element. Substitutional solid solutions are found between elements (e.g,. Si-Ge, Au-Ni, Cu-Ni) and compounds (e.g., Al_2O_3-Cr_2O_3, NaCl-KCl, MgO-FeO). In an *interstitial solid solution* the solvent atoms occupy the normal atom sites and the solute atoms occupy interstitial sites. Carbon in iron is an example of an interstitial solid solution.

EXAMPLE 3-2

The compound CdS can exist over a range of nonstoichiometric compositions in which there are not equal numbers of Cd and S atoms. If the density of the compound $CdS_{0.95}$ is 4.91 g/cm³, determine whether the structure has anion lattice vacancies or interstitial cation atoms. CdS has a cubic structure identical to that shown for GaAs in Fig. 2-26. The lattice parameter of $CdS_{0.95}$ is equal to 5.82 A.

Solution:

If the density of $CdS_{0.95}$ is greater than that of CdS, then there are cation interstitials; if the nonstoichiometric density is less than pure CdS, there are anion vacancies. To compute the density of CdS we take

$$\rho_{CdS} = \frac{(4 \text{ atom pairs/unit cell})(144.5 \text{ g/mole})}{(6.02 \times 10^{23} \text{ atom pairs/mole})/[(5.82)^3 \times 10^{-24} \text{ cm}^3/\text{unit cell}]} = 4.85 \text{ g/cm}^3$$

Thus, since $\rho_{CdS} < \rho_{CdS_{0.95}}$, cation interstitials are present.

3-3 POINT DEFECTS IN AMORPHOUS SOLIDS

When describing point defects in amorphous materials, it is convenient to consider separately the two classes of amorphous materials, i.e., the network structures and the long-chained polymeric structures. In the following discussion we shall limit our attention to point defects in some relatively simple amorphous structures, noting that similar defects are found in even the most complex arrangement of atoms.

Structurally, the simplest network material is SiO_2, which is a random arrangement of silica tetrahedra (see Fig. 2-30). In pure SiO_2 each silicon atom is covalently bonded to four oxygen atoms and each oxygen atom is covalently bonded to two silicon atoms. Thus, each oxygen atom serves as a bridge between two silicon atoms. Imperfections arise in the three-dimensional network of silica tetrahedra as other oxides, such as Na_2O, are added. For example, if Na_2O is added, the Na^+ ions can be accommodated into holes in the structure as shown in Fig. 3-6. The increased

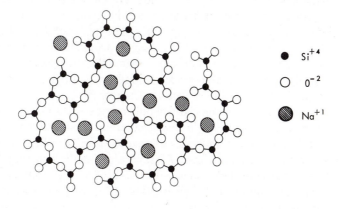

FIG. 3-6   Schematic of a sodium-silicate glass

oxygen content, however, means that the O:Si ratio increases and some of the Si-O-Si bridges are broken up into ionic Si-O groups containing non-bridging oxygen. When the O:Si ratio increases to 2.5, the number of bridging oxygens per silicon will have decreased to three. As additional Na_2O is added, the network structure becomes increasingly broken down, and when the O:Si ratio is equal to four, we have discrete SiO_4^{-4} ions with no three-dimensional network structure. Consequently, at high Na_2O contents we cannot form a continuous network structure, and the structure usually becomes crystalline. The reason *network modifiers* such as Na_2O are added to SiO_2 in commercial glasses is that as the number of strong covalent Si-O bonds is broken, both the melting temperature and viscosity at a given temperature decrease and the glass can be fabricated at lower temperatures. For example, a 20% addition of Na_2O to amorphous SiO_2 decreases the viscosity at 1400°C by about 10^{10}. Most glasses have an O: Si ratio of about 2.5. Oxides other than Na_2O which modify the network structure include practically all the alkali and divalent oxides, K_2O, Li_2O, CaO, MgO, BaO, ZnO, etc.

Whereas the addition of network modifiers such as Na_2O to SiO_2 lowers the melting temperature and increases the tendency for the glass to crystallize, or *devitrify*, there are many oxides, such as B_2O_5, GeO_2, Al_2O_3, and V_2O_5, that have the opposite effect. These oxides, which contribute to the formation of a network structure and resist devitrification, are known as *glass formers*. Very simply, the metallic ions in glass-formers have a high bond energy with oxygen (appreciable covalent component) and are incorporated into the three-dimensional network structure of the glass rather than fitted into holes in the network structure as are the Na ions in Fig. 3-6. The addition of B_2O_5 to SiO_2 would therefore tend to propagate the network structure rather than break it down into ionic groups. Generally, oxides that are glass-formers readily exist as glass when in the pure state. In addition to network modifiers and glass formers, there is a group of oxides, such as TiO_2, ZnO, and ZrO_2, that behave as network modifiers in some glasses and as glass-formers in others. These oxides are known as *intermediates*.

Other impurity atoms are also found in network structures, their presence usually altering the color or the transparency of the material. For example, dissolved oxides of transition metal ions such as copper, iron, and manganese add color to glass, while small amounts of crystalline materials such as CaF_2, SnO_2, and TiO_2 make glass opaque. We shall consider the influence of these point defects on the optical properties of glass in some detail in Chapter 15.

Imperfections in long-chained polymetric structures can be conveniently divided into two groups: defects in the structure of the individual molecules, and defects occurring when the molecules are packed together. We can analyze imperfections in the individual molecules by first considering the structure of a perfect molecule. Each long-chained molecule is made up of a large number of repeating units or *mers*. For example, polyethylene (Fig. 2-27), which is a member of the vinyl compounds, has a repeating unit of

$$\begin{array}{c} H \quad H \\ | \quad | \\ -C-C- \\ | \quad | \\ H \quad H \end{array}$$

Other vinyl compounds have a repeating unit of

$$\begin{array}{c} H \quad H \\ | \quad | \\ -C-C- \\ | \quad | \\ H \quad X \end{array}$$

where X may be a single atom, Cl in the case of polyvinyl chloride, or a complex monovalent side group,

$$
\begin{array}{c}
O \quad \diagup O \\
\diagdown\!\!/ \\
C \\
\mid \\
H\!-\!C\!-\!H \\
\mid \\
H
\end{array}
$$

in the case of polyvinyl acetate. When these repeating units combine together to form a long molecule, the side groups line up in one of the three ways shown in Fig. 3-7. The *isotactic* arrangement has all the side groups on the same side of the chain; the *syndiotactic* arrangement has the side groups regularly alternating from one side to the other; and the *atactic* arrangement of the side groups is completely random. The exact arrangement of the side groups is of profound importance in

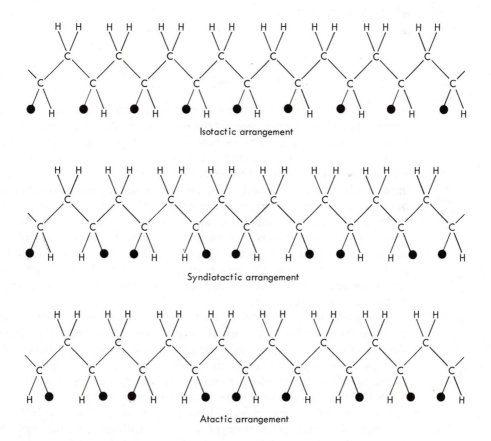

FIG. 3-7   Arrangement of side groups (dark circles) in a vinyl polymer. Each hydrogen pair or side group is oriented such that one atom is above and the other is below the plane of the page.

determining whether a polymer will be crystalline or amorphous. For example, a polymer with a large, bulky side group and an atactic arrangement will generally be amorphous. Conversely, polymers with isotactic arrangements can crystallize easily even if the side group is large.

Polymers with the same composition but with different atomic arrangements are called *isomers*. Because of their different structures isomers can have vastly different properties. For example, the monomer of natural rubber is

$$
\begin{array}{ccc}
CH_3 & H & \\
| & | & \\
H & C = C & H \\
\diagdown \diagup & & \diagdown \diagup \\
C & & C \\
\diagup \diagdown & & \diagup \diagdown \\
H & H &
\end{array}
$$

while that of gutta-percha is

$$
\begin{array}{cccc}
H & & H & H \\
| & & | & | \\
-C-C{=}C-C- \\
| & | & & | \\
H & CH_3 & & H
\end{array}
$$

Both isomers have the same chemical composition and differ only in that the CH_3 group is found on opposite sides of the molecule. In the natural rubber monomer the molecule is physically bent to allow room for the CH_3 group and the hydrogen atom on the double-bonded carbon atoms. This causes the molecular chains in natural rubber to be bent and tangled rather than long and straight. Consequently, natural rubber is never found in the crystalline state. Gutta-percha, in contrast, has a relatively straight molecular chain; the CH_3 group and hydrogen atom are on opposite sides of the molecule; and it crystallizes easily. The crystalline structure of gutta-percha results in a relatively rigid nonflexible solid, while the bent and tangled molecules of natural rubber can easily "stretch out" when a load is applied, making the material very flexible and elastic. Polymers with a structure similar to that of natural rubber are known as *elastomers* because of their large, reversible elastic deformations.

A common defect in polymer chains occurs when the polymer branches off and loses its linear form. A schematic illustration of branching in polyethylene is shown in Fig. 3-8. Although only one branch is shown, most polymers will often have a large number of branches per chain. Branching can be of two types: *tree branching*, where the branches are long (>10 monomer units) and *feather branching*, where the branches are only a few monomer units long. Practically all polymers have one or both of these types of branching. As we shall discuss in Chapter 10, polymers with branches are stronger than those without, since the branches make it more difficult for the molecules to slide past one another.

In polymeric systems as well as in crystalline systems it is possible to have solid solutions of different components. This solution may take the form of dissolving organic molecules into the spaces between polymer chains or to incorporate two

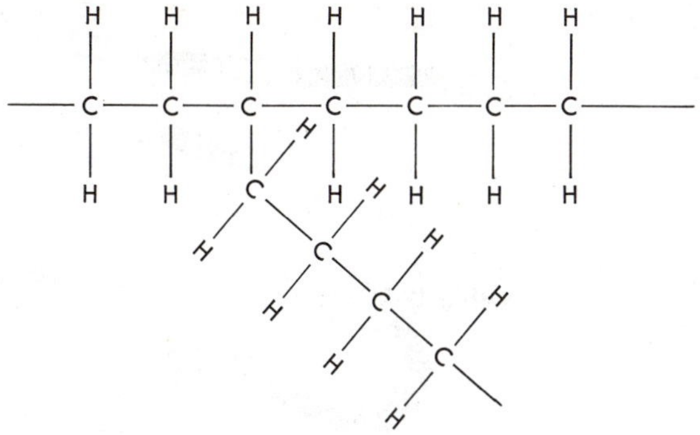

FIG. 3-8   Schematic branching in polyethylene

or more individual monomers into the same chain. In the former case the dissolved molecule is usually called a *plasticizer* because its presence inhibits crystallization of the polymer and promotes a soft, tough, flexible noncrystalline structure. Glycerol acts as a plasticizer when added to cellulose, producing the flexible plastic, cellophane.

A polymer chain consisting of two different types of monomers is called a *copolymer*. A schematic illustration of different copolymer structures is shown in Fig. 3-9. If the copolymer has the two different monomers distributed randomly or alternating along its length, it is called a *regular copolymer*. *Block copolymers* have chains made up of a long sequence of one monomer, followed by a long sequence of another monomer, and *graft copolymers* have a chain made up of one monomer unit and branches of another monomer unit. Whereas regular copolymers usually have properties quite different from the constituent monomers, block and graft copolymers normally retain some of the properties of each of the basic monomers. For example, it is possible to improve the dyeing properties of a hydrophobic fiber such as Acrilan (a commercial name for one of the acrylic fibers made from polyacrylonitrile) by grafting on branches of hydrophilic monomer units (e.g., cellulosic fibers such as cotton and rayon) without loss of the Acrilan fiber tensile strength.

Many polymers do not retain their linear molecular nature when formed into a solid object. Instead, they form a *cross-linked* or *network* structure. This is shown in Fig. 3-10 for the case of polyisoprene (natural rubber) cross-linked with sulfur atoms. The cross-links prevent the molecules from sliding past one another and greatly increase the mechanical strength. For example, natural rubber has a tensile strength of 300 psi, while rubber cross-linked with sulfur (vulcanized rubber, which contains $\approx 5\%$ sulfur) has a strength of 3000 psi. When cross-linking takes place at elevated temperatures through the application of pressure (during fabrication) the polymer is said to be *thermosetting*. The cross-linked network extending throughout the final structure is thermally stable and cannot easily be made to deform. In contrast, most linear polymers can be made to soften by raising the temperature and are referred to as *thermoplastic*. Phenol-formaldehyde (Bakelite) is

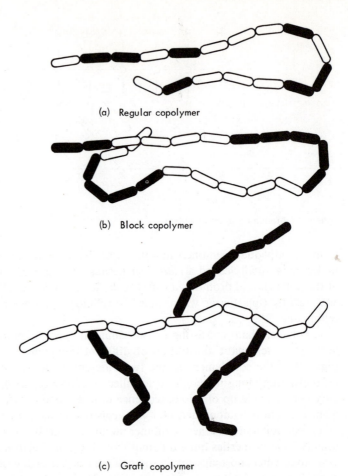

(a) Regular copolymer

(b) Block copolymer

(c) Graft copolymer

FIG. 3-9   Schematic representation of polymer solutions. Each unit represents one monomer.

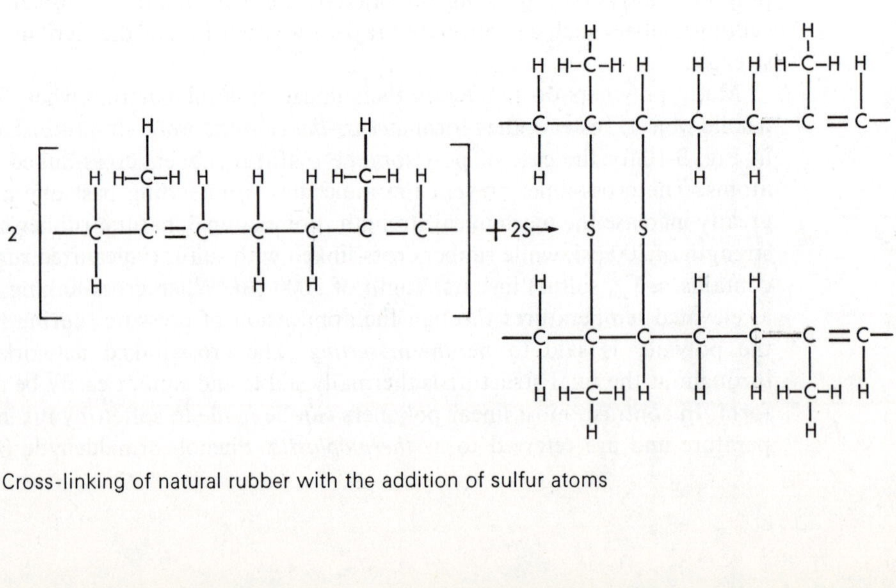

FIG. 3-10   Cross-linking of natural rubber with the addition of sulfur atoms

an example of a thermosetting polymer, while polystyrene and polymethyl-metha-crylate (plexiglass or Lucite) are examples of thermoplastic polymers.

EXAMPLE 3-3

The monomer of natural rubber (C_5H_8) is shown on page 76. If cross-linking during vulcanization with sulfur occurs only by breaking the double carbon bond (Fig. 3-10), what fraction of monomers are cross-linked in a rubber containing 5 weight percent sulfur?

Solution:

Since two sulfur atoms are required to cross-link two natural rubber monomers, we need to know the atomic fractions of sulfur and C_5H_8 monomers.

The atomic weight of one monomer of natural rubber is

$$5(12) + 8(1) = 68$$

while that of sulfur is 32. Thus

$$\text{Atomic fraction of sulfur} = \frac{0.05/32}{0.05/32 + 0.95/68} = 0.10$$

while the atomic fraction of C_5H_8 is 0.90. Hence, one of every nine, or 11% of the monomers, are cross-linked.

3-4 LINE DEFECTS IN CRYSTALLINE SOLIDS

Line imperfections, or *dislocations*, in crystalline solids are defects that produce lattice distortions centered about a line. To visualize the geometry of these defects it is convenient to examine one way in which dislocations might be formed. Consider a material with a simple cubic crystal structure (Fig. 3-11). Let us slice this crystal along the plane *ABCD* as shown in Fig. 3-11(b). Now displace the atoms on the top side of the cut relative to those on the bottom side by a distance equal to the lattice parameter, making sure that the direction of displacement is parallel to both the plane of the cut and a cube edge ($\langle 100 \rangle$ direction). If we then stick the crystal back together again, we find that we have a distorted crystal where the center of the distortion is along the line *AB*. Correspondingly, this line *AB* is a *dislocation line*. A similar situation exists if we slide the crystal in the manner shown in either Fig. 3-11(c) or 3-11(d), although it is clear that the nature of the atom displacements along the dislocation line depends very critically on exactly how we displace the atoms relative to the line *AB*. There are three possible cases: (1) If the atoms are displaced in a direction perpendicular to *AB* we call the resulting defect an *edge dislocation*, since the defect resembles the edge of an extra half plane of atoms. (2) If the atom displacement is parallel to *AB* we have a *screw dislocation*. (We shall show below that the atoms along the dislocation line are arranged in a spiral, or screw, orientation.) And (3) if the atoms are displaced at some other angle to *AB* we have a *mixed dislocation* or a dislocation with mixed edge and screw components.

Figure 3-12 shows the atomic structure of an edge dislocation in more detail. As mentioned above, it is simply the edge of an extra half plane of atoms. Normally the symbol ⊥ is used to represent a *positive edge dislocation* (extra half

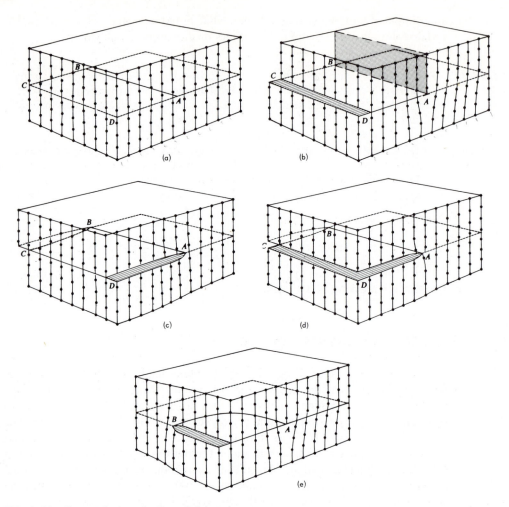

FIG. 3-11   Geometrical production of dislocations

plane above the slip plane *ABCD* in Fig. 3-11), and ⊤ is used to represent a *negative edge dislocation* (extra half plane below the slip plane).

The screw dislocation *does not* have an extra half plane associated with it. Rather, the atom planes perpendicular to the screw dislocation are distorted in such a way as to form a spiral ramp of atomic planes with the dislocation as the axis of the spiral, as shown in Fig. 3-13. Screw dislocations can be either right-handed or left-handed, depending on the nature of the spiral. We will refer to right-handed screw dislocations by the symbol S and left-handed screw dislocations by Ƨ.

The atom arrangement about a mixed dislocation is shown in Fig. 3-14(a). Where the atom displacements are nearly parallel to the dislocation line, it has predominantly a screw character, and when the displacements are nearly per-

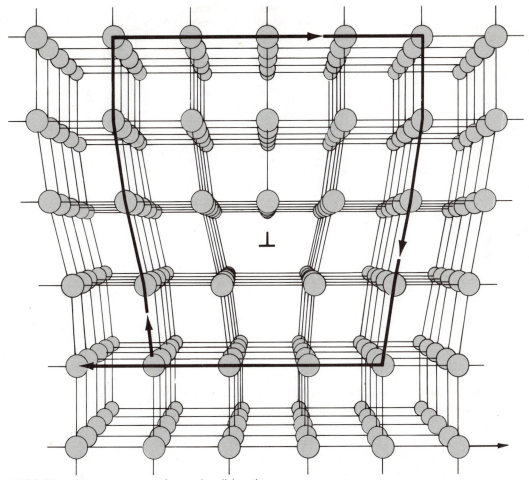

FIG.3-12   Atom arrangement in an edge dislocation

pendicular to the dislocation, it has predominantly an edge character. Between the edge and screw segments the dislocation line has atom displacements characteristic of a mixture of edge and screw components. The relative amounts of each depend on the angle between the dislocation line and the direction of atom displacements.

It is not necessary for the dislocation to end at the surface of the crystal. We could just as well have produced the dislocation by making a cut in the interior of a crystal and displacing the atoms on either side of the cut. The resultant dislocation loop is shown in Fig. 3-14(b). The character of the dislocation varies continuously around the loop, changing from pure edge to pure screw, etc. Segments on opposite sides of the loop are of the same type but of different sign. Dislocation loops can have any arbitrary shape, the only requirement being that the loop close

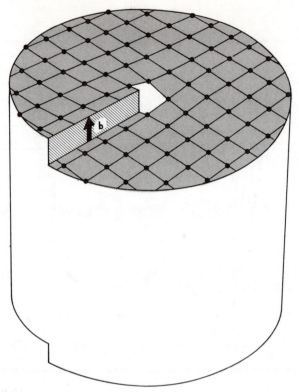

FIG. 3-13 Atom arrangements in a screw dislocation. Vector **b** represents the direction and magnitude of atom displacement.

upon itself or intersect another dislocation or a surface. *Dislocations cannot end within a crystal.*

We have introduced the concept of dislocations by allowing one portion of a crystal to slide or slip relative to another. It is now a simple matter to show the importance of dislocations in the deformation of crystalline materials. The plane in which a dislocation moves through the lattice is called a *slip plane* (Fig. 3-15). As the dislocation moves, one part of the crystal is displaced relative to the other. When the dislocation has passed all the way through the crystal, or swept out its entire slip plane, the portion of the crystal above the slip plane has shifted one atomic distance relative to the portion below the slip plane. In other words, the motion of the dislocation has caused the crystal to change its shape, or permanently deform. This is true for any type of dislocation, including a dislocation loop. Note that as the loop expands, even though the different dislocation segments move in different directions, they all produce the same shear displacement.

From the foregoing discussion it is evident that we can describe a dislocation as marking the boundary between the regions where slip has occurred and where it has not. On either side of the dislocation the crystal lattice is essentially perfect,

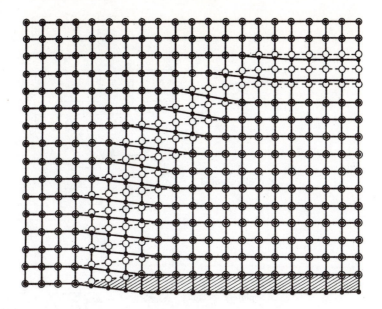

● Atoms below plane of the page

○ Atoms above plane of the page

(a)

Screw component

Edge component

b

Edge component

Screw component

(b)

FIG. 3-14   (a) Top view of atom arrangements in a mixed dislocation, (b) dislocation loop
in a crystal. Vector **b** indicates the direction of atom displacement.

but in the immediate vicinity of the dislocation the lattice is severely distorted.
For a positive edge dislocation, the presence of the extra half plane causes the atoms
above the slip plane to be put in compression, while those below the slip plane are
put in tension. Consequently, the edge dislocation will have a stress field around

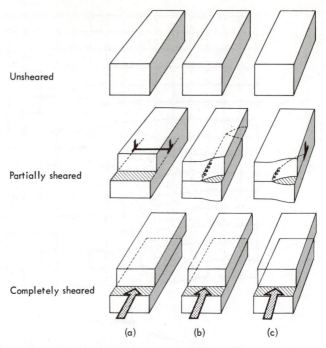

Unsheared

Partially sheared

Completely sheared

(a)   (b)   (c)

FIG. 3-15 Shear produced by dislocation motion: (a) edge dislocation, (b) screw dislocation, (c) mixed dislocation

it that is compressive above the slip plane and tensile below the slip plane. The screw dislocation has no dilatation (change in volume due to tensile or compressive stresses) around it, and all stresses are pure shear. Mixed dislocations have a stress field that is a combination of the individual edge and screw dislocation stress fields.

For convenience we usually associate with each dislocation a vector **b**, called the *Burgers vector*, which specifies the direction and amount of slip associated with the dislocation. We can determine **b** by allowing the dislocation to move and noting the magnitude and direction of the resulting lattice displacement, or we can use a simple geometrical construction known as a *Burgers circuit*. To use the Burgers circuit we mark out a circuit around the dislocation line where each step of the circuit is a lattice translation vector; the circuit would close upon itself if the crystal were perfect. This is shown in Fig. 3-16 for both edge and screw dislocations. If a dislocation is contained within the cirucit, then the circuit will not form a closed loop and the Burgers vector of the dislocation is the crystallographic vector that connects the start and finish of the circuit. In this way the Burgers circuit is a measure of the atom displacement associated with the dislocation. Note that for a pure edge dislocation the Burgers vector is *normal* to the dislocation line, and for a pure screw dislocation the Burgers vector is *parallel* to the dislocation line. For a mixed dislocation the Burgers vector is at an angle of between 0 and 90° to the dislocation.

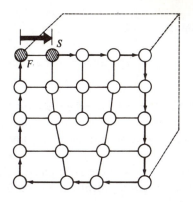 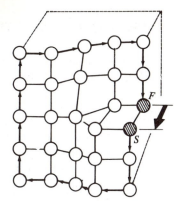

FIG. 3-16   Burgers circuits for edge and screw dislocations

EXAMPLE 3-4

Suppose a Burgers circuit is constructed around two parallel edge dislocations, where one is a positive edge and the other a negative edge. Would the Burgers circuit indicate the presence of a dislocation?

Solution:

The graphical construction is shown in the accompanying diagram. Note that the Burgers circuit closes upon itself, indicating a perfect crystal. This is because the net sum of the Burgers vectors contained within the circuit is zero (one positive and one negative). We would have obtained the same result if we had constructed the circuit around parallel right-handed and left-handed screw dislocations.

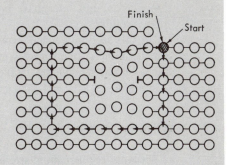

The density of dislocations is expressed as length of dislocation line per unit volume, or cm/cm³. The dislocation density present in any crystalline material depends very critically on the exact thermal-mechanical history of the material. For example, the dislocation density of silicon used in solid-state electronic devices may be as low as 1–10 cm/cm³, while the density in a high-strength steel may be as high as 10^{12} cm/cm³. In later chapters, when we discuss mechanical properties in more detail, we shall examine the variables that determine the dislocation density.

EXAMPLE 3-5

Suppose a material with a simple cubic crystal structure (1 atom per unit cell) has a dislocation density of 10^8 cm/cm³. If the lattice parameter is 4 A and the dislocations lie along $\langle 100 \rangle$ directions, what fraction of the atoms are on dislocation lines?

Solution:

If there is one atom every 4 A along a dislocation, then along 10^8 cm of dislocation there are $10^8/4 \times 10^{-8}$ or 2.5×10^{15} atoms. Now in 1 cc there are $1/(4 \times 10^{-8})^3 = 1.56 \times 10^{22}$ unit cells, and since there is only one atom per unit cell, there are 1.56×10^{22} atoms/cc. The fraction of atoms along dislocations is therefore

$$\frac{2.5 \times 10^{15}}{1.56 \times 10^{22}} = 1.6 \times 10^{-7}$$

3-5 INTERFACIAL IMPERFECTIONS IN SOLIDS

The several different types of interfacial or planar imperfections in solids can be grouped into the following categories:

1. Interfaces between solids and gases, which are called *free surfaces*
2. Interfaces between regions where there is a change in the electronic structure but no change in the periodicity of atom arrangement, known as *domain boundaries*
3. Interfaces between two crystals or grains of the same phase where there is an orientation difference in the atom arrangement across the interface; these interfaces are called *grain boundaries*
4. Interfaces between different phases called *interphase boundaries* where there is generally a change of chemical composition and atom arrangement across the interface

Grain boundaries are peculiar to crystalline solids, while free surfaces, domain boundaries, and interphase boundaries are found in both crystalline and amorphous solids.

In this section we shall consider only interfaces associated with atom arrangements, namely, free surfaces, grain boundaries, and interphase boundaries. Interfaces caused by changes in the electronic structure will be discussed in Chapters 13 and 14.

Free Surfaces

Because of their finite size all solid materials have free surfaces. The arrangement of atoms at a free surface differs slightly from the interior structure because the surface atoms do not have neighboring atoms on one side. Usually the atoms near the surface have the same crystal structure but a slightly larger lattice parameter than the interior atoms.

Perhaps the most important aspect of free surfaces is the *surface energy* associated with surface of any solid. The source of this surface energy may be seen by considering the surroundings of atoms on the surface and in the interior of a solid. To bring an atom from the interior to the surface we must either break or distort some bonds, thereby increasing the energy. The surface energy is defined as the increase in energy per unit area of new surface formed. In crystalline solids the surface energy depends on the crystallographic orientation of the surface; those surfaces that are planes of densest atomic packing are also the planes of lowest surface energy. This is because atoms on these surfaces have fewer of their bonds broken or, equivalently, have a larger number of nearest neighbors within the plane of the surface. Typical values of surface energies range from a few hundred to a few thousand ergs/cm^2; generally the stronger the bonding in the crystal, the higher the surface energy. Some examples are given in Table 3-2.

TABLE 3-2

Surface Energies for Various Solid Materials Measured in Inert Atmospheres

Material	Temperature (°C)	Surface Energy* (ergs/cm²)
Soda-lime-silica-glass	650	300
Al_2O_3	1850	905
MgO	25	1000
Silver	750	1140
TiC	1100	1190
Gold	≈ 900	1400
Copper	≈ 1000	1430
Iron	1450	2100

*Values reported for crystalline materials are values averaged over all crystallographic orientations.

EXAMPLE 3-6

When a very thin wire is hung under its own weight at high temperatures, it is observed that the wire actually becomes shortened in length. What is the driving force for this process?

Solution:

The driving force is the reduction in surface area, or the reduction in surface energy; i.e., the sample tends to assume a spherical shape to minimize its surface area to volume ratio. To observe this phenomenon it is necessary to have a thin wire in order to maximize the surface-to-volume ratio so that the compressive force due to the surface energy is greater than the tensile force due to the weight of the wire.

Surface energies can be reduced by the adsorption of foreign atoms or molecules from the surrounding atmospheres. For example, the surface energy of mica freshly cleaved in a vacuum is much higher than the same surface cleaved in air. In this instance oxygen is adsorbed from the air to partially satisfy the broken bonds at the surface. Impurity atom adsorption makes it almost impossible to maintain atomically clean surfaces. As a result, surface properties such as electron emission, rates of evaporation, and rates of chemical reactions are extremely dependent on the presence of any adsorbed impurities; these properties will be different if the measurements are made under conditions giving different surface adsorption.

EXAMPLE 3-7

Calculate the total surface energy of 10 g of gold at 900°C if the gold is in the form of spherical particles 2 μ in diameter. The density of gold is 19.3 g/cc.

Solution:

The volume of a particle 2 μ in diameter is $\frac{4}{3}\pi(10^{-4})^3$ cm³, and the weight of such a gold particle is $\frac{4}{3}\pi(10^{-4})^3(19.3)$ g. In 10 g of gold there are $10/\frac{4}{3}\pi(19.3)10^{-12}$ particles with a total surface area of $4\pi(10^{-4})^2(10/\frac{4}{3}\pi(19.3)10^{-12})$ cm². Taking the average surface energy of gold as 1400 ergs/cm² (see Table 3-2), the total surface energy is

$$\frac{4\pi(10^{-8})}{\frac{4}{3}\pi(10^{-13})19.3}(1400) \text{ ergs} = 2.18 \times 10^7 \text{ ergs}$$

or 5×10^{-2} cal/g.

Grain Boundaries

Grain boundaries separate regions of different crystallographic orientation. The simplest form of a grain boundary is an interface composed of a parallel array of edge dislocations, as shown in Fig. 3-17(a). This particular type of boundary is called a *tilt boundary* because the misorientation is in the form of a simple tilt about an axis parallel to the dislocations. The misorientation angle θ of a tilt boundary is given by

$$\tan \theta = b/D \qquad (3\text{-}1)$$

where b is the Burgers vector and D is the distance between dislocations. Grain boundaries can also form from a planar array of screw dislocations. As shown in Fig. 3-17(b), a crossed grid of two sets of right-handed screw dislocations produces a *twist boundary* where the misorientation is a simple rotation about an axis perpendicular to the plane of the dislocations. Tilt and twist boundaries are referred to as *low-angle boundaries* because the angle of misorientation is generally less than 10°. Many low-angle boundaries have both tilt and twist components and correspondingly are composed of complicated arrays of edge and screw dislocations.

When a grain boundary has a misorientation greater than 10° or 15°, it is no longer practical to think of the boundary as being made up of dislocations because the spacing of the dislocations would be so small that they would lose their individual identity. The atomic arrangement in such a *high-angle grain boundary* is shown in Fig. 3-18. The grain boundary represents a region a few atomic diameters wide where there is a transition in atomic periodicity between adjacent crystals or grains.

Grain boundaries have an interfacial energy because of the disruption in atomic periodicity in the vicinity of the boundary and the broken bonds that exist across the interface. The interfacial energy of grain boundaries is generally less than that of a free surface because the atoms in a grain boundary are surrounded on all sides by other atoms and have only a few broken or distorted bonds.

Solids with grain boundaries are referred to as *polycrystalline*, since the structure is composed of many different crystals, each with a different crystallographic orientation. Figure 3-19 shows the grain boundary structure of pure polycrystalline iron and polycrystalline polyethylene. In the case of iron the grain boundary structure has been revealed by preferential chemical attack or *etching* at the grain boundaries, while the grain structure in polyethylene is revealed by the use of polarized light. The grain structure is usually specified by giving average grain diameter or using a scheme developed by the American Society for Testing and Materials (ASTM). In the ASTM procedure the grain size is specified by a number n where

$$N = 2^{n-1} \qquad (3\text{-}2)$$

with N equal to the number of grains per square inch when the sample is viewed at $100\times$ magnification. For example, for $n = 8$ there are 128 grains/in.^{2} at $100\times$ and 1.28×10^6 grains/in.^{2} at $1\times$. If the grains are approximately square in cross section this corresponds to an average grain dimension of 8.8×10^{-4} in.

In polycrystalline samples the individual grains usually have a random crystallographic orientation with respect to one another, and the grain structure is

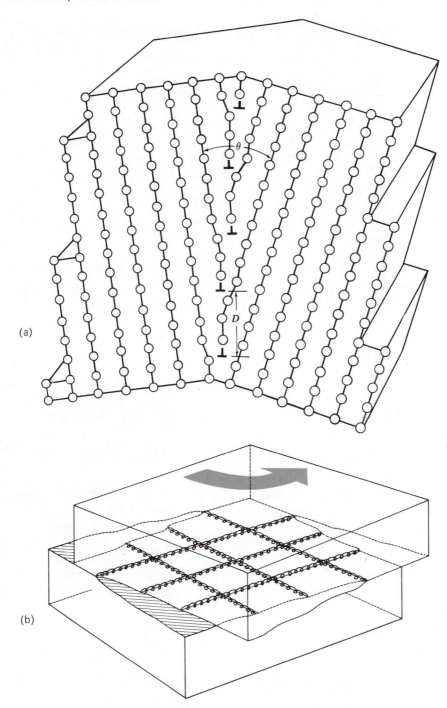

FIG. 3-17 (a) Edge dislocations in a low-angle tilt boundary, (b) screw dislocations forming a low-angle twist boundary

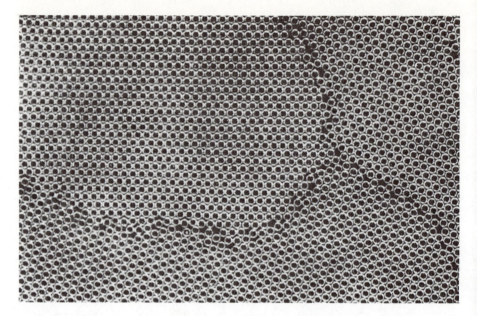

FIG. 3-18   Bubble raft model of a high-angle grain boundary. Photo courtesy J. F. Nye from L. Bragg and J. F. Nye, *Proceedings of the Royal Society*, London, **190**, 474 (1947). Used with permission.

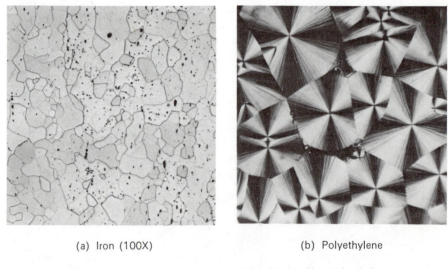

(a) Iron (100X)   (b) Polyethylene

FIG. 3-19   Grain boundaries in (a) polycrystalline iron and (b) polycrystalline polyethylene. Photos courtesy C. Young and R. Koch, Stanford University, and Prentice-Hall, Inc., from *Organic Polymers* by Turner Alfrey and Edward F. Gurnee (1967). Used with permission.

referred to as being *randomly oriented*. In some instances, however, the grains all have about the same orientation within a 5° spread. In this instance the material is said to have a *preferred orientation* or *texture*.

EXAMPLE 3-8

An iron–3% silicon solid solution alloy is used in electrical transformers because of its excellent magnetic properties. To take full use of these magnetic properties, which vary with crystallographic direction, this material is processed into polycrystalline sheet stock with a strong texture. Each grain has a {110} plane nearly parallel to the plane of the sheet and a ⟨100⟩ direction along the length of the sheet. Sketch the orientation of the unit cell with respect to the sheet.

Solution:

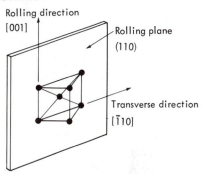

Rolling direction
[001]

Rolling plane
(110)

Transverse direction
[$\bar{1}$10]

Interphase Boundaries

We begin our discussion of interphase boundaries with a formal definition of what we mean by a *phase: A phase is a homogeneous, physically distinct and mechanically separable portion of the material with a given chemical composition and structure.* Phases may be substitutional or interstitial solid solutions, ordered alloys or compounds, amorphous substances, or even pure elements. In addition, a crystalline phase in the solid state may be either polycrystalline or exist as a single crystal.

Solids composed of more than one element may and often do consist of a mixture of a number of phases. For example, a dentist's drill, something painfully familiar to all, consists of a mixture of small single crystals of tungsten carbide surrounded by a matrix of cobalt (Fig. 3-20). Here the cobalt forms a continuous polycrystalline phase, while the tungsten carbide is present as a discontinuous phase. Polyphase materials such as the dentist's drill are generally referred to as *composite* materials. Composite materials have great importance in the engineering world because they have many attractive properties that set them apart from single-phase materials. For example, the dentist's drill has good abrasive characteristics (due to the hard carbide particles) and good toughness and impact resistance (due to the continuous cobalt matrix). Neither the tungsten carbide nor the cobalt has both abrasion resistance and impact resistance, yet the proper combination of the two phases yields a composite structure with the desired properties. In later chapters when we discuss the properties of materials (especially the mechanical properties), we shall see the great importance of composite materials.

The nature of the interface separating various phases is very much like a grain boundary. Boundaries between two phases of different chemical composition and different crystal structure are similar to high-angle grain boundaries, while boundaries between different phases with similar crystal structures and crystallographic

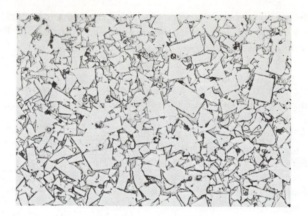

FIG. 3-20 Micrograph of tungsten carbide (small white particles) in a cobalt matrix (500 ×). Photo courtesy W. L. Silence, Stellite Division of Cabot Corp., Kokomo, Indiana. Used with permission.

orientations are analogous to low-angle grain boundaries in both energy and structure.

The concept of a solid consisting of a continuous phase and a discontinuous phase (or phases) leads to a simple classification of all the various types of composite materials. Table 3-3 gives this classification, which is based on the structure, whether amorphous or crystalline, of the continuous and discontinuous phases.

TABLE 3-3

Classification of Composite or Multiphase Materials

Continuous Phase	Discontinuous Phase (or Phases)	Examples
Crystalline	Crystalline	All metallic systems such as cast iron, steel, soft solder, etc.; most natural rocks such as granite and marble
Crystalline	Amorphous	None of practical significance
Amorphous	Crystalline	Most man-made ceramics such as building bricks and electrical insulator porcelain, concrete, partially crystalline polymers, some polymer–crystalline particle composites
Amorphous	Amorphous	Fiberglass, asphalt, wood, hydrated cement, other gels

Continuous Crystalline Phase–Discontinuous Crystalline Phase

We have already given one example of this type of multiphase solid, i.e., the cobalt–tungsten carbide structure shown in Fig. 3-20. Nearly without exception all multiphase solids with a metallic continuous phase fall in this category. Cast iron, for example, consists of a continuous matrix of an iron plus $\approx 3\%$ silicon

solid solution with discontinuous graphite particles. Practically all naturally occurring rocks also have continouus and discontinuous phases that are crystalline.
The structure of some granites, for example, has quartz (crystalline SiO_2) as the
continuous phase, with mica $(Al_2K)(Si_{1.5}Al_{0.5}O_5)_2(OH)_2$ and feldspar* as the
discontinuous phases.

The exact morphology (form and structure) of the phase distribution is very dependent on the way the solid was formed, i.e., its previous thermal history. Rapid
cooling from the melt usually results in a fine dispersion of the discontinuous phase,
while slow cooling yields a coarse structure. Rapid cooling also may not allow
enough time for one or more of the phases to crystallize. This is the reason most
natural rocks, which were formed while the earth's crust slowly cooled, are crystalline, while such rocks as lava and obsidian, which solidified relatively rapidly,
have a continuous phase (glass) that is amorphous.

Continuous Amorphous Phase–Discontinuous Crystalline Phase

There are three general classes of materials under this heading: (1) man-made
ceramics, (2) concrete, and (3) polymer–crystalline particle composites. We shall
describe each of these in some detail.

Many man-made ceramic materials are produced by mixing together a number
of minerals, forming an object of the desired shape and then heating the object
to produce a highly viscous liquid phase. Upon subsequent cooling this liquid
phase binds the other solid particles together. Because the cooling cycle is relatively
rapid (less than a day), the liquid phase usually does not crystallize but remains
amorphous. In a sense these man-made ceramics are similar to lava. If they had
been cooled slowly enough the continuous liquid phase would have crystallized.
Figure 3-21(a) shows the microstructure of an alumina ceramic in which Al_2O_3
crystalline particles are bonded with a glassy matrix. Structures qualitatively similar to that shown in Fig. 3-21(a) are found in many partially crystalline polymers.
In these instances a portion of the polymer has crystallized and a portion is still
amorphous [see Fig. 3-21(b)]. Although the chemical compositions in the crystalline and amorphous regions are the same for the polymer, the properties of each
phase are vastly different due to the different structure (see Chapter 10).

Concrete is a mixture of crushed stone or gravel, sand, and hydrated Portland
cement.† A typical microstructure is shown in Fig. 3-22. The hydrated cement forms
the continuous phase and bonds the crushed stone and sand, which act as a filler
material. Each of the stone or sand particles is usually crystalline and may in itself
be composed of more than one phase. The hydrated cement consists mainly of
a calcium silicate *gel*.‡ We shall describe the structure of this gel in some detail
in the next section.

---

*Feldspars are anhydrous alumina silicates containing K^+, Na^+, and Ca^{2+} ions.

†Portland cement is composed primarily of tricalcium silicate ($3CaO \cdot SiO_2$) and dicalcium
silicate ($2CaO \cdot SiO_2$), with smaller amounts of tricalcium aluminate ($3CaO \cdot Al_2O_3$), brown
millerite ($4CaO \cdot Al_2O_3 \cdot Fe_2O_3$), lime ($CaO$), and periclase ($MgO$).

‡A gel is a mixture of two phases—solid and liquid or solid and gas—in which the particle size
of the solid phase is on such a fine scale that the gel behaves as an amorphous solid.

(a) (3500X)   (b)

FIG. 3-21   Micrographs of two-phase materials containing one crystalline phase and one amorphous phase: (a) Al_2O_3 ceramic and (b) partially crystalline polymer (Isotactic polystyrene viewed under polarized light). Photos courtesy L. J. Anderson, Stanford University, and Prentice-Hall, Inc., from *Organic Polymers* by Turner Alfrey and Edward F. Gurnee (1967). Used with permission.

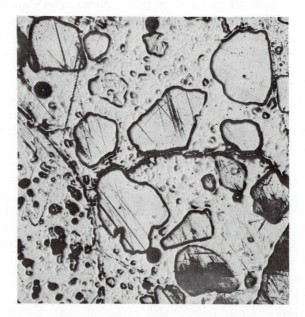

FIG. 3-22   A typical microstructure of concrete (35 $\times$). Photo courtesy American Concrete Institute from S. P. Shah and S. Chandra, *Journal of the American Concrete Institute,* **65**, 770 (1968). Used with permission.

There are a large number of multiphase materials produced by adding *fillers* in the form of small particles or fibers to a polymeric matrix. These fillers are added to alter the properties of the polymer, and usually the fillers that produce the most dramatic change in properties are those that form bonds or effectively cross-link the polymer chains. *Thermosetting* polymers are filled with substances like wood flour, pure short-fiber cellulose, powdered mica, short glass fibers, and asbestos.* These materials greatly enhance the dimensional stability, impact resistance, tensile and compressive strengths, abrasion resistance, and thermal stability of the polymer. *Thermoplastic* polymers are sometimes filled with large amounts (over 50% by weight) of mineral solids such as crushed quartz, limestone, or clay. In these instances the polymer (a hydrocarbon resin, for example) acts primarily to bond the inorganic particles together. The filled polymer has very poor tensile properties but good compressive strength, abrasion resistance, and dimensional stability.

An important example of a filled polymer is the case of rubber. For many applications, even vulcanized rubber does not exhibit satisfactory strength or abrasion resistance. It has been found that these properties can be enhanced by the addition of certain fillers to the rubber before vulcanization. Carbon black (amorphous carbon) is outstanding in this respect for both natural and synthetic rubbers. The particles of carbon black form a large number of weak bonds between the polymer chains in addition to the widely spaced strong cross-link bonds produced by vulcanization. These weak bonds restrict the short-range movement of the polymer molecules and consequently increase the strength.

Continuous Amorphous Phase–Discontinuous Amorphous Phase

Common materials in this group include the gels, such as cement and asphalt, wood, and fiberglass. In hydrated Portland cement the reaction of water with the dicalcium silicate and tricalcium silicate present in powdered cement forms a complex hydrated product that is a continuous network of amorphous or partially crystalline calcium silicates. In this network there are pore spaces between the individual silicate particles as well as large pores remaining from the excess water content required to form the cement and reduce its viscosity sufficiently so that it can be easily mixed. Asphalt is different from cement in that the network structure in cement is rigid, while asphalt has a flexible network structure. The structure of asphalt consists of high molecular weight hydrocarbons forming a loose three-dimensional structure surrounding an oily liquid phase. The structure and strength of cement is thermally stable, while asphalt behaves similar to a thermoplastic polymer, softening as temperature is increased.

Wood and fiberglass are both composite materials consisting of a relatively stiff continuous phase, bonding together particles of a higher strength second phase. Wood is composed of cellulose and the structurally complicated polymer, *lignin*. The lignin is the continuous phase bonding together the stronger cellulose molecules. In fiberglass an epoxy resin is used to bond together long, randomly oriented

---

*Although not all these substances are crystalline, we mention them in this section for convenience in order not to have to differentiate between crystalline and amorphous fillers.

glass fibers. In both wood and fiberglass the tensile strength is provided by the strong fibers (cellulose and glass, respectively), while the rigidity of the structure results from the continuous phase.

3-6 BULK DEFECTS

Bulk defects are generally found in materials for one of two reasons—either the defect has been introduced in the production of the material (e.g., the production of iron from iron ore) or it has been introduced in the fabrication of the material by such processes as casting, forging, rolling, etc. Below we shall briefly describe a few of these different types of bulk defects.

The most common bulk defects introduced during the production of materials are unwanted second-phase particles that are usually referred to as *inclusions*. Inclusions are found in nearly all materials in a wide variety of shapes and sizes. Common inclusions in ordinary glassware are small particles of clay or pieces of refractory brick which either originate from the raw materials used in production of the glass or come from the containers used to hold the molten glass. Such materials, being slow to dissolve, can appear as included particles in the final product. In metallic materials the most common example of an inclusion is an oxide particle, although sulfides, hydrides, etc., are sometimes also found. These particles usually originate during the initial refining operations when the metal is molten, and since they are generally insoluble in the metal they may be entrapped within the metallic matrix during the process of solidification. Figure 3-23 shows some inclusions in a sample of common structural steel. In our discussion of the mechanical properties of materials (Chapters 6–10) we shall show how inclusions can have a great influence on the mechanical strength of a material.

Bulk defects introduced during the fabrication of materials can take a variety of forms. Several of the more common defects are listed below:

1. Casting defects (shrinkage cavities and gas holes)
2. Working or forging defects (cracks)
3. Welding or joining defects

Casting defects usually originate from one of two causes: (1) the volume contraction accompanying solidification (shrinkage cavities) or (2) the evolution of gases during the solidification process (gas holes). Shrinkage cavities generally tend to be found in the center of castings where solidification occurs and hot molten material is not available to fill the void created by the volume change from liquid to solid. An example of shrinkage cavities is shown in Fig. 3-24. Gas holes in castings may range from fine porosity to large cavities. They are caused by gases, dissolved in the molten material, that are evolved and entrapped during solidification.

The most common bulk defects introduced during working or forging (plastically forming an object) or during a welding or joining operation are usually in the form of cracks. During forming operations cracks originate in areas where the material is severely deformed (such as a sharp bend) or where a previous defect

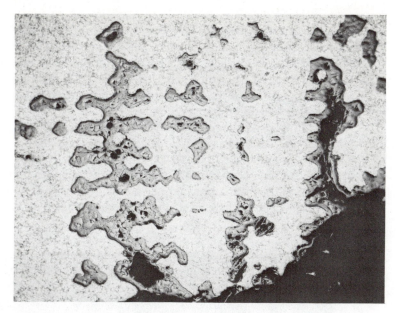

FIG. 3-23 Micrograph showing inclusions in structural steel. Photo courtesy Reinhold Publishing Co., from *Defects and Failures in Pressure Vessels and Piping* by Helmut Thielsch © 1965 by Litton Educational Publishing Inc. Reprinted by permission of Van Nostrand Reinhold Company. Used with permission.

FIG. 3-24 Micrograph showing shrinkage cavities in a steel casting. The length of this casting is ≈12 inches. Photo courtesy Reinhold Publishing Co., from *Defects and Failures in Pressure Vessels and Piping* by Helmut Thielsch (1965). Used with permission.

exists (such as a shrinkage cavity or a surface irregularity such as a notch, etc.). These cracks, even if they are very small (less than 0.1 cm), can have a very deleterious effect on the ultimate strength of the material (see Chapter 8). In addition, small surface cracks or scratches also form an important class of bulk defects, especially for the case of brittle materials like glass. As mentioned earlier, surface scratches only 10^{-4} cm deep can lower the fracture strength of glass by a factor of 100 from the strength of unscratched glass.

Cracks in welds commonly occur because large thermal stresses may be set up in the region of the weld as the weld metal rapidly cools and solidifies. These thermal stresses occur because the base material exerts a restraining effect on the weld metal as it thermally contracts following solidification. Since weld metals are usually more brittle than the base metal, the cracks usually form in the weld and may later propagate into the base metal under the application of an applied load. Figure 3-25 illustrates a crack in a weld.

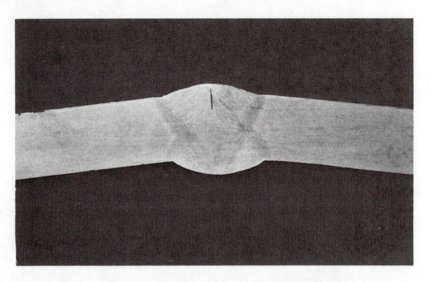

FIG. 3-25   Micrograph showing a crack in a weld. Photo courtesy Reinhold Publishing Co., from *Defects and Failures in Pressure Vessels and Piping* by Helmut Thielsch (1965). Used with permission.

PROBLEMS

3-1   In pure copper at room temperature there is about one atom vacancy every 10^{10} atom sites. At the melting temperature (1083°C), there is about one vacancy every 10^3 sites. What fraction of the volume change that takes place when the crystal is heated from room temperature to its melting temperature is due to the presence of vacancies? The linear thermal expansion coefficient of copper is 16.5×10^{-6}°C^{-1}.

3-2 Suppose lithium is dissolved interstitially in silicon. Would you expect the lithium atoms to be donors or acceptors? Would a lattice vacancy in silicon be a donor or an acceptor?

3-3 The nonstoichiometric compound $Fe_{0.93}O$ has a sodium chloride structure with a unit cell edge of 4.292 A. If the density is 5.658 g/cc, determine whether the structure has cation lattice vacancies or interstitial oxygen atoms.

3-4 Explain how you would expect the heat of fusion to vary with the average molecule length for a long-chain crystalline polymer.

3-5 It is possible for two elements to form a complete series of substitutional solid solutions over the whole composition range. Is this behavior possible for two elements that form an interstitial solid solution?

3-6 After polyethylene is exposed to a small dose of electromagnetic radiation of sufficient energy to break carbon-hydrogen bonds, the temperature at which polyethylene begins to soften increases. This effect is attributed to an increase in the density of cross-links following irradiation. Show schematically how such cross-links might form between two chains.

3-7 Assuming that there are no iron atom vacancies, find the maximum amount of carbon that could be dissolved interstitially in an FCC arrangement of iron atoms. Take the carbon atoms to occupy only the largest interstitial sites. If your answer is different from the experimentally observed value of 2 weight percent, give a possible explanation.

3-8 For a metal with an FCC crystal structure, sketch the atom arrangement around an edge dislocation with a {110} slip plane and with a Burgers vector parallel to ⟨110⟩. Let the length of the Burgers vector be equal to the smallest lattice translation vector in the ⟨110⟩ direction.

3-9 Suppose an edge dislocation in NaCl has a Burgers vector parallel to ⟨110⟩ and a {110} slip plane. Sketch the {100} face of several adjacent NaCl unit cells and show that if such a dislocation has only one extra {110} half plane there will be an interfacial defect associated with the dislocation (a plane across which like ions will be nearest neighbors). Also show that if the dislocation has two extra {110} half planes there is no interfacial imperfection.

3-10 What happens if a positive edge dislocation meets a negative edge dislocation on the same slip plane with a parallel Burgers vector of the same length? Show that the same thing happens when right-handed and left-handed screw dislocations on the same plane with parallel Burgers vectors meet.

3-11 Show with the aid of sketches that an edge dislocation may move in a direction normal to its slip plane when vacancies or interstitials are absorbed on the end of the extra half plane.

3-12 Substitutional solute atoms have a strain field about themselves, the sign of the strain depending on the size of the atom compared to a matrix atom. This strain field of a substitutional atom can interact with the stress field of an edge dislocation. Draw a schematic picture of an edge dislocation, and indicate the position around the dislocation where large and small substitutional atoms would be found. Where would interstitial atoms be found?

3-13 A low-angle tilt boundary has an energy (strain energy associated with the dislocations) less than the energy of an equivalent number of randomly oriented edge dislocations. By considering the dilatational stresses about an edge dislocation, show why this might be expected.

3-14 Which are the planes of densest atomic packing in the BCC, FCC, and HCP crystal structures?

3-15 A polycrystalline sample was heated to a high temperature for a period of time. Careful measurements of the grain size before and after the thermal treatment reveal that the average grain size increased as a result of the treatment. What is the driving force for this grain growth phenomenon?

3-16 Two small gold balls are placed in intimate contact and then heated to a temperature just below the melting point. After a period of time it is noticed that the balls have fused together and there is a small "neck" of gold joining them at the original point of contact. What is the driving force for this phenomenon? Give two atomic transport mechanisms by which this neck could have formed.

3-17 In experiments with liquids that readily form bubbles (e.g., soap, beer, etc.) it is noticed that when several bubbles impinge, the angle between adjoining interfaces at a line of contact (three interfaces intersecting on a common line) is invariably 120°. This is due to a mechanical equalization of surface tensions or surface energies. The same effect occurs in polycrystalline solids, although here the situation is made more complicated by the presence of grain boundaries with different energies. Find the angles between three intersecting grain boundaries that have energies of 1000, 1000, and 750 ergs/cm², respectively.

3-18 If natural rubber is kept at a temperature of −20°C for a long period it will partially crystallize. In addition, if some natural rubber is stretched to about three times its normal length and held at −20°C the process of crystallization is much more rapid. Why does stretching the rubber aid the crystallization process?

BIBLIOGRAPHY

T. ALFREY and E. F. GURNEE, *Organic Polymers*, Ch. 1–2. Englewood Cliffs, N.J.: Prentice-Hall, 1967.

F. W. BILLMEYER, *Textbook of Polymer Chemistry*, Ch. 1–3. New York: Interscience, 1957.

R. M. BRICK, R. B. GORDON, and A. PHILLIPS, *Structure and Properties of Alloys*, Ch. 1, 3, 5. New York: McGraw-Hill, 1965.

G. O. JONES, *Glass*, Ch. 1–3. New York: John Wiley, 1956.

W. D. KINGERY, *Introduction to Ceramics*. New York: John Wiley, 1960.

H. F. MARK, "Giant Molecules," *Scientific American* **204** (1961), p. 80 (also eight other articles in same issue).

J. A. PASK, *An Atomistic Approach to the Nature and Properties of Materials*, Ch. 5–8. New York: John Wiley, 1967.

W. T. READ, JR., *Dislocations in Crystals*. New York: McGraw-Hill, 1953.

H. G. VAN BUEREN, *Imperfections in Crystals*. Amsterdam: North Holland Publishing Co., 1960.

J. WEERTMAN and J. R. WEERTMAN, *Elementary Dislocation Theory*. New York: Macmillan, 1964.

4
Equilibrium

4-1 INTRODUCTION

In the previous chapter it was pointed out that all real materials are *imperfect*, *impure*, and *inhomogeneous*. The purpose of this chapter is to provide insight into why such heterogeneities exist in materials and to establish methods for predicting when and to what extent such deviations from perfection are stable. The entire chapter is devoted to a description of *thermodynamic equilibrium*. This particular state is the stable state for a material and is of critical importance in the science of materials because it is the condition that all materials strive to achieve.

In this chapter we shall study the thermodynamic principles that are needed to understand how and why equilibrium is achieved in materials. We shall develop the concept of *free energy*, which is of critical importance in describing equilibrium, and we shall show that a knowledge of equilibrium thermodynamics enables the explanation of why some materials naturally tend to be heterogeneous. We shall also demonstrate that the principles of equilibrium provide a basis for changing and controlling the internal structure of engineering materials. In later chapters we shall see that the ability to change or control the internal structure of materials leads to the ability to control their engineering properties.

4-2 MECHANICAL, THERMAL, AND CHEMICAL EQUILIBRIUM

Before considering the subject of equilibrium in materials, it is instructive to examine the meaning of equilibrium in some simple thermodynamic systems.*

Consider a ball with mass m that is placed exactly at the bottom of the valley shown in Fig. 4-1. It is evident that the position shown represents a minimum in

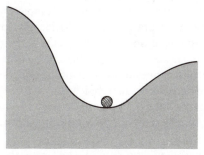

FIG. 4-1   A representation of mechanical equilibrium

the potential energy of the ball. If the ball is displaced to either side of its initial position, there is a gravitational force that acts to return the ball to the position of minimum potential energy. When the ball rests at the bottom of the valley, the net forces on it are zero, and the system is said to be in *mechanical equilibrium*. We see from this example that equilibrium in this case may be described either by the condition of zero net force or by a minimum in the potential energy. We shall see later that the concept of minimum energy used here is most useful in describing equilibrium.

Another simple form of equilibrium may be found by studying the exchange of heat between two bodies. Suppose that two bodies are initially at the same temperature and are in thermal contact. Consider what would happen if a small amount of heat were transferred from one body to the other. The temperature of the body receiving the heat is slightly increased, while the temperature of the other body is slightly decreased. As a result of the induced temperature difference, heat tends to flow back from the hotter body into the colder body so as to achieve an *isothermal* i.e., constant temperature, condition. The system is in *thermal equilibrium* only when the temperature is the same everywhere in the system.

Chemical equilibrium is defined as the condition that the potential energy of a given atom or molecule is the same in every part of the system. Where there is more than one phase in the system, the compositions and amounts of the component phases remain fixed and do not change with time when chemical equilibrium is achieved. We shall see later that it is not necessary for the phases to have the same composition to be in chemical equilibrium. In fact, except for systems composed

---

*A thermodynamic system is simply a region of space on which we wish to focus our attention. In the case of materials we frequently take 1 gram mole of substance to be the system.

of pure elements, phases that are in equilibrium have different chemical compositions. What is required is that the change in potential energy when an atom or molecule is moved from one phase to another be zero. When this condition is reached the compositions of the phases will remain invariant with time.

To illustrate the concept of chemical equilibrium in a single-phase system, let us suppose that two gaseous mixtures having exactly the same chemical composition are contained in two separate containers. Let us connect the two containers with a hollow tube and allow the gases to intermix completely. Now consider what would happen if one were to separate the gases artificially so that more molecules of a particular kind are found in one container than in the other. Because of the random motion of the molecules in the gas, the chemical imbalance would soon disappear since the molecules in question would diffuse equally into both containers. Single-phase *chemical equilibrium* is achieved when this condition is satisfied. In this case it is easy to imagine that the potential energy for a given molecule is the same in all parts of the system because the composition, and hence the environment, is everywhere the same.

Complete *thermodynamic equilibrium* in a system is reached only when *mechanical equilibrium*, *thermal equilibrium*, and *chemical equilibrium* are simultaneously achieved. Much of our attention will be directed toward an understanding of the features of chemical equilibrium.

Before turning our attention to a more quantitative description of equilibrium, it is instructive to describe in qualitative terms how mechanical equilibrium, thermal equilibrium, and chemical equilibrium may be satisfied in a material. Let us consider the problem of forming a substitutional solid solution of elements *A* and *B*, imagining that there is a lattice on which the *A* and *B* atoms reside. First, *mechanical equilibrium* is achieved in this system when the pressure is uniform throughout. The condition of uniform pressure is analogous to the condition of a ball resting at the bottom of a valley. Secondly, *thermal equilibrium* is obtained when the temperature in the solid solution is everywhere the same. Finally, *chemical equilibrium* is achieved when the potential energies for both *A* and *B* atoms are equal in all parts of the system. Since we are dealing with a solid solution and not a gaseous mixture, it is necessary that we examine the meaning of chemical equilibrium more carefully.

Chemical equilibrium in a solid solution involves two different considerations. The first is that the atoms will tend naturally to become completely intermixed. This *tendency toward mixing* was just illustrated for the case of gaseous mixtures. A second consideration deals with energy differences that depend on how the atoms are mixed or arranged on the lattice. In the case of a gas it may be assumed that since the molecules are so far apart, energies of the molecules are not strongly dependent on the types of nearest neighbor molecules. The case of a solid solution is considerably different in this respect. The atoms interact very strongly, and the energy of a given atom may be very dependent on the types of near neighbor atoms that it has. Therefore, while there is a tendency for *mixing* in a solid solution just as there is in a gaseous mixture, there is the added consideration of the *energies* of the atoms in different atomic environments. If the atoms in a solid solution do not interact with each other or if the *A-B* interaction is very similar to the *A-A* and *B-B* interactions, the tendency toward mixing would dominate the chemical equilibrium state, and the system would become a random solid solution. If, on the other hand, the *A-B* bond energy is very high relative to the energy of the *A-A* bonds and the *B-B* bonds, the *energy* considerations would dominate and the system would consist of regions containing mostly *A* atoms segregated from regions with nearly all *B* atoms. In the latter case the *A* and *B* atoms would tend to segregate into separate phases. At equilibrium the potential energy of an *A* atom in the *A*-rich phase would be equal to the potential energy of the *A* atom in the *B*-rich phase, in spite of the difference in compositions of the two phases. The potential energy of the *B* atom is also the same in both phases.

This simple example illustrates a very important fact about chemical equilibrium, namely that thermodynamic equilibrium in a material depends on a consideration of both the *energy* of the system and the *state of mixing* of the system.

4-3 THE FIRST LAW OF THERMODYNAMICS

Before we examine the conditions for equilibrium in materials it is necessary that we study briefly the thermodynamic quantities that are needed to describe equilibrium. Let us first consider the *internal energy* of a crystalline solid. Our model of a solid is one in which the atoms are considered to be tiny mass points fixed at their mean positions by springs. At a finite temperature the atoms vibrate about their average position with a frequency of about 10^{13} sec^{-1}. The amplitude of vibration of a particular atom is related to the maximum potential energy stored in the springs that hold that atom in place. When the solid absorbs heat, the increase in the internal energy is associated with an increase in the average amplitude of atomic vibration. Raising the temperature of the solid amounts to storing more energy in each vibrating atom (Chapter 2).

Let us now examine the meaning of internal energy in a gaseous system. When a gas is contained in a fixed volume and absorbs heat from the surroundings, the increment in the internal energy of the gas is simply equal to the heat energy absorbed. The energy absorbed is stored in the increased kinetic energies of the component molecules. Therefore, when a gas is heated to a slightly higher temperature, we may consider the individual molecules to be banging about with increased vigor. We may think of temperature as a measure of the motion of the molecules of the system.

Consider the case of an ideal gas* that absorbs an amount of energy ΔE from its surroundings. From the kinetic theory of gases it is possible to show that the average energy of a free molecule is simply $\frac{3}{2}kT$, where k is Boltzmann's constant and T is the absolute temperature. It follows from this that the internal energy of

---

*An ideal gas is simply a gaseous mixture of atoms or molecules that obeys the equation $pV = NRT$, where p, V, and T are, respectively, the pressure, volume, and temperature, N is the number of gram moles (1 gram mole = 6.023×10^{23} molecules) of the substance, and R is the molar gas constant 1.987 cal/mole °C.

a gas depends *only on the temperature* and is expressed as

$$E = \tfrac{3}{2}nkT \qquad (4\text{-}1)$$

where n is the number of molecules in the sample in question. This relation is used in Example 4-2 to describe the events that accompany the expansion of an ideal gas. Now we may turn our attention to the first law of thermodynamics.

The first law of thermodynamics refers to the exchange of energy between a system and its surroundings. This important law is expressed as

$$dE = dQ - dW \qquad (4\text{-}2)$$

where dE represents a small change in the internal energy* of a system which absorbs an amount of heat dQ and produces an amount of work dW. To demonstrate the applicability of the first law of thermodynamics, let us briefly consider what happens when a system absorbs heat from its surroundings.

From the first law of thermodynamics we see that the change in *internal energy*, dE, that accompanies the absorption of a small amount of heat, dQ, depends on whether any work is simultaneously done by the system on its surroundings. If a solid absorbs heat under the condition of constant pressure, it is free to expand, and the work of expansion is simply $p\,dV$, where p is the pressure exerted by the surroundings and dV is the change in volume. In this case the change in internal energy is expressed as $dE = dQ - p\,dV$, since the net work or energy that remains in the solid is clearly the difference between the heat energy absorbed and the work produced. On the other hand, when a system is constrained and is not free to expand (such as in the case of a gas in a container of fixed volume), the change in internal energy is simply equal to the heat absorbed.

Because the condition of constant pressure is so commonly satisfied in practice, we define a separate thermodynamic quantity, called the *enthalpy* and given the symbol H, which is equal to the heat absorbed at constant pressure. Enthalpy is defined as

$$H \equiv E + pV \qquad (4\text{-}3)$$

where V is the volume of the thermodynamic system in question and E and p are the *internal energy* and *pressure*, respectively. It is evident from this definition and from the first law that $dH = dQ$ when the pressure is constant. This term allows us to define precisely the heat capacity of a system at constant pressure as

$$C_p = \frac{\partial Q}{\partial T}\Big)_p \equiv \frac{\partial H}{\partial T} \qquad (4\text{-}4)$$

Alternatively, when heat is absorbed at constant volume we have for the heat capacity

$$C_V = \frac{\partial Q}{\partial T}\Big)_V \equiv \frac{\partial E}{\partial T} \qquad (4\text{-}5)$$

---

*The symbol E will be used to refer only to the internal energy of a system. This expression does not include the potential energy or the kinetic energy that the system might have with respect to a fixed reference. Some texts use the symbol U for the internal energy and E for the total energy.

In this book our attention is focused exclusively on the properties of condensed phases (solids and liquids) that are free to expand or contract as they are heated. Properly speaking, therefore, we should be concerned with the *enthalpy* of a system as opposed to its *internal energy*. It should be noted, however, that there is very little difference between the internal energy E and the enthalpy H in the case of solids and liquids. The reason for this is that solids and liquids are almost incompressible, and the pV term is nearly constant regardless of the actual constraints imposed on the system. Therefore, we shall hereafter refer only to the *internal energy*.

EXAMPLE 4-1

The heat of fusion (heat absorbed during melting) for pure aluminum is 2530 cal/mole, and upon melting, this metal undergoes a volume expansion of about 5.1%. Calculate the change in internal energy and the change in enthalpy that occur when 1 mole of aluminum is melted at a pressure of 1 atm.

Solution:

First, since the heat is absorbed at constant pressure we may immediately state that the change in *enthalpy*, ΔH, is given simply by the quantity of heat absorbed: 2530 cal. The atomic weight of aluminum is 26.9 g/mole, and the density is approximately 2.7 g/cm³, so that the volume of one mole of solid aluminum is given by

$$V = \frac{M}{\rho} = \frac{26.9}{2.7} = 9.95 \text{ cm}^3$$

On melting, the volume expansion ΔV is

$$\Delta V = 9.95(0.051) = 0.51 \text{ cm}^3$$

and the work of expansion done on the atmospheric pressure is

$$\Delta W = p\,\Delta V = 1 \text{ (atm)} \, 1.01 \times 10^6 \left(\frac{\text{dynes/cm}^2}{\text{atm}}\right) 0.51 \text{ (cm}^3\text{)} = 5.1 \times 10^5 \text{ ergs}$$

or

$$\Delta W = 5.1 \times 10^5 \text{ (ergs)} \, 2.38 \times 10^{-8} \text{ (cal/erg)} = 0.012 \text{ cal}$$

From the first law of thermodynamics [Eq. (4-2)] we see that the change in *internal energy* ΔE is given by

$$\Delta E = \Delta Q - \Delta W = 2530 - 0.012 = 2530 \text{ cal}$$

This computation indicates that the work term is so small in this case that it may be neglected, and we have $\Delta E = \Delta H = \Delta Q$ to a very close approximation.

4-4 IRREVERSIBILITY AND THE SECOND LAW OF THERMODYNAMICS

In our description of chemical equilibrium we pointed out that equilibrium involves a consideration of both the *energy* of a system and the tendency for *mixing*. We shall now examine each of these considerations in detail. Since the concept of minimum energy is one that we have already discussed, it will not be necessary to elaborate on that aspect of thermodynamic equilibrium.

While it is easy to see that a system should tend to adopt a condition of minimum energy (balls tend to roll downhill), the natural tendency of systems to become mixed up and more random is not immediately evident. Actually, the idea that a system should seek a condition of minimum energy is familiar to us only because of our everyday experiences. It does not follow from elementary laws of physics such as the principle of conservation of energy; it comes only from the second law of thermodynamics, to which we now direct our attention.

In the first law of thermodynamics the change in internal energy of a system that exchanges heat and work with its surroundings depends only on the sign of the heat absorbed and the work done. If the work and heat exchange is reversed, the change in internal energy is *exactly* reversed. In this sense the first law of thermodynamics is reversible and pays no attention to, nor makes any predictions about, the most probable direction of the change. The first law simply demands that the energy be conserved in any arbitrary change. The second law of thermodynamics, on the other hand, deals directly with the direction of spontaneous changes.

The inadequacy of the minimum energy principle to account for the directions of spontaneous changes in materials may be illustrated with a simple example. Suppose that a crystalline solid is heated to slightly above its melting temperature. Since the liquid phase is stable above the melting temperature, the solid will *spontaneously* absorb heat (heat of fusion) and transform (melt) to the liquid phase. In this case we see from the first law of thermodynamics that the internal energy of the system is *increased* by the amount of heat energy absorbed less the work done on the surroundings through expansion. When the liquid is cooled to slightly below the melting temperature, the process is reversed and the internal energy is correspondingly decreased. We see from this that the direction of the spontaneous reaction can certainly not be predicted by the change in the internal energy alone. In one case the internal energy *spontaneously increases*, while in the other case it *spontaneously decreases*.

Since much of our interest in the second law of thermodynamics comes from a desire to describe the tendency toward mixing in materials, it will be instructive for us to consider some simple problems of mixing. We have already pointed out that when two ideal gases are brought into contact, the gases tend naturally to diffuse together because of the random motions of the individual molecules. It may be noted that the two gases seek a *more probable* condition by intermixing. Once the gases have mixed, it is highly unlikely that all the molecules would ever become perfectly segregated as before. Certainly the probabilty that the molecules move in exactly the right way to become segregated is vanishingly small. On the other hand, the probability that the gaseous mixture remains randomly mixed is very high indeed, for almost all the possible movements of the molecules will again produce a random mixture. This example demonstrates the relation between the probability of a particular state or arrangement of atoms and the direction of a spontaneous change. We may draw from this example the general rule that systems tend toward a more probable state. This interpretation of the second law of thermodynamics is particularly useful in the case of materials because of our interest in the problem of mixing. The concept of probability plays an important part in the quantitative description of spontaneous mixing, as we shall soon see.

The tendency toward mixing may be expressed in terms of a compound quantity that we shall call the *mixing energy*. This is the product of the absolute temperature of the system (T) in units of °K and the *entropy* of the system (S) in units of cal/mole · °K (or ergs/atom · °K). We now focus our attention on the meaning of *entropy* and the form of the *mixing energy* term.

The change in entropy of a system is given by

$$\Delta S = k \ln \left(\frac{\mathfrak{W}_1}{\mathfrak{W}_2}\right) \tag{4-6}$$

where $\mathfrak{W}_2$ and $\mathfrak{W}_1$ are the numbers of distinguishable states in which the system may be found before and after the change, respectively. In this expression $\mathfrak{W}$, which is sometimes called the *randomness*, commonly refers to the number of distinctly different ways that the atoms of the system can be arranged. From Eq. (4-6) we see that when a spontaneous change takes place isothermally and the system tends toward a more probable state ($\mathfrak{W}_2 < \mathfrak{W}_1$) the entropy change is positive. The change in mixing energy that we associate with the mixing process is $T \Delta S$, where ΔS is the entropy change of the system during mixing. From the sense of the spontaneous change we see that the mixing energy term $T \Delta S$ is positive and tends toward a maximum value as equilibrium is approached.

In concluding the description of mixing it is worthwhile to point out that the *entropy* of a system may be defined in a different way. The change in the entropy dS of an isothermal system at temperature T that absorbs an amount of heat dQ in a reversible fashion* is defined as

$$dS = \frac{dQ}{T}\bigg)_{\text{rev}} \tag{4-7}$$

The following example shows how the form of Eq. (4-6) may be linked to this definition of entropy.

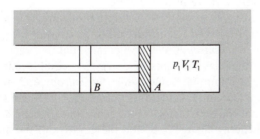

FIG. 4-2 Expansion of an ideal gas

EXAMPLE 4-2   Consider an ideal gas initially at temperature T_1 and pressure p_1 and contained in a volume V_1 (piston in position A in Fig. 4-2). As the gas expands isothermally it absorbs heat from the surroundings and produces work on the piston, which moves from A to B. Derive an expression for the change in entropy that accompanies the expansion.

---

*In a reversible change the system makes only infinitesimally small departures from equilibrium during the change.

Solution: Since the internal energy of an ideal gas depends only on the temperature [see Eq. (4-1)]

$$\left(\frac{\partial E}{\partial V}\right)_T = 0$$

it follows from the first law of thermodynamics that the heat absorbed during the expansion must exactly equal the work produced. It is this equality that allows us to compute the entropy change.

The work done by the expanding gas is

$$\Delta W = \int_{V_1}^{V_2} p \, dV = \int_{V_1}^{V_2} \frac{NRT_1}{V} \, dV = NRT_1 \ln \frac{V_2}{V_1}$$

where N is the number of moles of gas initially contained in the volume V_1. The heat absorbed by the gas is

$$\Delta Q = \Delta W = NRT_1 \ln \left(\frac{V_2}{V_1}\right)$$

and the total entropy change is

$$\Delta S)_{\text{total}} = \frac{\Delta Q}{T_1} = NR \ln \left(\frac{V_2}{V_1}\right)$$

The entropy change per mole is given by

$$\Delta S)_{\text{mole}} = R \ln \left(\frac{V_2}{V_1}\right)$$

and, since $R = N_0 k$, the entropy change per molecule is

$$\Delta S)_{\text{molecule}} = k \ln \left(\frac{V_2}{V_1}\right)$$

We see from this simple example that the form of Eq. (4-6) is reasonable and that as a gas expands to completely fill volume available to it, the entropy is increased. Again we may state that the entropy change undergoes a positive change as the system tends toward a more probable state.

We are now prepared to give a complete description of equilibrium in materials. As we have discussed, there is need to consider both the *internal energy* of the system and the mixing tendency (*entropy*). We have pointed out that each of these may be expressed by a term that has the units of energy. At equilibrium the internal energy tends to assume a minimum value, while the *entropy* and mixing energy term $T \Delta S$ tend to seek a maximum value. It is possible for us to construct a composite criterion to account for both factors at the same time. Specifically, we define the sum of the internal energy change, ΔE, and the negative of the mixing energy change, $-T \Delta S$, as the *free energy change*, ΔG, for the system:

$$\Delta G = \Delta E - T \Delta S \tag{4-8}$$

From this definition it is clear that since both the *internal energy* and the negative of the *mixing energy* tend toward minimum values as equilibrium is approached, it follows that at equilibrium the *free energy* reaches an absolute minimum. We see finally that the principle of minimum energy is applicable to materials systems, but we must be careful to specify that it is the *free energy* that seeks a minimum value.

It should be noted immediately that there are often circumstances under which one or the other of the terms in Eq. (4-8) dominates and that the approach to equilibrium is accompanied by either an *increase in the internal energy* or a *decrease in the entropy*. The melting example mentioned earlier is just such a case. In this instance a solid spontaneously melts and reaches its equilibrium state when it is raised above the melting temperature. This occurs in spite of the fact that the internal energy of the system is increased.

We shall now turn our attention to the application of the thermodynamic principles we have just described. In each case we shall be asking the question, "What is the equilibrium state for a material?" And in every case we shall answer that question by finding the condition for which the free energy is minimum.

4-5 EQUILIBRIUM CONCENTRATION OF DEFECTS IN CRYSTALS

One of the most direct applications of the principles of equilibrium involves the thermodynamic prediction that pure perfect crystals are unstable and tend to absorb defects from their surroundings. We will see shortly that this applies to impurities (foreign atoms) as well as to vacant lattice sites and other point imperfections. Because this fact forms the basis for the control of the internal structure of solids, it will be studied in some depth here.

Equilibrium Concentration of Impurities in Crystals

In our study of the equilibrium concentrations of impurities in crystals, we shall focus our attention on a particular alloy system—the germanium-gallium alloy system. As we shall see in Chapter 12, this alloy system is important because of its interesting electrical properties.

Let us take our system to be 1 mole of atoms in the form of a single crystal. Initially, all of the atoms in the system are germanium atoms. We shall investigate the effects of placing gallium atoms as impurities into the system. The gallium atoms are known to be substitutional impurities, and we shall require that when a gallium atom is added to the system the germanium atom that it replaces be removed from the system. In this way the number of atoms in our system will always be the same and is equal to Avogadro's number, N_0.

We are interested in describing the equilibrium state of the alloy system. As discussed in the previous section, equilibrium is achieved when the free energy of the system is at a minimum. Therefore, we must find that condition for which the free energy is lowest. We must consider both the internal energy and the mixing entropy as indicated in Eq. (4-8).

We suppose that when the system contains no gallium impurities it has a reference free energy, G_0. When n_i gallium atoms are placed into the system the change in internal energy may be expressed as

$$\Delta E = n_i \, \Delta E_i \tag{4-9}$$

where n_i is the number of impurity atoms dissolved and ΔE_i is the internal energy change associated with each impurity. One can easily imagine that ΔE_i comes from such factors as difference in size between the impurity atom and the solvent atom.

We now come to the question of the mixing entropy. The problem is to calculate the number of distinguishably different ways the atoms in the system can be arranged. Notice that when there are no impurities in the system, all possible arrangements of atoms on the crystal lattice (diamond cubic for the case of germanium) are identical. This means that when two germanium atoms exchange positions in the solid, the system appears exactly the same as before. The randomness of the system can be greater than one only if we allow some of the gallium atoms to enter the system. If one gallium atom is introduced into the system, there are N different places it might reside, and the randomness of the system is, therefore, equal to N. If we let two impurities enter the system, there are N places in which the first one might reside, and for each of these positions there are $N - 1$ places for the second impurity. This means that the total number of distinguishable ways of arranging a system having two gallium impurities is $N(N - 1)/2$, where the factor $\frac{1}{2}$ is introduced to account for the fact that putting the first impurity at the ith site and the second at the jth site is the same as putting the first one at site j and the second one at site i. From this simple statistical reasoning it can be shown that the number of different ways of arranging the atoms of the system when n_i impurities are introduced is

$$\mathcal{W} = \frac{N(N - 1)(N - 2) \cdots (N - n_i + 1)}{1 \cdot 2 \cdot 3 \cdot 4 \cdots n_i} \tag{4-10}$$

Multiplying both numerator and denominator by $(N - n_i)!$ yields

$$\mathcal{W} = \frac{N!}{n_i!(N - n_i)!} \tag{4-11}$$

Having calculated the randomness of the system with n_i impurities, we may now express the change in the mixing entropy with the use of Eq. (4-6). Using this result and Eqs. (4-8) and (4-9), we may express the free energy as

$$G = G_0 + \Delta E_i n_i - kT \ln \frac{N!}{n_i!(N - n_i)!} \tag{4-12}$$

This expression represents the free energy of the system as a function of the number of impurity atoms. The equilibrium state of this system is now obtained by finding the minimum value of the free energy. Mathematically, we simply require that

$$\frac{\partial G}{\partial n_i} = 0 \tag{4-13}$$

The physical interpretation of Eq. (4-13) is that the free energy of the system is lowered when impurities are added, and that when the addition of more impurities to the system no longer causes the free energy to be lowered the equilibrium state is achieved.

It can be shown using Eq. (4-12) that when Eq. (4-13) is satisfied (at equilibrium) we have

$$n_i = N \exp\left(-\frac{\Delta E_i}{kT}\right) \tag{4-14}$$

as an expression for the equilibrium number of impurity atoms in our system.

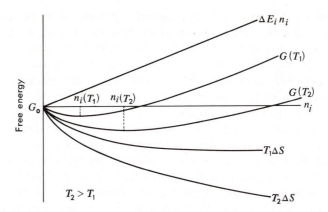

FIG. 4-3   Internal energy, entropy, and free energy of a solid solution with n_i impurity atoms

This procedure is illustrated in Fig. 4-3, where the terms in Eq. (4-12) are plotted as a function of n_i. Notice that the equilibrium number of solute atoms is defined by the minimum in the free energy curve. When the temperature is raised from T_1 to T_2, the curve denoting the entropy contribution to the free energy is increased as shown by the curve in Fig. 4-3. It is easy to see that increasing the temperature increases the importance of the entropy term and leads to a higher equilibrium concentration of solute atoms. This effect is expressed by Eq. (4-14), which is shown graphically in Fig. 4-4. Notice that when the temperature is increased, the fraction

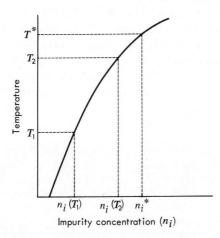

FIG. 4-4 Equilibrium concentrations of impurity atoms as a function of temperature

of sites that contain solutes (n_i/N) increases exponentially. The curve in Fig. 4-4 corresponds to the temperature variation of the free energy minimum of our system. Notice that $n_i(T_1)$ corresponds to the equilibrium value of n_i at temperature T_1 as illustrated in Fig. 4-3.

We also call the curve in Fig. 4-4 a *solubility curve* or a *solvus line*. This curve indicates that the number of solute atoms that can be dissolved by the system is zero at 0°K and increases exponentially at higher temperatures. When we have a system with a fixed number of solute atoms, say n_i^*, the state of aggregation of these atoms obviously depends on the temperature. Above T^* the system reaches a minimum free energy by dissolving all the solute atoms, while below T^* all the atoms will not go into solution. Later in this chapter we shall turn our attention to diagrams of the type shown in Fig. 4-4 when we treat the equilibrium between different phases in materials.

It is clear from Eq. (4-14) and Fig. 4-4 that the solubility of a particular component in a given crystalline matrix depends sensitively on the internal energy change ΔE_i associated with the introduction of each impurity atom. Generally speaking, when the impurity atom dissolves substitutionally, as in the Ge-Ga example, ΔE_i is determined in part by the difference in size of the solvent and solute atoms. For example, if the sizes of the solute and solvent atoms are nearly the same, there is almost no lattice distortion caused by introducing the impurity, and consequently ΔE_i is small. On the other hand, if there is a large size difference between the solute and solvent atoms, ΔE_i can be expected to be large. Therefore, by using Eq. (4-14) we may predict that substitutional impurity atoms that are either significantly larger or smaller than the solvent atoms will have a low solubility, while impurity atoms of about the same size as the solvent atoms will be highly soluble. It is known empirically that two metals with the same valence and the same crystal structure can often dissolve completely in each other if their atomic sizes are within about 15% of each other; they are usually only partially soluble when the size difference is greater than about 15%.

The previous discussion illustrates an important fact about materials, namely that they have an inherent tendency to become impure. Notice that in spite of the fact that the internal energy of the system increases when solutes are dissolved, the entropy of mixing is always sufficiently large to make the impure system the stable one. It might be noted that one of the fundamental difficulties in purifying materials comes from the fact that they tend naturally to be impure. The process of purification must necessarily work against the natural tendency for materials to reach an equilibrium state.

EXAMPLE 4-3

A silicon wafer is in contact with a large quantity of pure antimony. Because the Sb atom is so much larger than the Si atom, the strain energy associated with the introduction of each Sb atom into the silicon crystal is 1 eV. Calculate the temperature at which the equilibrium composition would be 0.01 atomic percent Sb.

Solution:

The temperature dependence of the solubility of impurity atoms is expressed by Eq. (4.14). Thus, the temperature (T_c) at which the equilibrium impurity content becomes 0.01 atomic percent is obtained from the following relation:

$$\frac{n_i}{N} = 10^{-4} = \exp\left(-\frac{\Delta E_i}{kT_c}\right)$$

$$T_c = -\frac{\Delta E_i}{k \ln(n_i/N)} = -\frac{(1.0)(1.6 \times 10^{-12})}{(1.38 \times 10^{-16}) \ln(10^{-4})}$$

$$T_c = 1260°K = 987°C$$

EXAMPLE 4-4

The equilibrium solubility of nitrogen in iron at 500°C is approximately 0.2 atomic percent. Calculate the solubility at 300°C.

Solution:

Using Eq. (4.14) we can write:

$$T_1 \ln\left[\frac{n_i(T_1)}{N}\right] = -\frac{\Delta E_i}{k} = T_2 \ln\left[\frac{n_i(T_2)}{N}\right]$$

If we let $T_1 = 500°C = 773°K$ and $T_2 = 300°C = 573°K$, the solubility at 300°C is expressed as

$$\ln\left(\frac{n_i(573)}{N}\right) = \frac{773}{573} \ln(0.002) = -8.37$$

$$\frac{n_i(573)}{N} = 2.3 \times 10^{-4}$$

Thus, the solubility at 300°C is 0.023 atomic percent nitrogen.

Vacancies and Other Point Defects in Crystals

In the previous section we showed that a perfect crystal is unstable and tends to dissolve impurities. In the same way crystals tend to dissolve vacancies. Consider the perfect crystal shown in Fig. 4-5. Notice that if atom (a) moves to the surface

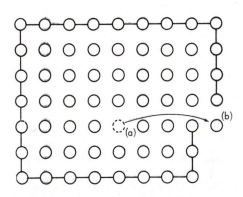

FIG. 4-5   Formation of a vacant lattice site

(site b), a vacancy is formed in the crystal. The vacancy can then migrate into the body of the crystal by changing positions with the atoms of the system. Because of the local distortions and electronic changes associated with the formation of the vacancy, the internal energy of the crystal increases when vacancies are formed. As in the case of impurities, we can write the total internal energy change associated with n_v vacancies as

$$\Delta E = \Delta E_v n_v \qquad (4\text{-}15)$$

where ΔE_v is the internal energy change per vacancy. Because the presence of the vacancy provides new ways for arranging the atoms of the system, the entropy is increased when vacancies are formed. Since the mixing entropy for vacancies is the same as the mixing entropy for impurities, it follows that the equilibrium state of the system is obtained in exactly the same way as it was for impure crystals. The result is that at equilibrium the number of vacancies in a system containing N atoms is

$$n_v = N \exp - \left(\frac{\Delta E_v}{kT}\right) \tag{4-16}$$

This result is the same as given in Eq. (4-14) and can be viewed as the solubility expression for vacancies. The formation energies for vacancies in several metals are shown in Table 4-1 along with the equilibrium vacancy concentrations at various temperatures.

TABLE 4-1

Formation Energies of Vacancies for Several Elements and Equilibrium Concentrations of Vacancies at Various Temperatures

Element	ΔE_v (eV)	T_m* (°C)	Equilibrium Vacancy Concentration (vacancies/cm³)			
			25°C	300°C	600°C	T_m*
Ag	1.1	960	1.5×10^4	1.5×10^{13}	3×10^{16}	7.8×10^{17}
Al	0.76	660	1.0×10^{10}	1.2×10^{16}	2.4×10^{18}	5.0×10^{18}
Au	0.98	1063	1.5×10^6	1.5×10^{14}	1.5×10^{17}	1.2×10^{19}
Cu	1.0	1083	1.1×10^6	1.4×10^{14}	1.4×10^{17}	9.0×10^{18}
Ge	2.0	958	< 1	1.3×10^5	1.3×10^{11}	8.2×10^{13}
K	0.40	63	2.1×10^{15}	Liquid	Liquid	1.3×10^{16}
Li	0.41	186	4.7×10^{15}	Liquid	Liquid	1.4×10^{18}
Mg	0.89	650	4.4×10^7	6.4×10^{14}	3.5×10^{17}	5.7×10^{17}
Na	0.40	98	4.0×10^{15}	Liquid	Liquid	1.0×10^{17}
Pt	1.3	1769	8.7	2.7×10^{11}	2.0×10^{15}	4.2×10^{19}
Si	2.3	1412	< 1	3.1×10^2	2.5×10^9	8.0×10^{15}

*Melting temperature

EXAMPLE 4-5

Measurements of the electrical resistivity of gold wires have indicated that the equilibrium vacancy concentration decreases two orders of magnitude when the temperature is reduced from 900°C to 523°C. Calculate the internal energy change associated with the formation of each vacancy in gold, and compute the fraction of lattice sites that are vacant at the melting temperature (1063°C).

Solution:

Since the equilibrium vacancy fraction at 523°C is two orders of magnitude lower than the value at 900°C, we may write:

$$\left(\frac{n_v}{N}\right)_{523°C} = 0.01 \left(\frac{n_v}{N}\right)_{900°C}$$

or

$$\exp\left(-\frac{\Delta E_v}{kT}\right)_{523} = 0.01 \exp\left(-\frac{\Delta E_v}{kT}\right)_{900}$$

Taking logarithms, we get

$$-\frac{\Delta E_v}{kT_{523}} = \ln(0.01) - \frac{\Delta E_v}{kT_{900}}$$

from which the energy of formation for each vacancy can be calculated:

$$\Delta E_v = \frac{k \ln (0.01)}{(1/T_{900}) - (1/T_{523})} = \frac{1.38 \times 10^{-16} \ln (0.01)}{(\frac{1}{1173} - \frac{1}{796})}$$

$$\Delta E_v = 1.57 \times 10^{-12} \text{ erg/vacancy}$$

$$\Delta E_v = \frac{1.57 \times 10^{-12}}{1.6 \times 10^{-12}} = 0.98 \text{ eV/vacancy}$$

The vacancy fraction at 1063°C is simply expressed as

$$\frac{n_v}{N} = X_v = \exp \left[-\frac{1.57 \times 10^{-12}}{(1.38 \times 10^{-16})(1336)} \right]$$

$$X_v = \exp (-8.5) = 2 \times 10^{-4}$$

Thus, at the melting point of gold about two sites in every 10,000 are vacant.

4-6 PHASE EQUILIBRIA IN ONE-COMPONENT SYSTEMS

We shall begin our study of phase equilibria with the general definition of a *phase*. As was noted in Chapter 3, *a phase is a homogeneous, physically distinct, mechanically separable portion of a material with a given chemical composition and structure*. In order to illustrate this definition we shall cite a few common examples of multiphase systems. Ice cubes in a glass of water represent a two-phase system (ice and liquid water), unless we choose to include the vapor above the glass in our system (in which case it would be a three-phase system). Other common examples include the mixture of oil and water (two phase) and the two-phase mixture of ice and a solution of scotch and water found in a cocktail. All these examples illustrate the various forms that phases may take in multiphase systems. Just as oil and water represent two distinct liquid phases, two regions of a solid with different compositions or crystal structures represent two distinct solid phases. We saw in Chapter 3 that one of the most important structural features of engineering solids is that they are very often composed of more than one solid phase.

Our attention is now directed to the meaning of the term *component*. The number of components in a system is simply the number of different kinds of atoms or molecules in the system. If we have an alloy consisting of two elements, such as iron and carbon in the case of steel, we say there are two *components* in the system. In the case of a compound such as H_2O, however, there is but one component in the system, namely the water molecule, H_2O. Thus, if the elements of the system always appear in the form of a compound we count only the molecule as the component and not the elements themselves.

The classical example of a one-component system is water. The conditions that define the stability of the common phases in water (solid, liquid, and vapor) are shown in Fig. 4-6. According to that diagram (which is called a *phase diagram*) the vapor phase is stable at high temperatures and low pressures, while the solid phase is stable at low temperatures and high pressures. Accordingly, the liquid phase is stable at intermediate pressures and temperatures. We may now ask the question, "Why is a particular phase stable at a given temperature and pressure?" The answer is, of course, that the stable phase is the phase that has the lowest free energy. This means that at low temperatures and high pressures the free energy of

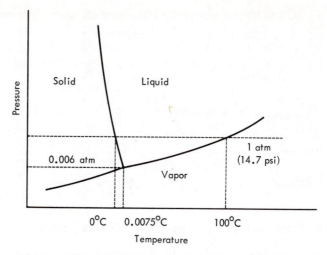

FIG. 4-6   Schematic representation of the pressure-temperature phase diagram for water

a system containing water molecules is lowest when the molecules are bonded together in the form of a solid.

A phase diagram such as the one shown in Fig. 4-6 is of great practical importance, for it allows us to predict the state of our system for any particular temperature and pressure. Suppose, for example, that we are interested in having all three of the phases in water coexist in a single container for very long times. We see immediately from Fig. 4-6 that this can happen only at a pressure of 0.006 atm and

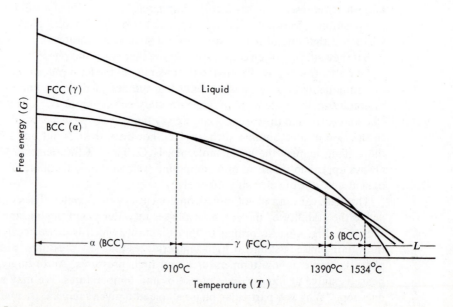

FIG. 4-7   Free energy vs. temperature for the phases that occur in pure iron

at a temperature of 0.0075°C. This point is called the *triple point*, and it is considered to be an *invariant* thermodynamic condition.

The existence of more than one solid phase in a one-component system is known as *polymorphism*. It is one of the key factors that allow us to change the properties of materials by heat treatment. Polymorphism in iron is particularly important because it forms the basis for obtaining the properties of steels. The free energy curves for the condensed phases (solid and liquid) in iron at 1 atm pressure are shown in Fig. 4-7. We see from this figure that at low temperatures (below 911°C) the α phase (BCC) is stable. In the temperature range 911°C to 1390°C the γ phase (FCC) is stable. The BCC phase again becomes stable just below the melting temperature. As will be shown in Chapter 9, it is the transformation from the γ phase (FCC) to the α phase (BCC) that ultimately provides the important properties of steels.

EXAMPLE 4-6

Aluminum oxide melts at 2300°K, and the heat of fusion is 26 kcal/mole. Assuming that the internal energy and the entropy changes that take place in going from the solid to the liquid state are independent of temperature, calculate the free energy that would be released if molten Al_2O_3 were to freeze at 1800°C (say, because of undercooling).

Solution:

At the melting temperature we know that the free energy of the liquid (L) and solid (S) phases are equal because neither is more stable than the other. (They coexist under conditions of equilibrium.) Thus, we may write

$$\Delta G_{S \to L} = \Delta E_{S \to L} - T_M \Delta S_{S \to L} = 0$$

where T_M denotes the melting temperature.

This relation allows us to compute the entropy of fusion as follows:

$$\Delta S_{S \to L} = \frac{\Delta E_{S-L}}{T_M} = \frac{26,000}{2300} = 11.3 \text{ cal/mole-deg}$$

At 1800°C (2073°K) the change in the free energy associated with the $S \to L$ transition (neglecting differences in heat capacity) is expressed as

$$\Delta G_{S \to L} = \Delta E_{S \to L} - T \Delta S_{S \to L}$$

and has the value

$$\Delta G_{S \to L} = 26,000 - (2073)11.3 = 2600 \text{ cal/mole}$$

If molten Al_2O_3 were to freeze at 1800°C, the corresponding change in the free energy would be $\Delta G_{L \to S} = -2600$ cal/mole. The free energy change is negative since the solid phase is stable at 1800°C.

The Gibbs Phase Rule

The equilibrium between different phases in a one-component system depends on both the pressure and the temperature. According to the phase diagram for water shown in Fig. 4-6, the liquid and solid phases are in equilibrium only for a well-defined set of pressures and temperatures. For example, when the pressure is 1 atm, the liquid and solid are in equilibrium only if the temperature is held at 0°C. This example indicates that if we are to have two phases in equilibrium in this system, only one of the two variables—pressure or temperature—may be arbitrarily chosen. The other variable is given by the condition of equilibrium. In this sense we say that there is but one *degree of freedom* regarding our choice of the temperature and pressure.

The temperature and pressure are called *intensive* variables because they do not depend on the size of the system.* The other intensive variables that represent possible *degrees of freedom* are the concentrations of the components in each one of the phases in the system. For a single component system we have only the temperature and the pressure as possible degrees of freedom.

The relation between the number of phases in equilibrium and the number of degrees of freedom is expressed in a general way by the *Gibbs phase rule*. If we designate the number of components of a system as C, the number of phases in the system as P, and the number of degrees of freedom for the system in question as F, then it can be shown that

$$P + F = C + 2 \qquad (4\text{-}17)$$

This important equation, originally derived by J. Willard Gibbs, allows us to determine, for a given system, the number of degrees of freedom that may be arbitrarily specified.

Now let us consider the meaning of the phase rule in terms of the phase diagram in Fig. 4-6. First let us determine the number of degrees of freedom for a single phase system. In this case $P = 1$, so that for a one-component system ($C = 1$), $F = C + 2 - P = 2$. These two degrees of freedom are interpreted to mean that both the pressure and temperature may be arbitrarily chosen in a single phase region. For two-phase equilibrium, $P = 2$ and $F = 1$; one may arbitrarily select either the pressure or the temperature for the two-phase equilibrium, but not both. Once one of these variables is chosen (say, the pressure), the other (the temperature) is determined by the equilibrium phase diagram. Figure 4-6 also illustrates the fact that three-phase equilibrium in this system can be achieved only at a particular pressure, 0.006 atm, and temperature, 0.0075°C. This corresponds to zero degrees of freedom, $F = 0$.

Since multicomponent materials are almost always subjected to a constant pressure, usually 1 atm, it is appropriate to write the phase rule for the special case in which pressure is not considered to be a degree of freedom. In this case we would write $P + F = C + 1$ (pressure constant). This form of the phase rule will be used throughout the remainder of this chapter and in other parts of the book.

4-7 PHASE EQUILIBRIA IN MULTICOMPONENT SYSTEMS

We now begin a study of equilibrium in multicomponent systems. Although most engineering materials are composed of many more than two chemical components, the study of two-component (*binary*) systems provides sufficient insight to understand phase equilibria even in very complex alloys.

Extensive variables are dependent on the size of the system. These variables include such quantities as volume, mass, and the total number of moles of a particular component.

Let us begin by considering one of the simplest alloy systems, the copper-nickel system. We want to find the equilibrium state of this system for all possible temperatures and chemical compositions. We shall suppose that the pressure is constant and is equal to 1 atm. As in all problems dealing with equilibrium, our task is to find the states for which the free energy of the system is minimum. In the copper-nickel binary system we have only two phases to consider—the liquid phase and the solid solution phase (having the FCC structure). In general, our approach to this problem is to determine how the free energies of each of these phases depend on both temperature and composition. The equilibrium diagram that describes the system is simply a composite of the information gained from considering the free energies in detail.

Suppose we consider a system of N atoms in which there are $X_{Ni}N$ atoms of nickel and $X_{Cu}N$ atoms of copper. (X_{Ni} and X_{Cu} represent the atomic fractions of

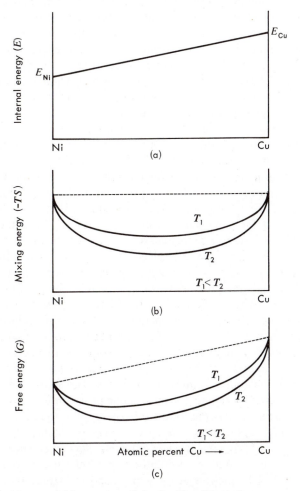

FIG. 4-8   Internal energy, entropy, and free energy of the Cu-Ni solid solution

the two components.) The internal energy of the solid solution can be treated in an approximate way simply by assigning different energies to the nearest neighbor atomic bonds. If we suppose that the bond between a nickel atom and a copper atom is the average of the nickel-nickel bond energy and the copper-copper bond energy, it can be shown that the internal energy of the alloy varies linearly with the atomic fraction of one of the components. This result is shown graphically in Fig. 4-8(a).

To obtain the free energy of the copper-nickel solid solution we must take account of the entropy of the solution as well as its internal energy. As was discussed in Section 4-4, the mixing entropy of a solid solution comes from the fact that there are many distinguishable atomic arrangements in a system that contains two types of atoms. The mixing energy $T\Delta S$ which is related to the state of mixing, is graphically illustrated for the copper-nickel solid solution in Fig. 4-8(b), which indicates that the mixing energy ($T\Delta S$) is maximum when there are equal numbers of the two types of atoms. This means simply that the number of distinguishable atom arrangements is maximum when the composition of the alloy is 50 atomic percent Cu and 50 atomic percent Ni. The free energy for the copper-nickel solid solution is obtained by adding the negative of the mixing energy to the internal energy. The result is illustrated in Fig. 4-8(c), where we see that the effect of increasing the temperature is to decrease the free energy of the alloy with respect to the free energies of the pure components.

Having established the form of the free energy of the copper-nickel solid solution, we can now turn our attention to the problem of predicting the equilibrium phases in this simple system. Suppose that the free energy of the liquid solution has a form similar to that for the solid solution. At very high temperatures ($T >$ 1453°C) we know that both copper and nickel are liquid. This condition is represented in Fig. 4-9(a), where the free energy of the liquid phase is lower than the free energy of the solid phase at all compositions. When the temperature is lowered to 1300°C, the free energy curves are repositioned in such a way [Fig. 4-9(b)] that the free energy of the solid phase of pure nickel (which freezes at 1453°C) is lower than the free energy of the liquid phase. Similarly, pure copper (which freezes at 1083°C) is liquid at 1300°C; hence, the free energy of liquid copper is lower than the free energy for the solid phase of copper. The free energy curves in Fig. 4-9(b) also indicate that the free energy of a dilute solid solution of copper in nickel is lower than the free energy of the liquid phase for that same composition. Thus, it is evident that a dilute solid solution of copper in nickel is more stable than the liquid solution at this temperature. By similar reasoning we see that a dilute liquid solution of nickel in copper is more stable than a solid solution with the same composition.

Generally speaking, the most stable phase for an alloy is determined by selecting the phase that has the lowest free energy. We shall now show that there are conditions for which the lowest free energy is obtained by allowing the alloy to be composed of *two separate phases* having *different* chemical compositions. Consider, for example, an alloy of 50 atomic percent Cu–50 atomic percent Ni at 1300°C. According to the free energy curves shown in Fig. 4-9(b), the free energy of the liquid solution is lower than the free energy of the solid solution for the composi-

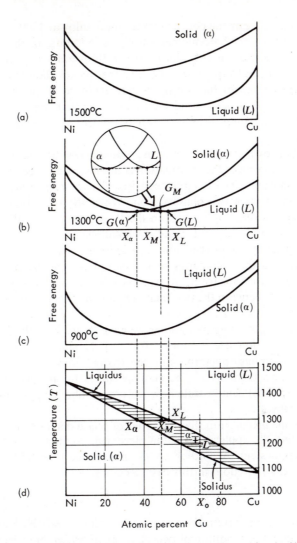

FIG. 4-9   Free energy curves at 1500°C, 1300°C, 900°C, and phase diagram for the Ni–Cu alloy system

tion. However, the lowest possible free energy for this alloy, and hence the most stable state of the alloy at this temperature and composition, is achieved by having *two separate phases* of *different compositions*. The free energy is given by the *line of common tangents* shown in Fig. 4-9(b). The two phases in question have the compositions X_α (solid solution) and X_L (liquid solution) as indicated in the figure. This result means that when the composition of a copper-nickel alloy falls between X_α and X_L, the equilibrium state of the alloy at 1300°C requires a mixture of the liquid and solid phases. Regardless of the exact overall alloy composition in this range (between X_α and X_L at 1300°C), the compositions of the component phases are

X_α and X_L. For the copper-nickel binary alloy at 1300°C, $X_\alpha = 36$ atomic percent Cu and $X_L = 52$ atomic percent Cu. In addition, in this composition range only the *amounts* of the two phases vary with overall composition.

When the temperature of the copper-nickel alloy is lowered to 900°C the free energy curves for the two phases are arranged in such a way [see Fig. 4-9(c)] that the free energy of the solid solution is lower than the free energy for the liquid solution for all compositions, and therefore the solid solution is stable for all compositions.

As shown in Fig. 4-9(d), the *phase diagram* for the copper-nickel binary system is a composite of the information contained in the free energy curves at the various temperatures. The diagram consists of two *single phase fields* (solid and liquid) separated by a two-phase region. When the temperature and composition of a copper-nickel alloy fall within the two-phase region, the equilibrium state of the alloy is a mixture of the liquid and solid phases. The compositions of the two phases are given by the intersection of an isothermal section (at the temperature in question) with the boundaries of the two-phase regions. The line that separates the liquid phase field from the two-phase region is called the *liquidus* line, while the boundary between the two-phase region and the solid phase field is called the *solidus* line.

In the preceding paragraphs we have shown how the free energy concept can be used to rationalize the form of one of the simplest phase diagrams. Actually, in practice, phase diagrams are constructed from experimental information about the alloys themselves. The cooling curves for the phase diagram shown in Fig. 4-11 (p. 126) demonstrate how features of phase diagrams can be obtained experimentally.

Some insight into the properties of phase diagrams may now be gained by considering the application of the phase rule to the case of two-phase equilibria in a binary system. According to the phase rule (constant pressure), the number of degrees of freedom in this case becomes:

$$F = C + 1 - P = 2 + 1 - 2 = 1$$

Thus, we may arbitrarily specify one intensive variable (say, temperature) and still maintain the condition of two-phase equilibrium. Now let us examine the meaning of this result in light of the Cu-Ni phase diagram shown in Fig. 4-9. Suppose we choose to fix the temperature at 1200°C. According to the phase diagram, two phases are stable as long as the overall composition of the alloy falls between 23 atomic percent Ni and 40 atomic percent Ni. For an alloy consisting of 30 atomic percent Ni–70 atomic percent Cu (denoted by X_0 in Fig. 4-9), and at a temperature of 1200°C, the compositions of the solid and liquid phases are given by the intersection of the 1200°C isotherm with the *solidus* and *liquidus* lines, respectively. From the phase diagram it is clear that the composition of the solid phase in equilibrium with the liquid at 1200°C is $X_\alpha = 40\%$ Ni–60% Cu, while the corresponding composition of the liquid is $X_L = 23\%$ Ni–77% Cu. It should be remembered that at this temperature the alloy achieves the lowest free energy when it is composed of a mixture of the two phases having compositions that differ from the overall composition X_0. We now see that once the temperature is chosen, the compositions of the component phases are fixed by the phase diagram and are not subject to arbitrary change.

The degrees of freedom that are possible in a multicomponent system include the concentrations of the components in each phase as well as the temperature. Let us again refer to Fig. 4-9 and consider the conditions for which we have two-phase equilibrium when the *liquid phase* has a composition of 50 atomic percent Ni–50 atomic percent Cu. Since the phase rule predicts that there is but one degree of freedom for this system, it follows that all other intensive variables, including the temperature, are fixed.

The isothermal section that cuts through the two-phase region is called the *tie line. The compositions of the phases that participate in two-phase equilibrium are given by the intersections of the tie line and the solubility limits of the single-phase regions.* This fact will hereafter be called the *tie-line principle.*

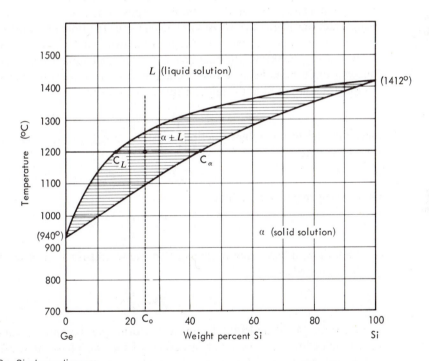

FIG. 4-10   Ge–Si phase diagram

EXAMPLE 4-7   Use the Si-Ge equilibrium diagram shown in Fig. 4-10 to predict the temperature at which the solid solution attains a composition of 60 weight percent Si–40 weight percent Ge by equilibrating with the liquid phase. Determine the composition of the liquid phase with which equilibrium is established.

Solution: A vertical section at 60 weight percent Si in the Si-Ge phase diagram intersects the solidus line at approximately 1280°C. The composition of the liquid phase that is in equilibrium with the solid solution at 1280°C is obtained by taking the intersection of the 1280°C isotherm with the liquidus line. The equilibrium liquid composition is about 30 weight percent Si.

So far in our discussion we have dealt only with the compositions of the phases involved in two-phase equilibria. It is possible to show that for an alloy consisting of two phases, the weight fractions of the two phases can be obtained from information contained in the phase diagram. Consider a two-phase alloy in the Si-Ge system (Fig. 4-10) with an overall composition C_0 (say, 25 weight percent Si). The compositions of the liquid and solid phases at 1200°C are denoted by C_L and C_α respectively (weight fractions). It is obvious from the phase diagram that the overall composition C_0 determines the amounts of the two phases in question. When C_0 is near C_L we have mostly liquid in our mixture (if $C_0 \leq C_L$, we have all liquid), and when C_0 is near C_α we have mostly solid phase in the mixture (if $C_0 \geq C_\alpha$ we have a single solid phase). When the overall composition falls between C_L and C_α the weight fractions of the two phases are given by the fractional position of the overall composition C_0 between the boundaries of the two-phase regions, C_α and C_L.

Let f_α and f_L be the weight fractions of the solid and liquid phases, respectively. It follows from a mass balance that the amount of Si in the two phases must equal the total amount of Si in the alloy. Thus, we may write

$$C_0 = C_\alpha f_\alpha + C_L f_L \qquad (4\text{-}18)$$

Since the sum of the weight fractions must be unity ($f_\alpha + f_L = 1$), it follows from Eq. (4-18) that

$$C_0 = C_\alpha f_\alpha + C_L - C_L f_\alpha \qquad (4\text{-}19)$$

On rearrangement, we see that

$$f_\alpha = \frac{C_0 - C_L}{C_\alpha - C_L} \qquad (4\text{-}20)$$

In addition, it can easily be shown that the weight fraction of the liquid phase can be expressed as

$$f_L = \frac{C_\alpha - C_0}{C_\alpha - C_L} \qquad (4\text{-}21)$$

Equations (4-20) and (4-21) represent a principle, commonly called the *lever rule*, which states that the weight fractions of the two phases that exist in a two-phase mixture are given by a simple lever relation between the overall composition and the compositions of the two phases.

EXAMPLE 4-8   Suppose that 1 kg of an alloy of 15 weight percent Si–85 weight percent Ge is held at 1100°C. Calculate the weight of Si in the liquid phase.

Solution: The composition of the liquid phase participating in the two-phase equilibria at 1100°C is given by the intersection of the 1100°C isotherm with the liquidus line: $C_L = 7$ weight percent Si. Isothermal intersection with the solidus line at 1100°C gives $C_\alpha = 26$ weight percent Si. Using the lever principle, the weight of liquid in this alloy is

$$W_L = (1 \text{ kg})f_L = \frac{C_\alpha - C_0}{C_\alpha - C_L} = \frac{26 - 15}{26 - 7} = 0.58 \text{ kg}$$

The weight of Si in the liquid phase is therefore $W_{\text{Si}}^{(L)} = (0.58)(0.07) = 0.04 \text{ kg}$.

We have now completed our study of two-phase equilibrium in binary systems. The principles we have described apply not only to simple isomorphous alloy systems but to more complicated multiphase systems as well. We will see shortly how both the *tie-line principle* and the *lever rule* apply to a wide variety of phase diagrams.

In the phase diagrams we have studied thus far, the solid forms of the two components have had the same structure and the alloys have been completely soluble in the solid state. The phase diagrams in Figs. 4-9 and 4-10 are called *isomorphous* equilibrium diagrams and are characterized by the fact that a maximum of two phases are in equilibrium at a given temperature and pressure. They are called *isomorphous* because the crystal structure of the solid state of this system does not change with composition.

4-8 INVARIANT REACTIONS

We now begin a study of equilibrium in alloys in which the two components are only partially soluble in the solid state. This condition arises if the two components have different crystal structures or if there are large differences in atomic size or electronic structure. A common feature of such alloy systems is the appearance of an *invariant reaction*, in which one or more phases react at a particular temperature to produce, on cooling, one or more new phases. Let us consider first the case of a *eutectic* reaction, in which a single liquid phase transforms, on cooling, to two separate and distinct solid phases:

$$L \xrightarrow[\text{cooling}]{} \alpha(s) + \beta(s)$$

During the course of a eutectic freezing reaction the liquid phase L is in equilibrium with the solid phases α and β (just as the liquid of a pure component is in equilibrium with the solid phase at the melting temperature). Thus, during a eutectic reaction we have the condition that three phases are in equilibrium. According to the phase rule (constant pressure), the number of degrees of freedom for three-phase equilibrium in a binary system is $F = 0$. This means that three phases can be in equilibrium only at one temperature. This particular temperature is called the *eutectic temperature* and is identified by an isothermal line in the equilibrium diagram.

An example of the eutectic reaction is found in the Pb-Sn binary phase diagram (see Fig. 4-11). This alloy system is of practical importance because Pb and Sn are the major alloying elements in most common solders. One of the most obvious facts about this system which is evident from the phase diagram is that the Pb-Sn alloy generally melts at lower temperatures than either pure Pb or Sn.

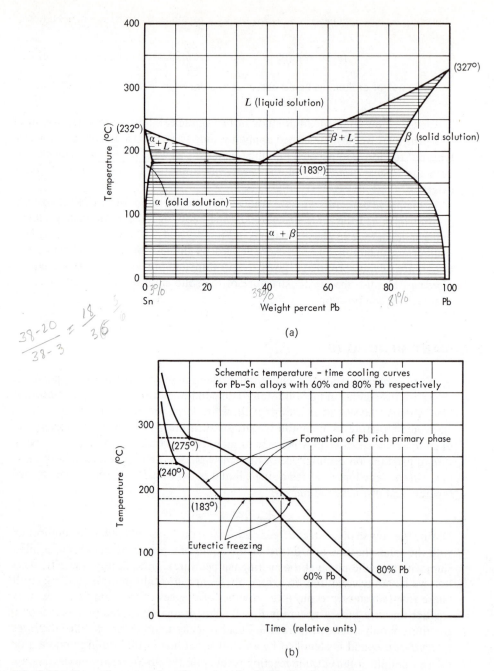

(a)

(b)

FIG. 4-11 Phase diagram and cooling curves for the Pb–Sn alloy system

In the Pb-Sn system the eutectic temperature is 183°C and the compositions of the phases that participate in the three-phase equilibrium are: C_α (Sn-rich solid solution) = 2.5 weight percent Pb, C_β (Pb-rich solid solution) = 81 weight percent Pb, and C_L (called the *eutectic composition*) = 38.1 weight percent Pb. In the eutectic reaction the liquid phase (C_L) transforms (freezes) at the eutectic temperature and produces two separate solid phases, α and β, with compositions C_α and C_β.

Two-phase regions in eutectic systems have the same thermodynamic properties as two-phase regions in isomorphous alloy systems. The tie-line principle and the lever rule are applied in the same way as before.

EXAMPLE 4-9

Use the Pb-Sn phase diagram to predict the compositions and the weight fractions of the phases of an alloy of 80 weight percent Pb–20 weight percent Sn at 250°C. How is it possible for both the liquid and solid phases to increase in Sn content when the temperature is lowered to 200°C?

Solution:

At 250°C the compositions of the phases are

$$C_\beta \text{ (Pb-rich solid phase)} = 87 \text{ weight percent Pb}$$
$$C_L \text{ (liquid)} = 67 \text{ weight percent Pb}$$

From the lever rule, the weight fractions of the two phases are

$$f_\beta = \frac{C_L - C_0}{C_L - C_\beta} = \frac{67 - 80}{67 - 87} = 65 \text{ weight percent } \beta$$

$$f_L = \frac{C_0 - C_\beta}{C_L - C_\beta} = \frac{80 - 87}{67 - 87} = 35 \text{ weight percent } L$$

When the temperature is reduced to 200°C the equilibrium compositions are

$$C_\beta = 83 \text{ weight percent Pb}$$
$$C_L = 45 \text{ weight percent Pb}$$

and the weight fractions of the two phases are

$$f_\beta = \frac{45 - 80}{45 - 83} = 92 \text{ weight percent } \beta$$

$$f_L = \frac{80 - 83}{45 - 83} = 8 \text{ weight percent } L$$

These results indicate that both the solid and liquid phase become more rich in Sn when the temperature is reduced. This apparent increase in the overall amount of Sn available is exactly compensated by the fact that at lower temperatures we have greater amounts of the phase having the lowest Sn content (the solid phase). The weight of Pb in 1 g of the alloy at 250°C is given by

$$W_{Pb}^{250°C} = f_\beta C_\beta + f_L C_L = (0.65)(0.87) + (0.35)(0.67) = 0.80 \text{ g}$$

while the weight of Pb at 200°C is given by

$$W_{Pb}^{200°C} = (0.92)(0.83) + (0.08)(0.45) = 0.80 \text{ g}$$

The important point in this calculation is that in spite of the compositional changes that take place when the alloy is cooled from 250°C to 200°C, the total amount of Pb in the system remains unchanged, as it should.

Now let us again turn our attention to the invariant eutectic reaction that takes place in the Pb-Sn alloy system. If an alloy having the eutectic composition is heated to a temperature slightly above the eutectic temperature, the alloy will become entirely liquid. Now if we attempt to cool the alloy by extracting heat to the surroundings, the liquid begins to transform to the two solid phases α and β. Since the three phases can be in equilibrium only at the eutectic temperature, the temperature remains unchanged until the alloy has transformed entirely to the two-phase solid state. In this sense the eutectic alloy freezes in a manner similar to the freezing of pure components. A cooling curve of temperature versus time would exhibit an arrest when the eutectic temperature is reached, as illustrated in Fig. 4-11. On cooling, the alloy remains at the eutectic temperature until the latent heat of fusion for the alloy has been removed. Once the alloy is entirely solid, further heat removal simply reduces the temperature of the two-phase solid mixture.

It is instructive at this point to consider other common types of invariant reactions that can take place in binary alloys. A listing of four of the most important invariant reactions is given in Table 4-2. *Eutectic* and *eutectoid* reactions are similar

TABLE 4-2

Invariant Reactions in Binary Alloys

Invariant reaction	Phase reaction	Phase diagram
Eutectic	$L \longrightarrow \alpha(s) + \beta(s)$ cooling	
Peritectic	$\alpha(s) + L \longrightarrow \beta(s)$ cooling	
Eutectoid	$\gamma(s) \longrightarrow \alpha(s) + \beta(s)$ cooling	
Peritectoid	$\alpha(s) + \gamma(s) \longrightarrow \beta(s)$ cooling	

in that they both involve the formation of two phases from one on cooling, while the *peritectic* and *peritectoid* reactions require that two phases react and produce a new single phase at the lower temperature. It might be noted that heating through a peritectic or peritectoid reaction is similar to cooling through the eutectic and eutectoid in the sense that a single phase reacts to form two new phases. The eutectic and peritectic reactions involve the liquid phase, while the eutectoid and peritectoid reactions are entirely solid state reactions. The nature of the construction of the phase diagram in each case is indicated in Table 4-2.

In all invariant reactions in binary alloys the temperature at which the reaction takes place is fixed and is not subject to change. In addition, the compositions of the component phases that participate in the invariant reactions are fixed by the points of intersection of the reaction isotherm and the respective single-phase fields. Consider the peritectic invariant reaction that is found in the Pt-Re phase diagram shown in Fig. 4-12. According to the phase diagram, this binary alloy exhibits a peritectic reaction at approximately 2450°C. The composition of the phases that participate in the peritectic reaction are: $C_L = 43$ weight percent Re, $C_\beta = 54$ weight percent Re, and C_α (called the peritectic composition) = 46 weight percent Re. At a temperature slightly above the peritectic temperature an alloy having the peritectic composition consists of a mixture of two phases: β (solid

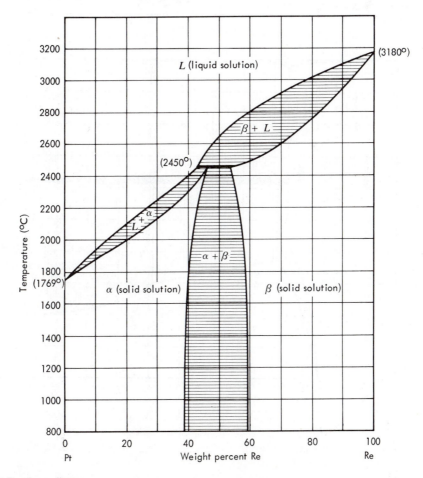

FIG. 4-12   Pt–Re phase diagram

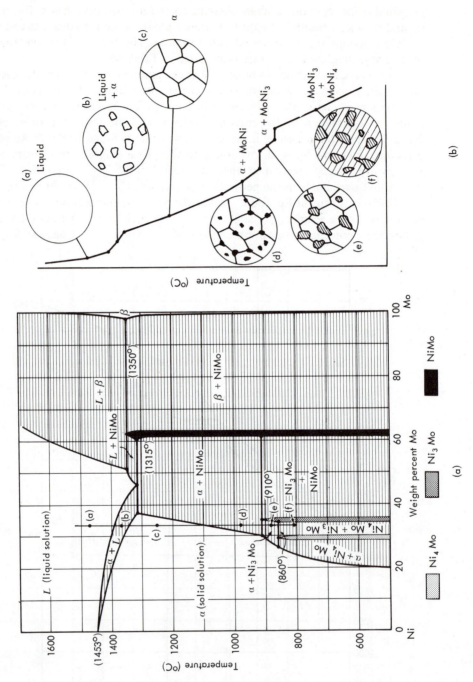

FIG. 4-13 Ni–Mo phase diagram and cooling curve for Ni–33%Mo alloy

solution) and L (liquid). According to the lever rule, the weight fractions of the two phases are

$$f_L = \frac{C_\beta - C_0}{C_\beta - C_L} = \frac{54 - 46}{54 - 43} = 73\% \text{ by weight}$$

$$f_\beta = \frac{C_0 - C_L}{C_\beta - C_L} = \frac{46 - 43}{54 - 43} = 27\% \text{ by weight}$$

It should be remembered that the compositions of the component phases must be expressed in weight fractions when making a calculation of the weight fraction of a particular phase.

In order for the peritectic reaction to occur, it is necessary that the β and liquid phases have the appropriate compositions (as given by the phase diagram) and that the temperature be equal to the peritectic temperature. It is not required, however, that the overall composition be equal to the peritectic composition. Suppose, for example, that we have an alloy of 50 atomic percent Re–50 atomic percent Pt (approximately 49 weight percent Re–51 weight percent Pt). According to the lever rule, at a temperature just above the peritectic temperature about half the alloy is liquid and the other half is solid (β, solid solution). Now if we try to cool the mixture by extracting heat, the peritectic reaction occurs and continues until one or both of the reacting phases have been used up. In the case of an alloy of 50 atomic percent Re, the phase diagram indicates that there is a certain amount of the β phase remaining after the peritectic reaction is complete. It is obvious that when the temperature is slightly below the peritectic temperature, the alloy consists of a two-phase mixture of β and α. The α phase is formed in the peritectic reaction, and the β phase constitutes the amount of excess β phase that existed prior to the peritectic reaction. It is equally clear that if the alloy has the peritectic composition, the amounts of β and liquid phases just above the peritectic temperature will be in the right proportion to react completely and form the single phase α.

In the cooling of an alloy through a peritectic reaction, the temperature remains constant as the invariant reaction takes place. As in the case of a eutectic reaction, the peritectic reaction causes a thermal arrest in a plot of temperature versus time as heat is extracted from the alloy at a fixed rate. The use of cooling curves for the clarification of the properties of phase diagrams is illustrated in Fig. 4-13 for the Ni-Mo alloy system.

Mon.

4-9 APPLICATION OF EQUILIBRIUM DIAGRAMS

Having developed all the relevant principles of phase equilibria, we are now prepared to describe the compositional changes that take place in an alloy when it is cooled or heated. We shall consider the case in which the rate of temperature change is sufficiently slow that thermodynamic equilibrium can be assumed at all times. As we shall see in the next chapter, it is not possible in practice to maintain equilibrium and still have phase changes occurring at a finite rate. We shall find, for example, that a reaction proceeds only if the temperature is somewhat below the equilibrium temperature. In spite of this complication, it is reasonably accurate to assume that equilibrium conditions are maintained when an alloy is heated or cooled very slowly.

Equilibrium Cooling

Consider the nickel-rich portion of the equilibrium diagram for the nickel-molybdenum system shown in Fig. 4-13. (This phase diagram is of technological importance since many of the alloys used for corrosion resistance in reducing environments are based on the Ni-Mo binary system.) Suppose we have a Ni-Mo alloy that contains 33% Mo by weight. Let us consider the compositional changes and phase changes that take place when the alloy is slowly cooled.

It should be noted before we begin that the isothermal lines in the Ni-Mo phase diagram correspond to the invariant reactions that are described in Table 4-2. As we describe the events that occur in the Ni-Mo system on cooling, we shall apply the principles of invariant reactions discussed in the previous section.

We begin by first heating the alloy to a sufficiently high temperature to obtain a homogeneous liquid phase ($T > 1390°C$). Now we cool the alloy slowly by allowing the heat to be transferred to the surroundings. Until the liquidus line is reached (at 1390°C), the state of the alloy is entirely liquid. The drawings accompanying the phase diagram in Fig. 4-13 illustrate the phase structure of the alloy. These drawings indicate what would be seen if the alloy were viewed with a microscope. Figure 4-13 also shows the cooling curve for an alloy of 33 weight percent Mo. The thermal arrests in the cooling curve correspond to the invariant reactions encountered for this alloy.

When the temperature falls below 1390°C, crystals of the nickel-rich solid solution begin to form and take the place of some of the liquid phase. At about 1375°C the weight fraction of the solid phase is approximately 50%, and the compositions of the respective phases are $C_\alpha = 27$ weight percent Mo and $C_L = 37$ weight percent Mo. The cooling curve changes slope in the two-phase region simply because the latent heat of fusion for the alloy must be removed as the alloy is cooled through the two-phase region. When the temperature falls below the solidus line the alloy is composed entirely of the solid phase α. As cooling proceeds from 1350°C to 1050°C the alloy remains in the form of the single solid phase.

When the limit of solubility of the α phase is reached (at 1050°C) a single phase having the overall composition is no longer stable. Accordingly, a new solid phase MoNi, which is called an *intermetallic compound*, is formed. The microstructure of the alloy at 950°C is illustrated in Fig. 4-13. As the drawing indicates, the new phase *nucleates* and *grows* either in the interior of the α grains or on the grain boundaries. The process by which this phase change takes place is called *precipitation* and is discussed at some length in Chapter 9.

When the temperature of the alloy reaches 910°C a peritectoid reaction begins to take place. In this reaction the MoNi particles react with the matrix phase α to produce a new intermetallic compound MoNi$_3$. The phase diagram indicates that the alloy is composed of approximately 50 weight percent MoNi$_3$ and 50 weight percent α just after the invariant reaction. On further cooling the alloy encounters another peritectoid reaction at 860°C. In this reaction the original matrix phase

FIG. 4-14 Al-Ge eutectic mixture (70×) Photo courtesy R. Koch, Stanford University. Used with permission.

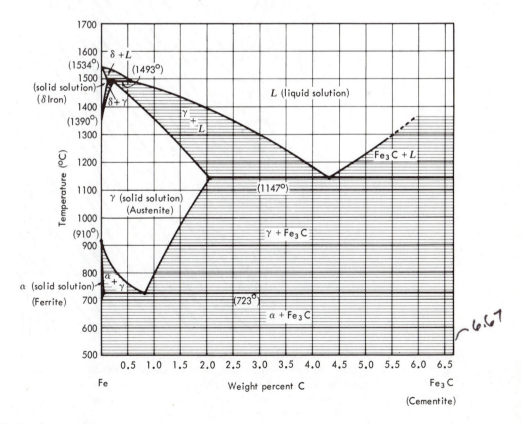

FIG. 4-15 Fe–C phase diagram

α is consumed by the reaction, and the alloy becomes a two-phase mixture of the intermetallic compounds $MoNi_3$ and $MoNi_4$.

In the alloy just described, the phases are said to have a *particulate* distribution. All of the two-phase mixtures can be characterized by particles of a secondary phase imbedded into the matrix of the primary phase. In the case of alloys that exhibit eutectic or eutectoid invariant reactions, the two-phase mixtures adopt either a *lamellar* form (alternating slabs of the respective phases) or a *rodlike* form (parallel rods of one phase in a matrix of the second phase). A eutectic mixture for the Al-Ge alloy is shown in Fig. 4-14. The two-phase colonies grow (at the expense of the parent phase) parallel to the component phases (i.e., parallel to rods and plates) simply because repeated nucleation of the two phases is not required in that direction.

Probably the most important example of a lamellar two-phase mixture is that which is encountered in carbon steels. The iron-carbon phase diagram illustrated in Fig. 4-15 exhibits a very important eutectoid reaction at 723°C and at a carbon content of 0.8 weight percent carbon. When the FCC solid solution (called *austenite*) is slowly cooled below the eutectoid temperature (723°C), a two-phase mix-

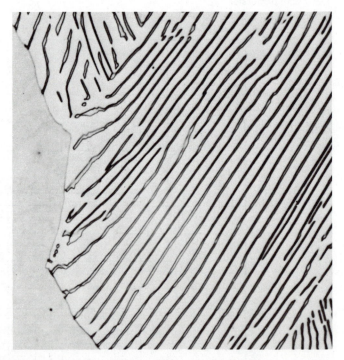

FIG. 4-16   Pearlite microstructure (1000×). Photo courtesy American Institute of Mining, Metallurgical and Petroleum Engineers, Inc., from *Decomposition of Austenite by Diffusional Processes* by V. F. Zackay and H. I. Aaronson (eds.) ("Some Observations on the Growth of Pearlite" by L. S. Darken and R. M. Fisher) (1962). Used with permission

FIG. 4-17   Al–Mg eutectic mixture (200×) Photo courtesy American Institute of Mining, Metallurgical and Petroleum Engineers, Inc., from A. S. Yue, *Trans. of the AIME* **224**, 1010 (1962). Used with permission.

ture of the BCC solid solution (called *ferrite*) and Fe_3C (called *cementite*) is formed. An optical micrograph of the phase structure in a eutectoid steel is shown in Fig. 4-16. The alternating layers of the two phases are characteristic of eutectoid and eutectic mixtures. The rodlike microstructures shown in Fig. 4-17 illustrate the two-phase eutectic mixture that is formed in the Al-Mg system.

In concluding our discussion of equilibrium diagrams, a brief comment about complex phase diagrams should be made. Many alloys form one or more *intermediate* phases, which appear as single-phase regions in the interior of the phase diagram. In some cases the intermediate phases are compounds that simply divide the diagrams into two separate parts. In the Ni-Mo phase diagram (Fig. 4-13), for example, the compound MoNi may be considered to separate the phase diagram into two separate parts. When the composition of a Ni-Mo alloy falls on the nickel-rich side of the compound, only that portion of the phase diagram is pertinent. In this sense MoNi may be viewed as a pure component that separates the Ni-Mo phase diagram into two distinctly different diagrams. In other alloy systems, such as in the Cu-Zn system, the intermediate phases are numerous and complex and do not divide the diagram into discrete simple parts. In this case it must be remembered that each two-phase region is composed of the single phases on either side of it and that the basic rules for invariant reactions apply as before.

Phase diagrams for ternary alloy systems are possible, but unless one is concerned with a precise determination of the compositional changes in an alloy, it is generally simpler and more practical to generalize from the pertinent binary phase diagrams.

4-10 OXIDE PHASE DIAGRAMS

The principles of phase equilibria discussed in the previous sections apply not only to metallic binary alloys but to alloys of oxides as well. In the case of oxide systems we can take the oxides that compose the system to be the components. Using this approach, we can represent a binary oxide system by M_wO_x–N_yO_z, where M and N are the metallic elements in each of the component oxides, O represents the oxygen, and w, x, y, and z (usually integers) define the exact chemical compositions of the oxides.

The Al_2O_3-Cr_2O_3 alloy system is described by the isomorphous phase diagram shown in Fig. 4-18. This alloy system is important because single crystals of the

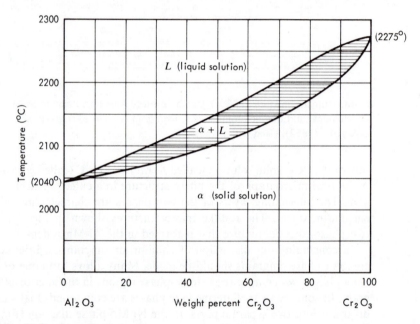

FIG. 4-18 Al_2O_3–Cr_2O_3 phase diagram

Al_2O_3-Cr_2O_3 solid solution (ruby crystals) can be made to emit coherent electromagnetic radiation (lasers). These unique optical properties will be discussed in Chapter 15.

The Al_2O_3-Cr_2O_3 phase diagram has been determined for a pressure of 1 atm; thus, the predictions of the phase rule for this system are exactly as they were for the Cu-Ni and Si-Ge alloy systems. The lever principle and the tie-line principle apply exactly as before.

EXAMPLE 4-10

A ruby crystal having the composition Al_2O_3 + 20 weight percent Cr_2O_3 is to be grown by slowly pulling the crystal from a liquid solution of Al_2O_3 and Cr_2O_3. The composition of the liquid bath is kept constant by making appropriate additions of Al_2O_3 and Cr_2O_3 to the liquid as the single crystal is grown. Determine the temperature at which the crystal must be grown, and find the composition of the liquid solution that will produce a crystal with the desired composition. Suppose that the total weight of the liquid solution is maintained at 1 kg and compute the weight of chromium in the liquid during the crystal-growing operation.

Solution:
From the Al_2O_3-Cr_2O_3 phase diagram in Fig. 4.18 we see that the solid solution containing 20 weight percent Cr_2O_3 is in equilibrium with the liquid phase at 2062°C and therefore must be grown at that temperature. Using the tie-line principle, we see that the composition of the liquid must be 90 weight percent Al_2O_3–10 weight percent Cr_2O_3.

The total weight of Cr_2O_3 in the liquid is $(0.1)(1 \text{ kg}) = 0.1 \text{ kg} = 100 \text{ g}$. The number of moles of Cr_2O_3 in the liquid is $100/M_{Cr_2O_3} = 100/[2(52) + 3(16)] = 0.657$. Thus, the number of moles of Cr in the liquid is $(0.657)(2) = (1.31)$. Finally, the weight of Cr in the liquid is $(1.31)(52) = 68.4 \text{ g}$.

Another important ceramic alloy system is the silica (SiO_2)–alumina (Al_2O_3) system, which the binary phase diagram in Fig. 4-19 describes. Alloys described

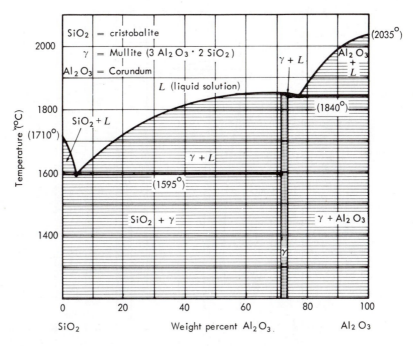

FIG. 4-19 SiO_2–Al_2O_3 phase diagram

by this phase diagram are important because they are used as refractory bricks in the construction of high-temperature furnaces. Generally speaking, silica-alumina alloys with high alumina content are more expensive and have better refractory (high-temperature) properties than high silica content alloys. The fact that alloys with high Al_2O_3 content have better high-temperature properties may be derived from the phase diagram, which shows that the liquidus temperature generally increases with Al_2O_3 content.

As shown in Fig. 4-19, the SiO_2-Al_2O_3 phase diagram exhibits two eutectic reactions and an intermediate phase which is called *mullite*. The fact that the oxides (SiO_2, Al_2O_3) that make up this system can be taken as the components is evident from the composition of the mullite solid solution. While the composition of the solution is not invariant, it may be described approximately by the chemical formula $3Al_2O_3 \cdot 2SiO_2$. Thus, if we call SiO_2 component A and Al_2O_3 component B, the intermediate phase (mullite) can be described by A_2B_3. It might be noted that the terms *cristobalite* and *corundum* simply describe the pure components SiO_2 and Al_2O_3, respectively. It should also be pointed out that since pure silica is polymorphic and forms different phases at low temperatures, the phase diagram in Fig. 4-19 applies only to the temperatures shown and cannot be extrapolated to lower temperatures.

The phase diagram shown in Fig. 4-20 illustrates the conditions for phase equilibria in the zirconia (ZrO_2)–lime (CaO) system. The intermediate phase which

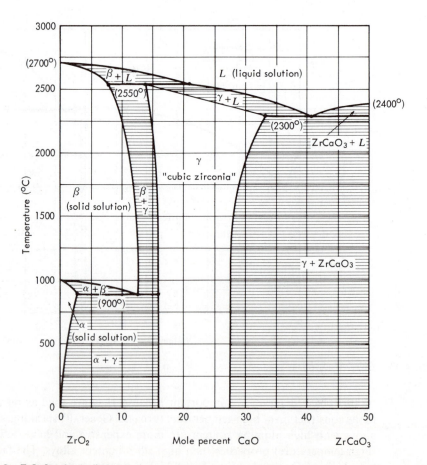

FIG. 4-20 ZrO_2–$ZrCaO_3$ phase diagram

appears at 20 mole percent CaO in that alloy and which is somewhat loosely called "cubic zirconia" is important because it is much stronger and more durable than pure monoclinic ZrO_2. It is common practice to "stabilize" the properties of zirconia by making additions of lime. Zirconia refractories which are "stabilized" with calcium are simply solid solutions of CaO-ZrO_2 that have the cubic structure.

PROBLEMS

4-1   Pure iron transforms on heating from the BCC phase (ferrite) to the FCC phase (austenite) at 910°C. The heat (enthalpy) required for the transformation is 215 cal/mole. Calculate the entropy change during the transformation. Assume that the atomic diameter is unchanged during the transformation and compute the work done by the atmospheric pressure as 1 mole of iron undergoes the transformation. Compute the change in internal energy and compare it with the enthalpy change. Is it accurate to say that the enthalpy change and internal energy change are almost the same? Would they be significantly different if the transformation were to take place at 100 atm? The atomic diameter of iron is 2.48 A.

4-2   Consider a sample of pure nickel in the shape of a coin. The area of the flat face is 1 cm², and the initial thickness is 0.1 cm. Suppose that a pressure of 40,000 psi (or 2720 atm) exerted on the flat face will cause the wafer to be flattened by 10%. Suppose further that during the deformation the internal energy of the solid remains unchanged. Calculate the temperature rise expected if the heat capacity of the solid is 6.1 cal/mole-°C.

4-3   It is known that lattice vacancies in silicon can seriously affect the electrical properties of semiconductors. Suppose that a transistor is being made with thin wafers that are produced by vapor deposition onto an inert substrate. The process is successful in producing single crystals of silicon, but the concentration of lattice vacancies is always very high ($\sim 10^{15}$ vacancies/cm³) just after deposition. Attempts to remove the vacancies by heating the wafers to just below the melting temperature have failed because the films cool so quickly after the anneal that the vacancy concentration is still too high. It is suggested that the wafers be heated to just below the melting temperature of silicon and that they be slowly cooled to a temperature where the equilibrium vacancy concentration falls below 10^{12} vacancies/cm³. Calculate the temperature to which the crystals should be slowly cooled. The formation energy for vacancies in silicon is 2.3 eV/vacancy. Explain why it is all right to rapidly cool the wafers once they have reached the critical temperature.

4-4   The crystals described in Problem 4-3 initially contain 10^{15} vacancies/cm³. Suppose that the crystals are cooled to the critical temperature (at which the equilibrium vacancy concentration is 10^{12} vacancies/cm³) and that the excess vacancies annihilate, causing an evolution of heat. Estimate the temperature rise that would be expected if the energy of each vacancy is 2.3 eV and the heat capacity is about 6.4 cal/mole-°C.

4-5   The formation energy of a vacancy in copper is about 1.2 eV. Estimate the percent volume change that takes place when the crystal is heated to its melting temperature. Consider only the expansion due to the absorption of vacancies. How does this expansion compare with the values that would be expected from the normal expansion coefficient, α (which relates to the change in the average distance between atoms)? ($\alpha_{Cu} = 16.5 \times 10^{-6}°K^{-1}$.)

4-6   An engineer claims to have observed five separate phases in a ternary (three-component) alloy at the same time. Is the observation accurate? Explain.

4-7   A crystal of silicon dissolves about 0.001 % of element X at 500°C. Estimate the concentration of the solute if equilibrium were established at 800°C. Assume that there is an infinite source of the element X.

4-8   The energy of formation of a vacancy in gold is 1.3 eV. Calculate the fraction of vacant lattice sites at 900°C and at room temperature. If a gold wire is quenched (rapidly cooled) from 900°C to room temperature, what is the fraction of excess vacancies present at the lower temperature?

4-9   How many parts per million of vacancies are there in gold at 900°C? (See Problem 4-8.) Assuming the vacancies to be distributed in a cubic array, what is the mean spacing between vacancies? The atomic density of gold is 5.9×10^{22} atoms/cm³.

4-10 Using the information and assumptions given in Example 4-6, show how the free energy change during freezing, $\Delta G_{S \to L}$, varies with temperature below the melting temperature. Assume that the rate of freezing is approximately proportional to the change in free energy (to be discussed in Chapter 5) during the transformation, and show that the freezing rate is zero at the freezing point and finite below the freezing point.

4-11 On the basis of information given in Fig. 4-6, show that ice should float in water.

4-12 The free energy curves in Fig. 4-7 indicate that the free energy of each phase decreases with temperature. Explain this observation in terms of the mixing energy.

4-13 Use the phase rule to predict the number of degrees of freedom for an alloy of Pb + 20 weight percent Sn at about 250°C. Interpret the degrees of freedom for this alloy at 183°C.

4-14 The phase rule (constant pressure) predicts that there is a single degree of freedom in a two-phase binary alloy. We commonly use this degree of freedom in selecting the temperature. In spite of this it is evident that the overall composition in the two-phase region may also be varied by adding more of one of the components. Is this inconsistent with the phase rule? Explain.

4-15 Draw the free energy curves for each of the phases in the Pb-Sn alloy system at 250°C, 200°C, and 150°C.

4-16 Draw the free energy curves for the phases in the Pt-Re alloy system at temperatures just above and just below the peritectic transformation temperature.

4-17 Describe the phases that appear as an iron–0.5 weight percent carbon alloy is slowly cooled from the melting temperature. Describe the compositions of the phases and the weight fractions of each phase at each of the following temperatures: 1470°C, 1100°C, 740°C, and 500°C. At what temperature will the austenite in this alloy have the highest carbon content? When the alloy is held at 700°C, what is the weight fraction of the ferrite phase that formed in the alloy before the eutectoid reaction had taken place?

4-18 Why would one expect alloys of iron and carbon, with carbon content greater than about 2.0 weight percent, to behave differently from alloys with lower carbon content?

4-19 Describe the phase changes that take place when a mixture of 50 weight percent silica (SiO_2) and 50 weight percent mullite ($3Al_2O_3 \cdot 2SiO_2$) is heated to 2000°C and slowly cooled to 1400°C. At what temperature is the ceramic entirely solid?

4-20 An engineer notes that since mullite has higher melting temperature than silica, it might be possible to increase the melting temperature of silica bricks by adding about 10% mullite by weight. Why is he wrong? Predict the temperature at which his bricks will begin to melt.

BIBLIOGRAPHY

R. M. BRICK, R. B. GORDON, and A. PHILLIPS, *Structure and Properties of Alloys*. New York: McGraw-Hill, 1965.

J. H. BROPHY, R. M. ROSE, and J. WULFF, *The Structure and Properties of Materials*, Vol. II. New York: John Wiley, 1964.

A. H. COTTRELL, *Theoretical Structural Metallurgy*. New York: St. Martin's Press, 1957.

L. S. DARKEN and R. W. GURRY, *Physical Chemistry of Metals*. New York: McGraw-Hill, 1953.

A. G. GUY, *Physical Metallurgy for Engineers*. Reading, Mass.: Addison-Wesley, 1962.

M. HANSEN, *Constitution of Binary Alloys*. New York: McGraw-Hill, 1958.

W. HUME-ROTHERY, J. W. CHRISTIAN, and W. B. PEARSON, *Metallurgical Equilibrium Diagrams*. London Institute of Physics, 1952.

W. D. KINGERY, *Introduction to Ceramics*. New York: John Wiley, 1960.

E. M. LEVIN, H. F. McMURDIE, and F. P. HALL. *Phase Diagrams for Ceramists*. Columbus, Ohio: The American Ceramic Society, Inc., 1956.

G. N. LEWIS and M. RANDALL (rev. by K. S. PITZER and L. BREWER), *Thermodynamics, 2nd Ed.* New York: McGraw-Hill, 1961.

W. F. LUDER, *A Different Approach to Thermodynamics*. New York: Van Nostrand Reinhold, 1967.

A. B. PIPPARD, *The Elements of Classical Thermodynamics*. New York: Cambridge University Press, 1966.

F. N. RHINES, *Phase Diagrams in Metallurgy*. New York: McGraw-Hill, 1956.

R. A. SWALIN, *Thermodynamics of Solids*. New York: John Wiley, 1962.

J. WASER, *Basic Chemical Thermodynamics*. New York: W. A. Benjamin, 1966.

C. ZENER, ed., *Thermodynamics in Physical Metallurgy*. Cleveland: American Society for Metals, 1950.

5
Kinetics

5-1 INTRODUCTION

In the previous chapter we discussed the principles of equilibrium and showed how the direction of a reaction is controlled by the change in free energy that accompanies it. Except for some cursory discussion, we did not consider the *rate* at which equilibrium is approached, i.e., *the rate of the reaction.* The rate of reaction is a very important consideration in materials science since many reactions never reach equilibrium and since properties of materials are, more often than not, determined by nonequilibrium, or *metastable*, states. As a simple example, let us consider the brick shown in Fig. 5-1, whose dimensions are l, m, and n, where $l > m > n$. Suppose that the brick rests on its end, such that the m and n dimensions lie in the horizontal (xy) plane. The center of mass of the brick is shown by the dot at its center. This brick is not in its most stable position. If a slight force F is applied along the y direction, for example, the brick will topple over to the position shown in Fig. 5-1(c), thereby lowering its center of mass, thereby lowering its potential energy, and thereby lowering its free energy by an amount ΔG. The configuration shown in Fig. 5-1(c) is the equilibrium, or most stable, position since it is the position of minimum free energy. Note that if the force had been applied along the x axis, the brick would have assumed another metastable configuration, with the l and n dimensions in the (xy) plane. The potential energy of the brick in this position, while lower than that which existed in the initial configuration, is still higher than that which exists in the equilibrium configuration. For the example shown here, there are, therefore, two metastable states and one stable state.

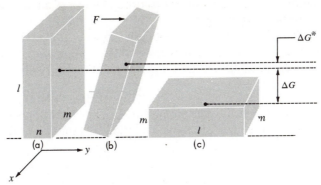

FIG. 5-1   (a) Metastable, (b) unstable, and (c) stable states for a brick

Why do we call configuration (a) metastable? The reason is that the brick can assume this configuration and maintain it for an indefinite period of time until sufficient force is applied to topple it over. This force is needed to raise the center of mass to the position shown in Fig. 5-1(b), and hence raise the free energy by an amount ΔG^*, before the brick can topple into configuration (c) without any further force being applied to it. Configuration (b) is therefore an *unstable configuration* since it cannot be maintained of its own accord (i.e., without an externally applied constraint).

Most kinetic processes can be broken down into at least two steps: (1) the transformation of a metastable state into an unstable state, which requires the application of some work (energy) to the system, and (2) the transformation of the unstable state to the stable state. This latter step is a spontaneous process and consequently does not require that work or energy be imparted to the system. While thermodynamics can predict that a reaction such as (a) $\rightarrow$ (c) will take place, kinetics determines the rate at which the metastable configuration becomes unstable, [e.g., the rate at which configuration (a) moves into configuration (b), and hence the rate of the reaction].

In the case of the brick described above, an external force is required to provide the work to raise the free energy of the system so that the unstable configuration (b) can be reached. The increase in free energy ΔG^* is known as the *activation free energy*, and the unstable configuration is referred to as the *activated state*.

5-2   REACTION RATE THEORY

In our study of the internal structure (atom arrangement) of materials we shall see that the fundamental step involved in changing from one atom arrangement to another, more stable, arrangement generally involves the discrete motion of individual atoms. Thus, to apply reaction rate theory to the study of atom arrangements we must be able to describe the motion of atoms. As we shall see later, the same concepts developed above to describe the equilibrium configuration of a brick can be used to describe atom movements; that is, the motion of an atom from one position to another can be described in terms of an activated state with a definite activation free energy.

Suppose that an atom is vibrating about position (a) (Fig. 5-2) and that the free energy of the system can be lowered from $G_{(a)}$ to $G_{(c)}$ if it moves into position (c). In order to do this, it must first surmount the energy barrier (b); i.e., a free energy

$$\Delta G^* = G_{(b)} - G_{(a)} \tag{5-1}$$

must be supplied to the atom. This energy can be supplied to the atom as *thermal energy* in the form of atomic vibrations. According to the principles of statistical

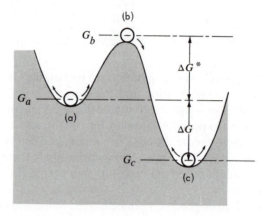

FIG. 5-2 (a) Metastable, (b) unstable, and (c) stable states for a materials system

thermodynamics the probability P that the atom will acquire sufficient thermal energy to surmount a barrier of height ΔG^* is

$$P = \exp\left(-\frac{\Delta G^*}{kT}\right) \tag{5-2}$$

Suppose that the atom is vibrating about point (a) with a frequency $v = 10^{13}$/sec. Thus, v is the frequency at which the atom will attempt to surmount the barrier. Consequently, the *rate* r at which atoms surmount the barrier is $r =$ (frequency of attempts) (probability that a given attempt is successful)

$$r = v \exp\left(-\frac{\Delta G^*}{kT}\right) = v \exp\left(\frac{\Delta S^*}{k}\right) \exp\left(-\frac{\Delta E^*}{kT}\right) \tag{5-3}$$

where ΔS^* is the entropy of activation and ΔE^* is known as the *activation energy*.

EXAMPLE 5-1   Suppose the probability that an atom has an energy greater than ΔG at 700°K is 10^{-7}. What is the probability that an atom has an energy greater than ΔG at 1200°K?

Solution:

$$P = \exp\left(-\frac{\Delta G}{kT}\right)$$

at 700°K

$$10^{-7} = \exp\left(-\frac{\Delta G}{k\,700}\right)$$

$$\frac{\Delta G}{k} = -700\ln(10^{-7}) = 1.28 \times 10^4 \ {}^\circ K$$

at 1200°K

$$P = \exp\left(-\frac{1.28 \times 10^4}{1200}\right)$$

$$= \exp(-9.4) = 8.2 \times 10^{-5}$$

Since the rate of atom motion, or more generally the rate of any reaction, depends primarily on ΔG^*, it is understandable that metastable states can exist. Suppose [Fig. 5-3(a)] that the free energy of a metastable state is G_1 and that the

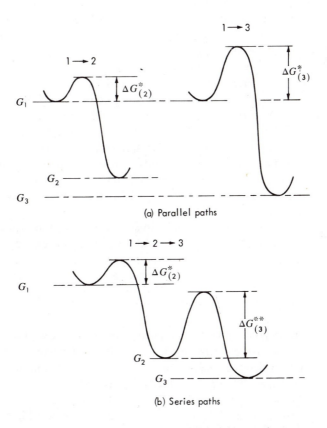

(a) Parallel paths

(b) Series paths

FIG. 5-3   Free energy changes associated with (a) parallel and (b) series reactions

state can transform either to another metastable state of free energy G_2 if an activation energy $\Delta G^*_{(2)}$ is supplied or to the equilibrium state of free energy $G_{(3)} < G_{(2)}$ if an activation energy $\Delta G^*_{(3)} > \Delta G^*_{(2)}$ is supplied. That is, we can have either

$$1 \longrightarrow 2 \qquad \text{or} \qquad 1 \longrightarrow 3$$

Thermodynamics predicts that eventually the system will reach state (3), but since $\Delta G^*_{(2)} < \Delta G^*_{(3)}$, kinetics predicts that the rate of reaction $1 \rightarrow 2$ will be faster than the rate of $1 \rightarrow 3$. Since the *time* required for the transformation is inversely proportional to the rate, in a given period of time more of the metastable product (2) will form than of the stable product (3). Thus, we can have the system transform to a metastable state (2) rather than to the equilibrium state (3) merely because the activation free energy from $1 \rightarrow 2$ is smaller than that from $1 \rightarrow 3$. Further, if the temperature is low enough so that with the available thermal energy the probability of transforming from $2 \rightarrow 3$ is essentially zero, then the system will remain in state (2) indefinitely. That is, *even though a state of lower free energy exists, if the activation free energy is large compared to the available thermal energy, then the reaction will not proceed at a measurable rate and the system will remain in the metastable state.*

We can generalize some of the above discussion as follows. When two alternative (parallel) reaction paths are possible, the decomposition of the initial state will occur along the fastest path and the rate of transformation will be determined by the rate of the fastest process. This is analogous to the situation in electric circuits where, if a given voltage V is applied across two parallel resistors of resistance R_1 and $R_2 > R_1$, most of the current flows through the resistor R_1. Alternatively, if two reactions occur in series [Fig. 5-3(b)] so that $1 \rightarrow 2 \rightarrow 3$, then if $\Delta G^*_{(2)} < \Delta G^{**}_{(3)}$,† the rate of $(1) \rightarrow (2)$ is much faster than the rate of $(2) \rightarrow (3)$; hence the rate of the overall reaction is controlled primarily by the rate of the *slowest* process. This is, of course, analogous to the situation in electric circuits where a given voltage V is applied across two resistors R_1 and $R_2 > R_1$ that are connected in series and where total current flow is determined primarily by the resistance R_2.

Consistent with the predictions of Eq. (5-3), those processes or reactions that depend on the rate of successful atom jumps are said to be *thermally activated* because the rate increases as the temperature (thermal energy) increases. The general expression for the rate of a thermally activated process is

$$\text{Rate} = A \exp\left(-\frac{\Delta E^*}{kT}\right) \tag{5-4}$$

where A is a constant for a particular process. Equation (5-4) is usually referred to as the *Arrhenius equation*, after Swedish Chemist Svante Arrhenius, who first demonstrated that the rate of many chemical reactions has this type of temperature dependence. Examples of thermally activated processes in solids that depend on atom movements include diffusion, oxidation, and phase transformations. These processes, all of which will be discussed in some detail later, can be described by an expression similar to the Arrhenius equation.

Because of the simple form of the Arrhenius equation it is possible to determine the activation energy ΔE^* for any thermally activated process by taking the logarithm of both sides of Eq. (5-4)

---

†$\Delta G^*_{(3)} \neq \Delta G^{**}_{(3)}$ because $1 \rightarrow 3 \neq 2 \rightarrow 3$ [see Fig. 5-3(b)].

$$\ln \text{(Rate)} = \ln (A) - \frac{\Delta E^*}{kT} \tag{5-5}$$

and then plotting the logarithm of the rate versus the reciprocal of the absolute temperature.

As an example of a thermally activated process, consider the following. A new class of charged particle detectors called *dielectric track detectors* has recently been discovered. The principle behind the operation of these detectors is based on the delineation of individual paths of heavily ionizing charged particles in a dielectric solid such as cellulose nitrate. Delineation is accomplished through the selective etching (chemical attack) of the radiation-damaged material along the particle's trajectory. That is, along the path of the charged particle there is a region of damage (displaced atoms, broken bonds, etc.) that, because of its higher internal energy, is more responsive to chemical attack than the undamaged matrix. Merely by measuring the size of the etch track information can be determined regarding the nature and energy of the charged particle.

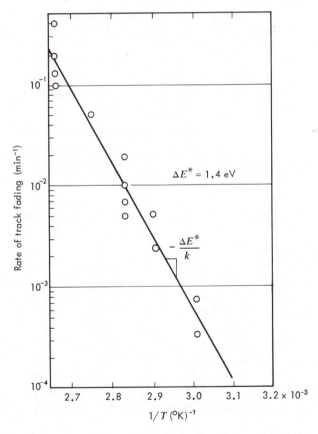

FIG. 5-4   Temperature dependence of the rate of track fading in cellulose nitrate (After E. V. Benton, Ph. D. dissertation, Stanford University, 1968. Used with permission.)

One of the limitations of this type of measuring device is that the damaged material along the path of the charged particle tends to recover or return to its initial perfect state over a period of time. This recovery process or atom rearrangement is thermally activated since it is controlled by the motion of individual atoms. As the amount of recovery increases, etching reveals a fainter and fainter track, until finally no track is produced when the material has fully recovered. The influence of temperature on the rate of track fading is shown in Fig. 5-4. The activation energy for the fading process, given simply from the slope of the curve, is about 1.4 eV (32,100 cal/mole), which is typical of processes involving atomic motion.

5-3 ATOMIC DIFFUSION IN SOLIDS

As discussed above, many of the thermally activated processes that occur in solids are diffusion controlled, and the rate of the process is determined by the diffusion rate of individual atoms from one atomic site to another. Although individual atoms can jump from one site to another in a random manner, in the presence of a concentration gradient* there will be a *net flow* of a particular species of atoms in one direction. It is this net flow, rather than the individual atomic movements, that is known as *diffusion*. Familiar examples of the many types of diffusion are the diffusion of heat that occurs when one end of a bar is heated and the cold end warms up after a period of time and the flow of electric current in a wire when an electric potential is applied across its ends. In each of these cases, the *flux* of matter or heat or electricity is given by the relation

$$\text{Flux} = (\text{conductivity})\,(\text{driving force}) \tag{5-6}$$

In the case of atomic diffusion, the conductivity is called the *diffusivity* or *diffusion coefficient* and is usually represented by the symbol D. The driving force for atomic diffusion is the *concentration gradient** that exists between one point and another, just as in heat flow the driving force is the temperature gradient.

As an example of diffusion in solid materials, consider the case where a piece of silicon and a piece of germanium are placed in contact with each other. The concentration profile at the time of contact, $t = 0$, is shown in Fig. 5-5. Silicon and germanium are completely soluble in each other in the solid state, as was shown in Fig. 4-10. Consequently, there is a tendency for silicon atoms to diffuse into the germanium and for germanium atoms to diffuse into the silicon, just as would be the case if two mutually soluble liquids were placed in intimate contact. Figure 5-5 shows that at time $t > 0$ increasing amounts of diffusion take place, and if we wait long enough, the silicon and germanium will dissolve completely in one another, the concentration gradient will vanish, and equilibrium will be achieved.

In practice there are two particular cases of diffusion to consider, *steady state* and *nonsteady state* diffusion. Steady state diffusion is diffusion that takes place under a concentration gradient that does not change with time. An example of

---

*A concentration gradient is essentially equivalent to a free energy gradient. In some cases the free energy gradient depends on factors other than the concentration gradient. Such cases will not be considered in this book.

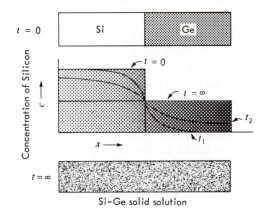

FIG. 5-5   Interdiffusion of silicon and germanium

steady state diffusion is the diffusion of a gas through a thin metal membrane where a high gas pressure is maintained on one side of the membrane and a low pressure on the other side. The situation where local concentration gradients and concentrations vary with time is referred to as nonsteady state diffusion. The example of the interdiffusion of silicon and germanium cited above is a case of nonsteady state diffusion.

Steady State Diffusion

Consider the diffusion of solute atoms along the x direction between two parallel atomic planes (Fig. 5-6) perpendicular to the plane of the paper and separated by

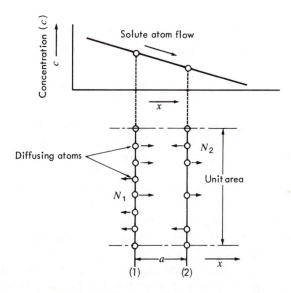

FIG. 5-6   Relation between atomic jumps and diffusion in a concentration gradient

a distance a. Suppose that the planes are of unit area and that there are N_1 solute atoms on plane (1) and N_2 solute atoms on plane (2), where $N_1 > N_2$. Here N has the units number/area. The atomic concentration of solute atoms in number/volume on plane (1) is $c_1 = N_1/a$ and on plane (2) is $c_2 = N_2/a$. Thus, a concentration gradient $dc/dx = (c_2 - c_1)/a$ exists along the x direction.

Now if the atomic jump frequency from each plane is r [see Eq. (5-3)], then in unit time $N_1 r/2$ atoms from plane (1) will jump to plane (2), and $N_2 r/2$ atoms will jump from plane (2) to plane (1). The factor of $\frac{1}{2}$ accounts for the fact that atoms may jump either in the positive or negative x direction. The *net flux* of diffusing solute atoms, J, from plane (1) to plane (2) is

$$J = \frac{1}{2}(N_1 - N_2)r = \frac{a}{2}(c_1 - c_2)r \tag{5-7}$$

where the units of J are number of atoms per unit area per unit time. Since the concentration gradient $dc/dx = (c_2 - c_1)/a$, we can rewrite Eq. (5-7) as

$$J = -\frac{1}{2}a^2 r \frac{dc}{dx} = -D\frac{dc}{dx} \tag{5-8}$$

where the diffusion coefficient $D = \frac{1}{2}a^2 r$ and has units cm²/sec. Equation (5-8) is known as *Fick's first law of diffusion*. This expression merely states that in the presence of a uniform concentration gradient the flux of diffusing atoms is proportional to the product of the diffusion coefficient (the atomic conductivity) and the concentration gradient (driving force). The minus sign indicates that atom flow occurs in the direction of negative concentration gradient.

Before examining the case of nonsteady state diffusion, let us examine in some detail the variables that influence the diffusion coefficient D. If we generalize the above discussion to allow atoms to jump along the y and z axes as well as along the x axis, then we should actually write

$$D = \frac{1}{6}a^2 r \tag{5-9}$$

since only one-third the jumps will be in the x direction. Now introducing Eq. (5-3) we have

$$D = \frac{1}{6}a^2 v \exp\left(\frac{\Delta S_D}{R}\right)\exp\left(-\frac{\Delta E_D}{RT}\right) = D_0 \exp\left(-\frac{\Delta E_D}{RT}\right) \tag{5-10}$$

where all the temperature-independent constants have been combined into D_0 and where ΔE_D and ΔS_D are the activation energy and activation entropy for diffusion per mole of diffusing species in units of cal/mole. Equation (5-10) reflects the thermally activated nature of the diffusion process; that is, as the temperature is raised, the increased thermal energy enhances the probability that an atom can jump to an adjacent site, and the diffusion rate increases.

The value of ΔE_D depends primarily on two factors: (1) the strength of the atomic bonds and (2) the atomic mechanism for diffusion. As bond strengths increase (usually accompanied by a rise in melting temperature, T_m) ΔE_D also increases. This is reasonable, for the higher the bond strengths, the greater the amount of thermal energy that must be provided to the atom to allow it to break its bonds (surmount the energy barrier) and jump to an adjacent site. Table 5-1 lists some typical *self-diffusion* activation energies, i.e., the activation energies for motion of an atom in a single-phase material, in the absence of a chemical concentration gradient.

TABLE 5-1

Typical Activation Energies for Self-Diffusion, ΔE_D

Material	Diffusing Species	Bond Type	Melting Temp. (°C)	ΔE_D (kcal/mole)
H_2	H_2	Van der Waals	−259	0.38
Ar	Ar	or hydrogen	−189	4.15
CH_4	CH_4		−184	3.2
Xe	Xe		−112	7.4
H_2O	H_2O		100	13.5
NaCl	Cl	Ionic	801	53
NaCl	Na		801	20
LiF	F		870	50
LiF	Li		870	43
Al_2O_3	Al		3500	110
Ge	Ge	Covalent	940	68
GaAs	Ga		1238	128
GaAs	As		1238	230
Si	Si		1412	118
Na	Na	Metallic	97.5	10
Al	Al		660	34
Cu	Cu		1083	47
Ni	Ni		1455	66

Diffusion in amorphous materials is also strongly dependent on the nature of the interatomic bonding. In polymeric materials where strong covalent bonds exist within individual molecules and where the molecules are bound together by weak secondary bonds, diffusion usually takes place by the motion of individial molecules. Generally the larger the molecules, the higher the activation energy. For example, the activation energy for the diffusion of CH_3 molecules in nylon is 26.8 kcal/mole, while that for C_4H_9 molecules is 37 kcal/mole. Diffusion of individual atoms in network solids such as ordinary glass also shows a marked dependence on the type of bonding the atom experiences. Silicon atoms, which have strong covalent bonds with oxygen, have a high ΔE_D and diffuse very slowly through glass. Sodium ions or small atoms such as hydrogen and helium, which are dissolved "interstitially" into the relatively open glass structure and experience weak bonds with the silicon and oxygen atoms, can diffuse rapidly through glass with low activation energies for diffusion. For example, the diffusion rate of Na^+ in a sodium-silicate glass is some 10^5 times that for Si at temperatures of $\approx 1000°C$.

The variation of ΔE_D with the atomic mechanism of diffusion can be illustrated with the aid of Fig. 5-7. Diffusion can take place by two different mechanisms, *interstitial diffusion* and *vacancy diffusion*. First, consider the case of solute diffusion in a dilute interstitial solid solution. The activation energy for diffusion is merely the energy that must be supplied to the interstitial atom to allow it to squeeze between solvent atoms as it moves from one interstitial site to another [Fig. 5-7(a)].

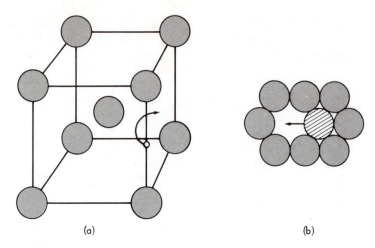

(a)   (b)

FIG. 5-7   Mechanisms of (a) interstitial and (b) substitutional diffusion

Because we are considering a dilute solution (the great majority of the interstitial sites are empty) we need not be concerned with the probability that the adjacent interstitial site is empty. That is, if the interstitial atom has sufficient energy to make the jump, we can assume that the jump occurs and need not be concerned about the possibility that another interstitial atom already occupies that particular site. This is very different from the case of diffusion in systems where the majority of atom sites are occupied. In this instance care must be taken to calculate the probability that an adjacent atom site is vacant and available to receive an atom. Consider the substitutional solid solution shown in Fig. 5-7(b). For the solute atom to move to the left two simultaneous events must occur: (1) the atom must have sufficient thermal energy to make a jump, ΔE_m, and (2) there must be a lattice vacancy adjacent to the jumping atom. As demonstrated in the previous chapter, the probability that a lattice site is vacant (equilibrim concentration of vacancies) is proportional to $\exp\left(-\Delta E_f / RT\right)$, where ΔE_f is the formation energy of vacancies. The total activation energy for this vacancy-controlled diffusion process is given by

$$\begin{pmatrix} \text{Probability that} \\ \text{atom jump is} \\ \text{successful} \end{pmatrix} = \begin{pmatrix} \text{probability that atom} \\ \text{has sufficient thermal} \\ \text{energy to jump} \end{pmatrix} \begin{pmatrix} \text{probability that} \\ \text{adjacent atom} \\ \text{site is vacant} \end{pmatrix}$$

$$= \exp\left(-\frac{\Delta E_m}{RT}\right) \exp\left(-\frac{\Delta E_f}{RT}\right) = \exp\left(-\frac{\Delta E_m + \Delta E_f}{RT}\right)$$

Thus, the activation energy for diffusion is composed of two parts, one related to atom motion, ΔE_m, and the other related to vacancy formation, ΔE_f.

Summarizing the above discussion, we can state that diffusion in pure elements, compounds, and substitutional solid solutions will occur by a vacancy mechanism, while solute diffusion in dilute interstitial solutions can take place without the presence of vacancies. In addition, the activation energy for diffusion by a vacancy mechanism is greater than that for diffusion of interstitials. As an example of this latter point, for diffusion of interstitial carbon in alpha (BCC) iron, $\Delta E_D = 20$ kcal/mole, while for self-diffusion of iron atoms in alpha iron, $\Delta E_D = 67$ kcal/mole. Because of the large differences in activation energies interstitial diffusion usually occurs rapidly at much lower temperatures than does vacancy-controlled diffusion. Some typical values of diffusion coefficients for interstitial and substitutional solutes are given in Table 5-2.

TABLE 5-2
Comparison of Diffusivities of Interstitial and Substitutional Solutes

	Solute	Solvent*	ΔE_D (kcal/mole)	D_0 (cm²/sec)	D (cm²/sec) 500°C	D (cm²/sec) 1000°C
Interstitial	C	Fe (BCC)	20	0.008	1.8×10^{-8}	3×10^{-6}
	N	Fe (BCC)	18	0.007	6×10^{-8}	5.7×10^{-6}
	H	Fe (FCC)	10	0.01	1.5×10^{-5}	1.9×10^{-4}
Substitutional	Ni	Fe (FCC)	66	0.5	1×10^{-19}	2.5×10^{-12}
	Co	Fe (BCC)	54	0.2	1.2×10^{-16}	9×10^{-11}
	Si	Fe (BCC)	48	0.4	1.2×10^{-14}	2.2×10^{-9}
	Al	Cu	39	0.07	5.6×10^{-11}	1.5×10^{-8}
	S	GaAs	92	4000	3.5×10^{-23}	1.6×10^{-12}
	Zn	GaAs	57	1.5×10^{-8}	1.3×10^{-24}	1.5×10^{-18}

*When the solvent can occur with more than one crystal structure, the appropriate structure is listed.

As in the general case of any thermally activated process, it is possible to determine a value of ΔE_D from the slope of a plot of the logarithm of the diffusivity versus the reciprocal of the absolute temperature. The value of D_0 can be obtained by extrapolating the plot to the intercept $1/T = 0$. Figure 5-8 is a plot of the dif-

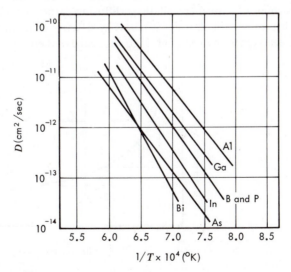

FIG. 5-8   Impurity diffusion coefficients in silicon

fusion coefficients of several impurity elements in silicon as a function of temperature. Alternatively, D_0 and ΔE_D can be obtained from solution of two simultaneous equations if the diffusion coefficient D is determined at two separate temperatures.

EXAMPLE 5-2

The diffusivity of lithium in silicon is 10^{-5} cm²/sec at 1376°K and 10^{-6} cm²/sec at 968°K. What are the values of ΔE_D and D_0 for diffusion of lithium in silicon?

Solution:

Let $T_1 = 1376$°K and $T_2 = 968$°K.

$$\log D_1 - \log D_2 = \frac{-\Delta E_D}{2.303R}\left(\frac{1}{T_1} - \frac{1}{T_2}\right)$$

$$(-5) - (-6) = \frac{-\Delta E_D}{2.303R}(0.725 - 1.033) \times 10^{-3}$$

$$\Delta E_D = \frac{(1)(2.303)(1.98)}{0.308 \times 10^{-3}} = 14{,}800 \text{ cal/mole}$$

To evaluate D_0 we simply use Eq. (5.10) and the value of D at *either* 1376 or 968°K. For example, since $D = 10^{-5}$ cm²/sec at 1376°K

$$D_0 = \frac{D}{\exp(-\Delta E_D/RT)} = \frac{10^{-5}}{\exp[-14{,}800/(1.98)(1376)]}$$

$$= 2.55 \times 10^{-3} \text{ cm}^2/\text{sec}$$

Nonsteady State Diffusion

Although Eq. (5-8) can be used to predict the steady state flux in the presence of a linear concentration gradient, it cannot be used to treat more interesting problems such as the mutual diffusion of silicon and germanium (Fig. 5-5) where the concentration gradient *changes* with time and position. This is the type of diffusion problem that is most often encountered in materials science, whether it be the hardening of a steel camshaft by diffusing carbon or nitrogen into the surface or the loss of certain alloy constituents by selective evaporation of a material operated in a deep space environment (high vacuum). In these cases, there will be an accumu-

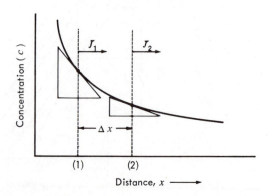

FIG. 5-9 Diffusion in a nonuniform concentration gradient

lation or diminution in the amount of the diffusing species over a period of time. Suppose that the concentration profile varies with diffusion distance x, as shown in Fig. 5-9. Consider two planes of unit area separated by a distance Δx. If the flux through plane (1) is J_1 and the flux through plane (2) is $J_2 < J_1$, then the rate of accumulation of atoms in the region Δx is simply the difference between the two fluxes, or $J_1 - J_2$. Now we can relate the rate of accumulation of atoms to the change in concentration with time as

$$\frac{dc}{dt} = \frac{J_1 - J_2}{\Delta x} \left(\frac{atoms}{cm^3 \cdot sec} \right) \tag{5-11}$$

As Δx becomes vanishingly small, Eq. (5-11) reduces to

$$\frac{dc}{dx} = -\frac{J_{(x+\Delta x)} - J_{(x)}}{\Delta x} = \frac{d}{dx}(-J) = \frac{d}{dx}\left(D \frac{dc}{dx} \right) \tag{5-12}$$

where we have used Eq. (5-8) to evaluate J. For those cases in which D is independent of composition (generally a good approximation),

$$\frac{dc}{dt} = D \frac{d^2c}{dx^2} \tag{5-13}$$

which is known as *Fick's second law of diffusion*. In physical terms this law states that the *rate of compositional change* is proportional to the *rate of change* of the concentration gradient rather than to the concentration gradient itself; this suggests why it is often very time consuming to reach final equilibrium (uniform composition) via diffusion. As equilibrium is approached and composition gradients are smoothed out, $d^2c/dx^2 \rightarrow 0$, and hence $dc/dt \rightarrow 0$. Thus, although the composition gradient, dc/dt, may be large at the start of a diffusion process (because d^2c/dx^2 is large), it is very small toward the end of the process.

The solution of Eq. (5-13) depends on the boundary conditions imposed by the particular problem of interest. As an example, let us consider the diffusion problem illustrated in Fig. 5-10. A semi-infinite slab (dimensions large compared with the

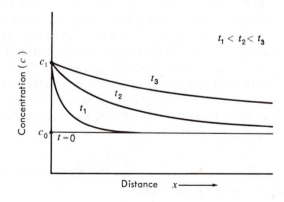

FIG. 5-10 Transient diffusion into a semi-infinite solid. The concentration gradient changes with time.

diffusion distance) of initial composition c_0 has, for all $t > 0$, the concentration at the $x = 0$ interface maintained at a value c_1. This surface concentration is greater than the uniform initial composition of the slab, and consequently diffusion occurs into the slab (in the x direction) from the surface. This hypothetical problem has several real analogies. For example, the manufacture of solid state electronic devices commonly involves the diffusion of electrically active impurities such as indium into a pure silicon substrate. This diffusion process can be carried out by placing a silicon wafer in an indium vapor at high temperatures. The indium vapor maintains a constant surface concentration of indium on the silicon wafer during the entire diffusion process.

The solution of Eq. (5-13) for the conditions described above is

$$\frac{c(x, t) - c_0}{c_1 - c_0} = 1 - \text{erf}\left(\frac{x}{2\sqrt{Dt}}\right) \qquad (5\text{-}14)$$

where $c(x, t)$ is the concentration at some point x in the slab at $t > 0$ and erf is the *Gaussian error function*, as tabulated in mathematical tables. Examples of the type of concentration curve predicted by Eq. (5-14) are shown in Fig. 5-10. Although Eq. (5-14) is somewhat unwieldy, we can take advantage of some simplifications suggested by its form. A particularly interesting aspect of this equation, and equivalently of the solutions to numerous other diffusion problems, is that $c(x, t)$ is completely described at all x and t by the error function of $(x/2\sqrt{Dt})$. This has two important consequences. First, if we are interested in a given composition, c', then

$$\frac{c' - c_0}{c_1 - c_0} = 1 - \text{erf}\left(\frac{x}{2\sqrt{Dt}}\right)$$

is constant, and therefore

$$\frac{x}{\sqrt{Dt}} = \text{constant} \qquad (5\text{-}15)$$

This means that for a given temperature the location of the composition c' will vary with the square root of time; that is, $x \propto t^{1/2}$, and the proportionality constant is related to the diffusion coefficient. The second important consequence of Eq. (5-14) is that the quantity $\sqrt{Dt}$ can be used as an order-of-magnitude estimate of the distance that an average atom will move during a diffusion cycle or, equivalently, of the distance over which appreciable diffusion will occur. This can be simply demonstrated with reference to Eq. (5-15) by considering the case where $c' = \frac{1}{2}c_1$ and where for simplicity we set $c_0 = 0$. That is, we solve for x and t so that the concentration of the diffusing species is equal to one-half the value at the interface $x = 0$. For these conditions

$$\text{erf}\left(\frac{x}{2\sqrt{Dt}}\right) = 0.5$$

and from mathematical tables

$$\frac{x}{2\sqrt{Dt}} = 0.5$$

or

$$x = \sqrt{Dt} \tag{5-16}$$

Thus, as mentioned above, this simple relationship can be used as rough estimate of the time and temperature needed to achieve appreciable diffusion or redistribution of solute atoms over a distance x. And since many other solutions to Eq. (5-13) also indicate that Eq. (5-16) can be used to predict the time required for diffusion to occur over a given distance, it is not only a very simple but also a very useful relationship.

EXAMPLE 5-3   Suppose that it is desired to diffuse indium into pure silicon such that the indium concentration at a depth of 10^{-3} cm from the surface will be one-half the surface concentration. How long should a silicon wafer be heated in contact with indium vapor at 1600°K in order to accomplish this diffusion?

Solution: From Fig. 5-8 we see that $D = 8 \times 10^{-12}$ cm²/sec at $T = 1600$°K ($1/T = 6.3 \times 10^{-4}$ °K^{-1}). Then from Eq. (5.16) we have

$$t = \frac{x^2}{D} = \frac{10^{-6}}{8 \times 10^{-12}} = 1.25 \times 10^5 \text{ sec}$$

5-4 DIFFUSION-CONTROLLED PROCESSES

The rates of many processes are controlled by atom diffusion. Of these processes, the most important include solid state phase transformations, solid–liquid phase transformations, oxidation, the high-temperature electrical conduction of ionic solids, the high-temperature plastic deformation of practically all materials, and sintering (the bonding between particles that occurs when they are placed in intimate contact and heated to high temperatures). All of these processes reflect the basic nature of the rate-controlling diffusion mechanism; i.e., they are thermally activated processes that occur rapidly only when the temperature is high enough to allow for rapid atomic mobility.

Although we will discuss most of these topics in some detail at a later stage, it is useful at this point to show how at least one of the processes is controlled by diffusion. In our discussion we shall make a special effort to show how a particular process can be analyzed in terms of the diffusion equations that were established in the previous section. For our example we shall consider the process of oxidation.

One of the most important of the diffusion-controlled processes is oxidation. All metals (except gold) tend to form oxides in air or in aqueous solutions at room temperatures, since the free energy of the oxide phase is lower than the sum of the separate free energies of the metal and oxygen. Consequently, a large amount of free energy must be supplied to reverse this process in refining naturally occurring ores to pure metals so that they can be used for engineering purposes. Similarly, metals will tend to oxidize and return to their equilibrium state. The rusting of iron is a typical example of a naturally occurring oxidation process. Oxidation that occurs in aqueous environments such as water or dilute acids is called *corrosion* and is a serious engineering problem.

Fortunately, oxidation is usually a slow process at ambient temperatures. The reason for this is that once the surface of a metal begins to oxidize (Fig. 5-11), subsequent oxidation can take place only when oxygen ions diffuse through the oxide and react with the metal ions at the metal-oxide interface or when metal ions diffuse through the oxide and react with oxygen ions at the oxide-oxidant interface. As the oxide thickens, the diffusion distance increases and the rate of oxidation decreases.

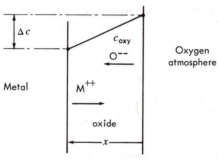

$$M^{++} + O^{--} = MO \qquad M + \tfrac{1}{2}O_2 = MO$$

FIG. 5-11 Oxygen diffusion through an oxide layer during oxidation

We can make a qualitative calculation of the oxidation rate in the following manner. Let us assume that oxidation proceeds by the diffusion of oxygen ions through the oxide to the metal-oxide interface. The calculation would be essentially the same if we assumed the metal ions were the diffusing species. The rate of increase of the oxide thickness dx/dt is then proportional to the flux of oxygen atoms reaching metal-oxide interface, or

$$\frac{dx}{dt} = a_1 J_{\text{oxy}} \tag{5-17}$$

where a_1 is a proportionality constant. Now noting that the difference in oxygen concentration between the oxide-oxidant and metal-oxide interfaces remains constant throughout the oxidation process and is equal to Δc, we can express the flux of oxygen atoms from Fick's first law as

$$J_{\text{oxy}} = -D_{\text{oxy}} \frac{\Delta c}{x} \tag{5-18}$$

where D_{oxy} is the oxygen diffusion coefficient and x is the oxide thickness. Combining these two expressions, we have

$$\frac{dx}{dt} = -a_1 D_{\text{oxy}} \frac{\Delta c}{x}$$

which upon integrating yields

$$x^2 = -a_1 D_{\text{oxy}} t \Delta c \tag{5-19}$$

This expression shows the oxide thickness to be proportional to $\sqrt{t}$, with the proportionality constant related to the diffusivity D. The minus sign in Eq. (5-19) is present because Δc is negative

Since D increases exponentially with temperature, oxidation is a particular problem when structural materials are exposed to high operating temperature, such as in jet engines and steam power plants. These considerations indicate that the rate of oxidation can be reduced by reducing the diffusivity of metal and/or oxygen ions in the oxide. Consequently, if the oxide that forms has a very high melting temperature (low diffusivity), the oxidation rate will be very slow. This is the reason that aluminum is oxidation resistant. The formation of a very thin protective layer of Al_2O_3, which is essentially impermeable to further diffusion, greatly retards the oxidation process. We shall discuss the kinetics of oxidation and corrosion in more detail at the end of this chapter.

5-5 KINETICS OF PHASE TRANSFORMATIONS

One of the most important areas in materials science is the area of microstructural control, which is the ability to obtain a desired electronic structure or microstructure to obtain a given set of electronic or mechanical properties. In some cases, such as the diffusion of indium into silicon to change the electronic structure, the change in properties can be achieved without changing the number of phases that are present. Usually this is not the case and, as in oxidation, new phases are formed by the reacting species. Many of the significant materials' processes, such as solidification, heat treatment of steel, and zone refining of semiconductor materials, involve a *change of phase*, or phase transformation.

The previous chapter showed how phase diagrams can be used to predict the amount and chemical composition of the various phases that should form at a given temperature in a given alloy. However, very little was said about the *size and shape* of these phases and the *rate* at which they form. These factors are primarily influenced by the kinetics of phase transformations, which is the subject of this section.

Consider two phases α and β, where α is the stable phase above a temperature T_E and β is the stable phase below T_E. For example, α could be the liquid phase, β the solid phase, and T_E the melting temperature. At $T = T_E$ the free energies of the two phases are equal,

$$G_\alpha = G_\beta$$

so that the change in free energy that accompanies the transformation $\alpha \rightarrow \beta$ is given by

$$\Delta G_V = G_\beta - G_\alpha = 0 \qquad (5\text{-}20)$$

The subscript V on ΔG indicates that we are considering a bulk or volume free energy, so that ΔG_V is the free energy change per unit volume of transformed material. Since

$$\Delta G_V = \Delta E_V - T\,\Delta S_V \qquad (5\text{-}21)$$

is equal to zero at the equilibrium transformation temperature T_E,

$$\Delta S_V = \frac{\Delta E_V}{T_E} \tag{5-22}$$

Since ΔS_V and ΔH_V do not vary significantly with temperature, if the transformation takes place at some temperature $T \neq T_E$, the free energy change that occurs is

$$\Delta G_V = \Delta E_V - T\left(\frac{\Delta E_V}{T_E}\right) = \Delta E_V\left(\frac{T_E - T}{T_E}\right)$$

$$\Delta G_V = \frac{\Delta E_V \, \Delta T}{T_E} \tag{5-23}$$

where $\Delta T = T_E - T$ is the degree of supercooling below (or superheating above) the equilibrium transformation temperature. If heat is evolved during the $\alpha \longrightarrow \beta$ transformation (e.g., solidification), then ΔE_V is negative and ΔG_V will be negative at all temperatures $T < T_E$. Consequently, the driving force for the reaction increases as the temperature decreases below T_E. [Note that if we were considering the reverse transformation $\beta \longrightarrow \alpha$, which absorbs heat (e.g., melting), then ΔE_V would be positive and ΔG_V would be negative at all temperatures $T > T_E$.]

Equation (5-23) tells us the driving force (free energy change) for a phase transformation, but it gives us no information about the rate at which the transformation actually takes place. For example, if the temperature is instantaneously lowered from $T > T_E$ to some $T < T_E$ where β phase *should form* ($\Delta G_V < 0$) by precipitating out of the α phase, this does not mean that the β phase will instantaneously appear. The formation of the β phase requires that the atoms align themselves in normal positions in the β lattice, and hence requires some local atomic rearrangement. This in itself suggests two reasons why the transformation cannot occur instantaneously over the entire volume of α phase. First, the probability that all the atoms would align themselves simultaneously into the β structure is extremely small. In fact, we shall see shortly that there is a free energy barrier to the formation of small β particles even at temperatures below T_E. Secondly, since atomic rearrangements are controlled by diffusion, this means that the formation of the β phase must be a time-dependent, thermally activated process. On this basis we

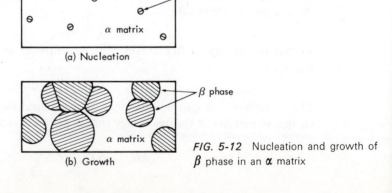

(a) Nucleation

(b) Growth

FIG. 5-12   Nucleation and growth of β phase in an α matrix

can subdivide the $\alpha \longrightarrow \beta$ transformation into two discrete steps. The first step is the formation of small nuclei of β in the α phase, a process called *nucleation* [Fig, 5-12(a)], and the second step is the *growth* of these nuclei [Fig. 5-12(b)]. The total rate of the transformation will therefore be dependent on how many β nuclei are formed and how fast they are growing. A complete analysis of transformation rates must therefore consider both the problems of nucleation and growth. As we shall see below, both of the processes are thermally activated, and hence very temperature-dependent.

Nucleation

As a first step in examining the kinetics of phase transformations we consider the calculation of the number of nuclei in a system. In order to perform this calculation we must first compute the total free energy change associated with the formation of a nucleus. This can be simply accomplished by examining the process of solidification. Let α be the liquid phase, β the solid phase, and T_E the melting temperature. Consider the formation of a small spherical β nucleus in the α matrix. (At this point we will not worry about how the nucleus has formed.) When the β nuclei form in the α phase, an interface will be created between the α and β phases [Fig. 5-12(a)]. Let γ be the surface free energy per unit area of this interface between nuclei and matrix.

The total free energy change ΔG_{tot} involved in making a spherical nucleus of the β phase, of radius r, is therefore

$$\Delta G_{tot} = \tfrac{4}{3} \pi r^3 \, \Delta G_V + 4\pi r^2 \gamma \tag{5-24}$$

The first term, the *bulk free energy change* involved in making the nucleus, is negative at $T < T_E$. The second term, the *surface energy change*, is always positive, since energy is always expended in making an interface. Figure 5-13 shows the

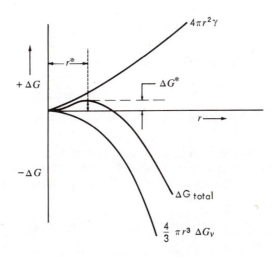

FIG. 5-13   Free energy changes associated with nucleation of a spherical particle of radius r

variation of these two terms and ΔG_{tot}, with radius r of the nucleus at $T < T_E$. We note that when r is less than a critical value, r^*, the surface term dominates and ΔG_{tot} increases with r, while for $r > r^*$, the volume term dominates and ΔG_{tot} decreases with increasing r. This means that there is a free energy barrier to the formation of a β nucleus. *A β nucleus must be larger than a critical size* $(r > r^*)$ *before it can continue to grow with a decrease in free energy.* Any β nuclei with $r < r^*$ will redissolve in the α matrix even at temperatures *below T_E*.

Because there is an energy barrier to the formation of a β nucleus we must somehow provide energy to the system in order to form a stable β particle. This energy is provided in the form of thermal energy. To calculate the number of nuclei in the system at any given temperature, we can treat the nuclei by the same general technique we used in Chapter 4 to estimate the density of vacancies. Let ΔG^* be the critical energy necessary to form stable nuclei $(r > r^*)$, and let s be the number of sites available for nuclei formation. To offset the energy increase in forming nuclei we have the mixing entropy that results from the various ways of arranging nuclei on their sites. Proceeding as in the case of vacancies, we find the number of nuclei, n^*, at any temperature is given by

$$n^* = s \exp\left(-\frac{\Delta G^*}{kT}\right) \tag{5-25}$$

To complete the calculation of n we must now evaluate ΔG^*; this can be accomplished by solving for r^* and then using Eq. (5-24) to obtain ΔG^*.

Since ΔG_{tot} has the maximum value ΔG^* at $r = r^*$, we can compute r^* by taking the derivative of ΔG_{tot} with respect to r and setting it equal to zero.

$$\frac{d}{dr}\left(\frac{4}{3}\pi r^3\right)\Delta G_V + \frac{d}{dr}(4\pi r^2)\gamma = 0 \qquad \text{at } r = r^*$$

yields

$$4\pi(r^*)^2 \Delta G_V + 8\pi r^* \gamma = 0$$

so that

$$r^* = \frac{-2\gamma}{\Delta G_V} \tag{5-26}$$

Introducing Eq. (5-23), we find

$$r^* = \frac{-2\gamma T_E}{\Delta E_V \Delta T} \tag{5-27}$$

The typical value of r^* is quite small, as we shall see in the following example.

EXAMPLE 5-4   The solid-liquid interfacial energy of pure silver is 126 ergs/cm². The latent heat of melting is 25 cal/g, and the melting temperature is 961°C. The density of solid or liquid silver is about 10.5 g/cm³. What is the value of r^* at 700°C?

Solution:

$$\Delta E_V = -25 \text{ cal/g} = (25)(10.5) = -262 \text{ cal/cm}^3$$

Since

$$1 \text{ cal} = 4.2 \times 10^7 \text{ ergs}$$

$$\Delta E_V = -1.1 \times 10^{10} \text{ ergs/cm}^3$$

$$\Delta T = 961 - 700 = 261°\text{C or } °\text{K}$$

$$T_E = 961 + 273 = 1234°\text{K}$$

Thus

$$r^* = \frac{-2\gamma T_E}{\Delta E_V \Delta T} = \frac{-2(126)(1234)}{(-1.1 \times 10^{10})(261)} = 2.24 \times 10^{-7} \text{ cm or 22 A}$$

Now using the above expression for r^* we can calculate ΔG^* from Eq. (5-24) as

$$\Delta G^* = \frac{16\pi\gamma^3}{3(\Delta G_V)^2} = \frac{16\pi\gamma^3}{3\Delta E_V^2} \frac{T_E^2}{\Delta T^2} \qquad (5\text{-}28)$$

The variation of r^* and ΔG^* with temperature T or degree of supercooling ΔT are shown in Fig. 5-14.

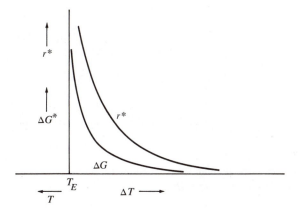

FIG. 5-14 Effects of undercooling on nucleation parameters r^* and ΔG^*

EXAMPLE 5-5

Calculate the value of ΔG^* for the critical-sized nucleus given in Example 5-4. If the density of nuclei sites is 10^{20} per cm^3, what is the density of nuclei at 700°C?

Solution:

From Eq. (5-28) and Example 5-4

$$\Delta G^* = \frac{16\pi\gamma^3 T_E^2}{3\Delta E_V^2 \Delta T^2} = \frac{16(3.14)(126)^3(1234)^2}{3(-1.1 \times 10^{10})^2(261)^2} = 6.1 \times 10^{-12} \text{ erg}$$

Now we can calculate the number of nuclei from

$$n^* = s \exp\left(-\frac{\Delta G^*}{kT}\right) = 10^{20} \exp\left(-\frac{6.1 \times 10^{-12}}{1.38 \times 10^{-16}(973)}\right) \text{ nuclei/cm}^3$$

$$n^* = 1.06 \times 10^{18} \text{ nuclei/cm}^3$$

Thus far we have considered only the equilibrium distribution of nuclei in a material and have paid no attention to the kinetics of nuclei formation, or the *rate of nucleation*. As nuclei form via the diffusive motion of atoms, we have two important thermal activation steps in the formation of a stable nucleus. First, we have to overcome the ΔG^* energy barrier, which arises solely from the bulk and surface free energy changes accompanying nuclei formation. In addition, we have the fact that atoms must diffuse and rearrange themselves to form a nucleus. Thus, the rate of nucleation, $\dot{N}$, will depend not only on the number of nuclei with sufficient energy to be stable [proportional to $\exp(-\Delta G^*/kT)$] but also on the ability of atoms to diffuse to the potentially stable nucleus [proportional to $\exp(-\Delta E_D/kT)$, where ΔE_D is the activation energy for diffusion]. Thus,

$$\dot{N} \propto \exp\left(-\frac{\Delta G^*}{kT}\right)\exp\left(-\frac{\Delta E_D}{kT}\right) \tag{5-29}$$

The variation of the nucleation rate with temperature is shown in Fig. 5-15. At temperatures just below T_E the rate of diffusion is rapid; but because of the small

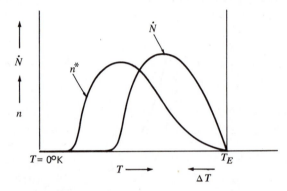

FIG. 5-15 Effects of temperature and undercooling on equilibrium nuclei concentration, n^* and on the nucleation rate, $\dot{N}$

undercooling ($T_E - T = \Delta T$ is small), we see from Eq. (5-28) that ΔG^* is large, and hence very few nuclei attain sufficient energy to reach the critical size r^*. At very low temperatures, where ΔT is large, ΔG^* is small; but because the diffusion rate is very slow and atoms can no longer move about and form nuclei, $\dot{N}$ is again low. At intermediate temperatures where diffusion rates are fairly rapid and ΔG^* is not too large, there is a maximum in $\dot{N}$. This maximum may occur at temperatures considerably below T_E.

Although the above discussion has considered only the liquid-solid transformation, the results are also applicable to solid-solid transformations. Generally, however, nucleation rates in solid-solid transformations are considerably slower than those found in the liquid-solid case. This is due to four separate factors:

1. Diffusion rates in solids are much slower than those in liquids; this means it is more difficult for atoms to rearrange themselves to form a nucleus.
2. Solid-solid surface energies are usually much larger than liquid-solid surface energies; this means there is an increase in ΔG^* for the nuclei forming in a solid parent phase.

3. ΔG_V for solid-solid transformations is smaller than for liquid-solid transformations, also increasing ΔG^* for the former case.
4. Because there are usually volume changes associated with the formation of a nucleus (the specific volumes of the α and β phases are different), there is a strain energy term that must be included in ΔG_{tot} for solid-solid transformations. This strain energy term is positive and hence increases ΔG^*. This term is not present in liquid-solid transformations, since the liquid can easily flow to accommodate any expansion or contraction due to the presence of the nucleus.

Because of the above effects it is clear why solid-solid transformations rarely ever approach equilibrium. The nucleation rates for these reactions are generally so sluggish that the retention of metastable phases is the rule rather than the exception. As we shall see in our later discussions this phenomenon proves extremely useful in many circumstances. That is, we are able to use the fact that solid-solid transformations occur very slowly to vary the internal structure of materials and hence change their properties.

Thus far we have considered only the ideal process of *homogeneous nucleation*, wherein spherical nuclei of β particles are formed in the α phase. In actual practice nucleation rarely occurs homogeneously, but rather occurs *heterogeneously* on surfaces. These surfaces are the walls of the container or foreign particles inadvertently or deliberately introduced into the melt during solidification or grain boundaries of the transforming phase in the case of solid-solid transformations. The reason for this effect can be seen in Fig. 5-16, taking the solidification transformation as

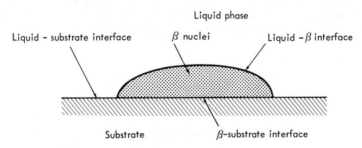

Liquid phase

Liquid – substrate interface β nuclei   Liquid – β interface

Substrate β–substrate interface

FIG. 5-16 Heterogeneous nucleation

an example. If γ (β–substrate) is lower than γ (liquid–substrate), then it can be shown that the *average* surface energy, γ_A, of the β nucleus decreases if the β nuclei form on the substrate. This is true even though the nucleus no longer has a spherical shape. And since the activation energy for nucleus formation ΔG^* is proportional to γ_A^3 [see Eq. (5-28)], a small decrease in γ_A can increase the nucleation rate at a given temperature. In addition, a smaller amount of undercooling is necessary to obtain a given rate of heterogeneous nucleation compared to homogeneous nucleation. Figure 5-17 illustrates the differences between homogeneous and heterogeneous nucleation.

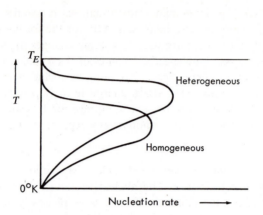

FIG. 5-17   Effects of temperature and undercooling on homogeneous and heterogeneous nucleation rates.

Growth of a Nucleus

The *growth* of a stable nucleus ($r > r^*$) will be determined by the rate at which atoms can move to and attach themselves to the nucleus. Since both these processes require atom movements, the growth rate is diffusion-controlled. Analogous to Fick's first law, there are two important factors that affect the diffusive motion of atoms to the nucleus: (1) the driving force for the transformation (equivalent here to the free energy change associated with the transformation that is proportional to ΔT, the amount of undercooling), and (2) the diffusion coefficient of the atoms, proportional to $\exp(-\Delta E_D/kT)$. Accordingly, the temperature dependence of the growth rate $\dot{g}$ will appear as shown in Fig. 5-18. At temperatures near T_E the driv-

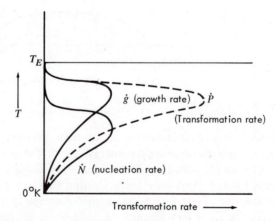

FIG. 5-18   Effects of temperature and undercooling on rates of nucleation, growth, and transformation

ing force is small, so that $\dot{g} \longrightarrow 0$ and at very low temperatures the diffusion co-efficient is small, so that again $\dot{g} \longrightarrow 0$. At some intermediate temperature there is a maximum in the growth rate. Because the diffusion rate is much more tempera-ture-sensitive than the driving force, the maximum in growth rate usually occurs at a higher temperature than the maximum in nucleation rate.

The *total rate of transformation* $\dot{P}$ will be dependent on both the rate of nuclea-tion $\dot{N}$ and the rate of growth $\dot{g}$ (see Fig. 5-18). At low temperatures and at tempera-tures just below the equilibrium temperature the total transformation rate is small because both the nucleation and growth rates are small. In the temperature range in between, the transformation rate increases to a maximum.

Since the *time* required for the transformation process to proceed to a certain extent (say, 10% of the total volume is transformed) is inversely proportional to the transformation rate, the transformation time will vary as shown in Fig. 5-19(a).

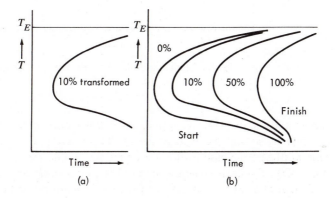

FIG. 5-19 Time-temperature-transformation curves for a hypothetical transformation

This curve is called a *C curve*. Actually, for any transformation, there will be a family of *C* curves that mark the start (0%), finish (100%), and intermediate points of the *isothermal transformation* [Fig. 5-19(b)], conducted at a constant tempera-ture. The minimum time intercept on the "start" curve is called the *nose* of the curve. The functional form of the time dependence of the transformation is com-plicated and is beyond the scope of this book. Since these curves contain time, temperature, and extent of transformation, they are commonly referred to as *T-T-T curves*. These curves are very important in certain metallurgical operations, such as the heat treatment of steel, and we shall discuss some specific examples in Chapter 9.

In practice, the isothermal transformation temperature or the *cooling rate* will play an important role in determining the size of the β grains (or particles). When the nucleation rate is low but the growth rate is high, only a relatively few nuclei will form before the transformation is complete, so that the β grains will be large. This would be the case for an alloy quenched from $T < T_E$ to T_1 (Fig. 5-20) and then isothermally transformed. Alternatively, when the nucleation rate

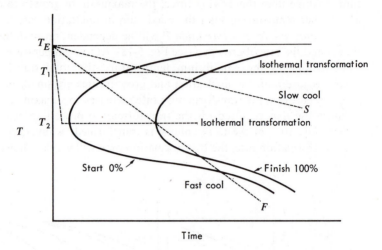

FIG. 5-20 Effects of cooling rate and temperature of transformation on the rates of transformation.

is high and the growth rate is low, many β nuclei form before the transformation is complete, so that the average size of the β phase grains (particles) will be small. This would be the case for an alloy that was isothermally transformed at $T = T_2$ (Fig. 5-20). Similarly, when a material is *continuously* and slowly cooled from $T > T_E$ (curve S, Fig. 5-20) most of the transformation occurs at high temperatures, where the nucleation rate is low and the β phase grains are then large. When the continuous cooling rate is fast (curve F) most of the transformation occurs at low temperatures, where the nucleation rate is high and the β phase grains are then small. As an application of this effect, in the following chapters we shall show that high mechanical strengths can be obtained when the size of microstructure is small. From the preceding discussions, this indicates that a high cooling rate tends to lead to high strength.

5-6 LIQUID-SOLID TRANSFORMATIONS

As an example of the application of kinetics to phase transformations, we shall first consider the process of solidification, or casting. To give some idea of the importance of solidification in materials production, we can cite the fact that almost all metallic components are melted and cast during some stage of the fabrication process. They are either cast directly into the shape of the component or as ingots that will be subsequently worked into shape.

From the economic point of view it would be desirable to build most components directly from castings since the expense of additional working and fabrication operations such as forging, extruding, annealing, joining, etc., would be eliminated. However, since the rate of growth of solid phases is much higher in the liquid phase than in the solid phase (because of the higher rate of diffusion in the liquid), the microstructure of as-cast materials is much coarser than that of heat-treated or mechanically fabricated and annealed materials. Consequently, as-cast parts are usually weaker than fabricated parts. Casting is therefore used as a primary fabrication process (1) when the structural component will not be subjected to high stessses during service (e.g., a cast iron engine block), (2) when high strength can be obtained by subsequent heat treatment (e.g., aluminum alloy pistons), (3) when the component is too large to be easily worked into shape (e.g., steel casting used in a steam turbine housing), or (4) when the shape of the component is too complicated to permit inexpensive subsequent fabrication (e.g., a zinc carburetor). In addition, all soldered, brazed, and welded joints contain an as-cast structure between two parts that are being joined.

Solidification is also important in nonmetallic systems. For example, many solid state electronic components, such as transistors, can be manufacutred using sophisticated solidification techniques, and in fact all the silicon and germanium single crystals used in the production of solid state electronic devices are prepared via solidification.

In the ensuing discussion we shall focus our attention on the influence of kinetics on solidification in three different types of alloy systems. In particular, we shall examine binary alloys with three different types of phase diagrams—isomorphous, eutectic, and peritectic. The results can be readily generalized to other systems.

Solidification of Isomorphous Systems

Consider the solidification of a system such as copper-nickel, which exhibits complete liquid and solid solubility. The equilibrium diagram is shown in Fig. 5-21. For illustrative purposes let us consider the solidification of an alloy that contains 60% nickel and 40% copper. If C_1 is the composition of the solid formed at temperature T_1 and the alloy is slowly cooled to T_2, the equilibrium diagram predicts that the composition of the solid should be C_2 and that all of the solid, both that formed at T_1 and that formed at T_2, should have this composition. This means that the solid formed at T_1 will have to change its composition by solid state diffusion. In this particular instance Cu atoms must diffuse into the solid from the liquid to change the composition of the solid from C_1 to C_2. However, in commercial practice, alloys are cooled at rates that are too high for equilibrium to be reached; that is, there is insufficient time for appreciable solid state diffusion to take place. Each layer that forms on a growing solid particle will therefore have a different chemical composition. Consequently, the *average* composition of the solid at T_2 will have a composition $C_2' < C_2$; i.e., a composition gradient will exist between the solid that formed at T_1, of composition C_1, and the solid formed at T_2,

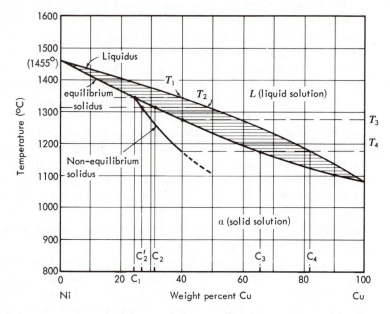

FIG. 5-21   Nonequilibrium cooling in the nickel-copper alloy system

of composition C_2. As further solidification takes place, the average composition of the solid will fall consistently below that predicted by the equilibrium solidus line. In Fig. 5-21 we have sketched out the average solid composition as a function of temperature. We can refer to this curve as the *nonequilibrium solidus line*, and it can be used in conjunction with the liquidus line to predict the fraction of solid or liquid present at any temperature using the lever rule. The actual position of the nonequilibrium solidus line depends on the cooling rate. As the cooling rate decreases, equilibrium is more closely approached and the equilibrium and non-equilibrium solidus lines are closer together.

An interesting aspect of this type of nonequilibrium solidification is that the Cu-Ni alloy does not become completely solid until a temperature, T_4, below the equilibrium solidus temperature T_3. That is, a portion of the solid will melt at a temperature lower than the melting temperature for an alloy with an identical but homogeneous chemical composition. The greater the amount of segregation, the more the melting temperature is lowered.

This type of segregation is known as *coring*. It can occur in any alloy that solidifies over a range of temperature. A cored microstructure is shown in Fig. 5-22. In order to remove the cored microstructure, which can have an adverse effect on properties due not only to the composition gradients but also to the lowering of the effective melting temperature of the alloy, it is necessary to *homogenize* the alloy by heating it for a sufficient period of time in the vicinity of temperature T_4 to allow diffusion to reduce the concentration gradients.

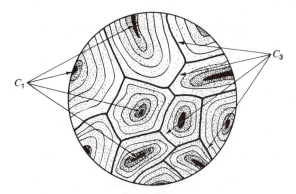

FIG. 5-22   Cored microstructure (see Fig. 5-21). Lines of constant composition are shown.

EXAMPLE 5-6   Suppose that the average size of the cored region in the copper-nickel alloy is 10^{-2} cm. How long will it take to remove this coring by annealing at 1100°C? For the diffusion of nickel in copper, $D_0 = 2.7$ cm²/sec and $\Delta E_D = 56.5$ kcal/mole.

Solution:

$$D = D_0 \exp\left(\frac{-\Delta E_D}{RT}\right) = 2.7 \exp\left(\frac{-56,500}{1.98(1373)}\right) \text{ cm²/sec}$$

$$D = 2.7 \times 10^{-9} \text{ cm²/sec}$$

The average distance over which diffusion must occur is 10^{-2} cm. Then, using Eq. (5-16),

$$t = \frac{x^2}{D} = \frac{10^{-4}}{2.7 \times 10^{-9}} \text{ sec} = 3.7 \times 10^4 \text{ sec}$$

Solidification of Eutectic Systems

Consider the eutectic phase diagram shown in Fig. 5-23. If an alloy of composition C_0 is cooled slowly from the liquid to the solid state, maintaining equilibrium all the while, then the alloy will not undergo the eutectic reaction and the microstructure will be single-phase α. However, let us examine the situation where the alloy is rapidly cooled. Suppose the cooling rate is rapid enough so that solid state diffusion does not have enough time to occur during the solidification process. Then coring will take place and the average composition of the solid will fall below the equilibrium solidus (imagine it to follow the nonequilibrium solidus in Fig. 5-23). Now we can ask what happens when the temperature reaches the eutectic invariant temperature T_E. At that temperature any liquid of composition C_E should undergo the eutectic transformation $L \longrightarrow \alpha + \beta$, and this is precisely what happens in this case. Because coring takes place, lowering the effective melting temperature of the alloy to T_E, the alloy will undergo the eutectic reaction and the microstructure will consist of a cored α phase plus some of the $\alpha + \beta$ eutectic structure. Note that the eutectic structure is clearly a nonequilibrium structure

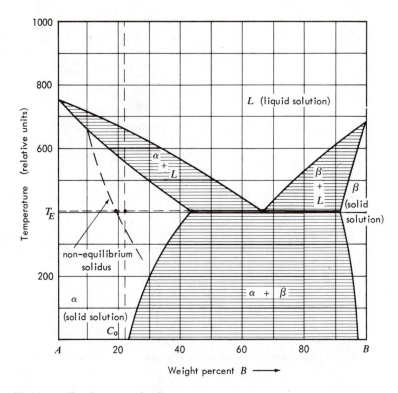

FIG. 5-23 Nonequilibrium cooling in a eutectic alloy system

that is present only because the kinetics have not been rapid enough to maintain equilibrium.

The distribution of the α and β phases in the eutectic structure is extremely dependent on the rate of the transformation (cooling rate). As pointed out in Chapter 4, the eutectic structure consists either of alternating plates, or lamellae, of α and β or of parallel rods of one phase symmetrically distributed in the other phase. That aspect of the structure which is critically influenced by cooling rate is the *size* of the microstructure (spacing between plates or rods). This is a result of the fact that the eutectic decomposition takes place by counterdiffusion of different types of atoms in the liquid. (*A* atoms diffuse to the α phase; *B* atoms to the β phase.) During a slow cool, atoms have time to diffuse over large distances, and, correspondingly, the spacing between plates or rods is large. Conversely, a rapid cool means short diffusion distances and small spacings. The eutectic structure will always tend toward the largest spacing allowed by the kinetics in order to minimize the α-β surface area (surface energy). The Al-Mg eutectic structure is shown in Fig. 5-24.

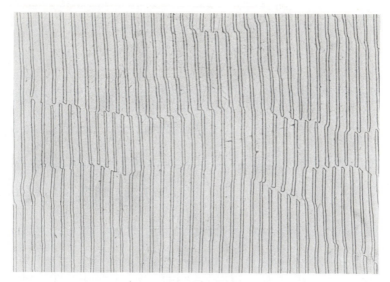

FIG. 5-24   Eutectic microstructure for the Al-Mg system (200×). Photo courtesy American
Institute of Mining, Metallurgical and Petroleum Engineers, Inc., from A. S. Yue,
Trans. of the AIME, **224**, 1010 (1962). Used with permission.

Solidification of Peritectic Systems

The peritectic reaction is much more sluggish than the eutectic reaction because
it involves diffusion in the solid state. Consider the solidification of an alloy with
composition C_0 in Fig. 5-25(a). If the alloy is slowly cooled to just above the peritectic temperature, then the structure will consist of solid of composition C_α and
liquid of composition C_L. Now since the peritectic reaction $L + \alpha \longrightarrow \beta$ requires
that liquid and α combine to form β, at the start of the peritectic reaction each α
particle will combine with liquid to form some β around the periphery of the α
particle [see Fig. 5-25(b)]. The reaction can then continue only by having B atoms
from the liquid diffuse through the β layer to combine with the α. Since the diffusion process is relatively slow, it is usually the rule rather than the exception for
some nonequilibrium α phase to be retained when cooling past the peritectic
temperature.

Thus far we have considered only the microscopic or atomistic aspects of solidification. We shall now briefly discuss the macroscopic aspects of this process in
order to understand the macrostructure of cast ingots.

Solidification begins with the formation of solid nuclei and the growth of these
nuclei into the liquid by the addition of atoms at the advancing interface. In commercial practice nucleation occurs heterogeneously, and therefore begins at temperatures just below the equilibrium melting temperature during continuous
cooling from above T_m. If the latent heat that is evolved during solidification is

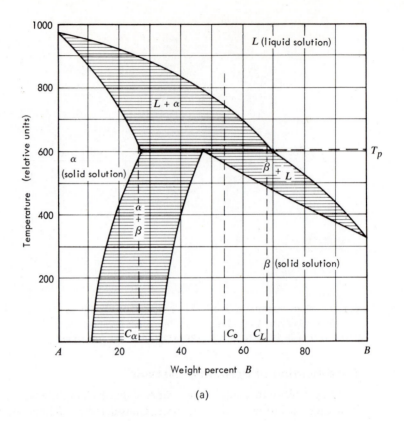

(a)

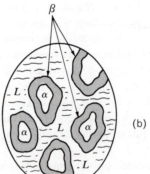

(b)

FIG. 5-25 Phases formed in a peritectic reaction: (a) phase diagram; (b) phase formation

not removed, the solid nuclei will heat back up to the melting temperature and solidification will stop. Consequently, the rate of solidification is determined primarily by the rate of removal of the latent heat of melting.

On a macroscopic scale, the structure of the as-cast ingot depends on the rate of nucleation and the rate of heat removal from the casting (e.g., on the temperature of the melt when it is poured into the mold and on the temperature of the mold walls). If the wall is cold and the casting is large, solidification will begin in the chill zone at the mold wall (region 1, Fig. 5-26) and proceed inward, parallel to the flow of heat out from the melt. The grains so formed (region 2, Fig. 5-26) are therefore elongated, or *columnar*, in shape. If they can reach the center of the casting before the temperature there drops low enough for nucleation to occur, the entire ingot structure will be composed of columnar grains. Usually this is not the case, since the center part of the casting begins to solidify before the columnar grains arrive there. Because the grains at the center of the casting are not forced to grow in any particular macroscopic direction, heat being removed isotropically, they are more equiaxed in shape (region 3, Fig. 5-26).

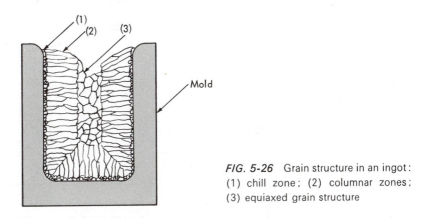

FIG. 5-26 Grain structure in an ingot: (1) chill zone; (2) columnar zones; (3) equiaxed grain structure

The relative amount of columnar versus equiaxed grains depends on such macroscopic factors as the rate of heat removal and the uniformity of cooling as well as the presence of solid nuclei in the liquid region. The latter are believed to be small crystals that are broken off (or dissolved off) the tips of the grains growing in from the sides of the mold. These crystal nuclei are carried into the center of the ingot by convective fluid flow in the liquid. A uniform but slow cooling rate accompanied by a quiescent liquid pool leads to a coarse grain structure in the interior of the ingot, whereas a fast cooling rate and a turbulent pool produce a fine grain structure, which is more desirable for good mechanical properties. In fact, a general rule about any solidification process is that *the faster the cooling rate, the finer the microstructure.*

5-7 SOLID-SOLID TRANSFORMATIONS

Very little will be said concerning the kinetics of solid-solid transformations since this topic will be discussed in more detail in Chapter 9 when we consider the influence of thermal treatments on the microstructure-mechanical property relationships. Here we shall briefly describe only one simple example of this class of transformation, the eutectoid transformation.

As mentioned earlier, solid-solid transformations rarely attain equilibrium because of the slow rate of solid state diffusion. Also, the size of the microstructural features is generally very small due to the short diffusion paths. Consider the solid state transformation of the alloy labeled C_0 in Fig. 5-27. As this alloy is cooled

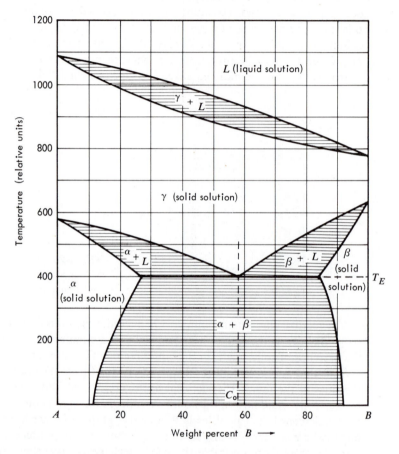

FIG. 5-27 Phase diagram representing a eutectoid reaction

below the eutectoid temperature, the reaction $\gamma \rightarrow \alpha + \beta$ should take place. Obviously, if we cool rapidly past the eutectoid temperature to room temperature, where diffusion rates essentially go to zero, the eutectoid reaction can be completely suppressed, and the microstructure will consist entirely of metastable γ. However, if we cool slowly, then the eutectoid reaction will take place. The eutectoid structure will be similar to the eutectic structures described earlier, i.e., either alternating plates of α and β or a rodlike distribution of one phase in the other. And, similar to the eutectic transformation, the spacing between rods or plates will decrease with increasing cooling rates. At rapid cooling rates, this spacing can be as small as a few hundred angstroms.

5-8 ENVIRONMENTAL DEGRADATION OF MATERIALS

In this section we consider the kinetic aspects of the degradation of materials in their operating environment. We shall limit our attention to the chemical and microstructural changes that take place as a result of the interaction of materials with their environment. Included in this discussion will be (1) the chemical reactions and structural changes commonly observed in polymeric metarials, (2) the degradation of metals in gaseous or dry environments (e.g., high-temperature *oxidation or sulfidation*), and (3) the electrochemical reaction of metals with liquid environments (e.g., *corrosion*). Absent from the following discussion is any consideration of the microstructural changes (phase transformations) brought about merely by a change in environmental temperature. The principles governing this last type of structural change have been adequately covered earlier in this chapter.

Polymeric Materials

One common form of degradation in polymeric materials is a breakdown in the long-chain polymer structure, leading to a decrease in molecular weight and an associated change in properties. There are two ways in which this decrease in molecular weight can occur. In *random degradation*, breaking, or *scission*, of the polymer chain occurs at random points along the chain, leading to a sharp decrease in molecular weight and an abrupt change in properties. Alternatively, the degradation may involve the successive release of monomer units from the end of the chain. This latter process, known as *chain depolymerization*, results in a gradual change in molecular weight, as the monomer units liberated from one chain may escape from the sample as a gas before other chains begin to depolymerize. These two forms of degradation may be caused by an increase in operating temperature or by exposure to radiation, oxygen, ozone, or other chemical species.

As an example, let us consider the reaction of natural rubber with ozone. This reaction is similar to that which occurs in many other polymers where double carbon bonds are available for chemical reactions. The breaking of the polymer chain occurs as follows (only one monomer unit is shown):

$$\begin{matrix} \text{H} & \text{CH}_3 & \text{H} & \text{H} \\ | & | & | & | \\ -\text{C}- & \text{C}= & \text{C}- & \text{C}- \\ | & & & | \\ \text{H} & & & \text{H} \end{matrix} + \text{O}_3 \Rightarrow \begin{matrix} \text{H} & \text{CH}_3 & \overset{\text{O}}{} & \text{H} & \text{H} \\ | & | & & | & | \\ -\text{C}- & \text{C} & & \text{C}- & \text{C}- \\ | & & & | & | \\ \text{H} & \text{O}& \!\!-\!\!\!-\!\! & \text{O} & \text{H} \end{matrix} \Rightarrow \begin{matrix} \text{H} & \text{CH}_3 \\ | & | \\ -\text{C}- & \text{C}= \text{O} \\ | & \\ \text{H} & \end{matrix} + \begin{matrix} & \text{H} & \text{H} \\ & | & | \\ \text{O}= & \text{C}- & \text{C}- \\ & & | \\ & & \text{H} \end{matrix} + \tfrac{1}{2}\text{O}_2$$

The reaction products normally form a layer on the surface and protect the polymer from further degradation. However, if the polymer is subjected to stress, this protective layer can be easily ruptured, causing increased attack and ultimate failure of the material.

Polymeric materials are also subject to *swelling* (the incorporation of a gas or liquid into the relatively open polymer structure). This dimensional instability in different environments can be troublesome when polymers are used for bearings or seals or in other applications where strict dimensional tolerances are required. For example, nylon (polyhexamethyleneadipamide) will absorb up to 1 weight percent H_2O and, correspondingly, can change its dimensions by about 1%, depending on the moisture content of the surrounding environment. Generally there is a loss in strength (i.e., softening) associated with swelling. Structural features that minimize swelling include high bond energies, high degree of crystallinity, cross-linking, and chemical dissimilarity between the polymer and penetrating molecules.

Metallic Oxidation

The process of oxidation is one in which a metal combines with an atom or molecule of a reactant and loses electrons. The oxidation of iron according to the reaction $Fe + \frac{1}{2}O_2 \longrightarrow FeO$ is typical of this process, although it is not necessary that the nonmetallic reactant be oxygen. Thermodynamics predicts that most metals should oxidize in air at room temperature. However, as was mentioned earlier, the kinetics of the oxidation process are such as to limit the oxide (or sulfide, etc.) formation to a thin region on the surface of the metal. Further oxide formation requires mass transport (diffusion) through the oxide layer, and at low temperature these diffusion rates are so slow that the oxidation process cannot continue. At high temperatures, where diffusion rates are high, the oxidation process occurs much more rapidly. Degradation through oxidation (sulfidation) limits the use of many materials in oxidizing (sulfidizing) environments. This is especially evident in applications such as jet engines or gas turbines, where both oxygen and sulfur may be present in the combustion products.

Figure 5-28 illustrates the oxidation of iron in air, showing that both ion and electron transport must occur through the oxide layer to sustain further oxidation. For many metals, ion diffusion through the oxide layer controls the rate of oxidation. In the Fe-O system, for example, the oxidation process $Fe^{++} + O^{--} \longrightarrow FeO$ occurs near the oxide-air interface, requiring that the Fe^{++} diffuse through the oxide. As shown in Eq. (5-19), if ion diffusion controls oxidation, a parabolic oxidation law is observed where the oxide thickness x is proportional to (time)$^{1/2}$.

Although parabolic oxidation is observed for many metals, there are other factors that can change the observed oxidation behavior, For example, if the oxide scale is not coherent and cracks or spalls off, the metal is not protected from the oxidizing atmosphere by the scale and oxidation will occur more rapidly than the parabolic law predicts. Also, the oxide may be volatile (as in the case of oxides of Mo) so that it vaporizes as it forms. Yet another important feature is the conduction of electrons through the oxide layer. As shown in Fig. 5-28, electron transport must accompany the ion diffusion. Most oxides are poor electrical conductors, and hence ion diffusion may be more rapid than electron transport, making the latter process rate-controlling. Each of the above features can change the oxidation kinetics and give rise to a different oxidation law. Below are listed the most common of these oxidation laws, with a short explanation of the appropriate rate-controlling mechanism.

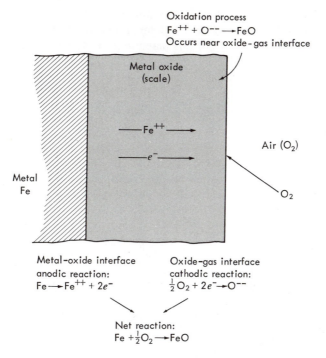

FIG. 5-28 Schematic illustration showing ion and electron flows associated with oxidation of iron

Parabolic oxidation ($x^2 = A_1 t$, where A_1 is the temperature-dependent rate constant). This behavior is observed when the oxide scale is coherent and ion diffusion is the rate-limiting process.

Linear oxidation ($x = A_2 t$, where A_2 is the temperature-dependent rate constant). Linear oxidation may occur when the transport of ions is more rapid than the chemical reaction involved in the oxidation step. In this case the reaction itself will be rate-limiting, leading to a time-independent oxidation rate. Linear oxidation also occurs when the oxide scale is nonprotective; that is, cracks or porosity in the oxide allow the oxidizing atmosphere to remain in contact with the metal and maintain a constant oxidation rate. Figure 5-29 illustrates an approximate linear oxidation rate for an oxide that continually fractures or spalls during growth. Each break in the curve represents a fracture in the oxide, with a parabolic growth between fractures. MgO on Mg is an example of a porous oxide film, while Nb and Ta form oxide films that continually fracture during growth.

Logarithmic oxidation [$x = A_3 \log (A_4 t + A_5)$, where A_3 is the rate constant]. Oxidation at low temperatures with very thin oxide films often follows logarithmic kinetics and is thought to be controlled by the diffusion of ions driven by large electric fields. The electric potential existing between the cathodic and anodic reaction areas (see Fig. 5-28) is about 1 V for most oxidation reactions. Thus, for very thin films (<50 A), voltage gradients of the order of 10^6 V/cm are established. Electric fields of this magnitude are sufficiently large to drive ion diffusion through the oxide. As the oxide thickens, the effect of the electric field diminishes and growth stops unless the temperature is raised high enough to promote thermally activated diffusion. Al and Cu both show logarithmic oxidation at low temperatures with limiting oxide thicknesses of ≈ 40 A.

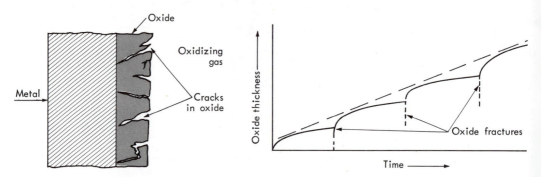

FIG. 5-29   Average linear oxidation behavior resulting from a parabolically thickening film that continually fractures, exposing fresh metal surfaces to the oxidizing gas

EXAMPLE 5-7

Copper oxidizes to Cu_2O, exhibiting parabolic behavior at temperatures above 500°C. The activation energy for the process (37.7 Kcal/mole) corresponds to the activation energy for diffusion of cuprous ions in Cu_2O. If 0.001 cm of Cu_2O forms after exposure to air for 5 min at 600°C, what will the oxide thickness be if a clean copper surface is exposed to air at 550°C for 10 min?

Solution:

From Eq. (5.19), $x^2 = At$, where A is a temperature-dependent constant of the form $A = A_0 \exp(-\Delta E/RT)$. If $x = 0.001$ cm after 5 min (300 sec), then

$$A_0 = \frac{x^2}{t \exp(-\Delta E/RT)} = \frac{(0.001)^2}{300 \exp\{-37,700/[1.98(873)]\}} = 9.35 \text{ cm}^2/\text{sec}$$

The oxide thickness after 10 min (600 sec) at 550°C (823°K) is

$$x = (At)^{1/2} = \left\{9.35 \exp\left[-\frac{37,700}{1.98(823)}\right]600\right\}^{1/2} = 7.5 \times 10^{-4} \text{ cm}$$

Metallic Corrosion

The corrosion of metals in liquid environments proceeds via an electrochemical mechanism; that is, the corroding metal has both cathodic and anodic areas. These areas may be permanently separated from each other or may consist of cathodic and anodic sites that are continually shifting. An example of this latter process is the simple dissolution of a pure metal in an acid.

Figure 5-30 illustrates the electrochemical nature of the corrosion process. The requirements for the process are: (1) anodes and cathodes must be present to form a cell, (2) the anode and cathode must be in electrical contact, and (3) the

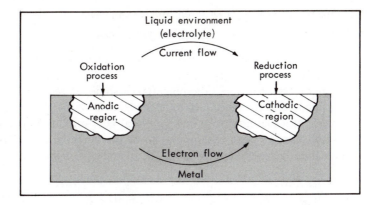

FIG. 5-30 Schematic illustration of metallic corrosion

liquid environment must serve as an electrolyte. At the anodic site an oxidation process occurs and the metal ion goes into solution via a reaction of the type

$$\underset{\text{(metal)}}{\text{Fe}} \longrightarrow 2e^- + \underset{\text{(ion in solution)}}{\text{Fe}^{++}}$$

At the cathode a reduction process occurs that generally results in the reduction of dissolved oxygen or the liberation of hydrogen gas. These two reactions can be written as

$$\underset{\text{(in solution)}}{\text{O}_2} + 4e^- + 2\text{H}_2\text{O} \longrightarrow \underset{\text{(in solution)}}{4\text{OH}^-}$$

$$\underset{\text{(in solution)}}{2\text{H}^+} + 2e^- \longrightarrow \underset{\text{(gas)}}{\text{H}_2}$$

The actual corrosion process occurs at the anode, where metal ions leave the metal surface and enter the solution. At the cathodic region there is no corrosion, and this is the area where electrons flow from the metal to the solution. On a single piece of metal anodic and cathodic regions can form due to localized differences in composition or localized differences in the environment. However, most commercial corrosion failures occur when two dissimilar metals are electrically connected in an electrolytic solution.

When two metals are electrically coupled and placed in a electrolyte, the property that determines which metal will be anodic and which will be cathodic is called the *electrode potential*. The electrode potential can be defined by referring to the following example. Suppose a piece of pure metal (Fe) is placed in water and the following reaction occurs:

$$\text{Fe} \rightleftarrows \text{Fe}^{++} + 2e$$

Fe^{++} ions will spontaneously enter the solution as long as there is a net decrease in free energy accompanying the process. The factors contributing to the free energy change are: (1) the chemical free energy change ΔG associated with the formation of F^{++} ions and (2) the removal of the Fe^{++} ions from the negatively-charged piece of iron. The iron has a negative charge because the electrons released during the oxidation process are not conducted away. At equilibrium the two contributions to the free energy must be equal, and we can write

$$\Delta G = Ze\phi \tag{5-30}$$

where ϕ is the potential bteween the piece of iron and the surrounding solution and Z is the ionic charge (2 for Fe^{++}). ϕ cannot be measured directly, but the difference between ϕ and the potential of some reference electrode can be readily measured. The reference electrode is taken to be the hydrogen electrode, which consists of a platinum wire in a 1 normal HCl solution in contact with hydrogen gas at 1 atm pressure. Taking ϕ_{ref} to be the potential of the hydrogen electrode, the standard electrode potential E_0 for any metal is defined as

$$E_0 = \phi - \phi_{ref} \tag{5-31}$$

The experimental apparatus for determining E_0 is shown in Fig. 5-31, and tabulated values for different metals are given in Table 5-3.

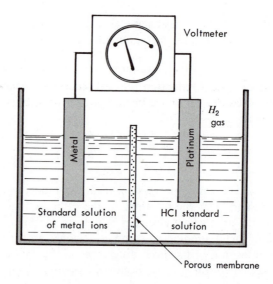

FIG. 5-31 Schematic illustration of apparatus used to determine standard electrode potentials

If two metals are in electrical contact the data in Table 5-3 tell us which metal will act as the anode and which will act as the cathode. The metal with the most positive E_0 will always be the cathode. The difference in electrode potentials for the two metals represents the driving force for current flow and accelerated corrosion of the anodic materials. For example, copper is cathodic with respect to aluminum, and if these two metals are in contact in an aqueous electrolyte, the aluminum will corrode preferentially. Generally the cathodic material corrodes very little or not all in this type of galvanic cell, while the anodic material corrodes much more rapidly than if the two metals were not in contact. The larger the difference in electrode potentials, the larger the effect.

TABLE 5-3
Standard Electrode Potentials

Metal Reaction	E_0(V)	
$Au = Au^{3+} + 3e^-$	+1.498	cathodic
$Pt = Pt^{++} + 2e^-$	+1.20	
$Pd = Pd^{++} + 2e^-$	+0.987	
$Ag = Ag^+ + e^-$	+0.799	
$Fe^{3+} + e = Fe^{++}$	+0.771	
$Cu = Cu^{++} + 2e^-$	+0.337	
$Sn^{4+} + 2e^- = Sn^{++}$	+0.15	
$2H^+ + 2e^- = H_2$	0 (arbitrary reference)	
$Pb = Pb^{++} + 2e^-$	−0.126	
$Sn = Sn^{++} + 2e^-$	−0.136	
$Ni = Ni^{++} + 2e^-$	−0.250	
$Co = Co^{++} + 2e^-$	−0.277	
$Cd = Cd^{++} + 2e^-$	−0.403	
$Fe = Fe^{++} = 2e^-$	−0.440	
$Cr = Cr^{3+} + 3e^-$	−0.744	
$Zn = Zn^{++} + 2e^-$	−0.763	
$Al = Al^{3+} + 3e^-$	−1.662	
$Mg = Mg^{++} + 2e^-$	−2.363	anodic

Two important examples of galvanic corrosion are shown in Fig. 5-32. Tin is commonly used as a protective layer on iron because tin has an electrode potential near iron and corrodes very slowly in many solutions. However, tin is cathodic to iron, and hence a small scratch in the tin plate exposing the iron will produce a galvanic cell where the anodic iron will corrode. An additional factor increasing the corrosion rate here is the relative area of the cathode to the anode. The small anode area corrodes very rapidly since the cathodic reaction occurs over a large area, thus allowing a large electron current to be removed from the anode. Whereas a tin coating must be continuous to protect steel from corrosion, a thin layer of zinc will provide corrosion protection for steel even if the coating is not continuous. This is because zinc is anodic with respect to iron; therefore this corrosion protection is commonly used. For example, galvanized steel sheet and pipe are merely steel coated with zinc, and sacrificial anodes of zinc and magnesium are commonly used on ship hulls or in water tanks to make the steel cathodic and corrosion-resistant. These sacrificial anodes, of course, must be replaced periodically.

The standard electrode potentials listed in Table 5-3 do not always predict the relative oxidation tendencies for metals, since this behavior does depend to some extent on the nature of the environment; that is, the positions of some of the metals may change somewhat, depending on exact conditions. For example, a very thin film of reaction product on a metal surface may render the metal corrosion-resistant. This is the reason chromium has such a profound influence on the corrosion behavior of stainless steels. Chromium is anodic with respect to iron; yet, if a small amount of chromium oxidizes and then combines with oxygen to form the chromate ion $CrO_3^=$, this ion can be adsorbed onto the metal surface and isolate the surface from further corrosive action. In this state the metal is said to

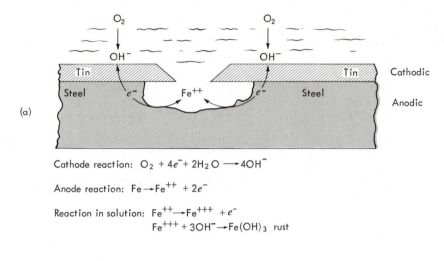

Cathode reaction: $O_2 + 4e^- + 2H_2O \longrightarrow 4OH^-$

Anode reaction: $Fe \longrightarrow Fe^{++} + 2e^-$

Reaction in solution: $Fe^{++} \longrightarrow Fe^{+++} + e^-$
$Fe^{+++} + 3OH^- \longrightarrow Fe(OH)_3$ rust

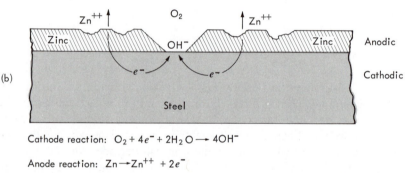

Cathode reaction: $O_2 + 4e^- + 2H_2O \longrightarrow 4OH^-$

Anode reaction: $Zn \longrightarrow Zn^{++} + 2e^-$

FIG. 5-32   Schematic illustration showing (a) accelerated corrosion in galvanic couple of tin and steel and (b) protection of steel by anodic zinc layer

be *passivated*. Because the corrosive tendencies of metals are environment-sensitive, it is generally necessary to evaluate their behavior in situations approximating actual service rather than relying on standard electrode potentials. A compilation of such data for the behavior of metals in sea water is shown in Table 5-4. The alloys grouped in parentheses behave similarly and generally are not subject to galvanic corrosion when in contact with one another.

There are several microstructural features that give rise to galvanic corrosion within a single piece of metal. Alloys with two or more phases may corrode readily if the phases differ in the galvanic series. If this is the case, the anodic phase will corrode rapidly and lead to deterioration of the metal. Grain boundaries also promote the formation of galvanic cells, since the bonding and free energies of the atoms at a grain boundary are different from atoms away from the boundary. The boundary region is generally anodic with respect to the rest of the grain and hence corrodes more rapidly. This is why fine grain sizes lead to higher corrosion rates. Preferential grain boundary attack may also occur if precipitates occur along grain boundaries, especially if the precipitation process locally changes the

TABLE 5-4

Galvanic Series in Sea Water

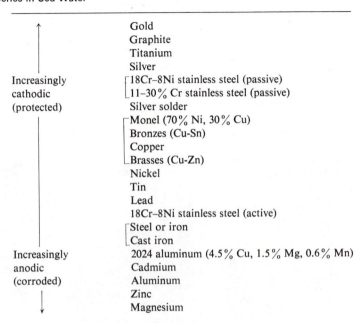

	Gold
	Graphite
	Titanium
	Silver
Increasingly	⎡18Cr–8Ni stainless steel (passive)
cathodic	⎣11–30% Cr stainless steel (passive)
(protected)	Silver solder
	⎡Monel (70% Ni, 30% Cu)
	Bronzes (Cu-Sn)
	Copper
	⎣Brasses (Cu-Zn)
	Nickel
	Tin
	Lead
	18Cr–8Ni stainless steel (active)
	⎡Steel or iron
	⎣Cast iron
Increasingly	2024 aluminum (4.5% Cu, 1.5% Mg, 0.6% Mn)
anodic	Cadmium
(corroded)	Aluminum
	Zinc
	Magnesium

chemical composition near the grain boundary. This can be a problem for chromium-bearing stainless steels if chromium carbides form along the grain boundaries, depleting the adjacent region of chromium. The low chromium content region adjacent to the grain boundaries is then anodic to the grain interior and corrodes rapidly. This problem can be alleviated through proper heat treatment to dissolve the chromium carbides or by adding other alloy elements that are stronger carbide formers than chromium.

Variations in the surrounding environment can also cause electrochemical cell formation and accelerated corrosion. One example is a *concentration cell* or galvanic cell set up due to variation in environment from one point to another on the surface. For example, variations in oxygen content in the electrolyte can produce anodic and cathodic regions along the surface of a metal. An area low in oxygen will be anodic to areas high in oxygen content. In practice, variations in oxygen content are usually associated with gasket surfaces, lap joints, surface deposits, crevices under bolt heads and rivet heads, etc. In each of these situation a stagnant region is produced where the oxygen content is lower than the surrounding environment. Examples are shown in Fig. 5-33. As a dramatic example of crevice corrosion, a stretched rubber band placed around a piece of stainless steel that is then immersed in salt water will cause sufficient localized corrosion to cut through the steel in a short period of time. In essence, the bulk of the material is cathodically protected at the expense of the small anodic region under the rubber band (or in the crevice), thereby leading to very high corrosion rates.

Effects similar to those produced by variations in oxygen concentrations occur

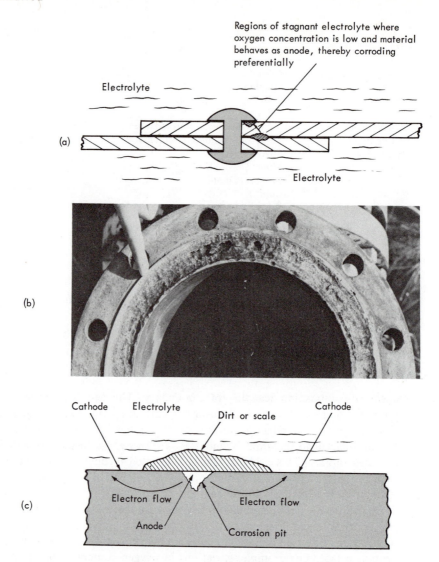

Regions of stagnant electrolyte where oxygen concentration is low and material behaves as anode, thereby corroding preferentially

Electrolyte

(a)

Electrolyte

(b)

Cathode   Electrolyte   Dirt or scale   Cathode

Electron flow   Electron flow

(c)   Anode   Corrosion pit

FIG. 5-33 Examples of crevice corrosion caused by local variations in oxygen concentration. Photo courtesy Mars G. Fontana, The Ohio State University. Used with permission.

when local differences in metal ion concentration exist. Surface regions in contact with high metal ion concentrations become anodic compared to regions in contact with low ion concentrations.

One further example of localized corrosion due to environmental changes exists in buried pipelines. Local variations in soil chemistry or oxygen content may lead to localized corrosion as long as the soil is damp enough to act as an electrolyte.

For instance, aerated soil with a high oxygen content will produce a cathodic region compared to compact soil with a lower oxygen content.

PROBLEMS

5-1 What are the values of ΔG^* and ΔG when a 2-lb ball resting in a valley 100 ft above sea level is rolled to the top of a hill 150 ft above sea level and then allowed to roll to another valley 25 ft above sea level? Express your answer in calories.

5-2 Suppose that two thermally activated reactions $1 \rightarrow 2$ and $1 \rightarrow 3$ are occurring simultaneously. If $\Delta G^*_{(1 \rightarrow 2)}$ is twice as large as $\Delta G^*_{(1 \rightarrow 3)}$ and the ratio of the reaction rates is equal to 100 at 500°C, what are the relative rates at 1000°C? Assume the pre-exponential factors to be equal for both reactions.

5-3 A certain thermally activated process takes 3 min at 900°C. How long will it take at 700°C if the activation energy is 40,000 cal/mole?

5-4 Consider the reaction $A \rightarrow B$. The kinetics of the reaction are described by the equation $dc_A/dt = Kc_A$, where K is the rate constant and c_A is the concentration of A. Show that the concentration of B, c_B, is given by the equation $c_B = 1 - \exp(-Kt)$. If c_B is equal to 0.3 at $t = 100$ sec, what is the value of c_B at 1000 sec?

5-5 Consider the thermally activated reaction $E \rightarrow F$. If the reaction is half completed after 100 sec at 300°C or after 20 sec at 500°C, what is the activation energy for the process?

5-6 The concentration of hydrogen in alpha iron is given by the equation

$$c = 42.7 p^{1/2} \exp\left(-\frac{6500}{RT}\right)$$

where c is in parts per million by weight and p is the external hydrogen pressure in units of atmospheres. The diffusivity of hydrogen in alpha iron is

$$D = 1.4 \times 10^{-3} \exp\left(-\frac{3200}{RT}\right) \text{ cm}^2/\text{sec}$$

Suppose that a thin iron membrane, 10^{-2} cm thick, separates a large reservoir of hydrogen at a pressure of 200 atm from another large reservoir at a pressure of 2 atm at 200°C. What will be the flux of hydrogen through the membrane in units of moles per square centimeter per second?

5-7 Suppose a piece of lead and a piece of tin are placed in intimate contact and held at a temperature of 150°C for a length of time sufficient to allow some diffusion to take place. Sketch the resulting composition profile across the diffusion couple. (*Hint:* see the lead-tin phase diagram, Fig. 4-11.)

5-8 Suppose that a piece of pure iron is carburized (carbon is allowed to diffuse into the iron) at 850 or 950°C for a very long time. Sketch the variation of carbon content with distance from the graphite-iron interface for both the 850 and 950°C diffusion treatments. (*Hint:* see the iron-carbon phase diagram, Fig. 4-15.)

5-9 The parabolic oxidation rate constant for pure copper varies with temperature as shown on the next page.

Rate Constant (gm$^2 \cdot$cm$^{-4} \cdot$sec^{-1})	$T\degree$C
5.3×10^{-11}	500
3.1×10^{-10}	600
1.6×10^{-9}	700

Determine the value of the rate constant at 650°C.

5-10 The oxidation rates of metals, whose oxides have a specific volume considerably larger than the metal itself, quite often show a linear rather than a parabolic oxidation rate. Suggest a reason why this might be so.

5-11 For small values of y (<0.5), the Gaussian error function (erf) can be approximated as erf $(y) \approx y$. Suppose a piece of silicon is exposed to aluminum vapor at 1300°C and the aluminum begins to diffuse into the silicon. How long will it take for the concentration of the aluminum at a point 0.01 cm from the surface to be 0.35 of that at the surface? The diffusivity of aluminum in silicon is given in Fig. 5-8.

5-12 The electrical conductivity of ionic crystals at high temperatures is principally due to the diffusion of charged ions. The conductivity σ varies with temperature T as

$$\sigma = \left(\frac{\text{const}}{T}\right) D$$

where D is the diffusion coefficient of the mobile ion. The conductivity of ZrO^2 is 3×10^{-4} (ohm-cm)$^{-1}$ at 700°C and 2×10^{-4} (ohm-cm)$^{-1}$ at 1100°C. What is the conductivity at 1000°C?

5-13 Suppose that nuclei formed during a phase transformation are cubic in shape rather than spherical. If a is the dimension of the cube, compute a value for a^* and ΔG^* in terms of γ and ΔE_v.

5-14 Suppose that a grain boundary AB exists in a solid phase α that is transforming to another solid phase β and that the energy of this boundary is 600 ergs/cm. If heterogeneous nucleation of β occurs on the boundary, such that part of the boundary is removed by the presence of the spherical nucleus

what is the value of r^* relative to the value for homogeneous nucleation? Take $\gamma_{\alpha-\beta}$ equal to 300 ergs/cm^2.

5-15 It is commonly observed that diffusion at low temperatures along grain boundaries or free surfaces is much faster than diffusion in the matrix. Suggest a reason why this might be so.

5-16 If a small amount of a divalent cation such as Mg^{++} is dissolved in NaCl, it is observed that the cation diffusion rate is much higher than in pure NaCl. Why?

5-17 A small droplet (1 mg) of liquid copper is cooled until it solidifies. Calculate the amount of undercooling (i.e., the temperature at which at least one stable nucleus forms) if $T_{\text{melt}} = 1083\degree$C, $\gamma = 200$ ergs/cm^2, density $= 10$ g/cm^2, and $\Delta E_v = -50$ cal/g

5-18 Can you suggest a way to refine materials by a solidification process making use of the phenomenon of "coring"? Explain.

5-19 A turbine blade is to be manufactured in the form of a single crystal that is grown by simply freezing the part from one end to the other, as shown in the accompanying diagram. The alloy is represented by composition C_0 on the phase diagram. The dashed line on the curve shows the average composition of the solidified part during freezing (diffusion is slow in the solid).

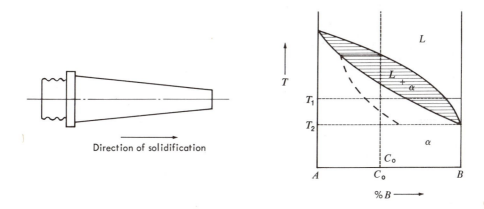

Direction of solidification

(a) The single crystal blade was service-tested at temperature T_1 and catastrophic fracture occurred. What happened?
(b) An engineer suggested that the blades could be annealed at T_2 prior to use at T_1. Is this suggestion sound in principle?
(c) If the diffusivity at T_2 is 10^{-9} cm^2/sec and the blade is 10 cm long, how long would the blade have to be annealed at T_2 to solve the problem? Is annealing at T_2 a practical solution to this problem?

BIBLIOGRAPHY

J. BURKE, *The Kinetics of Phase Transitions in Metals.* London: Pergamon Press, 1965.

J. BYRNE, *Recovery, Recrystallization and Grain Growth.* New York: Macmillan, 1965.

B. CHALMERS, *Physical Metallurgy*, Ch. 6, 8. New York: John Wiley, 1959.

A. H. COTTRELL, *Theoretical Structural Metallurgy*, Ch. 12, 14. New York: St. Martin's Press, 1957.

F. DANIELS and R. A. ALBERTY, *Physical Chemistry.* New York: John Wiley, 1961.

M. F. FINE, *Introduction to Phase Transformations in Condensed Systems.* New York: Macmillan, 1965.

M. FONTANA and W. GREENE, *Corrosion Engineering.* New York: McGraw-Hill, 1967.

W. D. KINERGY, *Introduction to Ceramics*. New York: John Wiley, 1960.

J. C. SCULLY, *The Fundamentals of Corrosion*. Oxford: Pergamon Press, 1966.

P. G. SHEWMON, *Diffusion in Solids*. New York: McGraw-Hill, 1963.

D. TURNBULL, "Phase Changes," *Solid State Physics*, Vol. 3. New York: Academic Press, 1956.

6
Introduction to the Mechanical Properties of Solids

6-1 INTRODUCTION

The application of mechanical forces to a solid body causes the body to change shape (deform) and, in some cases, to break (fracture). These responses define the mechanical properties that are studied in this chapter. Of special importance in this study are the *stress* and *strain*, which will be defined later. These quantities will be used to characterize the behavior of materials under different types of mechanical loading. Typical mechanical properties are the *elastic modulus*, which defines the reversible, time-independent elastic strain induced by a given applied stress, and the *ductility*, which defines maximum gross strain that the material can withstand without breaking. Some mechanical properties, such as the elastic moduli of crystals, are material constants, essentially independent of the conditions under which they are measured. Other properties, such as ductility, depend on external conditions of measurement, particularly the time interval over which the stresses are maintained, the rate and temperature at which the stresses are applied, and the microstructure of the material. For example, a bar of constructional steel with one particular microstructure will break after being strained 75% at 18°C, but at −253°C it requires only about 1% strain to cause fracture. In addition, if the microstructure is varied in a particular fashion, the steel will break after only a few percent strain at 18°C.

The design requirements of a structural component dictate the mechanical properties that the materials used in component construction must possess. Steel used in girders must have strength sufficient to support load without deforming excessively and ductility sufficient to prevent premature fracture in the vicinity of sharp corners (stress concentration). Materials, such as beryllium, used in gyroscope construction must have a high elastic stiffness so that they will be resistant to very small deflections and so that the gyro will retain its dimensional stability. Conversely, while the fiberglass used in the construction of a rocket motor case must be extremely resistant to fracture during the short time that the missile is being fired, long-term dimensional stability is not required.

In this chapter we shall describe the more important mechanical properties of solids, particularly those that are used in structural design and serve as a basis for calculation of the strength of structures, as described in courses on strength of materials, elasticity, and plasticity. These subjects deal with questions such as, "What is the maximum load that can be applied safely to a pressure vessel of given dimensions, having particular strength and fracture characteristics?" We shall illustrate that many of the mechanical properties of a material, such as strength, are *not* constants, as implied in some engineering handbooks, but are very sensitive to microstructure and environment.

In the following chapters we shall show how these properties depend on the way that atom arrangements respond to the application of stress. In Chapters 7 and 8 we examine the role of dislocations in the plastic deformation of crystalline materials and discuss the influence of microstructure on the motion of dislocations. In Chapter 9 we use the principles outlined in Chapters 4 and 5 to show how the microstructure of crystalline materials can be controlled to obtain a desired set of mechanical properties. Finally, in Chapter 10 we consider the mechanical behavior of amorphous materials.

6-2 ELASTIC DEFORMATION AND STRESS DISTRIBUTIONS

Elastic Stress-Strain Relations

When solid materials are subjected to small stresses they usually respond in an elastic fashion; that is, the strain produced by the stress is reversible (the strain goes back to zero when the stress is removed), and the magnitude of the strain is directly proportional to the magnitude of the stress. We discussed this problem briefly in Chapter 2, where we showed that application of an external force to a material produces atomic displacements in which the magnitude of the displacement was directly proportional to the force. This relationship between stress and strain (or force and displacement) is usually referred to as *Hooke's law* and can be written

$$\frac{\text{stress}}{\text{strain}} = \text{constant} \qquad (6\text{-}1)$$

where the constant is known as the *elastic modulus*. Equation (6-1) is generally valid only for small strains much less than 1% in crystalline materials.

We shall now describe the various relationships existing between stress and strain in the region where Hooke's law is valid. We begin with a formal definition of stress: Stress is the *force intensity* (force per unit area) acting on a material.

For example, if P is a uniform load acting on a cross-sectional area A of a structural member, then the component of load intensity or stress that acts across A is P/A. The stresses are *shear* (τ) when they act in the plane A and *normal* (tensile, $+\sigma$, compressive, $-\sigma$) when they act perpendicular to A.

Figure 6-1(a) shows a parallelepiped of material that has been deformed by the

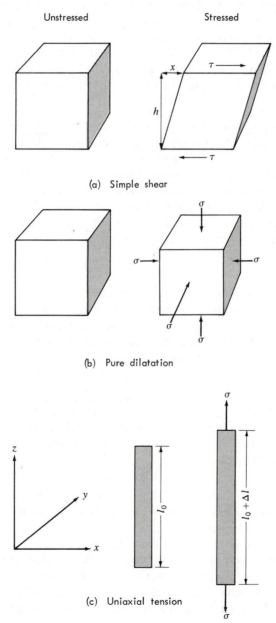

Unstressed   Stressed

(a) Simple shear

(b) Pure dilatation

(c) Uniaxial tension

FIG. 6-1 Example of simple stress states

application of a load P. The *elastic* shear strain γ is the amount of shear x divided by the distance h over which the shear has occurred, or $\gamma = x/h$. In pure elastic shear, Hooke's law describing the proportionality between stress and strain is written

$$\tau = G\gamma \tag{6-2}$$

where G is the *elastic shear modulus*. Note that pure shear does not produce any change in volume of the body, but does produce a change in shape.

Alternatively, suppose that a parallelepiped [Fig. 6-1(b)] is subjected to equal normal pressure $-\sigma$ (compressive stress) across its six faces such that the volume of the body V is reduced by an amount ΔV but the shape of the body remains the same. This type of deformation is called *pure dilatation*. The dilatation strain Δ is equal to $\Delta V/V$, and Hooke's law for elastic deformation is written as

$$\sigma = B\Delta \tag{6-3}$$

where B is called the *bulk modulus*. Note that pure dilatation does not produce any change in shape but does produce a change in volume.

In most instances the stress applied to a structure is neither pure dilatation nor pure shear, but is a mixture of the two. Uniaxial tensile or compressive loading is the most common example. When a tensile stress σ is applied along the axis of a rod of length l, the rod extends an amount Δl [Fig. 6-1(c)]; the tensile strain ϵ_z is the change in length divided by the initial length, $\Delta l/l_0$. Hooke's law describing the relation between elastic stress and elastic strain is

$$\sigma = E\epsilon_z = E\frac{\Delta l}{l_0} \tag{6-4}$$

where E is called the *modulus of elasticity* (*Young's modulus*).

The elongation in the axial direction is accompanied by a contraction in the two perpendicular transverse directions, and hence by a compressive strain $\epsilon_x = \epsilon_y$ in the transverse directions. The ratio of the induced transverse strains to the axial strain is known as *Poisson's ratio* and is denoted by the symbol v

$$v = -\frac{\epsilon_x}{\epsilon_z} = -\frac{\epsilon_y}{\epsilon_z} \tag{6-5}$$

The negative sign in Eq. (6.5) is present because the transverse strains ϵ_x and ϵ_y are negative. Typically, v varies between 0.25 and 0.35 for metals.

The modulus of elasticity is proportional to the shear and bulk moduli according to the relations

$$E = 2G(1 + v) \quad \text{and} \quad E = 3B(1 - 2v) \tag{6-6}$$

The elastic modulus and Poisson's ratio of some common materials are given in Table 6-1. As described in Chapter 2, these moduli provide measures of the cohesive forces between atoms.

The elastic modulus is only slightly influenced by small variations in internal structure such as small additions of alloying elements or the presence of defects such as vacancies, dislocations, or grain boundaries. For alloys that show complete solid solubility, the modulus usually varies linearly with composition following the rule of mixtures. Alloys that form intermediate phases have a much more

TABLE 6-1

Typical Room Temperature Values of the Elastic Modulus and Poisson's Ratio
for Several Materials

Material	Elastic Modulus, E (10^6 psi)	Poisson's Ratio, ν
Diamond	114	0.20
W_2C	90	—
W	56.4	0.28
Al_2O_3	50	—
TiC	45.7	—
Be	42	0.27
MgO	40	—
Ni	30	0.30
Fe	29	0.28
Si	29	—
Ge	23	—
LiF	19	—
Cu	17	0.34
SiO_2 (glass)	10	0.23
Al	10	0.34
Mg	6.3	—
NaCl	4.7	—
Wood (typical)	1.5	—
Rubber (vulcanized)	0.5	0.4
Polystyrene	0.4	0.4
Nylon	0.4	0.4
Polyethylene	0.02	0.4
Natural rubber	10^{-3}–10^{-2}	0.49

complex composition dependence of the elastic modulus. For example, in the
Mg-Sn system, both elements have an elastic modulus equal to about 6×10^6
psi at room temperature, while the compound Mg_2Sn has an elastic modulus greater than 10^7 psi. The general rule has been mentioned previously: The stronger the
interatomic forces (curvature of the potential energy well), the higher the modulus.

In addition to the composition dependence of the elastic modulus, there is also
a crystallographic variation in E; that is, if we measure E along different crystallographic directions in a single crystal we get different values. As described in
Chapter 1, this directional variation in properties is known as *anisotropy*. As an
example of anisotropy in a single crystal, the elastic modulus of iron varies between
41×10^6 psi and 19×10^6 psi, depending on the direction of measurement. The
values of E listed in Table 6-1 are average values taken from polycrystalline materials with a random orientation of the individual grains.

Finally, the elastic modulus decreases with increasing temperature. Generally,
the decrease is approximately linear with temperature up to about half the melting temperature, and then the modulus decreases more rapidly with further increase
in temperature. The magnitudes of the moduli at one-half the melting temperature
and just below the melting temperature are, respectively, ≈ 0.8 and ≈ 0.4 of the
value near absolute zero.

Simple Stress States

As mentioned above, most materials behave in a perfectly elastic fashion for only small strains. Deflections beyond these small elastic strains cause plastic or permanent shape changes or fracture. As we shall see below, the amount of permanent deformation produced by a given applied load depends primarily on the magnitude of the shear stresses induced by the load, while the conditions for fracture often depend on the magnitude of the tensile stresses produced by the load. Consequently, it is necessary to understand the state of stress set up in all parts of a structure under a given condition of loading. For example, the application of a uniaxial load P produces shear stresses τ on particular geometric planes, in particular directions, and hence produces shear strains in the body. The magnitude of the resolved shear stresses depends on the angles that the directions and planes make with the tensile axis.

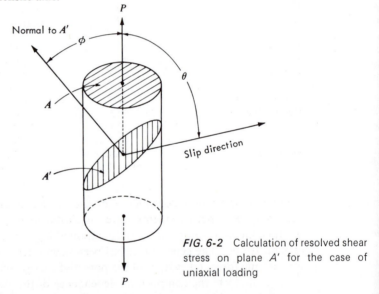

FIG. 6-2 Calculation of resolved shear stress on plane A' for the case of uniaxial loading

Consider the sectional plane A' (Fig. 6-2) whose normal makes an angle ϕ with the tensile axis. If A is the area of the circular cross section of the rod, then A' has an area $A/\cos\phi$. To compute the shear stress on A' acting along a direction that makes an angle θ with the tensile axis we note that the shear load resolved in this direction is $P\cos\theta$. Consequently,

$$\tau = \frac{\text{load}}{\text{supporting area}} = \frac{P\cos\theta}{A/\cos\phi} = \sigma\cos\phi\cos\theta \tag{6-7}$$

where $\sigma = P/A$ is the applied tensile stress. To compute the maximum value of τ, we note that for a fixed value of ϕ, the smallest possible value of θ is $(\pi/2 - \phi)$, so that

$$\tau = \sigma\cos\phi\cos\left(\frac{\pi}{2} - \phi\right) = \sigma\cos\phi\sin\phi \tag{6-8}$$

This function has a maximum value when $\phi = \pi/4$, so that

$$\tau_{max} = \sigma(0.707)^2 = \frac{\sigma}{2} \qquad (6\text{-}9)$$

Thus, the maximum resolved shear stress occurs on those planes that are 45° to the tensile axis in a direction that is co-planar with the tensile axis and the plane normal, and makes an angle of 45° with the tensile axis. Alternatively, we note that $\tau = 0$ across planes that are perpendicular to the tensile axis $\phi = 0$, and across planes that are parallel to the tensile axis $\phi = \pi/2$, since $(\cos \phi)(\cos \theta)$ is equal to zero when θ or $\phi = \pi/2$.

EXAMPLE 6-1   Consider the rectangular element of material subjected to a biaxial stress state shown in the accompanying diagram. Calculate the shear stress τ on the plane AA and find the orientation of the plane (value of θ) for which τ is a maximum.

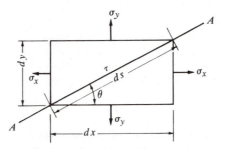

Solution: We can find τ by writing the equilibrium equation for the sum of the forces in the direction of τ (assuming the element has unit thickness):

$$\tau \, ds(1) + \sigma_x \, dy(1) \cos \theta - \sigma_y \, dx(1) \sin \theta = 0$$

Dividing by ds and noting that $dy/ds = \sin \theta$ and $dx/ds = \cos \theta$,

$$\tau = (\sigma_y - \sigma_x) \sin \theta \cos \theta = \left(\frac{\sigma_y - \sigma_x}{2}\right) \sin 2\theta$$

Here τ has its maximum value at $\theta = 45°$ and is zero for $\theta = 0$ and 90°.

In general, structural materials are subjected to stress states considerably more complex than uniaxial tension. Usually there are both normal and shear stresses applied to the material. Regardless of the complexity of the stress state, however, it can always be expressed equivalently by just three mutually perpendicular normal stresses. These normal stresses are referred to as the *principal stresses*. If we refer to the principal stresses as σ_1, σ_2, and σ_3, then the maximum shear stresses produced by these stresses are (see Example 6-1):

$$\frac{\sigma_1 - \sigma_2}{2}, \quad \frac{\sigma_1 - \sigma_3}{2}, \quad \frac{\sigma_2 - \sigma_3}{2} \qquad (6\text{-}10)$$

By convention $\sigma_1 \geq \sigma_2 \geq \sigma_3$, so that the maximum shear stress is

$$\tau_{max} = \frac{\sigma_1 - \sigma_3}{2} = \tau_2 \qquad (6\text{-}11)$$

For the case of uniaxial tension or compression, $\sigma_1 = \sigma$, and $\sigma_2 = \sigma_3 = 0$, so that $\tau_{max} = \sigma/2$, as given by Eq. (6-9). For hydrostatic compression (or tension) where $\sigma_1 = \sigma_2 = \sigma_3$, no shear stresses exist. This is important because permanent (plastic) deformation occurs only when shear stresses are present.

In fact, we can state that large-scale, permanent (plastic) deformation occurs when τ_{max} reaches a critical value, τ_Y, which is called the *yield strength*. We can thus establish a *plastic yield criterion* in which plastic flow will occur whenever

$$\tau_{max} \geq \tau_Y \qquad (6\text{-}12)$$

Consequently, it is possible to determine the maximum load that can be applied to a structure without causing permanent deformation if the relation between load and principal stress is known.

EXAMPLE 6-2

In a thin-walled pressure vessel subjected to an internal pressure p, a state of biaxial stress is produced and there is essentially no component of stress (σ_3) perpendicular to the walls of the vessel. If the wall thickness is t and the diameter is d, then the stresses $\sigma_1, \sigma_2,$ and σ_3 (Fig. 6.3) are given by

$$\sigma_1 = \frac{pd}{2t}, \quad \sigma_2 = \frac{pd}{4t}, \quad \sigma_3 = 0$$

What is the maximum internal pressure that can be applied to a steel vessel without causing plastic deformation if the vessel is 4 ft in diameter and its wall thickness is 0.10 in.? For this particular steel, $\tau_Y = 24,000$ psi.

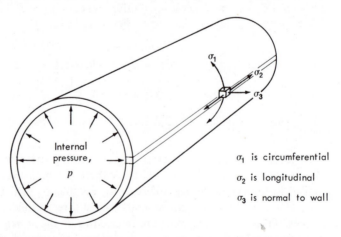

σ_1 is circumferential

σ_2 is longitudinal

σ_3 is normal to wall

FIG. 6-3   Stress distribution in a thin-walled pressure vessel due to internal pressure, p

Solution: Since $\sigma_3 = 0$, τ_{max} is given by

$$\tau_2 = \frac{\sigma_1 - \sigma_3}{2} = \frac{pd}{4t} = \tau_{max}$$

At yield

$$\tau_{max} = \tau_Y$$

so that

$$p_{max} = \frac{4t\tau_Y}{d} = \frac{4(0.1 \text{ in.})(24{,}000 \text{ psi})}{4(12 \text{ in.})} = 200 \text{ psi}$$

While the conditions for the onset of permanent deformation depend on the magnitude of the shear stress (Eq. 6-12), the conditions for fracture often depend on the magnitude of the tensile stress. Thus, a fracture criterion can be written where fracture will occur whenever a principal stress exceeds the fracture strength σ_F. Expressed mathematically, we have fracture whenever

$$\sigma_1 \geq \sigma_F \tag{6-13}$$

Consequently, the prediction of whether a material will yield or fracture is determined by whether Eq. (6-12) or Eq. (6-13) is satisfied. As the ratio of maximum tensile stress to maximum shear stress increases, the likelihood of fracture increases.

EXAMPLE 6-3

For the case of the pressure vessel shown in Fig. 6-3, what is the minimum value of fracture strength the material must possess so that the vessel will deform rather than fracture under internal pressure p? Take $\tau_Y = 24{,}000$ psi.

Solution:

If $\tau_Y = 24{,}000$ psi, then $\tau_{max} = 24{,}000$ psi at yield. From Eq. (6.11) $\sigma_{max} = 48{,}000$ psi. Consequently, σ_F must be at least 48,000 psi for yielding to occur instead of fracture.

Low values of the macroscopic fracture strength usually result from the presence of long, sharp cracks that have been accidentally introduced into the structure. These defects act as *stress concentrators* and raise the stress on a local level up to the point where atomic bonds are broken. To understand the meaning of stress concentration, let us consider a rectangular bar, of cross-sectional area A, that is loaded in uniaxial tension by a force P. The stress $\sigma = P/A$ is a force intensity and can be represented by lines of force shown in projection in Fig. 6-4(a). Now suppose that a crack is introduced into the body. Since load cannot be transmitted across a free surface, the lines of force must bend around the crack tip, as shown in projection in Fig. 6-4(b). The lines of force (stresses) are highly concentrated near the crack tip but become more evenly distributed at distances removed from the crack. The concentration of stress therefore decreases at increasing distances from the crack tip.

The elastic stress concentration factor is simply the ratio of the local stress σ_l near the crack to the gross applied stress $\sigma = P/A$. This factor is usually denoted by K, and hence

$$K = \frac{\sigma_l}{\sigma}$$

K depends on many factors, such as the length of the crack ($2a$), its tip radius of curvature (ρ), and the distance (r) from the crack tip at which K is to be evaluated. For example, right at the crack tip ($r = 0$)

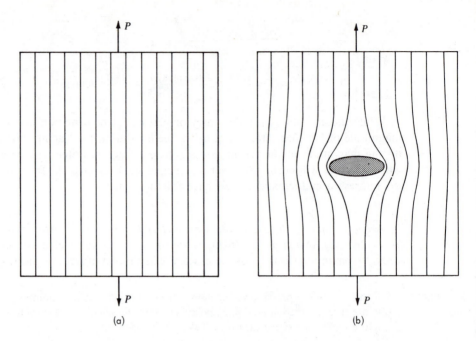

FIG. 6-4   Schematic diagram of stress distribution (a) in a homogeneous body and (b) in a body containing a void or crack.

$$K = 2\sqrt{\frac{a}{\rho}} \qquad (6\text{-}14)$$

so that the local stress increases as the crack becomes longer and sharper. K decreases with increasing distance from the crack tip.

6-3 STRESS-STRAIN RELATIONS UNDER UNIAXIAL LOADING

Evaluation of many of the important mechanical properties of solids is made by subjecting specimens to uniaxial, uniform loading. Metals and polymers are generally evaluated by pulling them in tension. Very brittle materials, such as concrete or ceramics, are usually tested under uniaxial compressive loading to prevent premature fracture and to simulate best the loading they will undergo in service. Before we consider the details of the stress-strain curve for any one material, it is instructive to examine briefly the wide variety of stress-strain curves observed for different materials.

We can group most observed stress-strain curves under one of three general headings: (1) materials that exhibit both elastic and plastic deformation, (2) materials that exhibit essentially no plastic deformation and fracture after only small elastic strains, and (3) materials that exhibit very large elastic strains. All metals and ceramic materials exhibit elastic behavior and, if they are not brittle and do not contain sharp notches, also undergo some plastic deformation before fracture. However, the stress-strain curves, and hence the tensile properties, of different materials vary widely. For example, Fig. 6-5 illustrates the room tem-

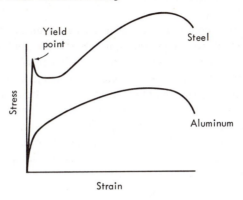

FIG. 6-5   Schematic illustration of the stress-strain characteristics of mild steel and aluminum

perature stress-strain curves of mild steel (Fe $+ 0.2\%$ C) and pure aluminum. Mild steel exhibits a sharp *yield point* on the stress-strain curve making the transition from elastic to plastic deformation abrupt, whereas the curve for aluminum exhibits a gradual transition. Practically all BCC transition metals behave in a fashion similar to the curve for mild steel, while nearly all other crystalline materials exhibiting some plastic deformation (FCC and HCP metals, ionics, etc.) behave in a fashion qualitatively similar to the aluminum curve.

For *brittle materials*, such as many ceramics, concrete, and some metals (e.g., cast iron, high-strength steel), fracture can occur in the elastic region and no macroscopic plasticity is observed (Fig. 6-6). The maximum elastic strain observed for these materials is generally less than 0.5%.

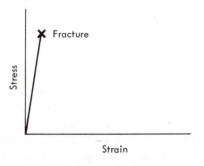
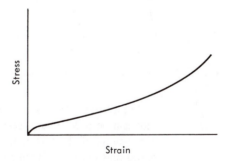

FIG. 6-6 Schematic illustration of the stress-strain relation for a brittle material (e.g., cast iron, ceramics, concrete)

FIG. 6-7 Schematic illustration of the stress-strain relation for an elastomer (long-chain polymer such as rubber)

Polymeric materials such as rubber often exhibit elastic strains greater than 100%. These materials also have a modulus E that is both strain- and time-dependent. As shown in Fig. 6-7, the stress-strain curve of rubber continually slopes upward as the strain increases. This is *not* plastic behavior, since the material will return to its original shape *if the stress is removed*. Instead, it results from the fact that the modulus increases with strain; the reasons for this are discussed in detail in Chapter 10.

The most common type of test used to measure mechanical properties is the *tensile test*, in which the relation between the applied stress and induced strain is determined on a test specimen that is being elongated at a particular rate at a fixed temperature. The test is conducted by placing a test specimen into a machine and elongating the specimen at a constant rate. The force required to produce a given elongation is recorded, and forces and elongations are converted to stress and strain by simple formulas given below. The test specimen contains a reduced cross section, or gauge length, over which the deformation takes place, since the stresses are higher there than in the wider parts of the specimen that are coupled to the machine (Fig. 6-8).

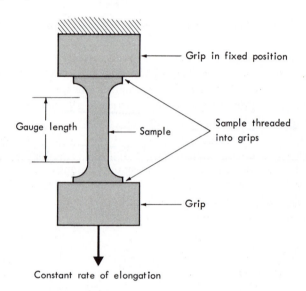

FIG. 6-8   Schematic illustration of specimen arrangement in tensile test

In plotting a stress-strain curve, there are two different systems of stress and strain that can be used. One system (*engineering* stress and strain) is based on the initial dimensions of the test sample, while the second system (*true* stress and strain) is based on the instantaneous sample dimensions. The engineering system is used for convenience when working at small strains but does not give a valid description of the actual behavior of the material at strains above $\approx 10\%$ where it becomes necessary to use true stress and true strain.

To describe both these systems we define:

l_0 = initial gauge length of undeformed specimen
A_0 = initial uniform cross-sectional area in gauge length
l = instantaneous gauge length after some deformation has occurred
A = instantaneous uniform cross-sectional area after some deformation has occurred.

The *engineering stress* is defined as $S = P/A_0$ and is based on the original cross-sectional area, whereas the *true tensile stress* $\sigma = P/A$ takes into account the fact that the load-bearing area decreases with increasing strain. In the elastic region (strains generally less than 1%) engineering stress and true stress are essentially the same. The *engineering strain e* is defined as the change in length divided by the initial length:

$$e = \frac{l - l_0}{l_0} = \frac{\Delta l}{l_0} \qquad (6\text{-}15)$$

whereas the *true strain* ϵ is the sum of all the instantaneous length changes *dl* divided by the instantaneous length *l* or,

$$\epsilon = \int_{l_0}^{l} \frac{dl}{l} = \ln\left(\frac{l}{l_0}\right) = \ln\left(\frac{l_0 + \Delta l}{l_0}\right) = \ln(1 + e) \qquad (6\text{-}16)$$

Since $\ln(1 + x) \approx x$ for $x < 0.10$, e and ϵ are essentially the same for strains less than 10%. Also, since plastic deformation occurs by a process of shear (Chapter 7), there is essentially no *volume change* in the specimen during deformation. After the specimen has elongated to a length *l*

$$Al = A_0 l_0$$

so that the true strain can be written in terms of cross-sectional area

$$\epsilon = \ln\left(\frac{A_0}{A}\right) \qquad (6\text{-}17)$$

EXAMPLE 6-4

The initial cross-sectional area of a tensile specimen is 0.20 in.² What is the cross-sectional area when the engineering strain is 0.3?

Solution:

$$\epsilon = \ln\left(\frac{A_0}{A}\right) = \ln(1 + e)$$

$$\frac{A_0}{A} = 1 + e$$

$$A = \frac{A_0}{1 + e} = \frac{0.2 \text{ in.}^2}{1.3} = 0.154 \text{ in.}^2$$

One of the reasons for preferring true strain over engineering strain when considering large deformations is that true strains are additive while engineering strains are not. Consider the following simple example. On two successive days the *same* aluminum tensile sample is given to a laboratory class with the instruction that the sample is to be strained 20%. After the second day the question is raised as to how much strain the sample has undergone. The calculations outlined in Fig. 6-9 show that on the basis of engineering strain the sample has undergone a total strain of 44% (unequal to the sum of the 20% increments), whereas on the basis of true strain the total strain *is* equal to the sum of the increments. The reason for the problem with engineering strain is that each strain increment is computed

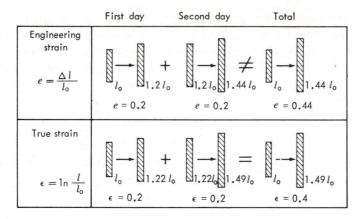

FIG. 6-9   Illustration of the additive nature of true strain.

on the basis of a different gauge length, whereas the overall strain is computed using the initial sample dimensions. This problem is alleviated when we consider true strains.

Also, when comparing tension and compression tests, the stress-strain curves will be identical only if we consider stresses and strains in terms of true stress and true strain. Since plastic deformation takes place by shear in both instances, it is expected that there should be no basic differences between the stress-strain characteristics, whether measured in tension or compression; thus, it is logical to use a system for stress and strain that gives indentical results.

EXAMPLE 6-5

Imagine that we deform two specimens of original length l_0. One sample is deformed in a tension test to double its length, $2l_0$, while the other is deformed in compression to one-half its original length, $\frac{1}{2}l_0$. Show that these are equivalent strains in terms of true strain but are different strains in terms of engineering strain.

Solution:

For true strain:

$$\text{Tension, } \epsilon = \ln \frac{2l_0}{l_0} = \ln 2 = 0.69$$

$$\text{Compression, } \epsilon = \ln \frac{l_0/2}{l_0} = \ln \frac{1}{2} = -0.69$$

For engineering strain:

$$\text{Tension, } e = \frac{2l_0 - l_0}{l_0} = 1.0$$

$$\text{Compression, } e = \frac{l_0/2 - l_0}{l_0} = -0.5$$

Let us now examine in some detail the stress-strain curve of a material that exhibits some plastic deformation. For our example we choose an aluminum alloy that is being strained at a temperature $T < 0.5T_m$ (where T_m is the absolute melting

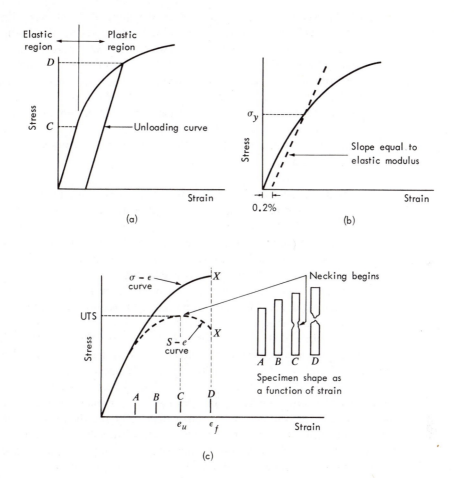

FIG. 6-10   Schematic illustration of the general characteristics of a tensile stress-strain curve for an aluminum alloy

temperature) and at a moderate strain rate ($\approx 10^{-2}$ per min). Figure 6-10 shows a typical stress-strain curve for such an alloy. Two regions are of interest.

1. *Elastic region*: At low stresses most materials exhibit elastic behavior in accordance with Hooke's law [Eq. (6-4)]. In this range the strains are reversible with stress; that is, if the specimen is unloaded when $\sigma = C$ [Fig. 6-10(a)], the material returns to its original shape. Also, it should be noted that the events leading to elastic behavior in crystalline materials are both *rate-independent* and *time-independent*. A strain ϵ will be produced by a stress σ, irrespective of the loading rate, and ϵ will not change if σ is then maintained constant for a long period of time under static loading. In polymers, the situation is more complicated and the modulus depends on the rate of loading. This behavior will be discussed in some detail in Chapter 10.

2. *Plastic region*: At higher stress levels the slope of the stress-strain curve becomes much lower and Hooke's law no longer describes the relation between stress and strain. This region is known as the *plastic region* and is characterized by the fact that the deformations become permanent, or *plastic* (although the sample is still elastically deformed). When the stress is removed from point *D*, for example, the material unloads elastically, along a line that is parallel to the elastic slope, but a permanent strain remains in the material. At temperatures below $0.5T_m$ plastic deformation is ordinarily time-independent; if a stress σ is maintained for a period of time under static loading, ϵ will not change.

The applied tensile stress required to induce plastic behavior is known as the *elastic limit* or *yield stress*, and is designated by the symbol $\sigma_Y = 2\tau_Y$. This stress is very important in structural design, because it marks the limit of reversible deformation and, more important, because it marks the limit at which small deformations are produced by small increases in stress. In *soft* materials, such as aluminum, the boundary between the elastic and plastic regions is not well defined. In fact, some very slight permanent deformation occurs even for very small stresses. In this case, the yield stress is taken to be the stress required to produce a specified, small value of permanent strain, usually 0.2% [Fig. 6-10(b)]. For *hard* materials, or those materials that show a sharp yield point (see Fig. 6-5), the yield stress is usually taken as the minimum, or plateau, in stress that occurs just after the yield point. This stress is referred to as the *lower yield stress*.

The increase in stress required to produce an increase in strain in the plastic region is called *strain hardening*, or *work hardening*, a phenomenon familiar to anyone who has ever bent a wire hanger or paper clip and then tried to bend it back to its original shape. The more a material is plastically deformed, the more difficult it becomes to plastically deform the material. The slope of the stress-strain curve in the strain hardening region, $d\sigma/d\epsilon$, is usually not constant, but varies with strain. In general, $d\sigma/d\epsilon$ varies from $E/400$ to $E/100$, and there are no generalized laws of plastic deformation (corresponding to Hooke's law) that apply for all materials.

As plastic deformation continues, the cross-sectional area decreases [Fig. 6-10(c)], but the load-carrying capacity of the specimen increases due to strain hardening. Eventually, an elongation is reached where the incremental increase in load-carrying capacity ($A\,d\sigma$) due to strain hardening becomes less than the incremental decrease in load-carrying capacity ($\sigma\,dA$) due to decreasing load-bearing area, and the specimen cannot withstand further increases in load. In short, we have a plastic instability. A small inhomogeneous strain in any region of the gage length will be accentuated and further increases in elongation will be confined to this localized region, causing a *neck* in the tensile sample [Fig. 6-10(c)]. Because deformation is confined to the neck region, the cross-sectional area rapidly decreases in this region, and the applied load decreases even though the material in the neck region continues to strain harden. That is, even though the true stress in the neck region increases, the load, and consequently the engineering stress, begins to decrease. The maximum load that the specimen withstands defines a common engineering property, the *ultimate tensile strength*, UTS $= P_{max}/A_0$. Maximum uniform engineering strain at this point is labeled e_u. Further elongation in the neck region eventually leads to fracture at point D. The engineering strain at fracture is e_f.

It must be emphasized that the true stress-true strain curve for the material is different [Fig. 6-10(c)] from the engineering stress-engineering strain curve: (1) because the true stress $\sigma = S(A_0/A)$ is always greater than S and always increases with increasing strain for tensile tests and (2) because the true strain $\epsilon = \ln(1 + e)$ is less than e for uniform elongation. For compression tests the opposite is true; i.e., $\sigma < S$ and $\epsilon > e$. It is important to note that even though $\epsilon < e$ for tensile tests, where elongations are uniform, the true fracture strain $[\epsilon_f = \ln(A_0/A_f)$, where A_f is the cross-sectional area at the point of fracture] is greater than e_f, since e_f is computed on the basis of uniform elongation. Similarly, the true stress at fracture, or fracture stress σ_F, is greater than the UTS. Since the true stress-true strain curve gives a more accurate picture of a material's *mechanical properties* than the engineering curves, *all stress-strain curves used in this text are true stress-true strain curves.*

In addition to yield strength, ultimate strength, and fracture stress, two other mechanical properties of interest, particularly in brittle materials, are *ductility* and *toughness.*

Ductility is a measure of the strain required to cause fracture. One way to express ductility is to give the percent reduction in cross-sectional area (RA) at fracture:

$$RA = \frac{A_0 - A_f}{A_0} \times 100 \qquad (6\text{-}18)$$

where again A_0 is the initial cross-sectional area and A_f is the cross-sectional area at the point of fracture. *Ductile materials* are those in which RA is large (greater than 50%); *semi-brittle materials* are those in which RA is small (less than 10%); *notch-brittle materials* are those in which RA is moderate in a simple tension test (say 30%) but is very small or zero when a notch or crack is introduced into the tensile specimen prior to testing.

The toughness of a material is a measure of the work per unit volume required to cause fracture. For a simple tensile test this work can be calculated in the following fashion:

$$\text{Work} = \int (\text{force, } P)(\text{increment of extension, } dl)$$

$$\text{Work per unit volume} = \int \frac{P\,dl}{Al} = \int_0^{\epsilon_f} \sigma\,d\epsilon \qquad (6\text{-}19)$$

where A and l are the sample dimensions. The toughness is therefore simply the area under the true stress-true strain curve (Fig. 6-11). Generally, tough materials have moderately high strengths and ductilities (for maximum area under the stress-strain curve). Materials with high strength and low ductility* have low toughness (Fig.

---

*Ductility generally decreases as the strength increases.

6-11). Thus, the materials with the highest strength generally do not have the highest fracture resistance (maximum energy or work required to cause fracture). This can be simply demonstrated for the case of diamond, which, although it has an extremely high fracture stress ($\sigma_f \approx 10^6$ psi), can be easily shattered by the blow of a hammer. We shall discuss the concept of toughness in more detail in Section 6-6 when we consider the fracture characteristics of materials.

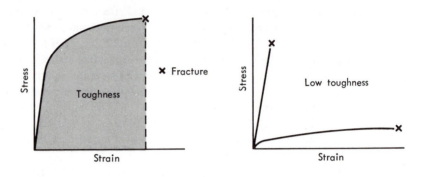

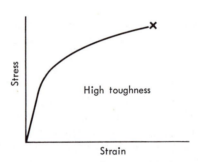

FIG. 6-11   Illustration of the determination of toughness from the stress-strain curve

EXAMPLE 6-6   Calculate the elastic strain energy in a sample pulled in tension to a stress of 40,000 psi. Take $E = 10^7$ psi.

Solution: From Eq. (6.19) and Hooke's law, we can write

$$\text{Work per unit volume} = \int_0^\epsilon \sigma\, d\epsilon = \int_0^\epsilon E\epsilon\, d\epsilon = \frac{E\epsilon^2}{2}$$

or, as $\epsilon = \sigma/E$,

$$\text{Work per unit volume} = \frac{\sigma^2}{2E} = \frac{(4 \times 10^4)^2}{2 \times 10^7} = 80 \text{ in.-lb/in.}^3 = 80 \text{ psi}$$

EXAMPLE 6-7   The stress-strain curves for many materials are observed to obey the following relation (in the plastic region):

$$\sigma = k\epsilon^n$$

where k is a constant and $n < 1$, usually varying from 0.1 to 0.5. If the stress-strain curve for a particular material is given by

$$\sigma = 10^5 \epsilon^{0.5} \text{ psi}$$

what is the work per unit volume absorbed in fracturing this material if the fracture strain is $\epsilon_f = 0.3$?

Solution:

$$\text{Work per unit volume} \int_0^{\epsilon_f} \sigma \, d\epsilon = \int_0^{\epsilon_f} 10^5 \epsilon^{1/2} \, d\epsilon$$

$$= 66.7 \times 10^3 \, \epsilon_f^{3/2} = 66.7 \times 10^3 (0.3)^{3/2} = 11,000 \text{ in-lb/in.}^3$$

$$= 11,000 \text{ psi}$$

The above discussion has dealt only with the general characteristics of the stress-strain curve. Some of the unique stress-strain characteristics of materials, such as the yield drop for BCC metals, and the time and strain dependence of the elastic modulus of polymers, will be discussed in detail in the following chapters. We shall now turn our attention to the influence of temperature on the strength of materials.

6-4 STRENGTH OF MATERIALS AT LOW AND MODERATE TEMPERATURES

Plastic deformation at low and moderate temperatures (below about $\frac{1}{2}$ the absolute melting temperature T_m) is usually characterized by the general rule that *the amount of plastic deformation depends only on the applied stress and not the time duration of the load.* This, as we shall see in the next section, contrasts with the situation at temperatures greater than about $\frac{1}{2} T_m$ where a constant stress will produce time-dependent strain.

Table 6-2 summarizes the room temperature tensile properties of some important engineering materials. It is apparent that a wide variation in properties exists between polymers and metals, between different metallic alloy systems (e.g., aluminum alloys as compared to copper alloys), and within a given base alloy system (e.g., low-strength steel as compared with high-strength steel). There are three basic reasons for these differences in properties:

1. The strength of a material is generally proportional to the elastic modulus (discussed more fully in Chapter 7) and, as shown in Table 6-1, there is a large variation in elastic modulus for different base materials.
2. The strength of a material depends very critically on the microstructure. Different thermal-mechanical histories can produce different microstructures for a material of given chemical composition, as we shall see in Chapter 9.
3. The strength of a material depends on the *ratio* of testing (or operating) temperature T to absolute melting temperature T_m. The *room temperature* properties in Table 6-2 were actually obtained at different T/T_m ratios for the different materials listed.

TABLE 6-2
Mechanical Properties of Some Engineering Materials

The data are for yield strength (σ_Y), ultimate tensile strength (UTS), yield strength divided by density (σ_Y/ρ), uniform engineering strain to fracture (e_u), and elastic modulus divided by density (E/ρ).

Material*	σ_Y (psi)	UTS (psi)	σ_Y/ρ (10^3 in.)	e_u (%)	E/ρ 10^6 in.
Aluminum Alloys					
1100 (99% Al)	5,000	13,000	51	40	102
2024 (4.5% Cu, 1.5% Mg, 0.6% Mn)	50,000	70,000	510	16	102
6061 (1% Mg, 0.6% Si, 0.25% Cu, 0.25% Cr)	21,000	35,000	214	22	102
7075 (5.5% Zn, 2.5% Mg, 1.5% Cu, 0.3% Cr)	72,000	82,000	730	11	102
Copper Alloys					
OFHC (99.95% Cu)	10,000	32,000	31	45	52.8
Bronze (10% Sn)	10,000	37,000	31	45	52.8
Brass (30% Zn)	11,000	44,000	34	66	52.8
Beryllium copper (1.9% Be, 0.2% Ni)	130,000	175,000	402	5	52.8
Iron Alloys (Steels)					
Armco iron (99+% Fe)	24,000	44,000	83	45	97
1020 (0.2% C, 0.5% Mn)	43,000	65,000	148	36	97
4340 (0.4% C, 0.7% Mn, 1.7% Ni, 0.8% Cr, 0.25% Mo)	200,000	220,000	694	12	97
Marage (18% Ni, 48% Mo, 8.5% Co, 0.7% Ti)	280,000	290,000	1000	10	97
304 Stainless (18% Cr, 8% Ni)	28,000	75,000	95	50	97
Titanium Alloys					
Titanium (99.9% Ti)	20,000	34,000	122	54	103
90% Ti, 6% Al, 4% V	163,500	175,300	1000	6.4	103
Nickel Alloys					
Nickel A (99.4% Ni)	20,000	65,000	62	40	93
Monel (67% Ni, 30% Cu)	35,000	75,000	109	40	93
Inconel (76% Ni, 16% Cr, 8% Fe)	40,000	90,000	124	40	93
Hastelloy X (9% Mo, 22% Cr, 18% Fe)	55,000	112,000	171	20	93
Other Materials					
Beryllium	—	90,000	—	10	672
Nylon	—	120,000	—	—	7.5
Polystyrene	—	8,500	—	—	13.3
Polymethylmethacrylate (plexiglass)	—	7,000	—	5	9.2

*Most of the materials listed have a microstructure, and hence mechanical properties that are strongly dependent on the exact thermal-mechanical history of the material. The properties in the table are typical of materials commercially available.

The effect of variations in modulus and melting temperature may be partially normalized by comparing values of σ_Y/E at various T/T_m, as shown in Fig. 6-12

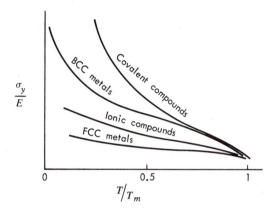

FIG. 6-12   Variation of modulus-compensated yield strength with homologous temperature

for a number of elements and compounds.* It is apparent that on this basis there are many similarities between different materials, although quite a range in properties can be found between materials with different crystal structure.

Perhaps one of the most striking features of the data shown in Fig. 6-12 is the difference in the temperature dependence of the yield stress for the FCC metals and most other materials. Although we shall not attempt to explain this temperature dependence (or lack of it) at this point, the presence of a strongly temperature-dependent yield stress gives rise to an effect known as *strain rate sensitivity*. That is, materials with a temperature-dependent yield stress also show a strain-rate-dependent yield stress; the higher the imposed strain rate, the higher the yield stress (Fig. 6-13). In fact, there is a definite equivalence between temperature and strain

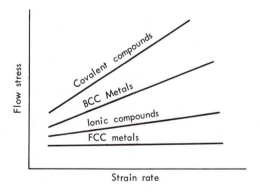

FIG. 6-13   Illustration of strain-rate sensitivity, or the variation of flow stress with strain rate at a given temperature

*We have not included data for alloys or polyphase materials since the microstructure of these materials is often temperature-dependent and would complicate the discussion considerably.

rate in determining the yield stress. This can be understood by considering deformation as a stress-aided, thermally activated process. At low temperatures we need a high stress to achieve a given strain rate, whereas at high temperatures, where more thermal energy is available, the stress need not be so high. Equivalently, at a constant temperature we require higher stresses the faster the strain rate (less time for thermal activation). Thus, we can write

High temperature + slow strain rate = low yield stress

and

Low temperature + fast strain rate = high yield stress

Materials that have a temperature-independent flow stress are also strain-rate insensitive; i.e., the flow stress is independent of the imposed strain rate.

At low temperatures strength is not the only requirement for selecting a material. For many applications, such as aircraft, submarines, or spacecraft, it is necessary to minimize the weight of the structure in order to maximize the weight of the payload. In these cases, it is necessary to choose materials on the basis of their strength-to-weight ratio, or *yield-strength-to-density ratio*, σ_Y/ρ. According to Table 6-2 it would be structurally more efficient (although perhaps cost more money) to construct a tension member from high-strength titanium rather than 4340 steel, even though the yield strength of 4340 steel is higher. In some cases, where cost is a criterion secondary to performance, great effort is spent in the development of materials with high strength-to-weight ratios.

A high strength-to-density ratio is also desirable for compression members such as struts or columns. However, under compressive loading long members often fail by *buckling* at loads that are less than those required to produce yielding. The buckling load is proportional to the *elastic* modulus E, so that high structural efficiency also requires materials of high modulus-to-density ratio (E/ρ). Table 6-2 lists some values of E/ρ for various materials. The high value of E/ρ for beryllium makes it extremely attractive, but at present it is brittle in bulk form and difficult to use reliably. One way of circumventing this problem is to combine the stiff, brittle material in the form of fibers with a softer, tough, ductile material to form a *composite*, such as fiberglass (glass fibers in a polymer resin matrix). We shall discuss these materials in Chapter 9.

In addition to the tensile test, another type of mechanical test used to evaluate the strength of materials at low temperatures is the *hardness test*. The *indentation hardness*, which is a measure of the material's resistance to penetration by a hardened steel ball or diamond indentor, can be used to evaluate the strength. A known load is applied to the indentor, and the diameter of the impression (Brinell test) or depth of the impression (Rockwell test) is recorded. The smaller the penetration, the harder the material. The hardness depends on both the yield strength of the material and on its strain hardening capacity, since relatively large strains (8–10%) are produced during indentation. In general, it is not possible to relate quantitatively various measures of hardness obtained by different tests on the same material, and hardness numbers *cannot* be directly employed in engineering design. However, for a given class of materials (e.g., low-carbon steel) the hardness num-

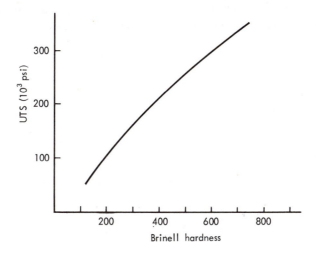

FIG. 6-14   Correlation of hardness with ultimate tensile strength (UTS) for steel

bers can be correlated with the tensile strength (Fig. 6-14). This is advantageous since hardness tests are easy to perform and require only very small amounts of material (pieces $\frac{1}{4}$ in. in diameter are generally sufficient). Therefore, they can be used for quality control purposes during production. Furthermore, the effect of heat treatment on mechanical properties can be evaluated very simply by annealing a small piece of material and measuring the hardness as a function of annealing time or temperature; it is not necessary to anneal large numbers of tensile specimens.

6-5 STRENGTH OF MATERIALS AT HIGH TEMPERATURES

At high temperatures ($T > 0.5T_m$) metals, ceramics, and polymers exhibit *time-dependent plasticity*. If a constant stress σ is applied and maintained constant despite any changes in cross-sectional area, the material will deform over a period of time irrespective of whether σ is greater than *or less* than the yield strength, as measured in a conventional tensile test at that temperature. This deformation is called *creep* and in many cases it is the limiting factor in the selection of materials for use at high temperatures. For example, creep deformations of turbine blades in a jet engine could cause the rotating blades to strike the walls of the turbine, causing the blades to bend or even fracture. Small creep deformations of nuclear fuel element cladding materials could lead to rupture of the cladding and the build-up of radioactivity in the cooling system that surrounds the fuel elements. Large-scale creep of high-pressure steam lines can cause these lines to crack and lead to plant shutdown. Consequently, most high-temperature designs are governed by the creep properties of the structural materials. The design criterion is usually based on a certain allowable amount of deformation over the desired lifetime of the structure (minutes for rocket nozzles, thousands of hours for jet engine turbine blades, and hundreds of thousands of hours for high-pressure steam lines).

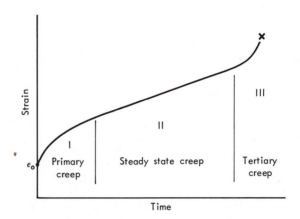

FIG. 6-15   Schematic illustration of creep curve showing time-dependent plastic strain

Figure 6-15 shows a schematic creep curve for a crystalline material and indicates that there are three stages to the creep process that follow the instantaneous strain ϵ_0 that is produced when the stress is first applied. During the first stage, the strain rate (the slope of the strain-time curve) decreases; it remains constant during the second or *steady state* stage and then increases rapidly until rupture occurs in the third stage. Most of the lifetime is spent in the steady state range, so that the creep rate in this range plays a dominant role in determining the lifetime of the structure.

Creep is a *stress-dependent*, *thermally activated process*, and hence is accelerated by increases in temperature or applied stress. Figure 6-16 shows the effect of stress and temperature on the steady state creep rate. Under creep conditions, rupture

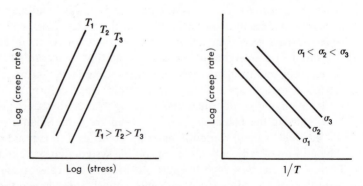

FIG. 6-16   Stress and temperature dependence of the steady state creep rate

occurs when the creep strain builds up to the ductility ϵ_f. Since the creep rate increases with stress and/or temperature, the rupture lifetime decreases as the stress or temperature is raised (Fig. 6-17).

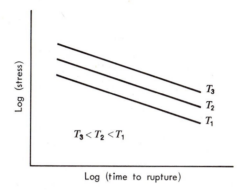

FIG. 6-17   Variation of time to rupture with stress and temperature

Amorphous, polymeric materials also exhibit a time-dependent form of deformation at high temperatures. However, in some instances, the time-dependent strain is recoverable after the load is removed. Figure 6-18 illustrates the strain-

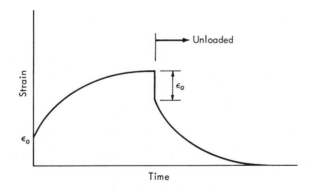

FIG. 6-18   Viscoelasticity or time-dependent elastic strain as exhibited by many polymeric materials

time curve for a polymer exhibiting time-dependent elastic strain. This type of retarded elastic response is known as *viscoelasticity*. In addition to viscoelastic responses, polymers also exhibit time-dependent plastic deformation at high temperatures (if the stresses are high enough), and, as with crystalline materials, the deformation rates increase with increases in either stress or temperature.

6-6 FRACTURE OF MATERIALS

The *notch toughness* of a material is another important mechanical property. Virtually all structural components contain notches or cracks of some sort, either those that have been introduced in design (e.g., a sharp corner or cutout), those introduced accidentally (e.g., in defective welds), or those that have formed by localized corrosion. These defects act as stress concentrators, which can cause brittle materials to break at applied stresses that are below the nominal yield strength. Figure 6-19 shows the effect of crack length on tensile fracture strength

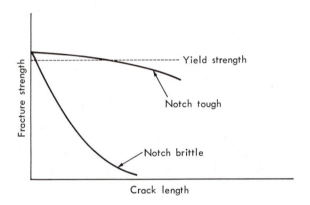

FIG. 6-19 Influence of crack or flaw size on fracture strength

in a tough material and in a notch-brittle material. In the former case, the presence of the crack has little effect, apart from reducing the load bearing area, and the load-carrying capacity is not reduced substantially. In the brittle material, the fracture strength drops rapidly as the crack size increases.

Previously we have defined the toughness as being the area under the true stress-true strain curve. This definition has little practical value for predicting the fracture behavior of materials containing cracks, since the stresses and strains in the immediate vicinity of the crack tip are not easy to calculate or measure. Basically, there are two methods that can be used to measure *toughness* for engineering purposes. First, it is possible to determine the toughness by measuring the fracture strength of a series of specimens containing cracks of different lengths and by constructing a curve such as that shown in Fig. 6-19, we can then relate the fracture stress to the crack length. In general, we find $\sigma_f \propto 1/\sqrt{a}$, where the proportionality constant is related to the elastic modulus and a parameter known as the *fracture toughness* (see Chapter 8 for a complete derivation). The fracture toughness is a measure of the work expended at the crack tip during crack propagation and generally decreases as the temperature decreases and as the yield strength increases. Thus, high-strength materials are the most sensitive to the presence of cracks and, correspondingly, have a low fracture toughness. Most ceramic materials and polymers are crack-sensitive and very brittle in the low and moderate temperature range ($T < 0.5T_m$).

A second approach to the measurement of toughness is based upon measurement of the energy required to break a standard size specimen that contains a notch. In the *Charpy impact test* a notched specimen is struck with a hammer (Fig. 6-20),

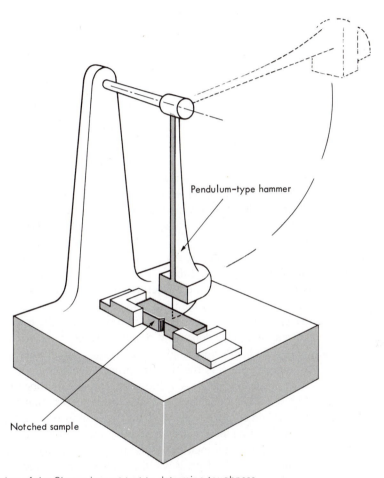

Pendulum-type hammer

Notched sample

FIG. 6-20   Illustration of the Charpy impact test to determine toughness

and the total energy absorbed in breaking the sample is recorded. Low values of impact energy are indicative of low values of fracture toughness and, hence, high notch sensitivity. Figure 6-21 indicates the effect of temperature on the impact energy of several classes of materials. Most low-strength FCC metals are tough at all temperatures. High yield strength materials ($\sigma_Y > E/150$) are always brittle. Ceramics, polymers, and BCC metals of low and moderate yield strength ($\sigma_Y < E/300$) undergo a *brittle-to-tough* transition as the temperature is raised. At low temperatures they are brittle, but as the temperature is increased, the toughness

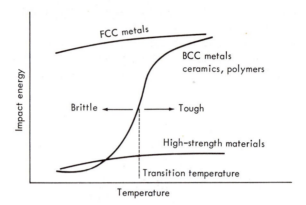

FIG. 6-21   Illustration of variation of impact energy with temperature for different materials

increases also. The *transition temperature*, where the impact energy is one-half the difference between maximum and minimum values, is very important in fracture-safe design with these materials. If the lowest temperature that the material will exhibit in service is above the transition temperature, then brittle fracture will not be a problem. Below the transition temperature, the toughness is so low that the fracture strength can be less than the yield strength when cracks are present, and service failure may occur. As mentioned in Chapter 1, cracks were observed in 25% of the World War II Liberty ships, and some of these ships actually broke in half (Fig. 1-6). Most of these fractures occurred on very cold nights when the temperature dropped below the transition temperature of the steel. As we shall see in Chapter 8, where we discuss fracture in more detail, the fracture toughness, like the yield strength, is very dependent on microstructure as well as on temperature.

Failure in both brittle and tough materials can also take place by a phenomenon known as *fatigue*, which is the fracture process that occurs in structures that are subjected to *alternating load*. A rotating drive shaft or torsion bar that supports a load (Fig. 6-22) is an example of a structural member that is subjected to alternating tension and compression. Many of the failures that occur in service are due to fatigue. Alternating loads cause small microcracks to form and grow very slowly (e.g., one-millionth of an inch per load cycle). Eventually, the crack will grow to the point where the applied stress is sufficient to cause the crack to propagate rapidly, and rapid fracture can occur (brittle material) or the remaining cross-sectional area cannot support the load and fails by simple overloading (tough material).

A typical fatigue fracture surface is shown in Fig. 6-23. The clamshell markings indicate that the fatigue crack grows progressively from the time it is initiated until fast fracture occurs. The markings on the fracture surface probably indicate small changes in the crack plane as the load distribution changes. Two separate fatigue cracks starting at points A propagated towards each other and met in the center of the shaft. The region marked B represents the last part of the shaft to fail. Failure in region B occurred because that particular region became overloaded.

The most important factor in determining the fatigue lifetime (number of cycles to failure, N_f) is the magnitude of the applied stress amplitude. Figure 6-24 shows

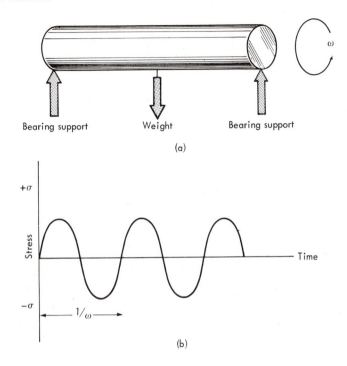

Bearing support   Weight   Bearing support

(a)

(b)

FIG. 6-22   (a) Example of structure subject to fatigue; (b) typical stress-time pattern in element of material

FIG. 6-23   Typical appearance of fatigue fracture surface. Photo courtesy A. J. West, Stanford University, Used with permission.

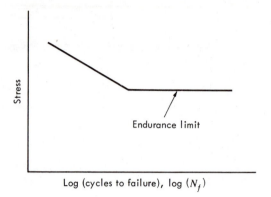

FIG. 6-24   Typical variation of fatigue life with stress amplitude

that N_f increases as the stress amplitude decreases. Below a certain stress level, i.e., about σ_Y in initially uncracked materials, the curve becomes horizontal and fatigue does not occur. This lower limiting stress is known as the *endurance limit*. Both the fatigue life at a given stress and the endurance limit are reduced by the presence of cracks or other stress concentrators (Fig. 6-25).

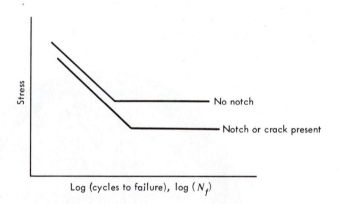

FIG. 6-25   Influence of crack or notch on fatigue lifetime

PROBLEMS

6-1   If yielding occurs at an axial stress $\sigma = \sigma_Y = 60,000$ psi in a uniaxial tensile test, what axial stress is required to cause yielding when a compressive stress of 10,000 psi is simultaneously applied in one perpendicular transverse direction and a tensile stress of 15,000 psi is being applied simultaneously in the other transverse direction?

6-2   Show that the resolved tensile stress across a plane that makes an angle ϕ with the tensile axis is $\sigma \cos^2 \phi$, where σ is the axial stress.

6-3   Suppose that yielding occurs in a single crystal of copper when a critical resolved shear stress $\tau_Y \doteq 10,000$ psi is produced on {110} type slip planes along a $\langle 111 \rangle$ type direction. If the tensile axis coincides with a $\langle 100 \rangle$ direction, what minimum axial stress must be applied to cause yielding on a {110} plane in a $\langle 111 \rangle$ direction?

6-4   Suppose that fracture occurs in a single crystal of MgO when a critical tensile stress $\sigma_F = 30,000$ psi is resolved across {100} planes, and that yielding occurs when a critical resolved shear $\tau_Y = 20,000$ psi is set up along $\langle 110 \rangle$ slip directions lying in {110} planes. Will the crystal deform plastically before fracturing when a stress is applied along $\langle 100 \rangle$ directions?

6-5   Suppose that plastic deformation occurs at the tip of a crack when the local stresses reach the tensile yield strength. If $\sigma_Y = 35,000$ psi, what axial stress is required to cause local yielding if a crack 0.5 in. long, having a root radius $\rho = 0.001$ in., is present?

6-6   What is the ultimate tensile strength if necking begins at a true strain $\epsilon = 0.25$ in a material whose stress-strain curve obeys the relation $\sigma = 120,000 \epsilon^{1/2}$ psi?

6-7   Derive a simple relationship between the tensile ductility ϵ_f and the percent reduction in area at fracture.

6-8   At the UTS, the load $P = A\sigma$ is a maximum. Show that the criterion for necking of a tensile specimen can be written $d\sigma/d\epsilon = \sigma$ and that if $\sigma = \sigma_0 \epsilon^n$, necking occurs when $\epsilon = n$.

6-9   What is the energy stored in an Al_2O_3 fiber 1 in. long and 0.001 in. in diameter when fracture occurs at a stress of 500,000 psi and no gross yielding has occurred? (See Table 6-1.) How high would you have to lift this Al_2O_3 sample so that the potential energy would equal the stored elastic energy?

6-10 Calculate the value of Poisson's ratio for the condition of constant volume.

6-11 A cylindrical sample 1 in. long and 0.5 in. in diameter is deformed in compression. A dial gauge indicates that the specimen has shortened by 0.152 in. when the load is equal to 15,000 lb. Calculate the *true plastic strain* and the true stress at this state.

6-12 Explain why a second hardness measurement made adjacent to a previous measurement indicates a higher hardness.

6-13 The modulus of elasticity is often measured by taking the slope of the stress-strain curve in the elastic region. This procedure is accurate at low temperatures but gives modulus values too low at high temperatures. Why?

6-14 Derive simple expressions relating engineering stress and strain to true stress and strain.

6-15 Suppose a crystalline material is subjected to a constant true strain rate tensile test at $T = 0.75T_m$. Sketch the stress-strain curve and compare it to the creep curve shown in Fig. 6-15.

6-16 Why does the macroscopic fracture stress decrease as the crack length increases in brittle materials? What influence does the radius of curvature at the crack tip have on the fracture stress?

BIBLIOGRAPHY

T. ALFREY, JR., *Mechanical Behavior of Polymers*. New York: Interscience, 1948.

A. H. COTTRELL, *The Mechanical Properties of Matter*. New York: John Wiley, 1964.

G. E. DIETER, *Mechanical Metallurgy*. New York: McGraw-Hill, 1961.

F. GAROFALO, *Fundamentals of Creep and Creep Rupture in Metals*. New York: Macmillan, 1965.

H. W. HAYDEN, W. G. MOFFATT, and J. W. WULFF, *The Structure and Properties of Materials*, Vol. III. New York: John Wiley, 1965.

F. A. McCLINTOCK and A. S. ARGON, *Introduction to the Mechanical Behavior of Materials*. Cambridge, Mass.: M.I.T. Press, 1962.

D. McLEAN, *Mechanical Properties of Metals*. New York: John Wiley, 1962.

R. E. SMALLMAN, *Modern Physical Metallurgy*, Ch. 5. London: Butterworths, 1962.

W. J. M. TEGART, *Elements of Mechanical Metallurgy*. New York: Macmillan, 1966.

7
Plastic Deformation in Crystalline Solids

7-1 INTRODUCTION

Plastic deformation in crystalline solids is inhomogeneous. It is inhomogeneous because it occurs by the shearing of whole regions of crystal past one another, rather than by continuous or homogeneous deformation in which *all* atoms are displaced the same amount from their equilibrium positions, as in elastic deformation. The shear displacements associated with plastic deformation occur primarily by the motion of dislocations (see Chapter 3). Therefore, to achieve an understanding of the mechanical properties of crystalline solids requires some knowledge of the properties of dislocations. We can get some idea of the importance of understanding the behavior of dislocations by briefly considering the three different aspects of the mechanical properties of crystals.

1. *Plastic Deformation.* Except for a few cases, plastic deformation and dislocation movement are mutually inclusive; one cannot occur without the other. This means that we can resolve essentially all permanent deformation into dislocation motion. This connection extends from a bent paper clip to a crushed automobile fender.

2. *Strength.* Since there is a unique relation between plastic deformation and dislocation motion, materials will support static loads without undergoing permanent deflection only if dislocations are prevented from moving. Fortunately, a great deal is known about preventing dislocation motion, and we

can, and do, prepare materials with different strengths by altering the mobility of dislocations (see Chapters 8 and 9).

3. *Fracture.* Dislocations are important in fracture processes because groups of dislocations can provide a sufficiently high concentration of stress to form a crack.

Because dislocations are so important in determining the mechanical properties of crystalline solids, this chapter is devoted to a thorough study of the behavior of individual dislocations and of the relationships existing between the motion of individual dislocations and plastic flow.

7-2 THEORETICAL STRENGTH OF CRYSTALS AND THE MOTION OF DISLOCATIONS

Consider a simple rectangular crystal lattice subjected to a shear stress τ as shown in Fig. 7-1(a). If the crystal deforms by the top plane sliding or shearing over the bottom plane, then we can make a simple calculation of the strength of the lattice. Let x be the displacement corresponding to the applied stress τ. Symmetry requires that τ be zero when $x = nb/2$, where $n = 0, 1, 2, 3, \ldots$ Also,

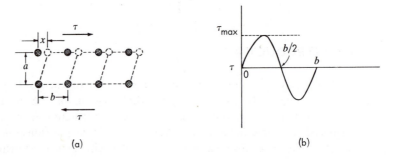

(a) (b)

FIG. 7-1   Periodic shear stress required to rigidly slip a perfect simple rectangular lattice; a model for computing the theoretical shear strength of a crystal

τ is positive (the lattice resists the applied stress) when $0 < x < b/2$ and negative when $b/2 < x < b$. The variation of τ with x, shown schematically in Fig. 7-1(b), is similar to a sine wave, and we may approximate it as such by writing

$$\tau = \tau_{max} \sin \frac{2\pi x}{b} \qquad (7\text{-}1)$$

where τ_{max} represents the strength of the lattice and can be estimated by noting that Eq. (7-1) reduces to

$$\tau \approx \tau_{max} \frac{2\pi x}{b} \qquad (7\text{-}2)$$

for small x (small strains). We also know that for small strains the crystals behave in an elastic fashion (Hooke's law) and $\tau = G\gamma$, where G is the shear modulus and γ is the shear strain. Combining this relationship with Eq. (7-2), we find at small strains

$$Gγ = G\frac{x}{a} = τ_{max}\frac{2πx}{b} \tag{7-3}$$

and since $a \approx b$,

$$τ_{max} \approx \frac{G}{2π} \tag{7-4}$$

or the theoretical shear strength is of the order of $G/2π$. Typical G values are around 10^7 psi, indicating theoretical lattice strengths of about 10^6 psi. This theoretical shear strength is the *maximum* that a crystal lattice can have. And since the theoretical strength is dependent only on the magnitude of the shear modulus, this means that solids with different shear moduli have different theoretical strengths. For example, the shear modulus of iron is about three times that of alminum, so that on an absolute scale iron *has the potential* to be three times as strong as aluminum.

Equation (7-4) has been verified for *perfect* crystals, i.e., crystals without dislocations. However, if dislocations are present in the crystal lattice, then the shear strength (stress at which permanent deformation is observed) is much lower, usually around $10^{-5} - 10^{-3}G$. Table 7-1 gives some typical values of the yield

TABLE 7-1

Yield Stress, $τ_y$, for Single Crystals tested at 20°C

Type of Bonding	Material	$τ_y$ (psi)	$τ_y/G$
Metallic	Cd	70	2.5×10^{-5}
	Ag	85	2.3×10^{-5}
	Cu	84	1.4×10^{-5}
	Cu*	88,500	1.5×10^{-2}
	Fe	2,000	1.7×10^{-4}
	Fe*	950,000	8.0×10^{-2}
Ionic	NaCl	284	1.4×10^{-4}
	SiO_2*†	up to 850,000	2.1×10^{-1}
	Al_2O_3*†	up to 3,000,000	1.3×10^{-1}
Covalent	Graphite†	up to 3,500,000	0.9×10^{-1}

*Essentially dislocation-free prior to testing.

†The inherent lattice resistance to dislocation motion is so high at room temperature that these materials fracture before they yield plastically.

strength for various pure crystalline solids, both with and without the presence of dislocations. Materials of engineering importance usually have a high density of dislocations present ($\approx 10^8$ cm/cm³) and usually have a yield strength well below the theoretical lattice strength.

The question then arises as to how dislocations permit the crystal lattice to deform at stresses far below the theoretical lattice shear strength. The answer to this question can be demonstrated simply with the aid of Fig. 7-2, which schematically illustrates an edge dislocation in a simple square lattice. The dislocation is centered at point A. What is the magnitude of the stress required (or how much work must be done) to move the dislocation to point B? We note that in moving from A to B each atom moves only a small distance. The work done in moving the atoms on the

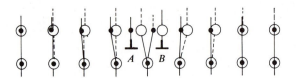

● Atom positions for dislocation centered at *A*

○ Atom positions for dislocation centered at *B*

FIG. 7-2   Local atom movements associated with the movement of an edge dislocation from one stable position in the lattice to the next

right side out of their low-energy positions is exactly compensated by the work recovered from the atoms on the left side as they fall back to low-energy positions. The net amount of work required to move the dislocation from *A* to *B* is then zero. Although this schematic representation simplifies the actual process, it has been found that in most pure materials with relatively nondirectional bonding (FCC and HCP metals) dislocations do indeed move very easily through the lattice at stresses in some instances lower than 100 psi. However, materials with either directional or ionic bonding (e.g., BCC metals, covalent materials, and other nonmetallic crystalline materials), have a higher lattice resistance to dislocation motion, and sometimes stresses well above 10^4 psi are needed before dislocation motion begins. Even in these instances, however, the stress to move dislocations through the lattice is much lower than the theoretical lattice strength.

The ease with which dislocations move through crystalline lattices has an analog in many other physical phenomena. For example, the motion of a positive edge dislocation can be likened to the motion of a caterpillar. A caterpillar does not attempt to pull itself along by dragging its entire body, just as a crystal does not shear by the block motion of one part of the crystal with respect to the other. Rather the caterpillar moves forward by forming a small region of compression at its tail (similar to the extra half plane of an edge dislocation) and then allowing this compressed region or hump to move along the length of its body. For each passage of a hump from tail to head, the caterpillar moves forward a small amount (Fig. 7-3(a)]. This is analogous to the displacement equal in magnitude to a Burgers vector associated with the passage of each dislocation. And as a dislocation allows a crystal lattice to shear easily, the caterpillar's hump allows it to move forward without the large frictional force that would be required to drag its whole body along. A worm also moves in a manner analogous to the passage of a dislocation in a lattice, but in this instance it is analogous to a negative edge dislocation [Fig. 7-3(b)]. The worm sticks its head forward a small amount and forms a tensile region (extra half plane below the slip plane); this tensile region then moves along the worm's body until it reaches the tail. The net result is that the worm moves forward by a small amount without dragging its whole body along the ground. A third example is to consider what happens when we vigorously shake a carpet by holding it on one end [Fig. 7-3(c)]. Each shake transmits a wave (positive edge

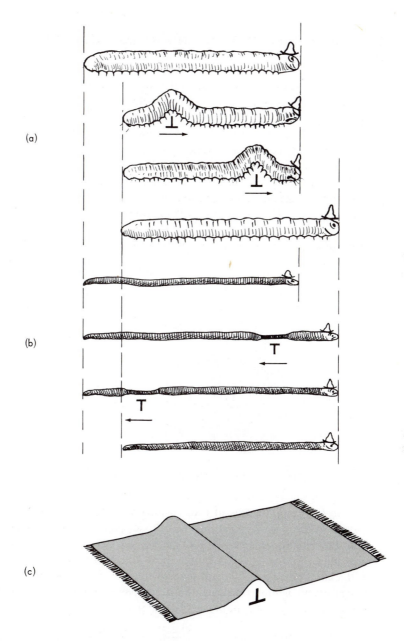

FIG. 7-3   Progressive movements of (a) caterpillars, (b) worms, and (c) rugs by the motion of dislocation-like defects

dislocation) through the rug, resulting in a net displacement. The easy passage of the wave along the length of the carpet permits very heavy carpets to be moved about quite easily, even though it would be very difficult to drag the rug by just pulling on one end.

7-3 ENERGY OF A DISLOCATION AND STABLE BURGERS VECTORS

Thus far we have only indicated that the presence of dislocations greatly reduces the strength of crystalline materials. We now turn our attention to the problem of predicting exactly what type (what Burgers vector*) of dislocations will be present in a given crystal structure. Our approach to this problem will be to calculate the energy of a dislocation as a function of Burgers vector and then show that those dislocations of miminum energy will be favored over all others.

Because the atoms about a dislocation are displaced from their normal lattice sites, all dislocations can be considered to have a stress field or strain field surrounding themselves. The elastic energy stored in this strain field represents what is also known as the *elastic energy* of the dislocation. We can calculate the elastic energy per unit length of a dislocation with the aid of Fig. 7-4. Consider a screw disloca-

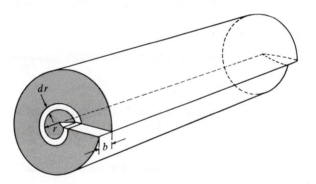

FIG. 7-4   Illustration of elastic distortion associated with a pure screw dislocation

tion to lie along the axis of a cylinder of radius r_1. This screw dislocation could have been produced by cutting the cylinder halfway through on a plane coplanar with the cylinder axis and then displacing the cut faces relative to one another by a distance b. Now let us calculate the strain energy in an annular ring of radius r, thickness dr, and unit length that is centered on the dislocation. The total elastic displacement in this ring (making a complete circuit around the ring) is merely equal to b, and the resultant shear strain in the ring is defined as

$$\text{Shear strain} \equiv \frac{\text{elastic displacement}}{\text{distance over which displacement occurs}} = \frac{b}{2\pi r} \qquad (7\text{-}5)$$

Knowing the shear strain, we can write the following expression for the strain energy per unit volume (see Example 6-6):

---

*In this section and throughout the book **b** is used to describe the Burgers vector and b stands for the magnitude of the Burgers vector.

$$\frac{\text{elastic strain energy}}{\text{unit volume}} = \frac{1}{2}(\text{stress})(\text{strain})$$

$$= \frac{1}{2}G(\text{strain})^2 = \frac{1}{2}G\left(\frac{b}{2\pi r}\right)^2 \tag{7-6}$$

where G is the shear modulus. The total elastic strain energy in the cylinder can be obtained by integrating over the volume of the cylinder

$$\frac{\text{elastic strain energy}}{\text{unit length of dislocation}} = \Gamma = \int_{r_0}^{r_1} \frac{1}{2}G\left(\frac{b}{2\pi r}\right)^2 2\pi r \, dr \tag{7-7a}$$

or

$$\Gamma = \frac{Gb^2}{4\pi} \ln \frac{r_1}{r_0} \tag{7-7b}$$

The lower cutoff in the integration, r_0, must be employed because Eq. (7-5) is only valid for $r > r_0 \approx 2b$. For $r < 2b$, the displacements are very large and the laws of linear elasticity (Hooke's law) no longer apply. In this region, commonly called the *dislocation core*, we must resort to more sophisticated theories to calculate the strain energy. The results of these calculations reveal that the strain energy associated with the dislocation core is *small* and that the elastic strain energy of the dislocation as defined by Eq. (7-7b) is a good measure of the total strain energy of the dislocations.

We can simplify Eq. (7-7b) by taking $\ln r_1/r_0 \approx 4\pi$. This corresponds to a value of r_1 of about 1 cm, which is a reasonable value for the size of a crystal containing a dislocation. Note that even if r_1 differs from 1 cm by a factor of 10 or 100, this changes the value of $\ln r_1/r_0$ by only a very small amount. Making this approximation, we find that Eq. (7-7b) reduces to

$$\Gamma \approx Gb^2 \tag{7-8}$$

This is a very important result, It indicates that *the energy per unit length of a screw dislocation is proportional to the square of its Burgers vector. Similar calculations for edge and mixed dislocations yield identical results.* From simple energy considerations, then, we might expect dislocations to have the smallest Burgers vectors possible. For example, a dislocation of Burgers vector 2b has an energy equal to $4Gb^2$, while two individual dislocations, each having Burgers vectors **b**, have a strain energy $Gb^2 + Gb^2 = 2Gb^2$. A dislocation of Burgers vector 2**b** can therefore reduce the total strain energy of the crystal by *splitting* into two individual dislocations. Alternatively, two dislocations of opposite sign will annihilate each other if they meet, thereby reducing the strain energy from $2Gb^2$ to 0.

With this knowledge we can predict the probable Burgers vectors to be found in the different crystal structures. In general, *the Burgers vector,* **b** *is the shortest lattice translation vector* of the crystal structure.* We require **b** to be a lattice translation vector so that the passage of a dislocation leaves a perfect crystalline structure behind and the requirement on the magnitude of **b** comes directly from Eq.

---

*Recall that a lattice translation vector is a vector connecting two Bravais lattice points. A vector connecting two atom sites may or may not be a lattice translation vector.

(7-8). In pure elements and solid solution alloys with only one atom per lattice point, the shortest lattice translation vector is parallel to the direction of closest atomic packing. This is not necessarily true for compounds or any solid solution where there is more than one atom associated with each lattice point. For example, in NaCl the stable Burgers vector is $a/2 \langle 110 \rangle$ even though the direction of closest atomic packing is $\langle 100 \rangle$ (see Fig. 7-5). A complete list of all the Burgers vectors

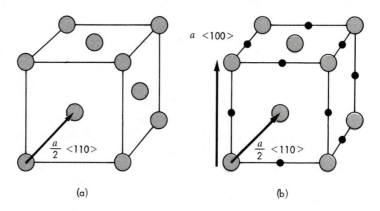

(a)   (b)

FIG. 7-5 The stable Burgers vector for the NaCl structure is $a/2 \langle 110 \rangle$ and not $a/2 \langle 100 \rangle$, as $a/2 \langle 100 \rangle$ is not a lattice translation vector. For clarity, only atoms on the exposed faces are shown.

TABLE 7-2

Slip Planes and Slip Directions for Common Crystal Structures

Crystal Structure	Examples	Possible Slip Planes	Slip Direction	Burgers Vector
FCC metals	Cu, Ag, Au, Al, Ni, brass, many stainless steels	{111}	$\langle 110 \rangle$	$\frac{a}{2} \langle 110 \rangle$
HCP metals	Be, Mg, Zn, Sn, Ti, Co, graphite	{001} {101} {112}	$\langle 110 \rangle$ $\langle 110 \rangle$ $\langle 113 \rangle$	$a \langle 110 \rangle$ $a \langle 110 \rangle$ $\mathbf{a} + \mathbf{c}$
BCC metals	Li, Na, K, Fe, most steels V, Cr, Mn, Cb, Mo, W, Ta	{110} {112}	$\langle 111 \rangle$ $\langle 111 \rangle$	$\frac{a}{2} \langle 111 \rangle$ $\frac{a}{2} \langle 111 \rangle$
Diamond	Diamond, Si, Ge	{111}	$\langle 110 \rangle$	$\frac{a}{2} \langle 110 \rangle$
NaCl	NaCl, LiF, MgO, AgCl	{110}	$\langle 110 \rangle$	$\frac{a}{2} \langle 110 \rangle$

Note: Some of the materials listed above are allotropic (can have different crystal structures depending on temperature). The room temperature form is listed here.

in the various crystal structures is given in Table 7-2. The length of most Burgers vectors is about equal to the lattice parameter, and typical values of b are $\approx 2 \times 10^{-8}$ cm.

EXAMPLE 7-1

Show that a dislocation in a BCC structure with a Burgers vector equal to $a \langle 110 \rangle$ is energetically unstable and will split into two $a/2 \langle 111 \rangle$ dislocations by the following reaction:

$$a\,[110] \longrightarrow \frac{a}{2}[111] + \frac{a}{2}[11\bar{1}]$$

Solution:

For the above reaction to take place the energy of the $a \langle 110 \rangle$ dislocation must be greater than the sum of the energies of two $a/2 \langle 111 \rangle$ dislocations. Since the energy of of a dislocation is proportional to b^2, we require

$$a^2[1^2 + 1^2 + 0^2] > 2\left(\frac{a^2}{4}[1^2 + 1^2 + 1^2]\right)$$

or

$$2a^2 > \tfrac{3}{2}a^2$$

The reaction is therefore energetically favorable.

7-4 SLIP PLANES AND SLIP SYSTEMS

In addition to producing displacements in definite crystallographic directions, dislocations also move on specific crystallographic planes. We refer to the plane on which a dislocation moves as its *slip plane*. Usually the slip plane corresponds to the plane that has the densest atomic packing. This is the result of two factors:

1. The slip plane must contain the Burgers vector, and the plane of densest atomic packing generally contains the direction of densest atomic packing.
2. The inherent lattice resistance to dislocation motion can be related to two geometrical parameters of the slip plane, the atomic smoothness of the plane and the distance separating adjacent planes. Smooth, widely separated planes offer considerably lower resistance to dislocation motion (atoms in adjacent planes sliding over one another) than do closely spaced, "rough" planes. The planes of densest atomic packing are those with the smoothest atomic structure and are also spaced further apart than any other set of crystallographic planes.

Slip planes for the various crystal structures are given in Table 7-2.

Every dislocation then produces slip in a specific direction (parallel to the Burgers vector) and moves on a specific slip plane. This combination is referred to as a *slip system*. Each crystal structure has a number of slip systems. For example, one specific slip system in an FCC metal would be $(111)[1\bar{1}0]$ or slip on a (111) plane in the $[1\bar{1}0]$ direction, while altogether there are twelve possible slip systems of this type (twelve possible $\{111\} \langle 110 \rangle$ combinations).

EXAMPLE 7-2   Suppose a BCC crystal deforms by slip on {110} planes in the ⟨111⟩ directions. How many slip systems of this type are there in the BCC structure?

Solution: There are six different {110} planes in the BCC structure: (110), (101), (011), (1̄10), (101̄), and (011̄). Contained in each of these planes are two ⟨111⟩ directions (e.g., the (110) plane contains the [11̄1] and [11̄1̄] directions as shown in the accompanying sketch). The total number of {110} ⟨111⟩ slip systems is thus 2 × 6, or 12.

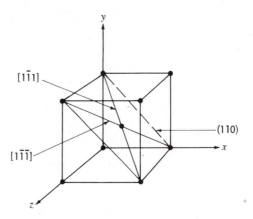

Slip on only one slip system cannot produce an arbitrary shape change for a crystal. It can only produce a shear strain in one direction, somewhat analogous to the shape change that occurs when a deck of playing cards is subjected to a shear stress and each successive card is displaced from the one below by a given amount [shown schematically in Fig. 7-6(a)]. A zinc single crystal oriented for this type of

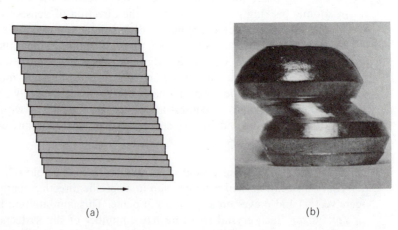

(a)   (b)

FIG. 7-6   (a) Schematic representation of slip, and (b) photograph of deformed single crystal of zinc. Photo courtesy American Society for Metals, from *Modern Research Techniques in Physical Metallurgy* ("Deformation of Single Crystals" by E. R. Parker and J. Washburn) 1953. Used with permission.

single slip is shown in Fig. 7-6(b). Note the slip steps on the surface. These steps are parallel to the slip plane, and each step represents the motion of thousands of dislocations on a given slip plane since all individual dislocations will produce a slip step of only ≈ 2 A.

To produce an arbitrary change in shape, a crystal must slip on a number of slip systems. Macroscopic slip is observed on a given system when the resolved *shear stress* [Eq. (6-8)] reaches the critical value for the onset of dislocation motion, i.e., a stress just high enough to overcome the lattice resistance to dislocation motion. This value of the resolved shear stress is called the *critical resolved shear stress* and is the same for all similar slip systems in a crystal. For some crystal structures it is possible to orient a crystal so that the resolved shear stress on all operative slip systems is zero. For the case of HCP crystals which deform on the $\{001\}\langle110\rangle$ slip systems this condition is attained for a tensile stress parallel to $\langle001\rangle$. If this is the case, dislocation motion does not occur and the crystal will not plastically deform at stresses below the theoretical lattice strength unless some other mode of plastic deformation occurs (see Section 7-7).

One important result of the necessity of a crystal to undergo slip on a number of different slip systems to achieve an arbitrary shape change is the difference in behavior between single crystals and polycrystals. A single crystal subjected to a shear stress can deform extensively with slip on only a single slip system. However, this is not the case for polycrystals. Since each grain in a polycrystal has a different crystallographic orientation, each will respond differently (slip in a different direction) when subjected to a shear stress. And if each grain deforms differently, then the region between grains (region around grain boundaries) is subject to complex shape changes if we demand coherency to be maintained between grains. Thus, unless the region around grain boundaries can undergo arbitrary shape changes, voids at grain boundaries will be opened up and the material will fracture at low strains. To achieve this arbitrary shape change, it is necessary to have five independent slip systems operative. Most metals have five independent slip systems, with the notable exception of those with hexagonal crystal structure. For this reason, many hexagonal metals (Be, Ti, etc.) have very poor ductility in polycrystal form, even though single crystals may be deformed extensively before fracture.

EXAMPLE 7-3

An NaCl single crystal is loaded in uniaxial compression with the [001] direction parallel to the compression axis. Dislocation motion is first observed on the (101)[10$\bar{1}$] slip system when the applied tensile stress is 500 psi. Calculate the inherent lattice resistance to dislocation motion.

Solution:

The resolved shear stress on the (101)[10$\bar{1}$] slip system is given by Eq. (6-7), where

$$\tau = \sigma \cos\theta \cos\phi$$

or $\tau = \sigma/2$ for (101)[10$\bar{1}$] slip since $\phi = \theta = 45°$. The shear stress necessary to overcome the inherent lattice resistance to dislocation motion is therefore 250 psi.

Aside from the shear strains associated with the glide of edge, screw, and mixed dislocations, it is possible for edge dislocations to produce tensile strains by moving in a direction perpendicular to the Burgers vector [Fig. 7-7(a)]. This type of motion

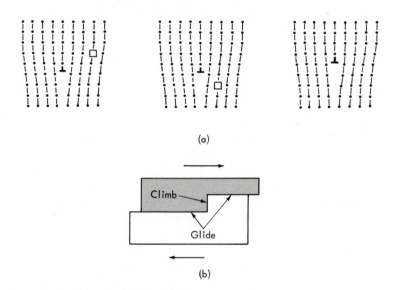

(a)

(b)

FIG. 7-7   Illustrations of edge dislocation climb. The □ represents a vacancy.

is called *dislocation climb* and involves the transport of atoms (or vacancies) away from or to the dislocation (edge of the extra half plane) by diffusion. Because dislocation climb is diffusion-controlled, it is important only at relatively high temperatures where the diffusion rates are fairly rapid. Motion of an edge dislocation showing both glide and climb motion is illustrated in Fig. 7-7(b).

7-5 RELATION BETWEEN DISLOCATION MOVEMENT AND PLASTIC FLOW

To demonstrate how we can relate the motion of individual dislocations to macroscopic plastic flow, let us consider the strain produced by the motion of a single dislocation. Suppose we have a crystal in the shape of a cube, where the edge length is L. Under the application of an appropriate shear stress, a dislocation is somehow produced on the left-hand side of the crystal and is allowed to move all the way through the crystal as shown in Fig. 7-8(a). The plastic shear strain associated with the passage of this dislocation is $\gamma_P = b/L$, where b is the Burgers vector. But suppose the dislocation had not passed all the way through the crystal but had gone only a distance x, where $x < L$. Then the overall shear strain would be somewhere between 0 and b/L, the exact value being proportional to the fraction of the slip plane area that the dislocation has traversed. In moving a distance x, the dislocation has swept out x/L of its slip plane and the resulting shear strain is given by

$$\gamma_P = \frac{x}{L}\frac{b}{L} \tag{7-9}$$

If instead of one dislocation we had considered N dislocations, all moving an aver-

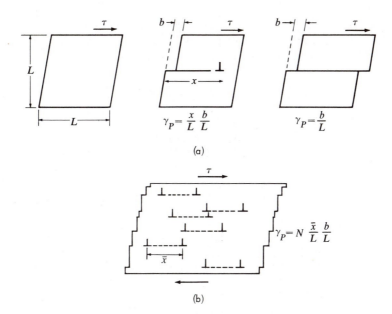

FIG. 7-8   Relation between shear strain and dislocation movement in glide

age distance $\bar{x}$ [Fig. 7-8(b)], the resulting strain would be

$$\gamma_P = \frac{N\bar{x}b}{L^2} \tag{7-10}$$

which can be rewritten

$$\gamma_P = \frac{NL\bar{x}b}{L^3} = \rho b\bar{x} \tag{7-11}$$

where NL is the total length of the dislocation line and ρ is the dislocation density, NL/L^3, in units of length per unit volume. *The shear strain is then the product of the dislocation density (the number of dislocations that have moved), the Burgers vector, and the average distance each dislocation has moved.* If the shear strain γ_P occurs over a time t, then we can write

$$\frac{\gamma_P}{t} = \frac{\rho b\bar{x}}{t}$$

or

$$\dot{\gamma}_P = \rho b\bar{v} \tag{7-12}$$

where $\dot{\gamma}_P$ is the shear strain rate and $\bar{v}$ is the average dislocation velocity. Equation (7-12) is a type of transport equation that often occurs in physics. It simply states that the strain rate is equal to the density of defects producing strain (dislocations) times the strength of each defect (the amount of strain associated with each defect), times the average velocity of the defects. In Chapter 12 we shall show that an identical equation exists for describing electrical conduction. For electrical conduction we write

$$J = nq\bar{v} \tag{7-13}$$

where the current density J (analogous to strain rate) is equal to the product of the number of charge carriers per unit volume, n, the strength or charge on each carrier, q, and the average velocity of each carrier, $\bar{v}$.

EXAMPLE 7-4

If a crystal with a dislocation density of 10^9 cm/cm^3 is deformed at a shear strain rate of 10^{-3} per sec,
(a) What is the average dislocation velocity?
(b) At what strain rate will the dislocation velocity approach the velocity of sound ($\approx 10^5$ cm/sec), the upper limit of the dislocation velocity?

Solution:

(a) $\dot{\gamma}_P = \rho b v$
   $\dot{\gamma}_P = 10^{-3}$ sec^{-1} = $(10^9$ cm/cm$^3)(2 \times 10^{-8}$ cm/sec$)v$
   $v = 5 \times 10^{-5}$ cm/sec

(b) For $v = 10^5$ cm/sec
   $\dot{\gamma}_P = (10^9$ cm/cm$^3)(2 \times 10^{-8}$ cm/sec$)(10^5$ cm/sec$)$
   $\dot{\gamma}_P = 2 \times 10^6$ sec^{-1}

Equations (7-11) and (7-12) are the fundamental expressions relating dislocation motion to plastic flow. However, before these expressions can be used to describe the mechanical properties of solids, we need to know something about the variables ρ, $\bar{x}$, and $\bar{v}$. All three of these quantities depend on stress, strain, temperature, and the internal structure of the solid under consideration.

To give an example, we might consider the Zn crystal shown in Fig. 7-6(b). The magnitude of the slip steps on the surface of this crystal indicate that several thousand dislocations have passed through the crystal on a given slip plane. Yet if we had examined the dislocation arrangement in the crystal prior to deformation, we would have found no more than a few dislocations on any one slip plane. Somehow a large number of dislocations must have been created and allowed to move during the deformation process. But where were these dislocations produced? We know they were not created from the perfect lattice because the maximum stress on the sample was only $\approx 10^{-5}G$, well below both the theoretical lattice strength and the stress required to create a dislocation in a perfect lattice. We are led to the conclusion that somewhere in the sample there are dislocation sources and that these sources are able to emit dislocations at very low stress. Apparently these sources continually generate dislocations during plastic flow, always replenishing the dislocation density, compensating for the dislocations that flow out the surface of the sample.

There is other indirect evidence that dislocations are generated during plastic flow. Suppose we have a polycrystalline sample, grain size equal to 10^{-2} cm, with a dislocation density of 10^8 cm/cm^3 (typical of a metal). Equation (7.11) tells us that the maximum strain such a dislocation density could produce is about 2%, taking $b = 2 \times 10^{-8}$ cm and $\bar{x}$ equal to the grain size. Yet we know from experience that many metals will deform much more than 2%. Indeed, bending a paper clip back and forth produces a strain far in excess of this amount. Once again we must conclude that dislocations are produced during flow. Some graphic evidence

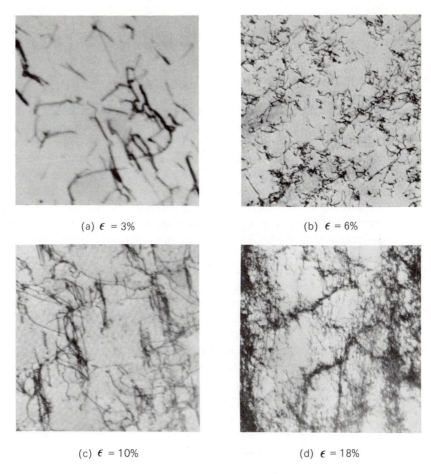

(a) $\epsilon = 3\%$

(b) $\epsilon = 6\%$

(c) $\epsilon = 10\%$

(d) $\epsilon = 18\%$

FIG. 7-9 Transmission electron micrographs illustrating the dislocation arrangements in deformed stainless steel (70% Fe, 18% Cr, 12% Ni) after elongations of (a) 3%, (b) 6%, (c) 10%, and (d) 18% at room temperature. Photos courtesy J. F. Breedis, AMF Inc. Used with permission.

of this fact is shown in Fig. 7-9, where we have used the electron microscope to examine the dislocation structure in stainless steel as a function of deformation. Clearly there is an increase in dislocation density with increasing strain.

The increase in dislocation density with plastic strain is usually found to obey a relationship of the form

$$\rho_{\text{total}} = \rho_0 + \alpha\gamma_P^m \qquad (7\text{-}14)$$

where ρ_0 is the initial or "grown in" dislocation density, about 10^3 cm/cm^3 in ionic and covalent crystals and 10^8 cm/cm^3 in most metals; α is a constant about equal to 10^8 cm/cm^3; and m generally has a value of ≈ 1.

In the following sections we shall examine the nature of dislocation sources in some detail. In particular, we shall focus our attention on the response of both a dislocation loop and a short segment of dislocation line to an applied shear stress.

7-6 DISLOCATION GENERATION

In considering the mechanisms of dislocation generation it is necessary to discuss two important properties of dislocations: (1) the concept of an effective force acting on a dislocation line and (2) the ability of a dislocation loop to expand under the application of an applied stress. Once we have established these two properties of dislocations we shall find that there are some very simple ways to generate dislocations within a crystal at very low applied stresses. We shall find that new dislocations can be created from existing dislocations and need not be created from a perfect lattice. In fact, our discussion will lead to the general statement that *when dislocations move, they invariably increase their total length, thereby increasing the dislocation density*.

The concept of a force acting on a dislocation originates from the fact that when we apply a stress to a crystal and a dislocation moves, the crystal deforms and the stress does work, lowering the potential energy of the system. The force on the dislocation can then be defined as the derivative of the energy change (work done) with respect to the dislocation position. We can calculate this force very simply in the following manner. Suppose a cube-shaped crystal containing an edge dislocation* is subjected to a shear stress τ, as shown in Fig. 7-10. If the dislocation

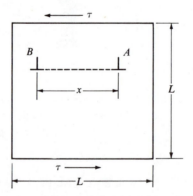

Fig. 7-10 Movement of a dislocation from *A* to *B* under an applied shear stress τ

moves a distance x, the top half of the crystal can be imagined to be displaced with respect to the bottom half by $(x/L)b$, where L is the length of the cube edge. The force acting on the top surface of the crystal due to the stress τ is τL^2 and, correspondingly, the work done by the applied stress when the dislocation moves is equal to

$$\tau L^2\left(\frac{x}{L}b\right) \quad \text{or} \quad \tau x b L$$

---

*The case of a screw or mixed dislocation is identical.

Since the energy change is just the negative of the work done, we can express the total force acting on the dislocation as

$$\text{Force} = \frac{-d(\text{energy})}{d(\text{position})} = \frac{d(\tau x b L)}{dx} = \tau b L$$

or

$$\text{Force per unit length} = \tau b \qquad (7\text{-}15)$$

Thus, the force per unit length acting on the dislocation is merely the product of the applied stress* and the Burgers vector. *This force acts in a direction perpendicular to the dislocation line or, equivalently, parallel to the direction of motion.*

Now we are in a position to consider what happens when a dislocation loop in a crystal is subjected to a shear stress. Consider the loop of radius r illustrated in Fig. 7-11. If there is an applied stress τ acting on the crystal, then the loop will

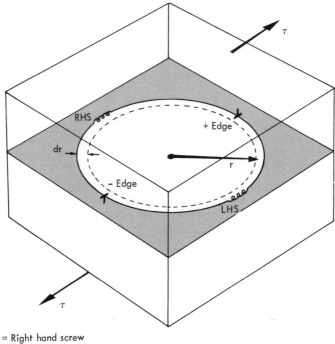

RHS = Right hand screw
LHS = Left hand screw

FIG. 7-11   Expansion of a dislocation loop under a shear stress τ

experience two opposing forces. One force, tending to collapse the loop, is due to the strain energy of the dislocation line, which can be decreased if the loop shrinks or shortens its total length. This strain energy makes the dislocation behave as if

---

*If the applied stress τ acts at some angle θ to the slip direction, then only the component of the applied stress parallel to the slip direction must be taken in evaluating the force on the dislocation.

it had a *line tension*, much the same as a stretched rubber band. This line tension T is defined as the change in energy with change in length of the dislocation and, following Eq. (7-8), we can write

$$T = Gb^2 \qquad (7\text{-}16)$$

where T is directed along the length of the dislocation. The force tending to expand the dislocation loop is due to the applied stress; i.e., the stress can do work if the loop expands. As long as the force on the dislocation due to line tension is greater than that due to the applied stress the loop will tend to collapse. And if the reverse were true, the loop would tend to expand.

For a given stress τ we can calculate the smallest diameter loop, which will expand in the following manner. Let the dislocation loop in Fig. 7-11 expand, increasing its radius by an amount dr. The increase in dislocation line length is $2\pi \, dr$. And since the energy per unit length of dislocation is Gb^2, the increase in elastic strain energy is $2\pi Gb^2 \, dr$. For the loop to expand without an overall increase in energy, the work done by the applied stress must be equal to or greater than $2\pi Gb^2 \, dr$. To calculate the work done by the applied stress we take the product of the force per unit length acting on the dislocation (τb), the length of the dislocation line ($2\pi r$), and the distance over which the force acts (dr). Equating the two energy changes, we have

$$2\pi Gb^2 \, dr = \tau b 2\pi r \, dr$$

or

$$\tau = \frac{Gb}{r} \qquad (7\text{-}17)$$

Equation (7-17) indicates that the stress necessary to expand a loop is inversely proportional to the radius of the loop, the proportionality constant being Gb. For any $\tau > Gb/r$, the loop expands, and for $\tau < Gb/r$, the loop contracts.*

Equation (7-17) is valid not only for loops but may also be used to describe the critical stress for bowing of dislocation segments. For example, consider the dislocation segment AB of length l shown in Fig. 7-12. We imagine this segment to be

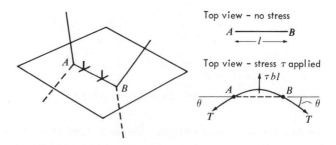

FIG. 7-12 Schematic illustration of dislocation bowing

---

*Here we assumed that the inherent lattice resistance to dislocation motion is negligible. If this is not the case, then this lattice friction force must be overcome before the dislocation can move.

pinned at its ends, possibly intersecting other dislocations on different slip planes as illustrated in the figure. Initially the line AB is assumed to be straight. Under the application of a stress τ, however, the segment begins to bow out, decreasing its radius of curvature. The dislocation bows out until the force in the forward direction due to the applied stress is just balanced by the line tension or, equivalently, until

$$\tau bl = 2T \sin \theta \tag{7-18}$$

When $\theta = 90°$ the force due to line tension is a maximum, and beyond this point the stress necessary to further bow the segment decreases. Evaluating Eq. (7-18) at $\theta = 90°$ to find the critical stress just large enough to fully bow out the dislocation segment, we find

$$\tau = \frac{2T}{bl} = \frac{2Gb^2}{bl} = \frac{2Gb}{l} = \frac{Gb}{r} \tag{7-19}$$

where $r = l/2$. Equation (7-19) is identical to (7-17), and hence we can consider a dislocation line of length l to behave in a fashion similar to a dislocation loop of radius $l/2$.

EXAMPLE 7-5

Calculate the stress necessary to move a dislocation through a matrix containing a fine dispersion of second-phase particles where the average spacing between the surfaces of adjacent particles is 400 A. Take $G = 10^{11}$ dynes/cm^2 and $b = 2 \times 10^{-8}$ cm.

Solution:

If the dislocation moves by bowing between particles, then the stress is given by $\tau = 2Gb/l$, where $l = 400$ A.

$$\tau = \frac{2(10^{11} \text{ dynes/cm}^2)(2 \times 10^{-8} \text{ cm})}{4 \times 10^{-6} \text{ cm}}$$

$$\tau = 10^9 \text{ dynes/cm}^2 = 14,500 \text{ psi}$$

Let us now consider what happens to the dislocation segment AB when $\tau > 2Gb/l$. As the segment bows out, the dislocation line essentially pivots around the pinning points A and B. Successive stages of this bowing process are shown in Fig. 7-13. When the bowing has reached sufficient proportions so that the two sides swing around and come into contact on the back side of the pinning points we observe a very curious phenomenon. The portions of the dislocation line that come into contact have the same Burgers vector but are of opposite sign, and consequently they annihilate one another. To give an example, if the initially straight dislocation segment AB was pure edge, the portions of the dislocation line that would come into contact would be left- and right-handed screw dislocations. When these two portions of the dislocation line annihilate, there remains a dislocation loop plus a dislocation segment identical to the initial segment AB. The bowing process has thus created a dislocation loop, and we can imagine the segment AB to act as a dislocation source. If a stress $\tau > 2Gb/l$ were maintained on the crystal, the source would continuously emit dislocation loops, which in turn would expand and move away. This type of source is known as a *Frank-Read* source, named after the two

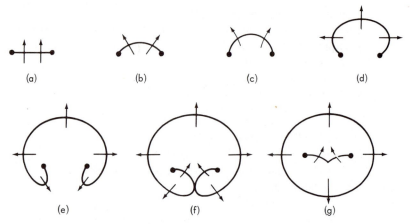

FIG. 7-13   Sequence of dislocation positions during the operation of a Frank-Read source

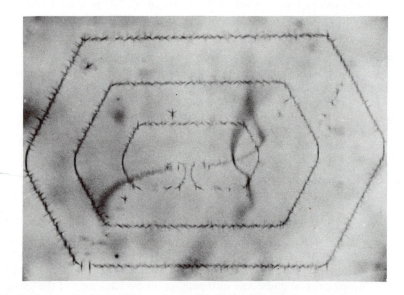

FIG. 7-14   A Frank-Read source in silicon. Photo courtesy The General Electric Company from *Dislocations and Mechanical Properties of Crystals* by J. C. Fisher et al. (eds), ("The Observation of Dislocations in Silicon" by W. C. Dash) 1957. Used with permission.

investigators who first postulated its existence. An example of a Frank-Read source in a silicon single crystal is shown in Fig. 7-14.

All crystals that initially have some dislocations present have potential Frank-Read sources. This is because the dislocations in the crystal occasionally intersect one another, forming a three-dimensional network. The points of intersection act as pinning points, and hence each dislocation in the crystal is qualitatively similar to the segment *AB* shown in Fig. 7-12. We can even make a simple calculation of

the yield strength of pure single-phase solids based on the spacing between dislocation intersections. This calculation assumes that yielding occurs when the stress is
large enough to fully bow out a dislocation segment; i.e., the yield stress is equal
to $2Gb/l$. For a random array of dislocations the value of l is given approximately
by $l = \rho^{-1/2}$, where ρ is the dislocation density. For annealed metallic crystals
typical values of ρ are $\approx 10^7$ cm/cm^3, giving $l \approx 3 \times 10^{-4}$ cm. With $b = 2 \times 10^{-8}$
cm, this calculation predicts the yield stress to be approximately $10^{-4}G$, which is
in agreement with the values given in Table 7-1.

**EXAMPLE
7-6**

If a crystal has a three-dimensional dislocation network with a dislocation density of
10^7 cm/cm^3, calculate the approximate number of Frank-Read sources per unit
volume and calculate the average stress required to activate a source. Take $G = 10^{11}$
dynes/cm^2 and $b = 2 \times 10^{-8}$ cm; assume that the inherent lattice resistance to
dislocation motion is zero.

Solution:

The average distance between dislocations is $\rho^{-1/2}$; therefore, there are $\rho/\rho^{-1/2}$, or
$\rho^{3/2}$, potential Frank-Read sources per unit volume. For the example above, $\rho^{3/2}$
$= 3.2 \times 10^{10}$ cm^{-3}. The stress necessary to activate a source is

$$\tau = \frac{2Gb}{l} = \frac{2Gb}{\rho^{-1/2}} = 2Gb\rho^{1/2}$$
$$\tau = 2(10^{11} \text{ dynes/cm}^2)(2 \times 10^{-8} \text{ cm})(10^7 \text{ cm/cm}^3)^{1/2}$$
$$\tau = 12.8 \times 10^6 \text{ dynes/cm}^2 = 185 \text{ psi}$$

The calculation of the yield strength for crystals with a large inherent lattice
resistance to dislocation motion (e.g., crystals with some covalent bonding) is
slightly more complex. In this instance we must add the lattice resistance stress
to the quantity $2Gb/l$ to obtain the yield strength. In many cases the lattice resistance stress is the dominant term, being much larger than $2Gb/l$. For example, Si
with a dislocation density of 10^2 cm/cm^3 ($l \approx 10^{-1}$ cm) has a room temperature
yield strength of over $10^{-3}G$ even though $2Gb/l$ is only about $10^{-7}G$.

Another way dislocation sources can be produced is by a process called *cross-
slip*. A dislocation is able to cross-slip if it can move from one slip plane to another as presented in Fig. 7-15. The simple geometrical requirement for a dislocation
to cross-slip is that the dislocation line and its Burgers vector must be contained
in two slip planes. This is possible only if the Burgers vector and dislocation line
are colinear. Thus, only pure screw dislocations, where the Burgers vector is parallel to the dislocation line, are capable of cross-slip. Figure 7-15 illustrates how the
cross-slip process can produce dislocation sources. A portion of a moving screw
dislocation has cross-slipped out of its original glide plane and, after moving a
short distance on this second slip plane, cross-slips back into a plane parallel to the
original slip plane.* The dislocation segment *AB* formed by this second cross-slip
process may then act as a Frank-Read source, being effectively pinned at its end

---

*The dislocation will tend to stay on the slip plane with the highest resolved shear stress, but
may cross-slip onto a plane with lower resolved stress when there is an obstacle in its path. We
shall discuss the nature of these obstacles in Chapter 8.

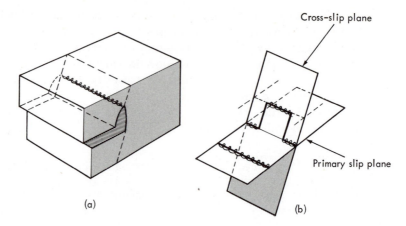

(a)   (b)

FIG. 7-15   Illustration of dislocation movements in cross-slip

points. This source may emit a number of loops (if the stress is large enough) any one of which may cross-slip again, producing new dislocation sources. Thus, as the dislocation moves along through the lattice it continually produces new dislocation lines, thereby increasing the dislocation density.

7-7 OTHER MODES OF DEFORMATION IN CRYSTALLINE SOLIDS

In addition to dislocation motion, there are three other modes of deformation that can occur in crystalline solids. Two of these deformation processes, *grain boundary sliding* and *directional diffusion creep*, occur only at high temperatures, while the third process, *twinning*, occurs at low temperatures. Generally, these processes occur *in addition* to dislocation motion.

Grain Boundary Sliding

Grain boundary sliding is a very inhomogeneous form of deformation produced by two grains sliding past one another along the plane of the common grain boundary (Fig. 7-16). Here the shear stress across the plane of the boundary is due to the applied stress. This mode of deformation is important only at low stresses and high temperatures (creep conditions). The contribution of grain boundary sliding to the total strain increases as the grain size decreases as there is more grain boundary area per unit volume. Generally, some dislocation motion within the grains accompanies any grain boundary sliding. This is because most grains have irregular shapes and, in order to keep from opening up large voids in the structure when grain boundary sliding occurs, there must be some grain shape change via dislocation motion.

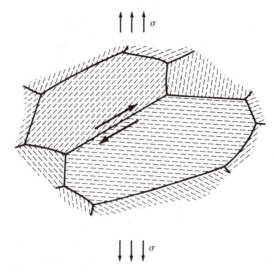

FIG. 7-16 Illustration of grain boundary sliding

Directional Diffusion Creep

At very high temperatures, greater than about 0.9 of the absolute melting temperature, and at stresses below the critical value for dislocation motion, crystals can deform by the directional diffusion of vacancies (Fig. 7-17). A net migration

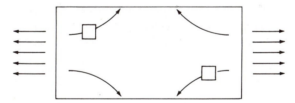

FIG. 7-17 Direction of vacancy migration at low stresses and high temperatures

of vacancies in one direction is equivalent to an atom flux in the opposite direction and can cause a macroscopic shape change. The driving force for this process is merely that the shape change allows the applied stress to do work and lowers the overall energy of the system. The vacancy flux is directed between good sources and sinks of vacancies—grain boundaries in the case of polycrystalline materials, and free surfaces for single crystals.

Twinning

Plastic deformation in crystals can also occur by twinning (Fig. 7-18), a process in which layers of atoms move in such a manner as to bring the deformed part of

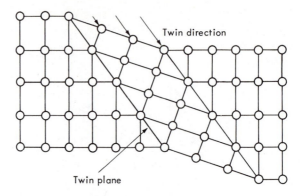

FIG. 7-18 Illustration of atom movement during twinning.

the crystal into a mirror-image orientation relative to the undeformed portion. The plane *AB* across which twinning occurs is called the *twinning plane*, and twinning, like slip, occurs in a specific direction called the *twinning direction*. Whether dislocation motion or twinning will be the predominant mechanism depends on which mechanism requires the lowest stress. The yield stresses for slip or twinning are not a constant for a given material, but vary with test temperature, strain rate, alloy content, and other extrinsic and intrinsic variables. In most materials having a BCC structure (refractory metals) the yield stress for slip increases sharply with decreasing temperature, whereas the twinning stress is relatively independent of temperature. At most temperatures $\tau(\text{slip}) < \tau(\text{twin})$, and slip is the preferred mode of deformation; but at very low temperatures ($T < T_t$, Fig. 7-19), the stress necessary for twinning is less than that required for slip, and twinning is the preferred mode of deformation. In FCC metals such as copper and aluminum the twinning

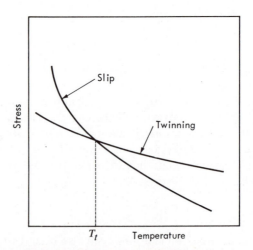

FIG. 7-19 Schematic illustration of the temperature dependence of the critical stress for slip and twinning in BCC crystals

stress is so much higher than the stress required for slip that twinning is generally not observed. HCP metals such as zinc, magnesium, and beryllium deform quite readily by twinning at most temperatures. Twinning is observed in only a few non-metallic solids.

PROBLEMS

7-1 What are the most stable (lowest energy) Burgers vectors for the body centered tetragonal, base centered orthorhombic, and triclinic Bravais lattices?

7-2 What is the most stable Burgers vector for the CsCl and GaAs crystal structures? (See Chapter 2 for atom arrangements.)

7-3 In the cubic crystal system, is the following dislocation reaction energetically favorable?

$$\frac{a}{2}[111] + \frac{a}{2}[1\bar{1}\bar{1}] \longrightarrow a[100]$$

7-4 The slip plane in FCC crystal structures is $\{111\}$, which is also the close-packed plane. Sketch a close-packed plane and show the slip direction $\langle 110 \rangle$. If an $a/2 \langle 110 \rangle$ Burgers vector dissociates according to the reaction

$$\frac{a}{2}[110] \longrightarrow \frac{a}{6}[21\bar{1}] + \frac{a}{6}[121]$$

(a) Show each of the reaction dislocations on your drawing.
(b) Is this dislocation reaction energetically favorable?
(c) What happens to the relative positions of two adjacent close-packed planes when one is shifted by a vector $a/6 \langle 112 \rangle$ with respect to the other?

7-5 Sketch an end view of an edge dislocation in the NaCl structure, and describe the atomistic processes involved in the climb of this dislocation.

7-6 Suppose a crystal with an edge dislocation is subjected to a stress as shown in the accompanying diagram. Calculate the climb force on the dislocation (force per unit length acting on the dislocation normal to the slip plane).

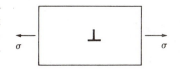

7-7 Suppose a copper crystal has a dislocation density of 10^{11} cm/cm^3. Calculate the strain energy associated with this dislocation density if $G = 4 \times 10^{11}$ dynes/cm^2 and $b = 2.5 \times 10^{-8}$ cm^2. If all this strain energy were converted to heat, how many degrees Celsius would the temperature of the copper rise? The heat capacity of copper is 0.09 cal/g°C, and the density is 8.96 g/cm.^{3}

7-8 Why can't a mixed dislocation cross-slip?

7-9 The inherent lattice resistance to the motion of dislocations in an FCC element is 50 psi. What is the *tensile* flow stress at which first dislocation motion begins if the tensile axis is parallel to the $\langle 100 \rangle$ direction?

7-10 Consult Table 7-2 and give those slip systems in a hexagonal close-packed crystal structure with nonzero resolved shear stress if a tensile stress is applied parallel to the *c* axis.

7-11 Suppose two orthogonal (mutually perpendicular) edge dislocations with parallel Burgers vectors intersect. Describe what happens to the dislocation lines, and discuss what influence this will have on their subsequent motion.

7-12 Repeat Problem 7-11 for the case of two orthogonal screw dislocations that intersect.

7-13 When dislocations glide through a matrix with dispersed second-phase particles, it is found that the first dislocation moving on a given plane meets less resistance to its motion than each subsequent dislocation moving on that same plane. Explain.

7-14 Some solute atoms have stress fields with only dilatational components, while others have both dilatational and shear components. Which type do you expect to be a more efficient barrier to dislocation motion? Why?

7-15 Some two-phase materials can have extremely small spacings between dispersed second-phase particles. If, in a particular material, the interparticle spacing is about $100b$, what will be the ratio of actual strength to theoretical lattice strength? Assume the actual strength can be estimated by calculating the stress necessary to bow a dislocation between two particles.

7-16 Show why it is generally necessary to have some dislocation motion accompanying any grain boundary sliding.

7-17 Describe the difference between slip and twinning.

7-18 Consider the nucleation and expansion of a dislocation loop in a crystal with an applied shear stress τ. Write an expression for the activation energy for loop nucleation, and show that the critical loop size r^* (there is free expansion for $r > r^*$) is $r^* = Gb/\tau$.

7-19 Suppose a BCC metal deforms via slip on both $\{110\}$ and $\{112\}$ planes in the $\langle 110 \rangle$ direction. How many possible slip systems are there?

7-20 Describe the motion of a dislocation in an amorphous material like glassy SiO_2.

BIBLIOGRAPHY

A. H. COTTRELL, *Theory of Crystal Dislocations*. New York: Gordon and Breach, 1964.

A. H. COTTRELL, *Dislocations and Plastic Flow in Crystals*, Ch. 1–3. London: Oxford University Press, 1958.

G. E. DIETER, JR., *Mechanical Metallurgy*. New York: McGraw-Hill, 1961.

D. K. FELBECK, *Introduction to Strengthening Mechanisms*. Englewood Cliffs, N.J.: Prentice-Hall, 1968.

H. W. HAYDEN, W. G. MOFFATT, and J. WULFF, *The Structure and Properties of Materials*, Vol. III. New York: John Wiley, 1966.

J. P. HIRTH and J. LOTHE, *Theory of Dislocations*. New York: McGraw-Hill, 1968.

W. T. READ, JR., *Dislocations in Crystals*. New York: McGraw-Hill, 1953.

J. WEERTMAN and J. R. WEERTMAN, *Elementary Dislocation Theory*. New York: Macmillan, 1964.

8
Strengthening Mechanisms

8-1 INTRODUCTION

In Chapter 6 we discussed some of the technologically significant properties of materials. Our treatment was macroscopic and somewhat descriptive in nature. In this chapter we begin to study the microscopic or atomic level structural imperfections that control the mechanical properties of crystalline materials. In particular we shall show that such mechanical properties as *yielding*, *strain hardening*, *creep*, and *fracture* are determined by the presence and behavior of such microstructural defects as dislocations, impurity atoms, second-phase particles, grain boundaries, cracks, and voids. A microscopic treatment of noncrystalline solids will be given in Chapter 10.

The object of this chapter is to establish the link between the structure of a material, including defects, and its mechanical properties. Knowledge of the relation between structure and mechanical properties will allow us to describe the microstructures that must be produced to achieve a given mechanical property. Chapter 9 deals with the application of this knowledge in the synthesis of mechanical properties in crystalline materials.

The approach in this chapter is to study the mechanical properties of relatively simple materials in which only a small number of factors (such as strengthening mechanisms) must be simultaneously considered. In this way we shall isolate the basic types of mechanical response of crystalline solids and develop sufficient insight into the operative mechanisms so that qualitative predictions of the mechanical properties of complex materials can be made. Again, Chapter 9 will deal with the integration of the ideas presented in this chapter, especially as they relate to the explanation and prediction of the mechanical properties of important structural materials.

8-2 DISLOCATION THEORY OF YIELDING

One of the most important mechanical properties of structural materials is the *yield stress*, or the stress at which significant plastic deformation begins to occur. In most engineering structures the yield stress represents an upper limit of stress for the designer. Consequently, the physical factors that establish the yield stress are of considerable interest and importance.

When the loads applied to a given material lead to stresses that are well below the yield stress, the behavior of the material in question or the engineering structure in which the material resides can be accurately predicted with the linear theory of elasticity. We sometimes choose or process materials in order to minimize the elastic deflections in a particular structure, but more often than not we are satisfied with the fact that the elastic deflections are small and that they can be accurately predicted. In the case of yielding, we have the double-edged problem of having large strains (or deflections) as well as a great deal of uncertainty about the expected deflection at a given stress. For these reasons some understanding of what actually happens during yielding is relevant to the general problem of structural design.

In this section we shall discuss a simple but accurate way to think about yielding in crystalline materials.* The model we shall use to describe yielding may be described as follows. At stresses well below the yield stress, dislocations do not move because they are pinned in place by their interaction with impurity atoms, second-phase particles, grain boundaries, and other dislocations, as well as by the direct interaction between the dislocation and the crystal lattice itself. At these low stresses the material deforms only elastically. When the stress level is high enough to cause dislocations to move, the material no longer behaves in a purely elastic manner. In this case we must take account of the fact that the material strains by two parallel mechanisms—*elastic deformation* and *plastic deformation*. The plastic deformation component causes a deviation from linearity in the stress-strain curve and near the yield stress begins to dominate the shape of the curve. In Chapter 7 [Eq. 7-12] we showed that the plastic strain rate $\dot{\epsilon}_p$† may be expressed as

$$\dot{\epsilon}_p = \tfrac{1}{2}\rho b \bar{v} \tag{8-1}$$

---

*Most structural materials of engineering importance yield by the movement and multiplication of dislocations.

†In the previous chapter it was emphasized that dislocations produce shear strain by their movement and respond only to shear stresses. However, in dealing with mechanical properties of materials it is common to use tensile stresses and tensile strains instead of the more fundamental shear stresses and strains. The orientation factor of $\tfrac{1}{2}$ is often used to convert shear strain to tensile strain ($\epsilon = \tfrac{1}{2}\gamma$) and tensile stress to shear stress ($\tau = \tfrac{1}{2}\sigma$).

where ρ is the density of mobile dislocations with Burgers vector b, and $\bar{v}$ is their average velocity.

When the stress in a tensile test becomes high enough to cause dislocations to move, the sample deforms both elastically and plastically. However, at the very start of the plastic deformation the number of dislocations that move is small, and they do not move at high velocities. As a result, even after some plastic deformation has begun, the elastic strain rate and the slope of the stress-strain curve is only slightly changed from the slope in the elastic region. As the tensile test proceeds, the dislocations multiply by mechanisms such as those discussed in Chapter 7 and also begin to move faster because they are being driven by higher stresses. The plastic strain rate soon becomes very large compared to the elastic strain rate and dominates the stress-strain relation beyond that point. A stress-strain curve for a crystalline solid is shown in Fig. 8-1. The figure also includes some of the features of the dislocation model for yielding just described.

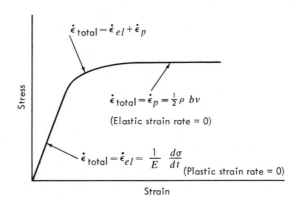

$$\dot{\epsilon}_{total} = \dot{\epsilon}_{el} + \dot{\epsilon}_p$$

$$\dot{\epsilon}_{total} = \dot{\epsilon}_p = \frac{1}{2}\rho \, bv$$

(Elastic strain rate = 0)

$$\dot{\epsilon}_{total} = \dot{\epsilon}_{el} = \frac{1}{E}\frac{d\sigma}{dt}$$

(Plastic strain rate = 0)

FIG. 8-1   Elastic and plastic strain rates in a tensile test

It is clear that the yielding phenomenon in crystalline solids ultimately involves a consideration of the mobility of the dislocations in a given material. Generally speaking, the dislocation mobility in a material is controlled by the mutual interactions between dislocations and other imperfections. Later sections in this chapter will deal with the various types of imperfections that impede dislocation movement and hence cause strengthening.

As discussed in the previous chapter, the dislocation density usually increases with strain. In the case of low dislocation density crystals it is safe to assume that all the dislocations produced are *mobile* dislocations. This is not necessarily the case for a material that has been heavily deformed or one in which there are precipitates or grain boundaries. In these cases some of the dislocations produced during straining may be stopped by other imperfections, so that all dislocations are not necessarily mobile. This is one of the reasons that yielding in complex materials is difficult to describe in terms of the properties of individual dislocations.

Our interest in the mobile dislocation density in a crystal and how it changes with strain stems from the fact that the mobile dislocation density can indirectly influence the velocity at which dislocations are required to move. Let us suppose that a crystal is being plastically deformed at a fixed strain rate $\dot{\epsilon}_p$. From Eq. (8-1) it is clear that the average dislocation velocity can be expressed as $\bar{v} = 2\dot{\epsilon}_p / \rho b$. It is clear that when the plastic strain rate is fixed, the average dislocation velocity decreases as the mobile dislocation density increases. This is one of the important features of yielding in relatively perfect crystals. In the early stages of yielding, the mobile dislocation density is low and the dislocations are required to move at high velocities; but as yielding proceeds, the dislocations multiply and the increase in ρ causes the average dislocation velocity to decrease.

In relatively perfect crystals (low densities of dislocations and precipitates) the dislocations velocity depends mainly on the applied stress σ. As can be seen

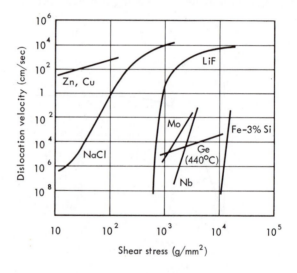

FIG. 8-2   Stress dependence of dislocation velocities for several materials

from the dislocation velocity data in Fig. 8-2, the relation between the dislocation velocity and the stress can be approximated by a relation of the form

$$\bar{v} = \left(\frac{\tau}{\tau_0} \right)^m \tag{8-2}$$

over much of the range of measurement. In this relation τ_0 is the shear stress that gives unit dislocation velocity (usually 1 cm/sec) and m is a parameter that defines the stress dependence of the dislocation velocity. Some typical values of τ_0 and m are given in Table 8-1.

We may now develop a simple model for yielding in relatively perfect crystals. If a tensile or compressive test is carried out at a constant strain rate, the stress-strain relation can be predicted qualitatively from a consideration of the terms in Eq. (8-1). By far the most important parameters in yielding are the initial mobile

TABLE 8-1

Values of τ_0 and m in the Expression $v = (\tau/\tau_0)^m$

Material	τ_0 (psi)	m
Zn	4.5	1
Cu	4.5	1
Mo	9.4×10^3	7
Nb	7.0×10^3	16
Fe + 3% Si	2.8×10^4	30
NaCl	2.1×10^2	8
LiF	1.7×10^3	25
Ge (440°C)	1.4×10^8	1

Note: All data are for room temperature unless otherwise indicated.

dislocation density ρ_0 and the parameter m that defines the stress dependence of the dislocation velocity. At the start of a test the applied stress is small enough so that the dislocations either do not move or move with negligible velocity. Consequently, from Eq. (8-1) we see that the plastic strain rate is zero, and we conclude that the sample deforms only elastically. The stress increases with strain according to Hooke's law until the stress becomes sufficiently large so that the average dislocation velocity, coupled with the initial mobile dislocation density, gives a plastic strain rate that is comparable in magnitude to the elastic strain rate. When this happens the stress-strain curve deviates from the linear relation obeyed at low stresses. As the stress continues to increase with strain, the average dislocation velocity increases even more, and the plastic strain rate becomes even more dominant. When the stress reaches a value such that the plastic strain rate $\frac{1}{2}\rho_0 b(\tau/\tau_0)^m$ is equal to the imposed strain rate, the slope of the stress-strain curve goes to zero, and a point of yielding is reached. For the case of highly perfect crystals and some BCC metals, this point is referred to as the *yield point* and is followed by a sharp drop in stress. Note that the stress level at the yield point depends on both the initial mobile dislocation density ρ_0 and the velocity stress exponent m. The high yield stresses associated with a low mobile dislocation density come from the fact that high stresses are needed to produce dislocation velocities high enough to maintain a fixed plastic strain rate. The low yield stresses associated with large values of m are simply due to the fact that low stresses are sufficient to produce a given dislocation velocity when the value of m is large.

The drop in the flow stress that follows the *yield point* is caused by the multiplication of dislocations. As dislocations multiply, the plastic strain rate exceeds the applied strain rate, and elastic strain rate becomes negative (the stress simply drops with strain). This behavior has a simple and straightforward meaning. As dislocations multiply, the mean dislocation velocity required to produce a given plastic strain rate decreases, and hence the stress required to move the dislocation at the appropriate velocities decreases.

The description of yielding and the yield drop we have given does not account for the fact that dislocations can and do interact with each other. As will be discussed in Section 8-5, this type of interaction gives rise to *strain hardening* and is manifested in a gradual rise in the stress-strain curve following yielding.

EXAMPLE 8-1

The stress-strain curve for a low-carbon steel exhibits a yield drop at the yield point [see Fig. 8-3(a)]. Estimate the shear stress at the upper yield point and the yield drop (in shear stress) that occurs when the mobile dislocation density increases by two orders of magnitude. The initial mobile dislocation density is 10^5 cm^{-2}, $\tau_0 = 2.8 \times 10^9$ dynes/cm^2, and $m = 20$. The total strain rate for the test is 10^{-3} sec^{-1} and $b = 2 \times 10^{-8}$ cm.

Solution:

After yielding, the plastic strain rate is very much larger than the elastic strain rate so that $\dot{\epsilon}_{\text{total}} \approx \dot{\epsilon}_p$ to a good approximation. At the upper yield point

$$\dot{\epsilon}_{\text{total}} = 10^{-3} = \tfrac{1}{2}(10^5)(2 \times 10^{-8})\left(\frac{\tau}{\tau_0}\right)^m$$

so that

$$\left(\frac{\tau}{\tau_0}\right)^m = 1$$

Thus,

$$\tau = 2.8 \times 10^9 \text{ dynes/cm}^2$$

When $\rho = 10^7$,

$$\left(\frac{\tau}{\tau_0}\right)^m = 10^{-2}$$

$$\tau = (2.8 \times 10^9)(10^{-2})^{1/20} = (2.8 \times 10^9)(0.794)$$

$$\tau = 2.22 \times 10^9 \text{ dynes/cm}^2$$

Hence, the stress drop is 0.58×10^9 dynes/cm^2 or 20.6%.

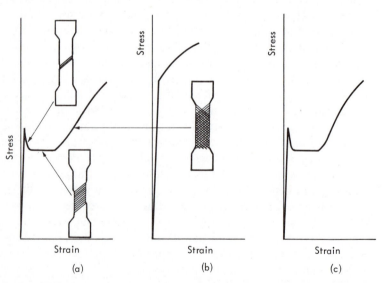

Strain

(a)

Strain

(b)

Strain

(c)

FIG. 8-3 Stress-strain relations for low-carbon steel: (a) as annealed; (b) just after (a); (c) as reannealed

8-3 THE YIELD POINT PHENOMENON

The dislocation model of yielding provides a simple and direct explanation of the yield point phenomenon that is commonly observed in BCC metals and alloys and that is particularly important in low-carbon steels. Figure 8-3(a) illustrates a room temperature stress-strain curve for a low-carbon steel that had been annealed and slowly cooled to the test temperature. The stress-strain curve shown in Fig. 8-3(b) is exhibited by the steel if a test is carried out immediately after the first test is completed. The sharp stress drop shown in the first stress-strain curve is called the *yield drop* and is notably absent in the second test. The yield drop can be made to reappear in the stress-strain curve [see Fig. 8-3(c)] by annealing the sample at a temperature above a few hundred degrees Celsius, and slowly cooling to the test temperature. This phenomenon is known as *strain aging*.

Another important feature of yielding in low-carbon steels is that deformation occurs inhomogeneously in the sample when it is strained slightly past the yield drop. The inhomogeneous deformation is associated with the plateau portion of the stress-strain curve and involves the propagation of bands of deformation along the length of the sample, as shown in Fig. 8-3(a). At large strains the deformation bands have propagated through the entire sample and uniform strain hardening occurs. The uniform deformation behavior is associated with the portion of the stress-strain curve that is roughly parabolic in shape.

The stress-strain relations for low-carbon steels can be explained with the dislocation model for yielding. As discussed in the previous chapter, the lattice distortions associated with interstitial impurity atoms such as carbon in iron give rise to strong interactions with the strain fields of dislocations. The carbon atoms tend to reside preferentially near dislocations because the distortion energy of the impurity atom is lowest when it is near the center of a dislocation. This means that if sufficient time is allowed for carbon atoms to collect at dislocations, most dislocations become pinned and are not free to move. The result is that the initial *mobile* dislocation density is greatly reduced by the annealing treatment. The *yield drop* or *yield point* shown in Fig. 8-3 is due to the low initial mobile dislocation density caused by the prior annealing treatment. As discussed before, when a material has a low dislocation density the flow stress decreases sharply just after yielding starts because the process of dislocation multiplication causes the average dislocation velocity and hence the flow stress to decrease. The yielding usually starts at a small notch or irregularity in the sample and subsequently propagates throughout the sample.

The stress required for plastic flow does not change appreciably with strain when the deformation is inhomogeneous. This is because the moving dislocations are concentrated mainly at the interface between the deformed and the undeformed regions, and the conditions of the interface do not change markedly as it moves from one end of the sample to the other. Once the plastic deformation has spread throughout the entire sample, further plastic deformation occurs in a more homogeneous fashion. The strain hardening that follows the inhomogeneous deformation is caused by dislocation-dislocation interactions, as will be discussed in Section 8-5.

The absence of the yield drop and the region of inhomogeneous deformation in the sample that had been prestrained past the yield point can be explained easily with the multiplication yield point theory. Prestraining introduces a high mobile dislocation density throughout the entire sample, so that the conditions for the yield point are no longer present. The effect of annealing the sample again at 200°C is to allow carbon atoms to diffuse to the dislocations and reduce the mobile dislocation density by pinning. The yield point reappears in the stress-strain curve because the mobile dislocation density is again small.

The yield point phenomenon and the inhomogeneous deformation that follows is of concern to users of sheet steel, especially automobile manufacturers. Automobile bodies and fenders are usually made of low-carbon steels, which exhibit the strain-aging phenomenon just described. It is important when forming a fender that the steel deform homogeneously so that the desired shape can be accurately produced. Because a low-carbon steel that has been annealed shows a yield point and deforms inhomogeneously, it is necessary to deform sheet steel by cold rolling before it is shaped into automobile parts. The cold rolling prestrains the steel past the region of inhomogeneous deformation so that subsequent bending operations occur smoothly.

8-4 SOME STRENGTHENING MECHANISMS

Having discussed the relation between the dynamical properties of dislocations and the mechanics of yielding, we are now prepared to describe the mechanisms by which crystalline solids may be strengthened. We are concerned mainly with the factors that control the speed of a dislocation at a given applied stress. We shall find that high yield strengths are achieved by forming obstacles to the movement of dislocations. These obstacles can take the form of impurity atoms, other dislocations, precipitates, or inert inclusions, such as oxides.

We shall limit our attention in this section to the mechanisms that control the yield strength of a material as opposed to the ultimate tensile strength. The ultimate strength depends significantly on strain hardening, which is discussed in the next section.

We begin our discussion of strengthening mechanisms by considering some of the factors that limit the motion of dislocations in relatively perfect crystals. There are essentially two different types of drag forces that act on dislocations moving in otherwise perfect crystals:* lattice vibration drag and drag due to bond angle distortions that occur at the core of a moving dislocation. The latter drag force is sometimes referred to as the *Peierls force*.

Lattice vibration drag operates in all crystals and is associated with the fact that a moving dislocation produces and interferes with sound waves (phonons) in the

---

*We imagine that the moving dislocations in question are the only defects in the crystal.

solid in which it travels. The drag forces are small compared to all of the other drag mechanisms we shall discuss and can be observed only in highly pure and perfect metal crystals.

The existence of the Peierls force in materials is directly related to the directionality of the bonding. In materials such as Si, Ge, and C, which have the diamond cubic structure, and InSb and GaAs, which have related structures, the bonding is largely covalent and hence is strongly directional. Accordingly, the core energies of dislocations in these materials are high because of the bond angle distortions at the dislocation core and are subject to significant change as the position of the dislocation is changed. In these materials the Peierls force represents the most significant barrier to dislocation motion at all temperatures, even up to the melting temperatures. As mentioned in Chapter 2, transition metals with the BCC structure also develop a directional bonding component. As a result the movement of dislocations in these metals at low temperatures is believed to be controlled by the Peierls force. At higher temperatures the atoms vibrate with sufficient amplitude that the effect of the Peierls force becomes nullified, and other factors such as impurities limit dislocation movement. Bonding in FCC and HCP metals and in ionic crystals is sufficiently nondirectional that drag due to the Peierls force is insignificantly small.

We now consider the role of imperfections in strengthening. Particular emphasis is placed on the imperfection strengthening mechanisms because the density and distribution of imperfections are subject to control. We shall see that an understanding of the following imperfection strengthening mechanisms provides a practical basis for controlling the strength of most crystalline solids (Chapter 9).

Solid Solution Strengthening

When an impurity atom is dissolved into a crystalline matrix, whether as a substitutional or interstitial solute, lattice distortion is produced and a stress field is set up near the impurity. This distortion occurs in the case of substitutional solute atoms when the impurity atom is either larger or smaller than matrix atoms. Distortions associated with interstitial solutes occur because the solute atoms are larger than the interstitial sites they occupy. The stress field set up by a solute atom interacts with the stress field of a moving dislocation, causing the motion of the dislocation to be impeded. This type of strengthening is known as *solid solution strengthening*.

The effects of solid solution strengthening are small if the solute atoms are randomly distributed in the lattice but may be large if the distribution of solute atoms is not random. For example, consider the case of a straight dislocation line and a perfectly random distribution of solute atoms. The force exerted on the dislocation from each solute atom depends on the exact position of the solute with respect to the dislocation line. Of importance here is the fact that the sign of the force depends on the atom position (the force may be either positive or negative). Thus, for a straight dislocation line and random solute distribution there is no *net* force on the dislocation, since, statistically, there are equal numbers of positive and negative forces (symmetric distribution of atoms about the dislocation). In reality, however, there is a net force on the dislocation, making it more difficult to move, since either (1) the solute distribution is not perfectly random or (2) the dislocation has some flexibility and can bow toward attractive forces and away from repulsive forces. With regard to the latter possibility of dislocation flexibility, it is important to note that dislocations are not free to assume any shape. Line tension makes it difficult for them to assume a small radius of curvature, and even for solute concentrations as low as $\approx 1\%$ the spacing between solutes is small compared to the expected curvature in the dislocation line.

Let us now consider a simple model that contains some of the important features of solid solution strengthening. Let the mean spacing between solute atom clusters in a slip plane be indicated by $\bar{l}$, as shown in Fig. 8-4. We shall assume

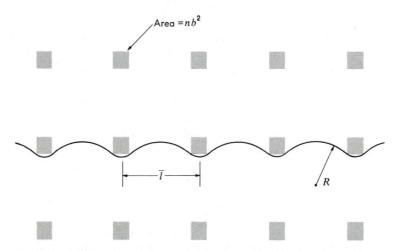

FIG. 8-4   Interaction between a moving dislocation and impurity atom clusters in its slip plane

that each cluster contains n solute atoms. A dislocation moving in the slip plane is held up by the impurity atom clusters, and dislocation bowing occurs between them. From our discussion of bowing in Chapter 7, recall that the radius of curvature r decreases as the applied stress increases and that as a result the line tension force F exerted on each solute atom cluster increases with increasing stress. Let us suppose, for sake of simplicity, that the dislocation is pulled past each cluster when the critical condition for bowing is achieved ($r = \bar{l}/2$). Using Eq. (7-17), we see that this critical condition requires that the shear stress be given by

$$\tau_s = \frac{2Gb}{\bar{l}} \tag{8-3}$$

As shown in Fig. 8-4, the area of the slip plane taken up by each cluster is approximately nb^2, so that the fraction of solute atoms or impurity concentration c may be expressed as

$$c \approx \frac{nb^2}{\bar{l}^2}$$

(8-4)

With this relation and with Eq. (8-3), we find that the contribution to the yield strength of a given alloy that comes from dislocation-solute interactions can be expressed as

$$\tau_S = K_s \sqrt{c}$$

(8-5)

where K_s is a constant equal to $2G/n^{1/2}$ for the simple model described. A model more realistic than the one just described would allow the dislocation to pass the impurity clusters before the critical radius is achieved. In that case K_s would signify the strength of the dislocation-solute–atom cluster interaction. The important features of Eq. (8-5) are that the solid solution strengthening component is directly proportional to the strength of the dislocation-solute interaction (as indicated by K_s) and proportional to the square root of the solute concentration. The data shown in Fig. 8-5 illustrate the composition dependence of the yield strength,

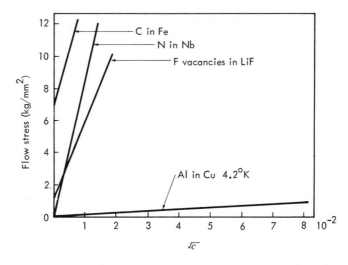

FIG. 8-5   Point defect concentration dependence of the flow stress for several materials

which is predicted by Eq. (8-5) for a number of different metallic and ceramic systems.

In solid solutions of FCC metals the hardening is often linearly proportional to concentration, indicating a different type of dislocation-impurity interaction. In addition to impurity atoms, vacancies (e.g., those produced by nuclear irradiation) can also interact with dislocations and provide strengthening.

EXAMPLE 8-2

The tensile yield stress of niobium with 50 ppm nitrogen (atomic fraction basis) is approximately 38,400 psi. Estimate the tensile yield stress of pure niobium and niobium with 500 ppm nitrogen. The shear modulus of niobium is $G = 15.6 \times 10^6$ psi, and K_s from experiment has the value of $G/12$.

Solution: The yield shear strength of the niobium-nitrogen alloy is

$$\tau_Y = \tau_0 + \tau_I = \tau_0 + K_s\sqrt{c}$$

where τ_0 is the yield shear strength of pure niobium (and not the shear stress required to produce unit dislocation velocity). At $c = 50$ ppm $= 50 \times 10^{-6}$,

$$\tau_0 = \frac{1}{2}(38,400) - \frac{G}{12}(50 \times 10^{-6})^{1/2}$$

$$\tau_0 = 19,200 - \frac{15.6 \times 10^6}{12}(50 \times 10^{-6})^{1/2}$$

$$\tau_0 = 10,000 \text{ psi}$$

Thus, the tensile yield strength for pure niobium is

$$\sigma_0 = 20,000 \text{ psi}$$

The yield stress for an alloy with $c = 550$ ppm is

$$\tau_Y = 10,000 + \frac{15.6 \times 10^6}{12}(500 \times 10^{-6})^{1/2}$$

and the tensile yield strength is

$$\sigma_Y = 2\tau_Y = 78,200 \text{ psi}$$

Precipitation Strengthening

Of great importance from a practical viewpoint is the strengthening achieved by precipitates of a second phase. In this case the solute atoms are not dissolved in the matrix but form separate second-phase particles that in turn are dispersed in the matrix. Most of the strengthening in ultrahigh-strength steels, aluminum, and titanium alloys results from this hardening mechanism.

It is instructive at this point to distinguish between *precipitation strengthening* and *dispersion strengthening*, since both involve strengthening by dispersing second phase particles into a crystalline matrix. *Precipitation strengthening* is a term that will be reserved for the case in which the second phase is formed by precipitation from solid solution. *Dispersion strengthening*, on the other hand, is a more general term and refers also to second-phase strengthening when the second phase is formed by a process other than solid state precipitation (such as the formation of oxide particles in a matrix by internal oxidation). Another term that will be used is *composite strengthening*, which refers to second-phase strengthening in the special case where the volume fraction is large and the second phase directly supports a significant share of the load.

When a moving dislocation encounters precipitates it will not, in general, be able to cut through them because the precipitates are generally stronger than the matrix. Consequently, the dislocation will have to bow between the precipitates (Fig. 8-6) and around them, leaving a dislocation loop around the the particle. The stress required for this process is approximately

$$\tau_p = \frac{2Gb}{l} \tag{8-6}$$

where l is the distance of closest approach between the particles; i.e., the bowing of the dislocation between the precipitates is exactly analogous to the bowing out of the dislocation segment illustrated in Fig. 7-12. When fine dispersions ($l \approx 50b$) are present, such as alloy carbides in steel, very large stresses must be applied before dislocation motion can occur, and the material exhibits a high yield strength.

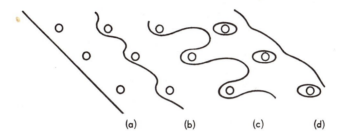

(a)   (b)   (c)   (d)

FIG. 8-6   Illustration of the process of pushing a dislocation through an array of precipitates

---

EXAMPLE 8-3

The metals tungsten and sodium have the same crystal structure (BCC), and their lattice parameters are not greatly different. Explain why a dispersion of strong precipitates with spacing $l = 10^{-4}$ provides a greater stength increase in tungsten than in sodium. Assume $a_{Na} = 4.29 \times 10^{-8}$ cm and $a_w = 3.16 \times 10^{-8}$ cm.

Solution:

The reason is that the strengthening produced by a dispersion of strong precipitates is governed by dislocation bowing, which depends on the elastic properties of the metal matrix. The shear modulus for tungsten is 1.51×10^{12} dynes/cm², while that for sodium is 0.026×10^{12} dynes/cm². Therefore, according to Eq. (8-6), the strength contributions due to the strong precipitates are

$$\tau_p \text{ (sodium)} = \frac{2(0.026 \times 10^{12})\frac{\sqrt{3}}{2}(4.29 \times 10^{-8})}{10^{-4}} = 1.93 \times 10^7 \text{ dynes/cm}^2$$

and

$$\tau_p \text{ (tungsten)} = \frac{2(1.51 \times 10^{12})\frac{\sqrt{3}}{2}(3.16 \times 10^{-8})}{10^{-4}} = 0.82 \times 10^9 \text{ dynes/cm}^2$$

Thus, the same microstructure produces a strength change in tungsten more than an order of magnitude greater than that produced in sodium.

EXAMPLE 8-4

Calculate the shear stress necessary to move a dislocation through a matrix containing a fine dispersion of spherical second-phase particles, where the average spacing between the center of adjacent particles is 500 A and the particles are 100 A in diameter. Assume that the dislocations do not cut through the particles and that the lattice resistance stress is zero. Take $G = 10^{11}$ dynes/cm² and $b = 2 \times 10^{-8}$ cm.

Solution:

The stress is given by $\tau = 2Gb/l$, where l is the distance between the surfaces of adjacent particles; i.e., $l = 400$ A.

$$\tau = \frac{2(10^{11} \text{ dynes/cm}^2)(2 \times 10^{-8} \text{ cm})}{(4 \times 10^{-6} \text{ cm})}$$

$$\tau = 10^9 \text{ dynes/cm}^2 = 14,500 \text{ psi}$$

It should be noted that there are some practical limits to the applicability of Eq. (8-6) to precipitation strengthening. These limits are that when precipitates are very small they are not sufficiently strong to block the movement of a dislocation and they can be sheared. Consider the precipitate shown in Fig. 8-7(a).

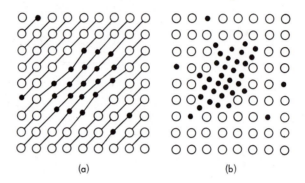

(a)   (b)

FIG. 8-7   Schematic illustration of (a) coherent and (b) incoherent precipitates

This precipitate has a chemical composition different from that of the matrix, but the precipitate and matrix have the same structure and share the same lattice; it is said to be a *coherent* precipitate. It seems clear, since the precipitate is but a few atoms thick, that it can be sheared by a dislocation moving in the matrix. On the other hand, the distortion field of coherent precipitate is obviously much larger than the precipitate itself, and hence it can interact with more distant dislocations than it could if it were *incoherent* [with separate structures or separate lattices; see Fig. 8-7(b).] Although Eq. (8-6) is generally applicable to precipitation strengthening, it cannot be applied when the precipitates are very thin.

The techniques used to form precipitates and provide precipitation strengthening in metallic and ceramic alloys involve the use of the principles of phase equilibria and kinetics. These ideas will be discussed in the next chapter.

Dispersion Strengthening

As mentioned before, dispersion strengthening refers to strengthening that arises from the presence of a second phase that is not formed by precipitation. The most common examples of this type of strengthening are found in metals and alloys that derive their strength from the presence of dispersed oxides. For example, TD-nickel and TD-nichrome* are high-temperature structural materials in which strength is derived from a dispersion of ThO_2 particles. These materials and this particular strengthening mechanism are important because such materials have superior strength at temperatures that correspond to the hottest sections of a gas turbine engine (1200°C).

There are two important features of the mechanism of dispersion strengthening.

The first is that the particles can act as obstacles to dislocation motion in the same way as was described for precipitates. In that case, Eq. (8-6) may be used to describe the strengthening process. Another contribution to dispersion strengthening is more complex and cannot be explained simply in a quantitative way. The dispersed particles usually act as prolific sources of dislocations, and consequently the crystalline matrix itself becomes strain-hardened. In fact, most dispersion-strengthened materials will not be unusually strong unless they have been mechanically processed (deformed at an intermediate temperature, such as 500–600°C for TD-nickel). This fact provides a strong indication that when the strain hardening contribution to dispersion strengthening is dominant, Eq. (8-6) cannot be generally used. Furthermore, the second-phase particles inhibit softening processes such as *recovery* and *recrystallization* (to be discussed in Chapter 9) and thereby permit the alloys to retain their strength even as the temperature approaches the melting temperature.

EXAMPLE 8-5

The tensile yield strength of pure polycrystalline silver is $\sigma_0 = 10,000$ psi. Estimate the strength of polycrystalline silver that contains 5 volume percent MgO in the form of inclusions 1000 A in diameter. For silver, $G = 0.436 \times 10^{12}$ dynes/cm² and $b = 2.9 \times 10^{-8}$ cm. The alloy is well annealed after the oxide inclusions are formed.

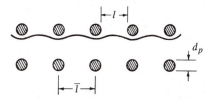

Solution:

The presence of MgO particles in silver provides dispersion strengthening through the mechanism of direct dislocation-particle interactions (the structure has not been strain hardened). The spacing between the MgO particles in the slip plane can be computed with reference to the accompanying figure.

$$\text{Particle volume fraction} = \text{area fraction} = \frac{\pi d_p^2/4}{\bar{l}^2} = 0.05$$

$$\bar{l} = \frac{\pi^{1/2}}{2(0.05)^{1/2}} d_p = 0.395 \times 10^{-4} \text{ cm} = l + 2d_p$$

$$\sigma_Y = \sigma_0 + 2\tau_p = \sigma_0 + \frac{4Gb}{l}$$

$$\sigma_Y = 6.8 \times 10^8 + \frac{4Gb}{l} = 6.8 \times 10^8 + \frac{(4)(0.436 \times 10^{12})(2.9 \times 10^{-8})}{(0.395 \times 10^{-4}) - (10^{-5})}$$

$$\sigma_Y = 2.40 \times 10^9 \text{ dynes/cm}^2 = 34,800 \text{ psi}$$

*The notation TD refers to "thoria dispersed."

Grain Size Strengthening

The imperfection strengthening mechanisms that we have discussed thus far may be classified as *extrinsic*, because they each involve the addition of either impurity atoms or aggregates of impurity atoms (precipitates and inclusions of a second phase). We now turn our attention to a strengthening mechanism that can be considered *intrinsic* in the sense that it can be induced without changing the composition of a given crystalline material. We shall show that the strength of a polycrystalline solid depends on the grain size and can be increased by treating the material in such a way as to achieve a fine grain size. In addition to grain size strengthening, *strain hardening*, which will be discussed in the next section, is an intrinsic strengthening mechanism because it can be produced in a given crystalline material by plastic deformation alone.

The yield strength of most crystalline solids increases with decreasing grain size, as shown in Fig. 8-8. It seems intuitively correct that fine grain solids should

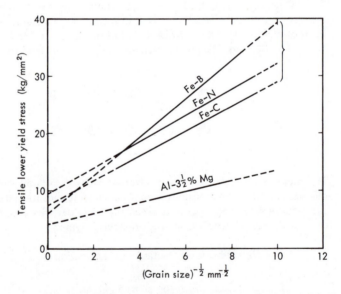

FIG. 8-8   Influence of grain size, *d*, on yield stress of various Fe alloys and an Al–3.5% Mg alloy

have high yield strengths because the grain boundary can act as an obstacle to dislocation motion. The following is a simple model to illustrate some of the physical conditions that cause grain boundary strengthening.

Consider the two adjacent grains in a polycrystalline sample shown in Fig. 8-9. Let us assume that yielding begins in grain *A* and takes the form of a slip band. In order for yielding of the polycrystal to occur throughout the sample it is necessary for the plastic strain to propagate from one grain to the next. This means

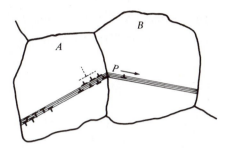

FIG. 8-9   Schematic illustration of slip propagation from one grain to the next

that the stress concentrations that are built up at the ends of the first slip band (at P) must be sufficient to start yielding in the second grain. The intensity of the stress at the tip of the slip band is obviously dependent on the applied stress and, less obviously, dependent on the length of the first slip band. Qualitatively, we may say that if the first slip band is very long (large grain size), a very large stress concentration will develop at the tip of the slip band because the shear load that would have been supported by the slip plane if yielding had not started would be transmitted to the end of the band and concentrated there. It is for this reason that materials with large grains usually have lower yield strengths. Once slip starts in one grain (the one most favorably oriented for yielding) the plastic flow can easily propagate from one grain to the next because of the high intensity of the stress at the ends of the long slip bands.

A quantitative description of the slip band model of yielding in polycrystalline solids leads to the following relation between yield stress σ_y and grain size d:

$$\sigma_Y = \sigma_i + k_y\, d^{-1/2} \tag{8-7}$$

where σ_i refers to the average yield strength of a single grain and is determined by all the operative strengthening mechanisms except grain boundary strengthening and k_y is a complex parameter that determines the effectiveness of grain boundaries in raising the yield strength. Equation (8-7) is usually called the *Hall-Petch* relation, after the workers who first observed the relationship experimentally. The data shown in Fig. 8-8 demonstrate that the Hall-Petch relation is obeyed experimentally for a number of metallic alloys.

In concluding the discussion of strengthening mechanisms in solids, consider the mechanisms by which a number of commercially significant structural materials are strengthened. These materials and strengthening mechanisms are summarized in Table 8-2.

EXAMPLE 8-6

Assume that strengthening of a low-carbon steel is achieved by grain size refinement and by fine carbide dispersions ($\sigma_i = \sigma_p$). Uniaxial tensile tests show that the tensile yield stress $\sigma_Y = 55{,}000$ psi when the grain diameter d is 0.32 mm and $\sigma_Y = 40{,}000$ psi when $d = 1.0$ mm. What is the average distance of closest approach between the particles? The shear modulus $G = 17.1 \times 10^6$ psi and $b = 2 \times 10^{-8}$ cm.

Solution:

From Eq. (8-7) we can write

$$(\sigma_Y^{(1)} - \sigma_i)\sqrt{d_{(1)}} = (\sigma_Y^{(2)} - \sigma_i)\sqrt{d_{(2)}}$$

where (1) and (2) refer to the two grain sizes.

$$(55,000 - \sigma_i)\sqrt{0.32} = (40,000 - \sigma_i)\sqrt{1}$$

Hence,

$$\sigma_i = \sigma_p = 18,500 \text{ psi}$$

Using Eq. (8-6)

$$l = \frac{4Gb}{\sigma_p} = \frac{4(17.1 \times 10^6)(2 \times 10^{-8})}{18,500}$$

$$l = 7.4 \times 10^{-5} \text{ cm}$$

It should be clear from our discussions of strengthening mechanisms that the strength of a crystalline solid depends sensitively on the microstructure of the material in question. In all cases we found that imperfection strengthening mechanisms become more significant when the mean spacing between imperfections is small. This general conclusion about the relation between strength and microstructure sets some guidelines for the design of high-strength materials. We shall see in the next chapter that methods can be devised to produce fine microstructures and hence high-strength structural materials.

8-5 STRAIN HARDENING AND RECOVERY

As mentioned in the previous section, strain hardening may be treated as an *intrinsic* strengthening mechanism in which other dislocations representing the principal obstacles to dislocation motion can be induced by plastic strain. We choose to treat this strengthening mechanism separately because it is so intimately related to the softening mechanism called *recovery*. Strain hardening is the *increase* in strength caused by an *increase* in the dislocation density produced by a given plastic strain, whereas recovery is defined as the *decrease* in strength that accompanies a decrease in the dislocation density or a rearrangement of the dislocations. We shall see that strain hardening and recovery in crystalline solids continually compete for the control of the structure and strength of crystals. This important competition will become particularly evident when we treat the phenomenon of high-temperature creep.

Strain Hardening

The phenomenon of strain hardening at low temperature (where recovery is either very slow or does not occur) is important in its own right. As already mentioned, strain hardening represents a very convenient strengthening mechanism. Most high-strength aluminum alloys, for example, are strain-hardened by rolling before they are put into service as structural materials. In addition, the very marked differences in the yield strengths of hot rolled and cold rolled low-carbon steel products are due to the effects of strain hardening. Strain hardening also provides a beneficial effect in terms of a safety factor for structural design. If a given structural component in a design is overloaded, it will plastically deform and strain-harden, thereby becoming stronger *in situ*. It should be noted, however, that since plastic strain is necessary to produce the strain hardening the structural material in question must have sufficient ductility to provide the necessary plastic strain.

TABLE 8-2

Principal Strengthening Mechanisms in Some Common Structural Materials

Material and Condition	Approximate Yield Strength (psi)	Operative Strengthening Mechanism
Aluminum Alloys		
Al-Cu-Mg; heat treated and cold worked (2024-T3)*	50,000	Strain hardening; precipitation strengthening
Al-Zn-Mg-Cu; annealed (7075-0)*	15,000	Solution strengthening
Al-Zn-Mg-Cu; heat treated (7075-T6)*	72,000	Precipitation strengthening
Copper Alloys		
Cu (electrolytic); annealed (commercially pure; 0.04% oxygen)	10,000	Strain hardening (dislocation-dislocation strengthening)
Cu (high conductivity); cold worked (oxygen-free; high conductivity; OFHC)	40,000	Strain hardening
Cu-Zn; annealed (brass)	35,000	Solution strengthening
Cu-Be; heat treated (beryllium copper)	140,000	Precipitation strengthening
Nickel Alloys		
Ni-Cr-Fe; cold worked (Inconel)	150,000	Strain hardening; solution strengthening
Ni-Mo-Fe; annealed (Hastelloy B)	40,000	Solution strengthening
Ni-Co-Cr-Mo-Ti-Al; heat treated (Udimet 700)	120,000 (at 1200°F)	Precipitation strengthening
Iron Alloys		
Fe; annealed (commercially pure; 0.01% C)	25,000	Solution strengthening; grain boundary strengthening
Fe-C; annealed (1020* steel; 0.2% C)	45,000	Solution strengthening; grain boundary strengthening
Fe-Ni-Mo-Mn-Cr-C; heat treated (4340* steel; 0.4%C)	200,000	Solution strengthening; grain boundary strengthening
Maraging steel 300 (special heat treatment)	290,000	Precipitation strengthening; grain boundary strengthening;
Cold drawn tungsten wire	540,000	Strain hardening; grain boundary strengthening

*Integers refer to composition and processing conditions for the structural material in question according to conventions established by industry and by professional societies.

Although the dislocation mechanisms responsible for strain hardening in crystalline solids are numerous and still the subject of research, it is possible to describe some of the important features of these mechanisms with the use of very simple models. Because dislocations have stress fields they are able to interact and exert forces on one another. For example, if the stress fields of two parallel dislocations tend to cancel when the dislocations approach, then the dislocations attract one another and there are forces acting on the dislocations, tending to move them together. Conversely, if the stress fields of parallel dislocations constructively interfere, the dislocations repel. Some typical dislocation interactions are illustrated in Fig. 8-10.

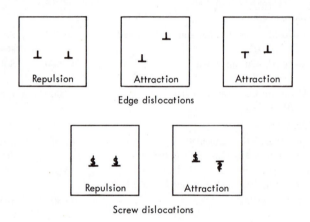

FIG. 8-10   Illustration of some common types of dislocation interactions

Dislocations on different slip systems can also interact (Fig. 8-11). In this case the moving dislocation, labeled *AA*, encounters other dislocations (*BB*) threading through its slip plane. In general, the interaction between *AA* and *BB* can be either

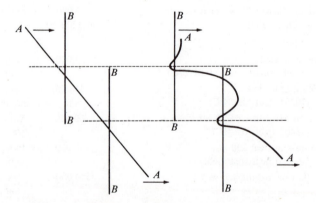

FIG. 8-11   Illustration of dislocation *AA* interacting with dislocations *BB* normal to slip plane

repulsive or attractive. If there is a repulsive force between the dislocations, then dislocation motion will be impeded, since the applied stress will have to do work by either bowing AA between the vertical dislocations or overcoming the stress field interaction and allowing AA to cut through BB. If the force between dislocations is attractive, then a similar situation exists, since in order to continue dislocation motion the stress must do work to pull AA away from BB.

EXAMPLE 8-7

Imagine that there are two parallel screw dislocations with identical Burgers vectors on different but parallel slip planes spaced a distance h apart. If one of these dislocations is fixed in position, calculate the value of the shear stress required to move the second dislocation past the immobile one. Take the value of the applied shear stress acting on the dislocation to be τ_a, and assume the lattice resistance to dislocation motion to be zero.

Solution:

From Eq. (7-5) we can calculate the radially symmetric shear stress about a screw dislocation as

$$\text{Shear stress} = G(\text{shear strain}) = \frac{Gb}{2\pi r}$$

Now we must find what component of this stress is acting on the moving dislocation. From the accompanying figure we see that

$$\frac{Gb}{2\pi r}\cos\theta = \frac{Gbx}{2\pi\sqrt{x^2 + y^2}\sqrt{x^2 + y^2}} = \frac{Gbx}{2\pi(x^2 + y^2)}$$

is the appropriate component and that this is a maximum (for constant y) at $x = y$. Thus, for the dislocations to pass one another, the applied stress τ_a must be greater than the stress due to the immobile dislocation: $\tau_a \geq Gb/4\pi h$.

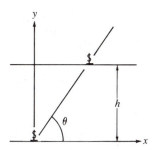

EXAMPLE 8-8

Calculate the shear stress required to move a dislocation through a forest of other dislocations (the forest dislocations are perpendicular to the slip plane of the moving dislocation) if the density of forest dislocations is 10^8 cm/cm³. Assume that the moving dislocation does not cut through the forest dislocations and that the lattice resistance stress is zero. Take $G = 10^{11}$ dynes/cm² and $b = 2 \times 10^{-8}$ cm.

Solution:

The average spacing between forest dislocations is $(10^8 \text{ cm/cm}^3)^{-1/2}$, or 10^{-4} cm. The moving dislocation must bow between the forest dislocations to continue motion, and the necessary stress is simply that required to completely bow out a dislocation segment 10^{-4} cm long; i.e.,

$$\tau = \frac{2Gb}{l} = \frac{2(10^{11} \text{ dynes/cm}^2)(2 \times 10^{-8} \text{ cm})}{(10^{-4} \text{ cm})}$$

$$\tau = 4 \times 10^7 \text{ dynes/cm}^2 = 604 \text{ psi}$$

In both of the preceding examples it is shown that the stress needed to drive dislocations past each other is proportional to the product of the shear modulus and the Burgers vector and inversely proportional to their separation distance. Since the separation distance l and the dislocation density are directly related through $\rho \approx 1/l^2$, it follows that the (tensile) stress needed to produce plastic flow in a strain-hardened crystal can be expressed as

$$\sigma_{\text{WH*}} = \alpha Gb \sqrt{\rho} \tag{8-8}$$

where α is a constant, usually about 0.2, which accounts for the orientation of the slip planes and other details of the strain-hardening process. The relation between flow stress and dislocation density predicted by Eq. (8-8) is illustrated in Fig. 8-12 for polycrystalline copper.

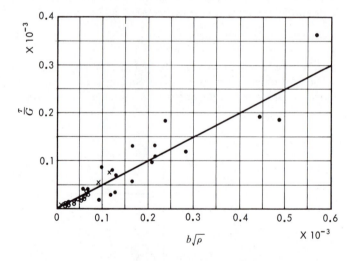

FIG. 8-12 Effect of dislocation density on the flow stress of polycrystalline copper

A complete account of the subject of strain hardening should indicate how dislocation density increases with strain. In general, this subject is beyond the scope of this book; however, it is worth noting the consequence of one of the more commonly observed relations between dislocation density and plastic strain. If the density varies linearly with strain, it follows immediately from Eq. (8-8) that the

---

*We use the symbol WH to denote *work hardening*, a term commonly used to describe strain hardening.

shape of the stress-strain curve is parabolic. This prediction is in qualitative agreement with the stress-strain relations observed for many metals and alloys.

**EXAMPLE
8-9**

A copper crystal of 5 cm³ contains 5×10^{10} cm of dislocation line. If $G = 0.75 \times 10^{12}$ dynes/cm² and $b = 2 \times 10^{-8}$ cm, calculate the tensile stress required to overcome the internal stress field of the dislocations.

Solution:

The dislocation density is

$$\frac{5 \times 10^{10} \text{ cm}}{5 \text{ cm}^3} = 10^{10}/\text{cm}^2$$

$$\sigma_{\text{WH}} \approx 0.2Gb\sqrt{\rho} = (0.2)(0.75 \times 10^{12})(2 \times 10^{-8})(10^5)$$

$$\sigma_{\text{WH}} \approx 3.0 \times 10^8 \text{ dynes/cm}^2 = 4400 \text{ psi}$$

Recovery

The strain energy stored in a deformed crystal is a manifestation of the number and arrangement of dislocations in that crystal. Since the free energy (dominated by the strain energy) can be decreased through dislocation annihilation and rearrangement, there is a thermodynamic driving force for these microscopic events to occur. Recovery is a process by which the internal strain energy and yield stress of a deformed crystal can be decreased through rearrangements and annihilations of the dislocations.

A crystal is strain-hardened when it is deformed at low temperatures. Subsequent annealing at various elevated temperatures allows recovery to take place and the hardness to decrease accordingly. Figure 8-13 illustrates the hardness

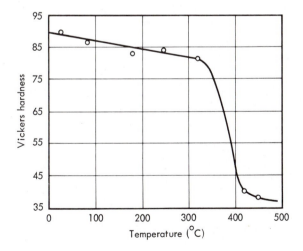

FIG. 8-13   Recovery of the Vickers hardness of polycrystalline copper as a function of temperature for a fixed heating rate

(on a Vickers hardness scale) of polycrystalline copper initially deformed 33% in tension at room temperature and subsequently heated to the various temperatures at a rate of 6°C/min. The gradual recovery that occurs between room temperature and 300°C is due to several different types of recovery mechanisms. The sharp drop in hardness at 400°C is caused by a phenomenon called *recrystallization*, which will be discussed in the next chapter.

The simplest recovery process is the annihilation of dislocations of opposite sign on the same slip plane [see Fig. 8-14(a)]. This process is relatively easy and will usually take place at low temperatures (room temperature for most metals).

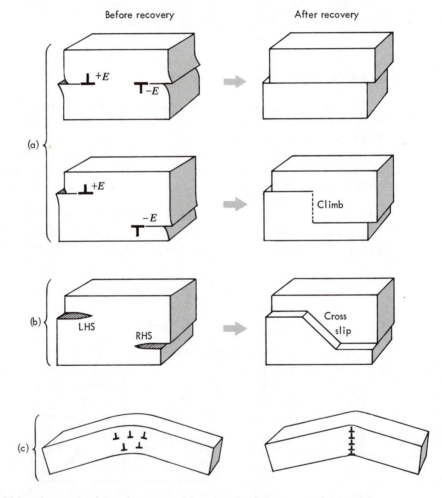

FIG. 8-14   Dislocation mechanisms of recovery: (a) and (b) annihilation; and (c) rearrangement

When dislocations of opposite signs are on different slip planes they cannot annihilate completely unless *cross-slip* (screw dislocations) or *climb* (edge dislocations) takes place [Fig. 8-14(b)]. These events are more difficult processes (they have higher activation energies), and they usually occur at elevated temperatures (above room temperature).

The edge dislocations illustrated in Fig. 8-14(b) can annihilate only if vacancies are created at the dislocation core and the vacancies are permitted to diffuse away from the dislocations. We shall see in the next section that this diffusion-controlled process plays an important part in high-temperature creep.

Recovery may also occur by the rearrangement of a given number of dislocations. Figure 8-14(c) illustrates a common type of dislocation rearrangement that occurs during recovery. The edge dislocations, which are initially positioned randomly, lower the strain energy of the crystal by stacking one above the other as shown. As discussed in Chapter 3, the final dislocation configuration is called a *low-angle grain boundary*. Each of these recovery mechanisms causes the internal stress (or strain energy) to be relaxed. These events are in opposition to those of strain hardening which cause the level of internal stress (or strain energy) to increase.

8-6 MECHANISMS OF DEFORMATION AT ELEVATED TEMPERATURES

The characteristics of *high-temperature creep* of crystalline solids were described in Chapter 6. We now turn our attention to the atomic mechanisms responsible for these high-temperature mechanical properties.

There are basically two different types of creep phenomena that can occur: *low-temperature creep*, sometimes called *logarithmic creep*, in which the plastic strain varies logarithmically with time, and *high-temperature creep*, or *steady state creep*, in which the plastic strain varies linearly with time after an initial transient stage. Low-temperature creep usually occurs at temperatures below $0.5T_m$, whereas steady state creep takes place above that temperature. The differences between low-temperature and high-temperature creep are shown in Fig. 8-15.

Let us consider the events that occur when a crystalline solid is deformed under creep conditions. The rate of deformation or *creep rate* is high at first. As plastic flow occurs, dislocations move and multiply and through their interactions cause strain hardening. This in turn causes the creep rate to diminish (see Fig. 8-15). Thus, creep deformation continues to occur, but at an increasingly slower rate. In low-temperature creep, the strain rate decreases because of strain hardening until it becomes too small to measure. In that event, for all practical purposes, creep will have stopped.

In the case of high-temperature creep, recovery processes such as dislocation annihilation and rearrangement counteract the effects of strain hardening. That

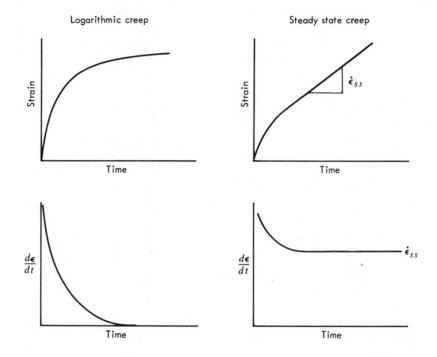

FIG. 8-15 Schematic strain-time relations for low-temperature (logarithmic) and high-temperature (steady-state) creep

is, while strain-hardening mechanisms cause the creep rate to decrease, recovery mechanisms permit the creep rate to accelerate. It is clear from the shape of the high-temperature creep curve (Fig. 8-15) that at the start of the test (decreasing strain rate) the effects of work hardening override the effects of recovery. However, after a stage in which the creep rate decreases with time, the creep rate reaches a steady state value. In this regime, the hardening effects of strain hardening are exactly compensated by the softening effects of recovery. The result is a deformation structure (number and arrangement of dislocations) that does not change with time. The constancy of the structure causes the creep rate to be constant. Thus, a steady state creep phenomenon arises naturally from the competition between strain hardening and recovery.

The description of steady state creep in terms of the effects of strain hardening and recovery leads to a simple but accurate quantitative account of steady state creep. At steady state the increment in strength due to strain hardening in a given period of time must be equal in magnitude to the decrement in strength caused by recovery. This condition may be expressed as

$$\left(\frac{d\sigma}{dt}\right)_{\substack{\text{strain}\\\text{hardening}}} = \left(\frac{d\sigma}{dt}\right)_{\text{recovery}} \tag{8-9}$$

or

$$\left(\frac{d\sigma}{d\epsilon}\right)_t \frac{d\epsilon}{dt} = \left(\frac{d\sigma}{dt}\right)_\epsilon \tag{8-10}$$

where $d\epsilon/dt$ is the steady state creep rate $\dot{\epsilon}_s$, $(d\sigma/d\epsilon)_t$ is the strain hardening rate h, and $(d\sigma/dt)_\epsilon$ is the recovery rate r. According to this simple analysis, the steady state creep rate is expressed as the ratio of the recovery rate to the strain-hardening rate at steady state

$$\dot{\epsilon}_s = \frac{r}{h} \tag{8-11}$$

This is the basic equation to which we shall refer when we discuss the atomic mechanisms that control high-temperature creep of solids. Specifically, we shall focus our attention on those factors that control recovery, since they are much more subject to structural control than are factors that change the strain-hardening rate.

It is clear from our discussion of creep thus far that the primary or transient stage of creep occurs when the material in question strain-hardens. Generally speaking, we find that relatively perfect crystals (for example, well-annealed metals) exhibit extensive primary creep, whereas solids that are first plastically deformed (cold worked) prior to testing may not show a transient creep stage at all.

We now turn our attention to the atomic mechanisms that control recovery and hence control high-temperature creep. As discussed in the previous section, the driving force for recovery is the reduction in strain energy that accompanies dislocation rearrangement and a decrease in the dislocation density. We also pointed out that recovery ultimately involves the climb motion of edge dislocations, which in turn depends on the process of self-diffusion. Considering all of these factors, it is possible to develop the following simple relation for the recovery rate r:

$$r = \left(\frac{d\sigma}{dt}\right)_\epsilon = A\left(\frac{\sigma}{E}\right)^p D_{SD} \tag{8-12}$$

where σ is the applied stress (strength), E is the elastic modulus, p is a constant, usually about 3 for pure metals, and D_{SD} is the self-diffusion coefficient. The diffusion coefficient appears in this relation because vacancy diffusion is required for dislocation climb, and the stress (or strength) compensated for elastic modulus appears because the rate of recovery is faster when dislocations are close together [high strength; see Eq. (8-8)].

In order to predict the steady state creep rate it is necessary to know both the recovery rate r and the strain hardening rate h. For our purposes it is sufficiently accurate to note that low-temperature (without recovery) stress-strain curves are often parabolic, with the consequence that the strain-hardening rate decreases with increasing stress according to

$$h = h_0\left(\frac{E}{\sigma}\right) \tag{8-13}$$

Now taking the ratio of the recovery rate to the strain-hardening rate according to Eq. (8-11), we have

$$\dot{\epsilon}_s = BD_{SD}\left(\frac{\sigma}{E}\right)^n \tag{8-14}$$

where $n = p + 1$ and $B = A/h_0$, both of which can be considered essentially constant for a given material. Using Eq. (8-14), we now turn our attention to the factors that influence the steady state creep rate.

Because the steady state creep rate of crystalline solids is proportional to the self-diffusion coefficient, it follows that the variation of the creep rate with temperature is dictated by the temperature dependence of D_{SD}. Since $D_{SD} = D_0 \exp(-\Delta E_D/RT)$ it follows that the activation energy for steady state creep is the same as that for self-diffusion. It is clear from this relation and Eq. (8-14) that

$$\ln \dot{\epsilon}_s = \ln\left[BD_0\left(\frac{\sigma}{E}\right)^n\right] - \frac{\Delta E_D}{RT} \tag{8-15}$$

and that a plot of $\ln \dot{\epsilon}_s$ vs. $1/T$ should reveal a straight line with a slope $= -\Delta E_D/R$. This relation is illustrated in Fig. 8-16 for a number of different aluminum alloys. It should be noted that since the steady state creep rate is controlled by diffusion it depends sensitively on the test temperature. Figure 8-16 shows that changing the

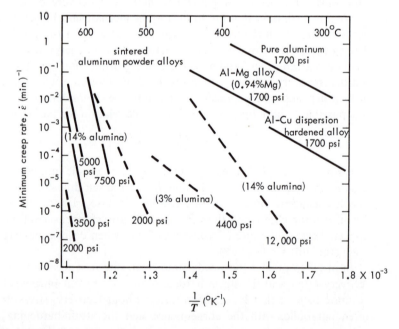

FIG. 8-16 Temperature dependence of steady state creep for a number of different Al alloys, after C. L. Meyers, J. C. Shyne and O. D. Sherby, *J. Aust. Inst. Metals,* **8,** 171 (1963). Used with permission.

temperature of aluminum from 300°C to 400°C can cause the creep rate at a given stress to change by as much as two orders of magnitude. The temperature dependence of creep in sintered aluminum powder (14% Al_2O_3) is even greater.

Equation (8-14) indicates that creep-resistant materials are those for which D_{SD} is small. We showed in Chapter 5 that D_{SD} is small at absolute temperatures, which are low compared to the absolute melting temperature T_M ($T/T_M \ll 1$). This means that a material will generally have better creep resistance if it is put into service at low temperatures relative to its melting temperature. It is for this reason that materials with high melting temperatures (oxides, refractory metals, and nickel- and cobalt-based superalloys) are used for high-temperature structural applications.

EXAMPLE 8-10

Polycrystalline copper creeps at a rate of 2.6×10^{-2} min^{-1} under a stress of 7000 psi at 560°C. If the activation energy for self-diffusion is 50,700 cal/mole, calculate the creep rate at 500°C at the same stress.

Solution:

Using Eq. (8.15) we can write

$$\dot{\epsilon}_s = B\left(\frac{\sigma}{E}\right)^n D_0 \exp\left(-\frac{\Delta E_D}{RT}\right)$$

Thus,

$$\ln\left(\frac{\dot{\epsilon}_1}{\dot{\epsilon}_2}\right) = -\frac{\Delta E_D}{R}\left(\frac{1}{T_1} - \frac{1}{T_2}\right)$$

Let

$$T_1 = 500°C = 773°K \qquad T_2 = 560°C = 833°K$$
$$\dot{\epsilon}_2 = 2.6 \times 10^{-2} \text{ min}^{-1}$$
$$\dot{\epsilon}_1 = 2.6 \times 10^{-2} \exp\left(-\frac{50,700}{1.98}\right)\left(\frac{1}{773} - \frac{1}{833}\right)$$
$$\dot{\epsilon}_1 = 2.29 \times 10^{-3} \text{ min}^{-1}$$

Equation (8-14) also predicts that the steady state creep rate is a sensitive function of the applied stress σ. Since

$$\ln\left(\frac{\dot{\epsilon}_s}{D_{SD}}\right) = \ln\frac{B}{E^n} + n\ln\sigma$$

it is clear that a plot of $\ln(\dot{\epsilon}_s/D_{SD})$ vs. $\ln\sigma$ should produce a straight line with a slope n. This graphical representation of the stress dependence of the steady state creep rate is shown in Fig. 8-17 for a number of different crystalline solids. Values of n range from 4 or 5 for the materials shown in Fig. 8-17 to about 40 for some dispersion-strengthened materials.

EXAMPLE 8-11

An alloy has a steady state creep rate of $\dot{\epsilon}_1$ when the stress is 10,000 psi and the temperature is 1000°C. The activation energy for creep in this alloy is 50,000 cal/mole and the stress exponent $n = 5$. It was discovered that certain modifications in the manufacturing procedure for this alloy would produce a more creep-resistant material. Specifically, as a result of the new procedure, the alloy will exhibit a creep rate $\dot{\epsilon}_1$ at 1000°C at an improved stress of 11,000 psi. The manufacturers of jet engines who use this material are more interested in advances in the operating temperature. Estimate the increase in the operating temperature that would result from this development.

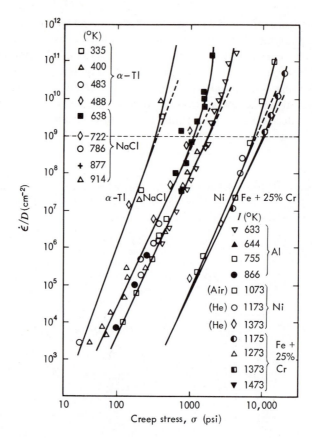

FIG. 8-17 Stress dependence of the steady state creep rate for various materials, after O. D. Sherby and P. M. Burke, *Prog. in Mat. Sci.,* **13,** 325 (1968). Used with permission.

(i.e., at 10,000 psi, what is the temperature at which the creep rate for the "new procedure" material is $\dot{\epsilon}_1$?).

Solution: Let $\sigma_0 = 10,000$ psi $T_0 = 1000°C\ (1273°K)$

$\sigma_1 = 11,000$ psi

For the new process

$$\dot{\epsilon}_1 = A'\sigma_1^5 \exp\left(-\frac{\Delta E}{RT_0}\right) = A'\sigma_0^5 \exp\left(-\frac{\Delta E}{RT_1}\right)$$

$$5 \ln\left(\frac{\sigma_1}{\sigma_0}\right) = \frac{\Delta E}{R}\left(\frac{1}{T_0} - \frac{1}{T_1}\right)$$

$$\frac{1}{T_1} = \frac{1}{1273} - \frac{5(1.97)}{(50,000)} \ln\left(\frac{11}{10}\right)$$

$$\frac{1}{T_1} = 0.784 \times 10^{-3} - 1.87 \times 10^{-5}$$

$$T_1 = 1308°K = 1035°C$$

Thus, the new process produces an increase of 35°C in the operating temperature.

Since all recovery processes involve the movement of individual dislocation, whether by glide or climb, it follows that all of the mechanisms that inhibit dislocation movement also inhibit recovery. All the imperfection strengthening mechanisms discussed in the previous section, with the exception of grain boundary strengthening, can contribute to high-temperature strength. The mechanisms that inhibit recovery by inhibiting dislocation motion in a number of different high-temperature structural materials are summarized in Table 8-3. As was the case for

TABLE 8-3

Strengthening Mechanisms for Some High-Temperature Structural Materials

Materials	Principal Strengthening Mechanisms
Refractory metals (Ta, W, Mo)	Intrinsic strength–Peierls force; grain boundary strengthening
Nickel-based superalloys	Precipitation strengthening (Ni_3Al precipitates)
Cobalt-based superalloys	Solid solution strengthening (W and Mo in solid solution)
TD-nickel and TD-nichrome	Dispersion strengthening (ThO_2 inclusions)
Sintered aluminum powder (SAP)	Dispersion strengthening (Al_2O_3 inclusions)
Oxides (aluminum oxide)	Intrinsic strength; precipitation strengthening

low-temperature strengthening mechanisms, the imperfection mechanisms for reducing recovery rates at high temperatures are more effective when the mean spacing between the defects is small. Specifically, this means that when an alloy is strengthened by precipitation or dispersion strengthening the spacing between the precipitates must remain small if the strengthening mechanism is to remain effective. For a fixed volume fraction of the second phase, this means that the individual particles must remain small. The achievement of this structural condition at high temperatures is one of the major problems in the development of alloys for high-temperature service. Since the surface-to-volume ratio of a spherical particle is $4\pi r^2/\frac{4}{3}\pi r^3 = 3/r$, the total surface area and surface energy (particle-matrix surface) associated with a microstructure are lowest when the particles are large. This means that large particles are thermodynamically more stable than small ones. At high temperatures rapid diffusion permits large particles to grow at the expense of small ones. Since particle strengthening is inversely proportional to particle size (or particle spacing), the yield and creep strengths tend to decrease with increasing temperatures. This tendency may be reduced through the use of particles that are insoluble in the matrix, such as Al_2O_3 in aluminum and ThO_2 in nickel, since the rate of particle coarsening depends on the solubility of the particle in the matrix.

In addition to recovery and strain hardening, two other modes of deformation contribute to high-temperature creep: *grain boundary sliding* and *directional diffusion creep*. These are discussed on pages 244-245. In the case of grain boundary sliding, deformation is very inhomogeneous since it is all concentrated at the grain boundary and often leads to the formation of cracks and premature rupture (Fig. 8-18).

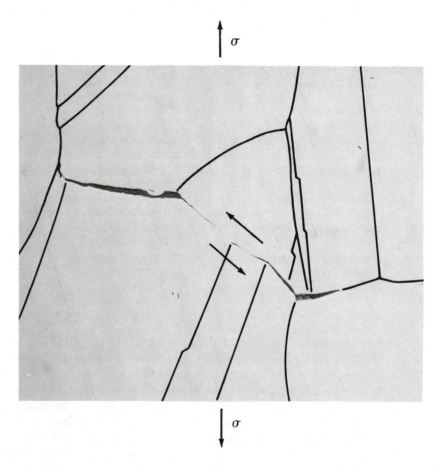

FIG. 8-18 Intercrystalline cracks formed in polycrystalline Ni + 6% W by grain boundary sliding. The positions of the grain boundaries and twin boundaries have been drawn on the photograph of the unetched surface. Photo courtesy D. K. Matlock, Stanford University. Used with permission.

8-7 THE MECHANICS OF FRACTURE

In Chapter 7 we discussed the process of plastic deformation in perfect crystals and showed that the theoretical shear strength of a perfect crystal was about $G/10$. We noted that plastic deformation can occur at stresses that are much below this value because of the presence of dislocations. Likewise, if we perform a similar calculation of the critical tensile stress σ_c required to break atomic bonds and

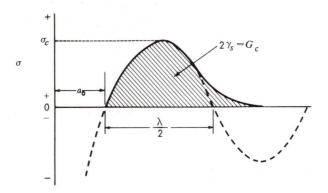

FIG. 8-19   Sine approximation for the force between atoms ; model for estimating the theoretical cleavage strength of a crystal, σ_c

separate atomic planes (Fig. 8-19), assuming again a simple sine law of forces between atoms, we find that

$$\sigma_c \approx \frac{\lambda}{2\pi}\frac{E}{a_0} \approx \frac{E}{10} \tag{8-16}$$

which is several orders of magnitude greater than the measured fracture strengths of most materials. Consequently, as in the case of a slip, it appears that defects must be present to cause fracture at applied stress levels that are much below the theoretical value. These defects are sharp cracks and notches.

Elastic Fracture

We now consider a method that may be used to calculate the fracture strength of a perfectly elastic (nonyielding) material (e.g., glass) that contains a sharp crack (with tip radius ρ about equal to the lattice spacing a_0). The crack shown in Fig. 8-20 advances when atomic bonds are broken at its tip. In order to extend the crack, a certain amount of work must be done to separate the atomic planes to three or four atomic distances, thereby forming two new crack surfaces. The shaded area of the atomic force–distance curve shown in Fig. 8-19 is the work done in creating two unit areas of surface, and hence is twice the *surface energy* of the material, $2\gamma_s$. For the two-dimensional crack shown in Fig. 8-20, having a total surface area of $4a$ per unit thickness, the total energy involved in forming its surface is

$$W_s = 4a\gamma_s \tag{8-17}$$

As the crack spreads, more surface is created and more work needs to be done. At the same time, however, the growth of a crack involves a decrease in the potential energy of the loading mechanism, since work is done by the applied forces

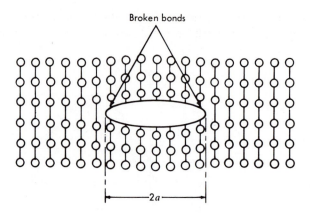

FIG. 8-20 Illustration of bond breaking at the tip of an elastic crack

when the crack surfaces are allowed to separate. There is also an increase in the strain energy of the body when the crack grows, but this increase is more than compensated by the work of the applied forces. The principles of elasticity can be used to show that the potential energy of the external forces, less the elastic energy for a body with a crack of length $2a$ under a tensile stress σ, is

$$W_e = -\frac{\sigma^2 \pi a^2}{E} \tag{8-18}$$

so that the total energy of a crack solid (including external forces) under an applied tensile stress is

$$W_{\text{total}} \approx -\frac{\sigma^2 \pi a^2}{E} + 4a\gamma_s \tag{8-19}$$

As in the case of the nucleation of a second-phase particle discussed in Chapter 5, the condition for unstable growth (W_{total} decreases as the crack extends) is obtained by taking the derivative of W_{total} with respect to a and setting it equal to zero. Thus, the condition of instability ($\sigma = \sigma_F$) is

$$\frac{dW_{\text{total}}}{da} = 0$$

or

$$-\frac{\sigma_F^2 \pi 2a}{E} + 4\gamma_s = 0$$

$$\sigma_F = \left(\frac{2E\gamma_s}{\pi a}\right)^{1/2} \tag{8-20}$$

so that the fracture strength decreases as the length of a pre-existing crack increases. Equation (8-20) is known as the *Griffith criterion* for fracture. Often this equation is written in the form

$$\sigma_F = \left(\frac{EG_c}{\pi a}\right)^{1/2} \tag{8-21}$$

where G_c is the *total work of fracture*.

Equation (8-21) can be interpreted as predicting the critical values of stress and crack length at which a crack will begin to propagate in an unstable manner. Evidently when $\sigma\sqrt{\pi a}$ reaches the value $\sqrt{EG_c}$, the crack will grow. In this context it is convenient to treat $\sigma\sqrt{\pi a}$ as a measure of the driving force for crack propagation. It is common practice to define this quantity

$$K = \sigma\sqrt{\pi a} \qquad (8\text{-}22)$$

as the *stress intensity factor*. We then consider that fracture occurs when the stress intensity factor K is equal to or greater than the *critical stress intensity factor*,

$$K_c = \sqrt{EG_c} \qquad (8\text{-}23)$$

usually called the *fracture toughness*. For practical applications of the fracture toughness concept, K_c is usually measured by determining the crack lengths and stresses at which artificially produced cracks actually propagate.

The theoretical value of the surface energy γ_s is about $Ea_0/20$. Table 8-4 gives the values of the measured fracture energies G_c for some ceramics (e.g., MgO); these are in good agreement with the theoretical values, indicating that fracture

TABLE 8-4

Theoretical and Measured Values of G_c for some Structural Materials

Material	$G_c = 2\gamma_s$ (Theoretical) (lb/in.)	G_c (Measured) (lb/in.)
Glass	0.020	0.08
Plexiglass	0.065	2.74
MgO	0.085	0.10
High-strength steel	0.13	300
High-strength aluminum	0.04	100
High-strength titanium	0.06	600

occurs by the elastic fracture process just described. The low value of G_c for glass and ceramics means that very small flaws can trigger an unstable fracture; consequently these materials are very brittle and weak when they contain cracks.

EXAMPLE 8-12

An inspection technique for finding cracks in a high-strength steel landing gear has sufficient resolution to find cracks that are 0.1 in. in length. Calculate the stress level we can expect the landing gear to support without breaking.

Solution:

From Table 8.4, $G_c = 300$ lb/in. for high-strength steels; $E = 30 \times 10^6$ psi for steel.

$$\sigma_F = \left[\frac{(30 \times 10^6)(300)}{\pi(0.1)}\right]^{1/2}$$

$$\sigma_F = 1.7 \times 10^5 \text{ psi}$$

Plastically Induced Fracture

Table 8-4 indicates that for plexiglass and almost all metals, even those that are often thought of as being brittle, the measured value of G_c is two to four *orders of magnitude* greater than the true surface energy term, $2\gamma_s$. The reason for this difference is that in metals the yield strength in a small volume near the crack tip (σ_Y) is much lower than the cohesive strength (σ_c), and the stresses concentrated at the crack tip therefore cause localized yielding before localized fracture (Fig. 8-21). Yielding extends over a region called the *plastic zone*, within which tensile

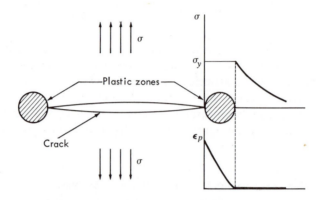

FIG. 8-21 Illustration of stresses and strains in the plastic zones at crack tips

stresses are of the order of the yield stress σ_Y. Since $\sigma_Y \ll \sigma_c$, the crack is able to advance only when a sufficiently high *plastic strain* is produced in the volume element at the crack tip.

In contrast to the case of elastic fracture, where most of the displacement perpendicular to the fracture plane is concentrated over a distance of the order of a few atomic spacings (10 A), plastic fracture involves large strains that extend over several grain diameters (about 10^{-3} cm) perpendicular to the fracture plane. Consequently, a much larger volume is deformed during plastic fracture, and the *work per unit area of fracture surface, G_c*, is increased proportionately. In polymers, the large value of G_c relative to $2\gamma_s$ is due to the work done at the crack tip in unravelling and stretching the long polymer chains.

The processes leading to fracture, either in the plastic zone at the tip of a crack or in an uncracked tensile specimen, are complicated, and the details are beyond the scope of this text. Basically the problem is one of determining how the stress level in the deforming element ($\sigma \approx \sigma_Y$) can be increased to the cohesive stress $\sigma_c \gg \sigma_Y$. The answer lies in the fact that on a *microscopic scale* plastic deformation is an inhomogeneous process and involves the formation of dislocation pile-ups or blocked slip bands at grain boundaries (Fig. 8-22) or at phase boundaries.

Under certain conditions (low temperatures and high strain rates) the tensile

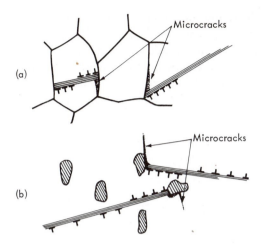

FIG. 8-22   Illustration of mechanisms of the formation of microcracks in solids: (a) grain boundaries; (b) phase boundaries

stress at the tip of a blocked slip band in BCC metals and ceramics* builds up to the cohesive stress and causes the formation of a microcrack, even though the *average stress* acting across the grains is of the order of the yield strength (Fig. 8-22). In BCC metals these microcracks tend to form on {100} planes called *cleavage planes*, across which the atomic bonds are weakest. Since cleavage occurs at relatively low strains, along well-defined cleavage planes, the cleavage fractures appear bright to the eye or under the microscope.

A second type of plastic fracture, which occurs in FCC and HCP metals and in BCC metals tested at moderate temperatures, involves the formation of voids at the crack tip (Fig. 8-23). Once formed, the voids act as centers of stress and

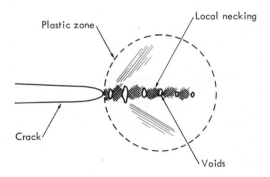

FIG. 8-23   Schematic illustration of mechanism of crack propagation under conditions of shear fracture

---

*FCC metals and many HCP metals do not fracture by cleavage because the stresses at the crack tip can be relieved by plastic deformation, which does not usually concentrate into bands.

strain concentration, and fracture ultimately occurs by local necking in the region of the crack tip. This fracture process is called *shear fracture* and produces a fracture surface with a dull, mottled appearance.

In general, the local strains required for shear fracture are independent of temperature, while the strains required to cause cleavage decrease with decreasing temperature (Fig. 8-24). Localized fracture occurs when the crack tip strains build

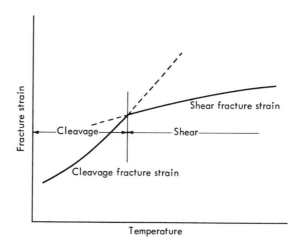

FIG. 8-24 Temperature dependence of cleavage and shear fracture strains; brittle-ductile transition

up to the critical value for cleavage or shear, and the fracture mode depends on which fracture criterion is satisfied first. Therefore, BCC metals and alloys, such as low-carbon steel, fracture by cleavage at low temperature and by shear at moderate and high temperatures. FCC and HCP metals fail by shear under most conditions.

The work of fracture, G_c, is proportional to the plastic strain required to induce fracture at a crack tip. Therefore, the transition from the cleavage mode (low strain to fracture) to the shear mode (high strain to fracture) in BCC metals and alloys is a *brittle-ductile transition*. This transition can be detected in a *Charpy impact test* as illustrated in Fig. 6-20. The fracture toughness shows a similar transition temperature. This is illustrated in Fig. 8-25 for both low-alloy/low-strength and high-alloy/high-strength steels. Rearranging Eq. (8-22), we see that at a given applied stress σ the critical crack length

$$2a_c = \frac{2}{\pi}\left(\frac{K_c}{\sigma}\right)^2 \tag{8-24}$$

and increases rapidly with the temperature in the brittle-ductile transition region. This rapid change occurs at temperatures just above a critical temperature called the *nil-ductility temperature*. In order to guarantee safe designs for low-alloy steel

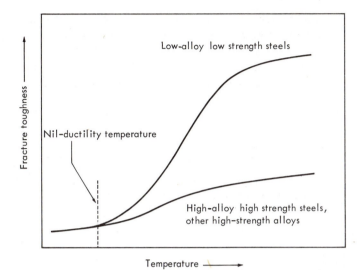

FIG. 8-25 Schematic illustration of the brittle-ductile transition in terms of the fracture toughness (K_c) for low-alloy/low strength steels and for high-alloy/high strength steels and other high-strength alloys (e.g., aluminum and titanium alloys)

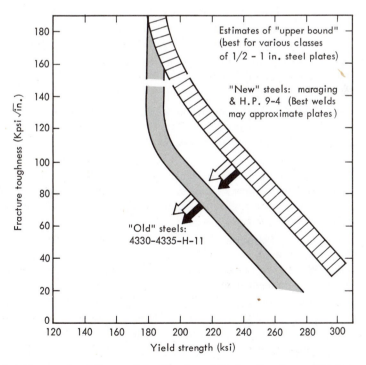

FIG. 8-26 Effect of yield strength on "upper-bound" value of fracture toughness (K_c) for various medium- and high-strength steels

structures such as bridges and pressure vessels, it is necessary to use steels for which the nil-ductility temperature is at least 120°C below the operating temperature.

In high-alloy/high-strength steels (e.g., 4340*), high-strength aluminum (e.g., 7075-T6), and titanium alloys (e.g., 6A1-4V) used in aerospace applications, a sharp brittle-ductile transition does not occur as the temperature is raised. At room temperature K_c is sufficiently low that very small cracks (2a $\approx$ 0.1 in.) introduced by fatigue or corrosion can cause brittle fracture in service. Figure (8-26) illustrates the unfortunate fact that most structural materials become more brittle as their strengths are increased. In consequence, materials with high yield strengths must be inspected very carefully to be sure that harmful defects are not formed. From Eq. (8-24) we see that for low values of K_c (i.e., for brittle materials) the critical crack size is small and the task of detecting the flaws by inspection difficult.

PROBLEMS

8-1 A crystal of LiF is tested in compression at a strain rate of 10^{-5} sec^{-1}. If the initial mobile dislocation density is 10^4 cm^{-2}, what is the compressive stress at which the slope of the stress-strain curve will be half the elastic modulus? For LiF, $E = 1.1 \times 10^{12}$ dynes/cm^2 and $b = 2 \times 10^{-8}$ cm. The stress dependence of the dislocation velocity is defined by $\tau_0 = 3.9 \times 10^7$ dynes/cm^2 and $m = 10$. This problem demonstrates that plastic flow starts at stresses well below the yield point. The conversion relation $\tau = \frac{1}{2}\sigma$ may be used for this problem.

8-2 A tensile specimen of tungsten containing a mobile dislocation density of 10^8 cm^{-2} is strained at a rate of 0.01 min^{-1}. Estimate the dislocation velocity at the upper yield point.

8-3 Explain why it is impossible for the mobile dislocation density of a crystalline solid to be lower than about 500 cm^{-2} at the yield point when the applied strain rate is 10^{-1} sec^{-1}. The shear wave (elastic wave) velocity for most crystals is about 10^5 cm/sec.

8-4 If the dislocation velocity exponent $m = 35$ (such as in Fe–3% Si steel), what is the ratio of the upper yield stress σ_{UY} to the lower yield stress σ_{LY} if the mobile dislocation density increases from 10^8 cm^{-2} to 10^{11} cm^{-2} when the lower yield stress is reached after the yield point?

8-5 The yield strength of a single crystal of iron is 10,000 psi when 0.01 atomic percent nitrogen is dissolved in the lattice and 15,000 psi when 0.04 atomic percent nitrogen is dissolved in the lattice. What is the yield strength when the nitrogen content is 0.03 atomic percent?

8-6 What is the dispersion hardening due to the presence of 5 weight percent Al_2O_3 particles, 10 μ in diameter, that are dispersed in pure aluminum? Al_2O_3 has a density of 4 g/cm^3, and aluminum has a density of 2.8 g/cm^3. The shear modulus of aluminum is $G = 4 \times 10^6$ psi.

---

*See Table 6.2 for more information on these alloys.

8-7 If the dispersion hardening due to the presence of Fe_3C particles in steel 1 μ in diameter is 20,000 psi, what is the mean spacing between particle centers? Estimate the carbon content of the steel.

8-8 Suppose that for a copper tensile specimen the dislocation density varies with true strain as $\rho = 10^{10}\epsilon$. If the entire stress-strain curve can be described by a power law $\sigma = \sigma_0\epsilon^n$, and necking occurs when $\epsilon = n$, what is the true stress at necking? For copper $G = 4 \times 10^{11}$ dynes/cm^2 and $b = 3 \times 10^8$ cm.

8-9 Suppose that the grains in a molybdenum tensile specimen are cubical in shape. If the grain size hardening is 10,000 psi when 625 grains are present in 1 mm^2 of cross section, what is the grain size hardening when 10,000 grains appear in 1 mm^2?

8-10 Suppose that a nickel-base alloy ($T_M = 1450°C$) can safely undergo a creep rate of 0.01 % per hour at an applied stress of 10,000 psi and a temperature of 600°C. If the stress is raised to 15,000 psi, what is the temperature at which creep rate reaches the same value? For this alloy $n = 5$ and $\Delta E = 70,000$ cal/mole.

8-11 Using the methods described in Chapter 7, show that

$$\sigma_c \approx \frac{\lambda}{2\pi} \frac{E}{a_0} \approx \frac{E}{10}$$

as given by Eq. (8-16).

8-12 A dilute alloy with the BCC structure contains 5×10^{15} foreign atoms per cubic centimeter. Initially, all the dislocations that are present are pinned (immobilized) by the collection of solute atoms at their cores. We assume that a dislocation is completely pinned when there is one solute at every lattice point along the length of dislocation. The initial dislocation density for this material is $\rho_0 = 10^6$ cm^{-2}. The lattice parameter is $a_0 = 2 \times 10^{-8}$ cm. Show the stress-strain curves after the following treatments:
(a) as received (initial condition);
(b) specimen prestrained 2%;
(c) specimen prestrained 2% and annealed for a prolonged time;
(d) specimen prestrained 20%;
(e) specimen prestrained 20% and annealed for a prolonged time.

8-13 A sample of polycrystalline iron is strained in a tensile test at the rate of 10^{-3} sec^{-1}. If the mobile dislocation density is 10^8 cm^{-2} (constant), what is the tensile flow stress for the material? Assume that the Burgers vector is 2×10^{-8} cm, $m = 20$, and $\tau_0 = 1422$ psi.

8-14 Mechanical tests are run on polycrystalline samples of copper at room temperature. It is found that at a strain of 0.002 the stress necessary for further plastic deformation is 10,000 psi. A stress of 15,000 psi is required at a strain of 0.04. What stress would be necessary for further plastic deformation if the sample had already undergone a strain of 0.14? Assume that the initial dislocation density is high.

8-15 The average dislocation velocity in a sample of iron can be expressed as $\bar{v} = (\tau/\tau_0)^{40}$, where τ is the average shear stress in the sample and τ_0 is 15,000 psi.
(a) Why must this expression be modified or discarded at high stresses?
(b) Estimate the stress at which a sample of iron would begin to deform plastically.
(c) Estimate the fractional change in the flow stress $\Delta\tau/\tau$ that would accompany a fractional increment in the strain rate on the sample $\Delta\dot{\epsilon}_p/\dot{\epsilon}_p = 0.5$.

(d) What assumptions about the dislocation density must be made to solve part (c) of this problem?

8-16 A crystal has a lattice friction stress of 10^7 dynes/cm^2. If the grown-in dislocation density is 10^9 cm/cm^3, calculate the stress necessary to initiate plastic flow. Take $G = 10^{11}$ dynes/cm^2 and $b = 2 \times 10^{-8}$ cm.

8-17 Would a high-strength iron alloy with a lattice friction stress of 100,000 psi and a dislocation density of 10^{12} cm/cm^3 be expected to yield or fracture if microscopic cracks are present, where the cracks are 10^{-2} cm long with a radius of curvature at their tip of 10^{-7} cm? Take

$$\gamma = 1000 \text{ ergs/cm}^2$$
$$G = 5 \times 10^{11} \text{ dynes/cm}^2$$
$$b = 2 \times 10^{-8} \text{ cm}$$

8-18 Explain why FCC metals generally do not behave in a brittle fashion at low temperatures while BCC metals do fracture in a brittle fashion.

BIBLIOGRAPHY

G. E. DIETER, JR., *Mechanical Metallurgy*. New York: McGraw-Hill, 1961.

F. GAROFALO, *Fundamentals of Creep and Creep Rupture in Metals*. New York: Macmillan, 1965.

J. J. GILMAN, *Micromechanics of Flow in Solids*. New York: McGraw-Hill, 1969.

H. W. HAYDEN, W. G. MOFFATT, and J. WULFF, *The Structure and Properties of Materials*, Vol. III. New York: John Wiley, 1965.

R. W. K. HONEYCOMBE, *The Plastic Deformation of Metals*. New York: St. Martin's Press, 1968.

F. A. McCLINTOCK and A. S. ARGON, *Mechanical Behavior of Materials*. Reading, Mass.: Addison-Wesley, 1966.

D. McLEAN, *Mechanical Properties of Metals*. New York: John Wiley, 1962.

D. PECKNER, ed., *The Strengthening of Metals*. New York: Van Nostrand Reinhold, 1964.

Strengthening Mechanisms in Solids. Cleveland: American Society for Metals, 1962.

W. J. M. TEGART, *Elements of Mechanical Metallurgy*. New York: Macmillan, 1966.

A. S. TETELMAN and A. J. McEVILY, JR., *Fracture of Structural Materials*. New York: John Wiley, 1967.

J. WEERTMAN and J. R. WEERTMAN, "Mechanical Properties I and II," in *Physical Metallurgy*, ed. R. W. Cahn. Amsterdam: North Holland Publishing Company, 1965.

9

The Relation Between
Mechanical Properties and
Microstructural Control

9-1 INTRODUCTION

In Chapters 4 and 5 we noted that the microstructure of a material depends upon its alloy composition and thermal history, and in Chapters 6–8 we noted that the mechanical properties of materials were determined by microstructure. In this chapter we synthesize the important aspects of all these phenomena and discuss the methods of *microstructural control* that are used to obtain optimum mechanical properties.

The practical methods of microstructural control are subject to certain constraints that are imposed by the processes of fabrication. For example, for most metallic materials a large ingot is worked into the desired shape by rolling, extruding, or drawing, as shown in Fig. 9-1. The energy expended in producing the large strains during forming is proportional to the flow stress. If the metal is hard and strong, as desired for service use at the *end* of the fabrication process, then more energy (i.e., cost) will be required for fabrication. Also, the large strains involved during forming would cause most hard materials to fracture. Consequently, while metals need to be strong for use in service, they need to be soft during fabrication. By using the principles of microstructural control, it is, in fact, possible to keep metals soft during fabrication and allow them to harden afterward.

The student may wonder why all metallic components are not made by simply melting and casting into the desired shape. In fact, many low-strength parts (e.g., cast-iron engine blocks) and complicated parts that are difficult to shape (e.g.,

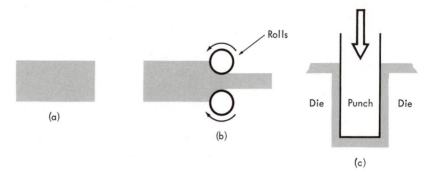

FIG. 9–1   Schematic illustration of working operations used to form metal parts: (a) initial ingot; (b) rolling; (c) stamping or punching

nickel-base alloy turbine blades) are made by casting. However, as-cast materials tend to have an inhomogeneous microstructure (Chapter 5) and consequently show great variability in mechanical properties. In addition, since these materials are cooled from the melt where diffusion is rapid, they tend to have a *coarse* microstructure (large grain size, large dispersed particles) and hence relatively low strength. High strength requires a fine microstructure (Chapter 8). This can best be obtained in worked (*wrought*) structures where a high dislocation density is generated and in structures that have undergone solid state phase transformations (small, closely spaced, second-phase particles).

There are numerous processes of microstructural control that, when combined with various fabrication processes, can allow the formation of high-strength, relatively inexpensive parts. We shall not be concerned with the details of specific processes. Instead, we shall focus our attention on the general principles, mentioning only how these principles apply to important commercial materials such as brasses, aluminum alloys, steels, ceramics, and fiber composites. In Chapter 10 we shall deal with similar topics for polymeric materials,

9-2 COLD WORKING, RECRYSTALLIZATION, AND HOT WORKING OF METALS

In the previous chapters we noted that the dislocation density in a material increases with increasing plastic strain. Since each dislocation has a certain strain energy associated with it, the overall elastic strain energy or free energy of the material increases with increasing plastic strain (increasing dislocation density). Because systems always tend toward the minimum in free energy there is a tendency for the dislocations either to annihilate or rearrange themselves into arrays where the overall strain energy is decreased. The rate at which these dislocation rearrangements occur is strongly temperature dependent. As the temperature is increased, thermal energy aids dislocation movement (both glide and climb, the latter being a diffusion controlled process) so that dislocations may begin to move in response to the stress fields of surrounding dislocations. That is, dislocations with attractive forces between them move together, and those with repulsive forces between them move apart. Accompanying these changes in dislocation arrangement is a corresponding decrease in the strength or hardness of the material. To demonstrate the influence of temperature on the strength of a cold worked material, consider the following example.

Suppose that a block of α brass (Cu–30% Zn) is rolled down from an initial thickness of 1 in. to a final thickness of $\frac{1}{2}$ in. (a 50% reduction in area) and that samples are cut from the deformed block. If each of these samples is heated (annealed) for a given period of time (say, one hour) at various temperatures and the hardness of each sample is then measured at room temperature, a curve such as that shown in Fig. 9-2 is obtained. The hardness of the cold worked metal decreases

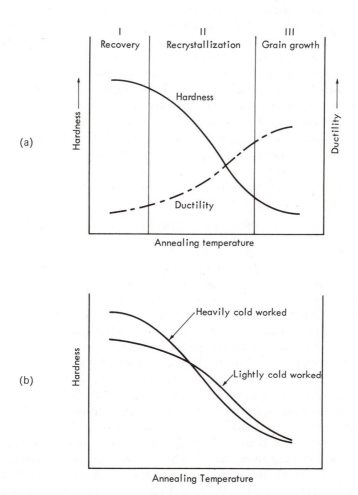

FIG. 9-2 (a) Variation of properties of cold-worked metals as a function of annealing temperature. (b) influence of amount of cold work on recrystallization behavior

as the annealing temperature increases. There are three distinct processes occurring during these annealing treatments: *recovery*, *recrystalization*, and *grain growth*.

The low-temperature softening results from *recovery*, a process in which the dislocations introduced during working are arranged into more stable configurations (lower strain energy). These arrangements are often called *subgrain boundaries* or *cell boundaries* [Fig. 9-3(a) and (b)]. Characteristic of these are the tilt and twist boundary described in Chapter 3. The formation of these dislocation arrangements produces only a slight decrease in hardness, since the dislocation density changes very little and the subgrain boundaries are still good barriers to further dislocation motion. One distinctive characteristic of the recovery process is that is does not involve any change in the grain structure. The only changes taking place are the dislocation arrangements within existing grains.

The rapid softening that occurs at higher temperatures is due to *recrystallization*, a process in which the dislocation density is reduced to a very low value. This process occurs by the following sequence of events. Small, strain-free (dislocation-free) grains are nucleated and grow in the deformed matrix. As these grains grow and consume the matrix, the dislocations in the matrix are absorbed and essentially annihilated at the boundaries of the newly formed grains. When the new grains impinge upon one another, the process of recrystallization is complete. Since dislocations are swept out of the deformed material and form the boundaries of the newly formed grains, the hardness (strength) decreases and the ductility increases as increasing numbers of strain-free grains are formed (i.e., as the recrystallization process proceeds).

If we consider recrystallization as a solid state transformation that can be described as

$$\text{crystal (deformed)} \longrightarrow \text{crystal (undeformed)}$$

we can then analyze it by the principles of nucleation and growth discussed in Chapter 5. As before, we must consider both the change in bulk free energy and surface energy when nuclei are formed. The bulk free energy ΔG_V that is driving the transformation is the strain energy associated with the dislocations. At the instant a strain-free (recrystallized) region is nucleated, a surface will be formed at the boundary between deformed and undeformed regions [Fig. 9-3(c)]. This interface has an energy γ that opposes the formation of the undeformed grain. Analogous to Eq. (5-25), the number of nuclei will be proportional to $\exp(-\Delta G^*/RT)$, where ΔG^* is the critical free energy required to form a stable nucleus. In contrast to the case of solidification, ΔG_V for recrystallization is essentially independent of temperature; ΔG_V depends on the amount of elastic strain energy, and this is not temperature-dependent except for the small temperature dependence of the elastic constants. Consequently, ΔG^* is independent of temperature, and the rate of recrystallization simply increases with increasing temperature, since both the number of nuclei and the rate of growth (which is proportional to the rate of atom motion) increase with increasing temperature. Also, since the driving force ΔG_V increases as the amount of stored elastic strain energy increases, recrystallization will occur faster, at a given temperature, as the amount of prior plastic strain increases [Fig. 9-2(b)]. Generally, by increasing

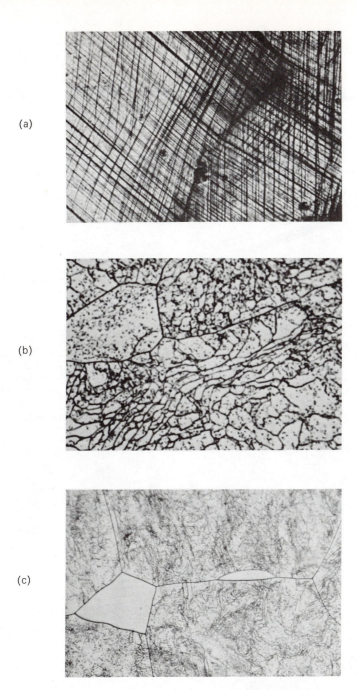

FIG. 9-3   Dislocation arrangement in (a) cold worked (150×), (b) recovered (1000×) and (c) partially recrystallized Fe + 3% Si after annealing at 800°C for 10 min. (500×). These are etch-pit micrographs, and each small dot represents the intersection of a dislocation with the free surface. The white appearance of the recrystallization nuclei in (c) indicates that they are essentially dislocation-free. Photos courtesy A. J. West and W. C. Harrigan, Jr., Stanford University. Used with permission.

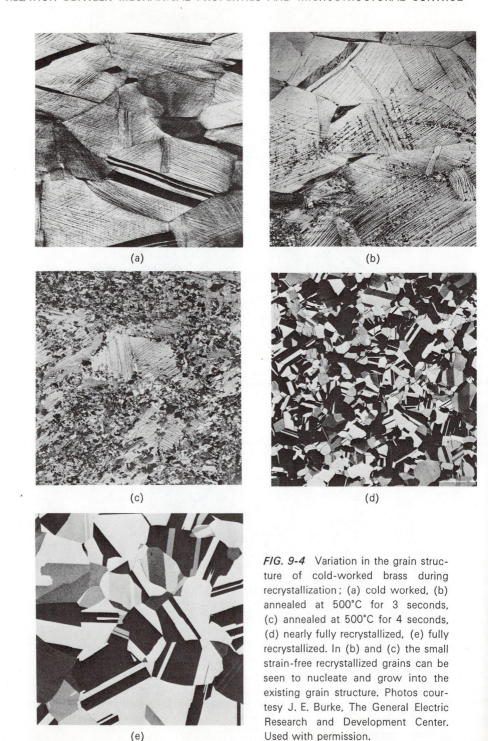

FIG. 9-4 Variation in the grain structure of cold-worked brass during recrystallization; (a) cold worked, (b) annealed at 500°C for 3 seconds, (c) annealed at 500°C for 4 seconds, (d) nearly fully recrystallized, (e) fully recrystallized. In (b) and (c) the small strain-free recrystallized grains can be seen to nucleate and grow into the existing grain structure. Photos courtesy J. E. Burke, The General Electric Research and Development Center. Used with permission.

the amount of prior plastic strain we increase the rate of nucleation more than the rate of growth; hence the as-recrystallized grain size decreases with increasing prior strain. Raising the annealing temperature has the reverse effect and results in a larger as-recrystallized grain size.

Figure 9-4 shows the microstructure of a sample of brass, initially cold worked and then held for various lengths of time at 500°C. Figure 9-4(b) and (c) show the recrystallization process at various stages of time, and Fig. 9-4(d) illustrates the grain structure completion of the recrystallization process.

Since the rate of recrystallization increases with increasing temperature, the time required to achieve a given state of recrystallization (e.g., hardness lowered by 50% of total possible decrease) decreases with increasing temperature. From curves such as those shown in Fig. 9-5 it is possible to determine an activation energy for recrystallization. This activation energy will depend on the activation energy for diffusion, ΔE_D, since atomic rearrangements are required for the nucleation and growth of the strain-free grains. ΔE_D is in turn proportional to absolute melting temperature, so that the minimum *recrystallization temperature* in a heavily cold worked metal (temperature at which recrystallization is 95% complete after 1 hr anneal) decreases with decreasing absolute melting temperature.

Recrystallization is generally inhibited by the presence of any defects that inhibit the motion of grain boundaries (e.g., small second-phase particles, inclusions, etc.). The purer the material, the lower the recrystallization temperature.

EXAMPLE 9-1

A heavily cold rolled brass sheet must be annealed for 2 min at 400°C before recrystallization is 50% complete. How long must the sheet be annealed at 300°C if the activation energy for recrystallization is 40,000 cal/mole?

Solution:

$$\text{Rate} = \frac{1}{\text{time for 50\% completion}} = \frac{1}{t} = A \exp\left(\frac{-\Delta E}{RT}\right)$$

where A is a constant.

$$\text{At } T_1 = 400°C = 673°K, \qquad t_1 = 2 \text{ min}$$
$$\text{At } T_2 = 300°C = 573°K, \qquad t_2 = ?$$

$$\frac{1/t_1}{1/t_2} = \frac{\exp\left(-\Delta E/RT_1\right)}{\exp\left(-\Delta E/RT_2\right)}$$

$$\ln\frac{t_2}{t_1} = \frac{\Delta E}{R}\left(\frac{T_1 - T_2}{T_1 T_2}\right) = \frac{40,000}{1.98}\left(\frac{100}{(673)(573)}\right) = 5.19$$

Thus,

$$\frac{t_2}{t_1} = 180, \, t_2 = 180 \times 2 \text{ min} = 360 \text{ min}$$

The final stage of softening [stage III in Fig. 9-2(a)] is due to the *growth* of newly recrystallized grains. The driving force for this process is the total grain boundary surface energy (total surface area), which can be reduced as the grains grow. The increase in grain size produces a decrease in yield strength (Chapter 8). This

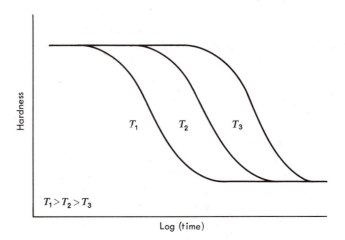

FIG. 9-5   Variation of time dependence of the hardness decrement, with annealing temperature due to recrystallization

process is also diffusion-controlled and hence proceeds faster as the temperature is increased.

Thus far we have considered the operations of working and annealing as two separate processes that are conducted in series. Certain parts are fabricated by alternate working and annealing operations, the number of anneals depending on how much strain is required to form the part and how much strain the material can withstand before fracture. For example, if a 60% reduction is required, but cracking will occur after a 20% reduction, then the two intermediate anneals must be performed during fabrication.

Alternatively, if a material is worked above its recrystallization temperature (*hot worked*), then recrystallization occurs simultaneously with plastic strain. Since the material remains soft and ductile during deformation, very large strains can be introduced without causing the formation of cracks. Large ingots are usually hot worked by rolling or forging to change the coarse, inhomogeneous microstructure into a homogeneous, fine microstructure and to reduce the ingot to a thickness that can be subsequently formed into the desired shape. It should be remembered that hot rolled products are essentially strain-free and consequently are softer than cold worked products. Therefore, the final fabrication step performed on ductile materials (e.g., low-carbon steel, brass and some aluminum alloys) is usually carried out by cold working to harden the part for use in service.

9-3 PRECIPITATION HARDENING

In Chapter 8 we noted that the strength of a material could be greatly increased if a second phase were present in the form of finely spaced, hard particles. We also showed that high strength required that the particle size be small. Also, in Chapter

5, we noted that when a phase transformation takes place (for example, the precipitation of β from α), the β particles will be small if the rate of nucleation is high and the rate of growth is low. This is the case when the transformation takes place at a low temperature where the amount of supercooling ΔT below the equilibrium transformation temperature is large and the rate of atomic diffusion is generally small. Consequently, it is very difficult to achieve a fine particle size (and high strength) in as-cast materials, which generally cool slowly from the melt and undergo phase transformations near the equilibrium transformation temperature. Fine particle size almost always necessitates rapid cooling rates and solid state transformations at relatively low temperatures.

The easiest way to achieve a fine dispersion of a second phase is through the precipitation of β particles from a supersaturated solid α phase according to the generalized reaction

$$\alpha \longrightarrow \alpha + \beta$$

Consider the phase diagram shown in Fig. 9-6. For an alloy of composition C_S,

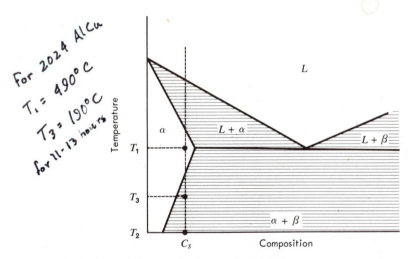

For 2024 AlCu
$T_1 = 490°C$
$T_3 = 190°C$
for 11-13 hours

FIG. 9-6   Phase diagram illustrating the precipitation-hardening process

following hot working or casting into shape and slow cooling back to room temperature, the β particles will be quite large and the alloy will be soft. We now desire to treat the material thermally to produce a fine dispersion of β particles. The first step in the precipitation-hardening process is to *solution anneal* the material by heating it to a temperature in the single-phase α region (temperature T_1 in Fig. 9-6) and then waiting until all the β phase dissolves in the α matrix. The alloy is then quickly cooled, or *quenched*, to temperature T_2 (usually room temperature). This rapid cooling prevents the precipitation of β particles, because the time required for cooling is less than the time required to start the transformation, as signified by the *T-T-T* or *C* curve shown in Fig. 9-7 and discussed in detail in Chapter 5.

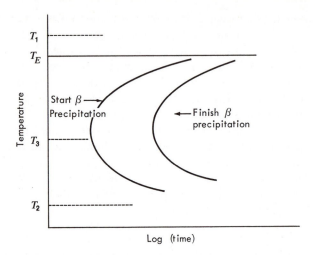

FIG. 9-7 T-T-T curve for the alloy C_S shown in Fig. 9.6. T_E is the equilibrium temperature at which β should first appear.

The alloy is relatively soft immediately after quenching. Any hardening that occurs is the result of solid solution hardening, due to the presence of B atoms dissolved in the α matrix. If the alloy is maintained at temperature T_2, the β phase will not form because the rate of diffusion is very slow. However, if we heat the alloy back up to temperature T_3, still in the two-phase region, then the diffusion rate will be rapid enough for the β phase to precipitate out at a measurable rate. Holding the alloy at temperature T_3 is referred to as *aging* the alloy. The sequence of events occurring during the precipitation (or aging) process is as follows.

First, B atoms start to cluster together by diffusion, forming nuclei of β phase precipitates. When only a few small precipitates are present, they will not offer much resistance to dislocation motion, and the strength of the alloy will not increase much. As the number and size of precipitates increase, more work is required to force the dislocations through them. Consequently, the yield strength increases as the time of aging increases (Fig. 9-8). Eventually, it becomes sufficiently difficult for the dislocation to cut through the β particles so that it is easier for the dislocations to bow between the particles (see Section 7-6) than to cut through them. As discussed in Chapters 7 and 8, when the dislocations bow between the precipitates, the flow stress is inversely proportional to the spacing between precipitates (i.e., the finer the dispersion of precipitates, the higher the flow stress). If the alloy is aged further, the particles will tend to grow, lowering the surface-to-volume ratio. This is accomplished by the dissolution of small particles of β and the diffusion of B atoms to the interface of a larger β particle. The spacing between particles then increases and, correspondingly, the yield strength decreases. An alloy that has been aged beyond the point of maximum strength is usually said to be *overaged*. For practically all precipitation-hardened alloys there is no significant overaging and softening occurring during long time service at ambient temperatures. This is

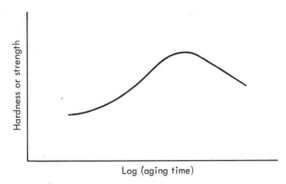

FIG. 9-8   Variation of hardness or strength with aging time in precipitation hardening

due to the slow rates of diffusion at ambient temperatures and also to the logarithmic time dependence of the thermally activated processes involved in nucleation and growth.

The time required for precipitation to achieve maximum strength depends on the temperature of aging (Fig. 9-9). For low aging temperatures it takes longer to reach

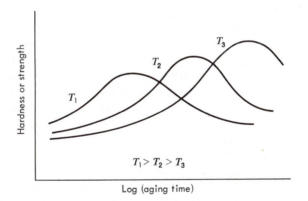

FIG. 9-9   Influence of aging temperature on precipitation-hardening characteristics

maximum strength than at high aging temperatures. Generally the maximum strength is higher for low aging temperatures since the low temperatures favor a finer second-phase dispersion (high nucleation rate and low growth rate).

The precipitation hardening process represents another example of microstructural control that allows a part to be soft during fabrication but then hardened for service use. The main requirement for precipitation hardening is decreasing solubility with decreasing temperature. To give some idea of the degree of strengthening obtained by precipitation hardening, the maximum strength increment obtained by precipitation hardening in aluminum alloys (aluminum-copper and aluminum-magnesium-zinc are the most common) is of the order of $G/75$. Thus, it is possible to raise the strength level to within 10 to 20% of the theoretical lattice strength. In addition to aluminum-base alloys, certain steels, nickel-base superalloys, and titanium alloys are strengthened by age hardening processes.

EXAMPLE 9-2

Suppose two sheets of an aluminum alloy that has been strengthened by both precipitation heat treatment and cold work are welded together. Describe the microstructural changes that accompany the welding process and explain how these changes will affect the strength of the alloy.

Solution:

Adjacent to the weld zone, where the metal is heated nearly to the melting temperature and then cooled rapidly (the rest of the sheet acts as a heat sink), the microstructure will be similar to a fully recrystallized, solution-treated material with fairly low strength. As the distance from the weld zone increases, the processes of recrystalization, recovery, and overaging occur to lesser extents. Each of these processes contributes to a lower strength for the alloy. At a large distance from the weld the material will still retain its high strength.

9-4 HEAT TREATMENT OF STEEL

We shall now discuss, in some detail, the heat treatment of steel (iron-carbon alloys). The reasons for choosing to discuss steel are numerous and do not stem solely from the fact that steels are probably the most important metallic materials used commercially. The phase transformations that take place in the iron-carbon system lend themselves nicely to a discussion of how to control microstructure and hence mechanical properties through control of thermal-mechanical history. Also, we shall see that there is an example of a diffusionless phase transformation (*martensite transformation*) in the iron-carbon system that has great significance when we consider the processing of ultrahigh-strength materials. Although the discussion will deal mainly with steel, it should be remembered that the general principles described also apply to many other materials,

Most steels contain less that 1.2 weight percent carbon and may contain considerably higher concentrations of other alloying elements. For example, stainless steels generally have about 18% chromium and 8% nickel. For purposes of discussion, however, we can imagine the phase equilibria of steel to be adequately represented by the binary iron-carbon phase diagram shown in Fig. 9-10. Iron-carbon alloys containing more than 2% carbon are known as *cast irons*, and although their properties can be controlled by heat treatment, this class of iron-carbon alloys is somewhat more specialized than the lower-carbon steels. For future discussion we shall focus our attention on the bottom left-hand corner of this diagram.

Practically all steel heat treatments involve heating into the γ (austenite) region and then cooling back to ambient temperatures. The important variable in the heat treatment is the *cooling rate*, for this determines not only the *size* of the microstructure but also the *nature* of the phases present. Examination of the iron-carbon

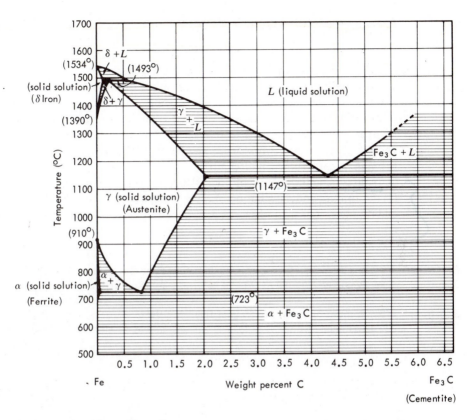

FIG. 9-10   The iron-carbon binary phase diagram

phase diagram indicates three *equilibrium* phase transformations to be of interest:

$$\gamma \longrightarrow \alpha$$
$$\gamma \longrightarrow Fe_3C$$

and the eutectoid transformation

$$\gamma \longrightarrow \alpha + Fe_3C$$

In Chapter 4 we showed that during equilibrium (slow) cooling the eutectoid transformation produces a platelike structure, called *pearlite*, in which the lamellae of α (ferrite) and Fe_3C (cementite) lie side by side (Fig. 4-17). In addition to these equilibrium transformations there is a *nonequilibrium* transformation of interest. At high cooling rates γ decomposes via the diffusionless transformation

$$\gamma \longrightarrow \alpha'$$

where α' (martensite) is essentially a supersaturated BCC iron-carbon alloy.

From the point of view of microstructural control and mechanical properties, most steels can be grouped into one of two categories—*ferritic-pearlitic steels* or *martensitic steels.*

Ferritic-Pearlitic Steels

Most of the steels used today, on a tonnage basis, are iron-carbon, *hypoeutectic* (carbon content below eutectoid composition) steels that contain small amounts of alloy additions to remove oxygen and sulfur during the steel-making operation and to refine the microstructure. These carbon steels are used in structural applications (bridges, I beams, pressure vessels) where moderate strength at low cost is desired. They are also used in thin sheets for "tin" cans and automobile fenders because they can be easily worked and because the strain hardening introduced by forming imparts sufficient strength.

In discussing the process of microstructural control, two factors are important: (1) the amount and distribution of each phase, as determined by the phase diagram, and (2) the size of each phase, as determined by the cooling rate from the austenite range.

EXAMPLE 9-3

What is the weight fraction and distribution of phases in an iron–0.2% carbon steel, slowly cooled from 900°C to just below the eutectoid temperature, T_{Eu}?

Solution:

The amount of α that precipitates from γ in the $\alpha + \gamma$ region, called *primary ferrite*, is determined by applying the lever rule just above the eutectoid temperature (723°C)

$$\%\alpha = \frac{0.8 - 0.20}{0.8 - 0.02} = 77\%$$

Therefore, the $\%\gamma$ available to transform to pearlite just below T_{Eu} is 23%. Just below the eutectoid temperature

$$\text{Total } \%\alpha = \frac{6.7 - 0.20}{6.7 - 0.02} = 97\%$$

Therefore, $\%\alpha$ in eutectoid structure $= 97 - 77 = 20\%$, and the total $\%\text{Fe}_3\text{C} = 100 - 97 = 3\%$.

Consider the cooling of a hypoeutectoid steel from the γ range through the $\alpha + \gamma$ range to the $\alpha + \text{Fe}_3\text{C}$ region. As discussed in Chapter 5, solid state nucleation tends to occur at the grain boundaries of the transforming phase, since the effective surface energy of the nuclei will be lower there. Therefore, the primary ferrite grains that appear at temperatures above the eutectoid temperature are nucleated at austenite grain boundaries and grow into the austenite. The remaining austenite, which transforms to pearlite below T_{Eu}, lies in between the primary ferrite grains. Consequently, the pearlite appears as patches, surrounded by primary ferrite, after the transformation is complete (Fig. 9-11).

The kinetics of the solid state precipitation of one solid phase from another are described by the T-T-T curve discussed in Chapter 5. A typical T-T-T curve for a low carbon (0.2%) steel is shown in Fig. 9-12. Austenite begins transforming to primary ferrite somewhere below the equilibrium transformation temperature, the exact temperature depending on the cooling rate. Likewise, the pearlite begins to form somewhere below the eutectoid temperature. We can get some qualitative informa-

FIG. 9-11   The microstructure of a hypoeutectoid steel showing primary ferrite (white) and pearlite (600×). From *Physical Metallurgy for Engineers* by Donald S. Clark and Wilbur R. Varney © 1952 by Litton Educational Publishing, Inc. Reprinted by permission of Van Nostrand Reinhold Company.

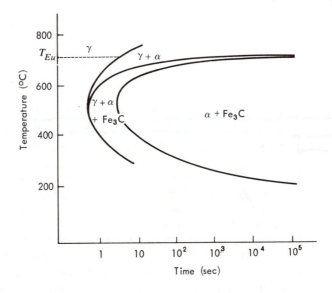

FIG. 9-12   Schematic *T-T-T* curve for Fe + 0.2% C steel

tion on the size of the microstructure by referring to our previous discussion on phase transformations. In Fig. 5-20 we noted that if a sample of an alloy was quenched from above the equilibrium transformation temperature to some temperature T_1, and another sample was quenched to a lower temperature $T_2 < T_1$, the sample transformed at a higher temperature T_1 would have the coarser microstructure.

We also noted that a fast, continuous cooling rate allowed the transformation to occur at low temperatures and produced a finer microstructure than did a slow, continuous cool. This situation applies to the case of steels. Steels cooled slowly have a coarser microstructure than steels cooled rapidly (Fig. 9-13). Since the yield

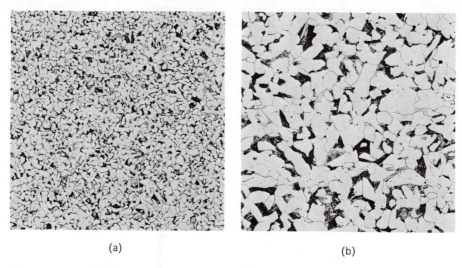

(a)   (b)

FIG. 9-13 Microstructure of 0.2% carbon steel: (a) fast cooling rate, (b) slow cooling rate (150 ×). From *Physical Metallurgy for Engineers* by Donald S. Clark and Wilbur R. Varney, ©1952 by Litton Educational Publishing, Inc. Reprinted by permission of Van Nostrand Reinhold Company.

strength increases with decreasing grain or microstructural size, the yield strength of a low-carbon steel increases with increasing cooling rate from the austenite range. This then offers a simple way to refine the microstructure and increase strength.

To facilitate microstructure refinement and further strengthening in low-carbon steels it would appear logical to look for some way to suppress the $\gamma \rightarrow \alpha$ and $\gamma \rightarrow \alpha + Fe_3C$ transformations so that they take place at lower temperatures. One way to achieve this goal is to use the knowledge that alloy additions can stabilize the presence of a given phase with respect to other phases. For example, austenite transforms to ferrite at 910°C in pure iron, but is stable down to the eutectoid temperature of 723°C when 0.8% carbon is present. Thus, carbon acts as an *austenite stabilizer* and allows austenite to remain in equilibrium with ferrite at lower temperatures as the carbon concentration increases. Now if we add an alloying element, such as nickel, which acts as an *austenite stabilizer*, to a low-carbon steel we can expect the eutectoid temperature to be decreased; consequently, the rate of austenite decomposition to ferrite and cementite will decrease. This decrease is caused by a reduction in the amount of supercooling at any given temperature $T < T_{Eu}$. This means that under conditions of continuous cooling the transformation will take place over a lower temperature range, where diffusion is slower, and hence the resulting microstructure (ferrite grain size, pearlite lamellae spacing) will be finer. In terms of the T-T-T curve, the addition of an austenite stabilizer has the effect of shifting the curve to the right so that it takes longer times at a given temperature to complete the transformation (Fig. 9-14). This leads to both an increase in yield strength and an increase in toughness (Chapter 8). In general, the yield strength of ferritic-pearlitic steels varies between 35,000 psi and 70,000 psi, depending on cooling rate and alloy content.

[handwritten margin notes: add. of ferrite stabilizer ... trans. take longer a given temp.]

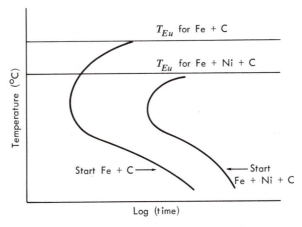

FIG. 9-14   Shift of the nose, or start, of the transformation curve due to the presence of an austenite stabilizer, Ni

EXAMPLE 9-4   Suppose a steel of eutectoid composition has been heat treated to give a fine pearlite spacing. Now if this steel is heated back up to a temperature just below the eutectoid temperature, indicate why the microstructure is unstable, the probable form the microstructure will assume, and what influence the change will have on the mechanical properties.

Solution: The microstructure is unstable because of the large amount of α-Fe$_3$C surface area (high surface energy). The structure will change in such a way to reduce the surface area/volume ratio for the two phases. This usually occurs by having the α-Fe$_3$C lamellae break up to form spherical particles of Fe$_3$C in an α matrix [see Fig. 1-1]. Because the microstructure coarsens in this change, the strength decreases along with a corresponding increase in ductility.

Martensitic Steels

From the above discussion it would seem that what is needed to produce very high-strength steel is to increase the cooling rate from the γ range to such a degree that an ultrafine pearlitic microstructure is produced. This concept is true in part; that is, very high-strength steel is produced employing rapid cooling, but it is not fine pearlite that provides the strength. Instead we find that, upon rapid cooling, the austenite decomposition produces microstructures that do not resemble pearlite at all.

Consider the austenite decomposition of a eutectoid steel (0.8% carbon) whose *T-T-T* curve is shown in Fig. 9-15. If a sample is quenched from the austenite range

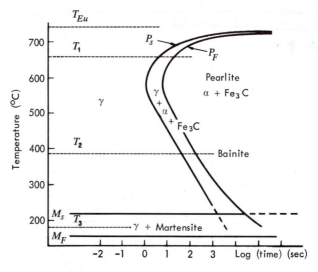

FIG. 9-15   *T-T-T* curve for eutectoid steel (Fe + 0.8% C)

to a temperature T_1 above the nose of the curve, and then isothermally transformed, pearlite will begin forming at the time indicated by the *pearlite start* curve P_s and all of the austenite will have transformed to pearlite at the time denoted by the *pearlite finish* curve P_F. Now, if the steel is transformed at a temperature T_2, below the nose of the curve (corresponding to a rapid cooling rate), the α and Fe_3C no longer form the platelike structure called pearlite. Instead, the Fe_3C phase appears as finely dispersed particles in the ferrite matrix, which has an elongated, lathlike appearance [Fig. 9-16(a)]. This mixture of α and Fe_3C is known as *bainite*. The reason that the pearlite structure does not form in this temperature range is that at these low temperatures the diffusion rates are so low, and the driving force for transformation is so high, that the carbon atoms do not have time to diffuse over sufficient distances to form an Fe_3C plate. Instead, they agglomerate over short distances and appear as fine particles. Since the resultant ferrite grain size is small as well, bainitic microstructures have very high strength and toughness.

Suppose now that the eutectoid steel is quenched to a lower temperature T_3 without intersecting the pearlite start curve. At this low temperature the diffusion rates are so low that Fe_3C *cannot* precipitate from the austenite lattice, even as fine precipitates. Consequently, the diffusion controlled, decomposition of austenite cannot take place. At the same time, T_3 is so far below the equilibrium transformation temperature T_{Eu} that the driving force for *FCC* austenite to transform to BCC ferrite is enormous. In fact the driving force is so large that it triggers off a diffusionless phase transformation in which the γ transforms to a body centered tetragonal structure which is merely supersaturated α (ferrite) with excess carbon. This supersaturated α is known as *martensite* or α'.

The diffusionless transformation by which γ decomposes to martensite takes place by a complicated shearing of the γ lattice. Each atom moves only a small distance relative to its neighbors, less than 1 atomic distance. Consequently, thermal activation in the sense of vacancy motion or solid state diffusion is not required for the formation of martensite.

Since martensite formation is a diffusionless transformation, it cannot be suppressed by quenching and, irrespective of time, a certain amount of martensite will form at a given temperature $T < M_S$, where M_S is the temperature at which martensite first appears, or the *martensite start temperature*. The amount of martensite that forms at a given temperature below M_S will increase with increased cooling below M_S. At $T = M_F$, all the austenite will have transformed to martensite, so that M_F is called the *martensite finish temperature*. Consequently, both austenite and martensite will exist at temperatures $M_F < T < M_S$. The reason that a temperature gap exists between M_S and M_F (i.e., that all the austenite doesn't convert to martensite at M_S) is that the formation of the first bit of martensite stabilizes the microstructure and decreases the driving force for further transformation. There is a volume change accompanying the $\gamma \rightarrow \alpha'$ transformation, and the strain energy associated with this difference in specific volume offsets the chemical driving force. The temperature must be decreased before the driving force is sufficient to cause further martensite formation.

Martensite appears very similar to bainite when viewed under the microscope [Fig. 9-16(b)]; it, too, has a very fine, needlelike appearance. The one difference, of course, is that Fe_3C has not precipitated from the body centered structure, so that martensite is a supersaturated single-phase structure. The very fine size of the martensite grains and the extremely high concentration of interstitial carbon atoms that are locked in the structure, and give a large, solid solution hardening contribution, combine to make martensite one of the hardest metallic structures yet produced. The hardness of martensite increases with increasing dissolved carbon content and is considerably greater than that of even fine grained, ferritic-pearlitic structures.

The specific volume of the BCT martensite is greater than that of the transforming FCC austenite, and the difference increases with increasing carbon content. This difference in specific volume introduces large residual strains in the newly formed microstructure which, combined with the high intrinsic hardness, make as-quenched martensite a very brittle material. Martensitic structures must there-

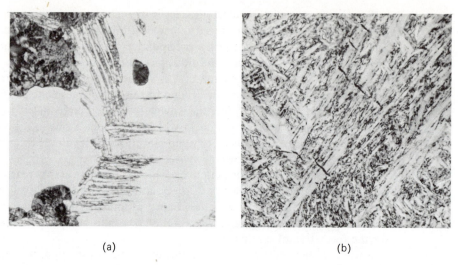

(a)   (b)

FIG. 9-16   Micrographs of (a) bainite (1250×) and (b) martensite showing very fine
microstructures produced by high cooling rates. From *Physical Metallurgy for
Engineers* by Donald S. Clark and Wilbur R. Varney, ©1952 by Litton Educational
Publishing, Inc. Reprinted by permission of Van Nostrand Reinhold Company.

fore be *tempered* before they can be reliably used. Tempering is a process designed
to remove the residual stresses that exist after martensite formation and soften the
martensite by precipitation of carbon as Fe$_3$C. Specifically, tempering consists
of annealing at temperatures between 200 and 723°C for times ranging from 10
min to 1 or 2 hr, depending on the size of the part and the desired hardness level.
Figure 9-17 illustrates a typical curve of hardness vs. tempering temperature. In

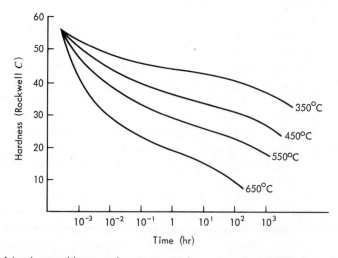

FIG. 9-17   Variation of hardness with tempering treatment for a quenched 0.35% C steel

general, residual stresses are relieved by microplastic deformation at temperatures around 200°C. Softening occurs above 200°C and is due to the precipitation of carbon as Fe_3C from supersaturated, tetragonal martensite according to the reaction

$$\alpha' \longrightarrow \alpha + Fe_3C$$
$$\text{martensite} \quad \text{ferrite} \quad \text{cementite}$$

When precipitation is completed (around 360°C), the BCC matrix is composed of very fine, ferrite grains that appear similar to bainite plates [Fig. 9-16(a)]. While the yield strength has been reduced because of the reduction in solid solution hardening of carbon atoms, the Fe_3C particles are still finely dispersed and provide a precipitation hardening effect. Hence, these steels possess relatively high strength. More important, this strength is combined with very high *toughness*, due to the removal of distortion, the decrease in yield strength, and the fine grain size, so that they possess an *optimum combination* of strength and toughness. Figure 9-18 shows the increase in toughness that accompanies the decrease in yield strength.

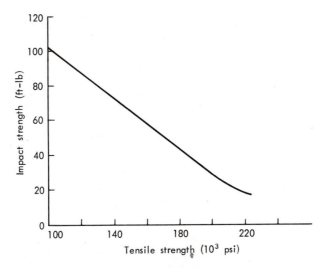

FIG. 9-18 Approximate impact energy-tensile strength correlation for quenched and tempered steels

Further increases in tempering temperature (or time) cause the ferrite grains and Fe_3C particles to grow. The latter assume the shape of spheres, since this minimizes their ratio of surface area to volume (see Example 9-4). These structures are referred to as *spheroidized* because of the spheroidal carbide particles.

We have seen above that the transformation of austenite to ferrite and cementite can proceed by either of two routes:

(1) $\gamma \longrightarrow$ pearlite $\longrightarrow$ ferrite + spheroidized carbides

(2) $\gamma \longrightarrow$ martensite $\longrightarrow$ ferrite + spheroidized carbides

Regardless of the route, the properties of the final products are the same. The properties of the intermediate structures depend upon grain size (cooling rate) and carbon content. Some pearlitic steels, for example, possess the same yield strength (but lower toughness) as quenched and tempered medium-carbon (0.3–0.6%) martensites. Typically, the yield strength of martensitic steels varies between 100,000 and 250,000 psi, depending on carbon content and tempering treatment.

Pearlitic-Martensitic Steels

It is possible to produce *mixed microstructures* in steels by varying the cooling steps from the austenite range. Suppose (Fig. 9-19) that an alloy is quenched to

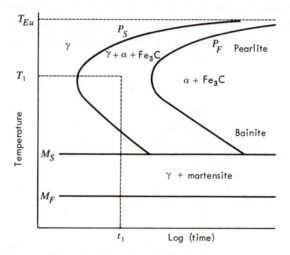

FIG. 9-19 Hypothetical *T-T-T* curve illustrating formation of mixed microstructure

T_1, isothermally transformed for a certain period of time t_1 until some pearlite has formed, and then quenched to below the M_F temperature. The austenite that did *not* transform at T_1 will be available to transform to martensite below M_S, and the microstructure will exhibit a mixture of martensite and pearlite. The longer the prior holding time at T_1, the greater the amount of pearlite formed there and the lower the amount of martensite. It is important to realize that the *pearlite itself remains unchanged* during the subsequent austenite-martensite transformation because pearlite is an equilibrium microstructure.

The strength of the mixed pearlite-martensite microstructure will depend on the relative amounts of the two types of constituents. To a first approximation we can estimate the ultimate strength of the mixtures UTS_{mix} by taking the weighted average of the two strengths (rule of mixtures)

$$\text{UTS}_{\text{mix}} \approx \text{UTS}_{\text{mart}}V_f + \text{UTS}_p(1 - V_f) \tag{9-1}$$

where UTS_{mart} and UTS_p are the ultimate strengths of the martensite and pearlite, respectively, and V_f is the volume fraction of martensite. In hypoeutectoid steels primary ferrite, pearlite, and martensite can all be formed by suitable heat treatment. The strength of these three phase mixtures can also be estimated by the rule of mixtures.

We noted previously (Fig. 9-14) that alloying additions that stabilize the austenite phase (carbon, nickel) shift the *T-T-T* curves to the right for both the austenite-ferrite and austenite-pearlite transformations. This ability to shift the *T-T-T* curves is a very important factor in high-strength martensitic steel design. The magnitude of the displacement of the *T-T-T* curve is referred to as the *degree of hardenability* of the steel. This terminology is very logical, as illustrated by Fig. 9-20. For a given

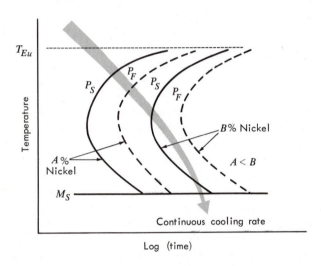

FIG. 9-20 Shift in the *T-T-T* curve due to alloying, which allows martensite to form at slower cooling rate

continuous cooling rate, more martensite (the harder phase) will form as the nickel content is increased from *A* to *B* and the *T-T-T* curve is shifted to the right. The nickel addition allows the steel to be harder for a given cooling rate and therefore imparts hardenability. Thus, hardenability is a property of steel that determines the depth and distribution of hardness introduced by quenching. The higher the hardenability, the slower the cooling rate needed to form a full martensitic structure. Note that nickel itself has no effect on the ultimate strength of martensite. The significant hardness increase arises because of the increased martensite content. Chromium and molybdenum, although they are not austenite stabilizers, produce similar effects. Increasing carbon content causes the ultimate strength and hardness to increase for two reasons: first, because carbon shifts the *T-T-T* curve to the right (see Figs. 9-12 and 9-15), and second, because the strength of martensite itself increases with increasing carbon content.

The hardenability imparted by specific alloying agents is a very important consideration in the choice of steels for high-strength, complicated parts. The reason is that a given hardness is primarily controlled by the amount of martensite in continuously cooled steels. If the hardenability is low, fast cooling is required to achieve large percentages of martensite. However, fast cooling causes the outside of a part to transform to the hard martensite structure before the inside of the part has a chance to cool down. When it reaches its M_S temperature and transforms to martensite, it will expand against an already hardened layer, and residual stresses, distortion, and even cracking will be produced. Therefore, parts must be cooled as slowly as possible from the austenite range (e.g., by oil quenching instead of water quenching), and this requires addition of alloy elements to impart hardenability. Since these elements are expensive, the cost of heat-treatable steel is much greater than that of nonhardenable, low-alloy, ferritic-pearlitic steel. Also, it must be remembered that the cooling rate at the center of a part is slower than at the outside. Consequently, the thicker the part to be hardened, the greater the hardenability (i.e., alloy content) must be to achieve the desired hardness for a given cooling treatment (air cooling, water quenching, etc.).

For some applications, such as rotating shafts, it is desirable to have a very hard outer surface to minimize wear, but a tough internal core. This may be produced by heating a ferritic-pearlitic steel at temperatures around 900°C in the presence of carbon (in methane gas, or in contact with graphite chips) and cooling rapidly to ambient temperature, a process known as *carburizing*. The carbon is allowed to diffuse into the outer layers of the part and causes the formation of hard martensite after cooling. The inner part, still with low-carbon content, retains the soft ferrite-pearlite structure.

Steels are generally classified by their alloy content—particularly the amount of nickel, chromiun, and molybdenum—present as hardenability agents. The last two of four digits specify the number of "points" of carbon. For example, the 1000 series steels have ferrite-pearlite microstructures, and a 1020 steel contains 0.20 weight percent carbon, while a 1040 steel contains 0.40 weight percent carbon. The 2000 series steels (e.g., 2315) are used in the ferritic-pearlitic condition or bainitic condition and contain about 3% nickel to refine the microstructure. The 3000 series steels (3140) contain nickel (1.4%) and chromiun (0.70%) and possess sufficient hardenability to allow the martensitic transformation in relatively thick structures at moderate cooling rates. The 4000 series steels (4340) contain more nickel (1.8%) and chromium (0.80%) as well as molybdenum (0.25%) and can be cooled at a slower rate from the austenitic range to produce martensite.

9-5 COMPOSITE MATERIALS

Composite materials are those in which individual phases are bonded together in such a manner that the average properties of the composite are determined by the individual properties of each phase. Generally the composite is constructed in order to gain advantageous characteristics from each of the component materials or to overcome disadvantageous characteristics of each. Materials containing small amounts of finely dispersed second-phase particles are normally not considered composites because the dispersed phase affects properties by *interacting* with the primary phase, rather than by contributing on its own. The term *composite* usually refers to materials in which the volume fraction of the minor phase is at least 10 or 15%.

Many common structural materials are composites of one form or other. Familiar examples are steels containing mixed microstructures, concrete (sand, gravel, hydrated cement), fiberglass (glass fibers in a polymer matrix), and wood (cellulose and lignin). The combination of specific materials to form a composite is particularly important with nonmetallic materials. Consider the case of tungsten carbide or titanium carbide, which are very hard, brittle ceramic materials that are used as die inserts for metal forming. The toughness of these materials is very low, of the order of the true surface energy. If a small crack is accidentally introduced into a large piece of one of these materials, brittle fracture can occur at very low stresses, since there is a continuous path for brittle crack propagation. However, if the tungsten carbide (WC) is dispersed in the form of fine particles in a tough, ductile metal matrix such as cobalt (Fig. 9-21), then even if a crack is introduced acci-

FIG. 9-21   WC in a cobalt matrix (10% cobalt) (500×). Photo courtesy W. L. Silence, Stellite Division, Cabot Corporation. Used with permission.

dentally, it can be stopped by the tough continuous phase. The strength and service reliability of WC can therefore be increased by dispersing it in a metallic matrix. Similarly, fiberglass composites combine the strength of glass fibers with the toughness and flexibility of a polymeric matrix.

Since we are interested in determining the bulk properties of composites in terms of the properties of the individual components, let us consider a simple example. Suppose two materials, such as boron and aluminum, are bonded together in the form of thin laminates and we desire to know the elastic properties of the composite. There are two extreme cases to consider.

1. Suppose that a stress σ is applied transverse to the laminates, as in Fig. 9-22(a), such that the laminates are loaded in *series*. Let V_f be the volume fraction of boron so that $(1 - V_f)$ is the volume fraction of aluminum. The elastic modulus of composite E_{comp} can be computed by dividing stress σ

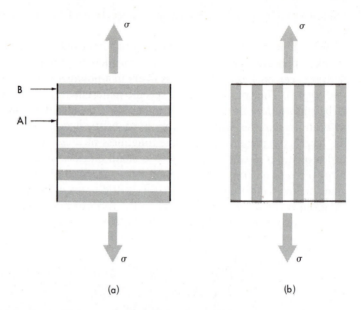

FIG. 9-22   Composite structure illustrating (a) isostress and (b) isostrain loading of an aluminum-boron laminate

by the total strain ϵ_{total} (ϵ_{total} is the sum of the strain in each phase multiplied by its volume fraction). If E_{B} is the elastic modulus of boron (50×10^6 psi) and E_{Al} the elastic modulus of aluminum (10×10^6 psi), then

$$\epsilon_{\text{total}} = \frac{\sigma}{E_{\text{B}}} V_f + \frac{\sigma}{E_{\text{Al}}}(1 - V_f) \tag{9-2}$$

Since

$$\sigma = E_{\text{comp}} \epsilon_{\text{total}}$$

$$\frac{1}{E_{\text{comp}}} = \frac{V_f}{E_{\text{B}}} + \frac{(1 - V_f)}{E_{\text{Al}}}$$

or

$$E_{\text{comp}} = \frac{E_{\text{B}} E_{\text{Al}}}{V_f E_{\text{Al}} + (1 - V_f) E_B} \tag{9-3}$$

This case is referred to as *isostress* deformation.

2. Alternatively, suppose that the load is applied parallel to the laminates, as in Fig. 9-22(b). If the laminates are well bonded together, such that the strains ϵ in each laminate are the same (*isostrain*), the elastic stresses in each laminate will vary according to the modulus of the laminate. The total stress σ_{total} is given by the sum of the loads carried by each phase. Thus,

$$\sigma_{\text{total}} = E_{\text{B}} \epsilon V_f + E_{\text{Al}} \epsilon (1 - V_f) \tag{9-4}$$

so that the composite modulus $E_{\text{comp}} = \sigma_{\text{total}}/\epsilon$ is given by

$$E_{\text{comp}} = E_{\text{B}} V_f + E_{\text{Al}}(1 - V_f) \tag{9-5}$$

Figure 9-23 shows how the composite modulus varies with volume fraction for these two extreme cases. Many composites, such as pearlite-martensite mixtures, deform by a combination of isostrain and isostress, particularly when one phase is discontinuously distributed in a continuous matrix. In this case, the composite modulus falls in between the two curves predicted by Eqs. (9-3) and (9-5).

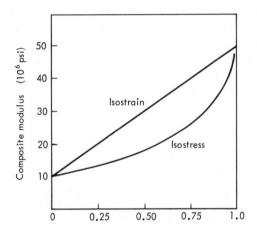

FIG. 9-23   Variation of aluminum-boron composite modulus with type of loading

One of the areas of greatest importance in the field of composite materials is that of fibrous composites, briefly mentioned in Chapter 1. The fibrous composite is one of two fundamentally different approaches aimed at achieving high strength in materials. The alternative approach is based on limiting the motion of dislocations. This latter approach has been discussed extensively in connection with solution and particle hardening and involves immobilizing the matrix dislocations with finely spaced barriers. There is a definite limit to this approach, in that to raise the yield strength to the theoretical limit, dislocation must be completely immobilized, and this leads directly to small fracture strains and brittleness.

The principle of fiber composites does not rely on restricting plastic flow but, to the contrary, uses plastic flow in the matrix to load the oriented fibers, thereby taking advantage of their high strength. Recall that fibers of many materials can be produced with very high tensile strengths (see Table 1-1). The high strength of the fiber is usually due to the fact that (1) it is dislocation-free, (2) it is very difficult for dislocations to move (strong covalent bonds), or (3) it has been extensively cold worked. The fibers cannot be used directly because of their small size, generally less than 0.01 cm in diameter, and the fact that they are generally brittle and fracture with little or no plastic deformation (low fracture toughness). However, if the fibers are combined with a tough, ductile matrix, a useful engineering material with both high strength and toughness can result.

We shall now briefly describe the theory of uniaxial fiber reinforcement based on a simple physical model. The most significant result of this discussion will be learning that it is *not* necessary to have continuous fibers to achieve large increases in strength.

As mentioned above, the strength of fiber composites depends upon successful transfer of the load from the ductile matrix to the strong fibers. In order to achieve substantial loading of the fibers, they must be introduced in the form of long fibers arranged parallel to one another and the loading direction. Assuming this to be the case, consider the response when a load is applied. Normally the matrix will have a lower modulus than the fiber, and, therefore, even when both are elastic, differences in the axial displacements will produce tangential stresses τ and corresponding tensile loading of the fibers (Fig. 9-24). Once the matrix has yielded plastically, its effective modulus $d\sigma/d\epsilon$ will be even lower. In this case, tensile loading of the fibers occurs rapidly in such a manner that they take up the major part of the load.

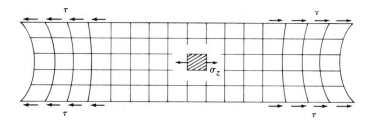

FIG. 9-24 Build-up of tensile stresses σ_z in a fiber due to tangential surface stress τ

Continuous Fibers

Provided a critical volume fraction is exceeded, the tensile strength or fracture strength σ_c of a composite of parallel continuous fibers is given accurately by a rule of mixtures [Eq. (9-1) or (9-5)]:

$$\sigma_c = \sigma_f V_f + \sigma'_m(1 - V_f) \tag{9-6}$$

where V_f is the volume fraction of fibers, σ_f is the tensile strength of the fibers, and σ'_m is the tensile stress in the matrix when the fibers fail. Similarly, the effective composite modulus E_{comp}, after yielding of the matrix, is given by

$$E_{\text{comp}} = E_f V_f + \left(\frac{d\sigma}{d\epsilon}\right)_\epsilon (1 - V_f) \tag{9-7}$$

where E_f is the modulus of the fibers and $(d\sigma/d\epsilon)_\epsilon$ is the slope of the stress-strain curve of the matrix material at the appropriate strain.

At low volume fractions, the composite strength is not given by Eq. (9-6) because fracture of the fibers does not lead to immediate failure if the matrix can work-harden sufficiently. Therefore, Eq. (9-6) applies only if the composite strength σ_c given by Eq. (9-6) is higher than the same volume fraction of matrix without the fibers. That is,

$$\sigma_c > \sigma_u(1 - V_f) \tag{9-8}$$

where σ_u is the ultimate strength of the matrix material. Combining Eqs. (9-6) and (9-8) yields the minimum volume fraction V_{min} for which Eq. (9-6) is applicable:

$$V_{min} = \frac{1}{1 + \sigma_f/(\sigma_u - \sigma'_m)} \tag{9-9}$$

Of course, for significant increases in strength, volume fractions much higher than V_{min} must be used.

Discontinuous Fibers

If the fibers are not continuous, then the fibers and matrix are not strained the same amount and Eq. (9-6) no longer applies. To determine the efficiency of the fiber in strengthening the composite we must calculate the extent to which the load is transferred to the fibers. Assuming that there is no loading from the fiber ends, the tensile stress σ_z in a discontinuous fiber builds up approximately as shown in Fig. 9-25. Since the strain in the fiber cannot exceed the composite strain ϵ_c,

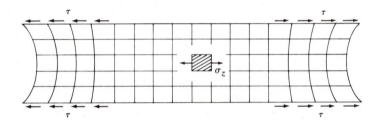

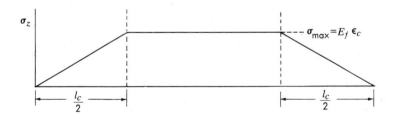

FIG. 9-25 Variation of tensile stress σ_z with position in a discontinuous fiber. l_c is the critical fiber length necessary to insure fibers are loaded to failure.

the maximum attainable stress in the fiber is $\sigma_{max} = E_f \epsilon_c$. The minimum fiber length l, to assure that the fibers will be strained to the strain of the composite, can be computed as follows:

$$\sigma_{max} = E_f \epsilon_c = \frac{2\pi r (l/2)\tau}{\pi r^2}$$

$$l \geq \frac{r E_f \epsilon_c}{\tau} \tag{9-10}$$

where we have merely assumed the stress at the center of the fiber to be equal to σ_{max} and calculated the length of the fiber necessary to achieve this stress through loading by the surface stresses τ.

To attain maximum strength it is desirable to load the fibers to their fracture stress σ_f (or fracture strain ϵ_f). The critical fiber length required to insure that fibers can be loaded to failure ($\epsilon_c = \epsilon_f$) is then

$$l_c = \frac{r \sigma_f}{\tau} \tag{9-11}$$

where τ is the tangential stress applied by the matrix and is either the interface strength or the matrix shear strength, whichever is smaller. It is important to note here that the critical fiber length increases with fiber radius and strength, but decreases as the matrix shear strength increases.

EXAMPLE 9-5   Suppose discontinuous tungsten fibers are imbedded in a copper matrix. If the fibers are 0.001 in. in diameter and have a fracture stress of 200,000 psi, what is the critical fiber length required to insure that the fibers are loaded to their fracture strength? Assume that the tangential stress between matrix and fiber is 10^4 psi.

Solution: From Eq. (9-11) we have

$$l_c = \frac{r \sigma_f}{\tau} = \frac{(10^{-3}) 2 \times 10^5}{2 \times 10^4} = 10^{-2} \text{ in.}$$

Therefore, the fibers must have a length ten times the diameter to insure maximum loading.

Examining Fig. 9-25 reveals that the average stress in the fiber will always be less than the fracture strength σ_f. Therefore the strength of a discontinuous fiber composite must be less than that of a continuous fiber composite of the same volume fraction of fibers. More specifically, Eq. (9-6) must be modified to

$$\sigma_c = \bar{\sigma} V_f + \sigma'_m (1 - V_f) \tag{9-12}$$

where

$$\bar{\sigma} = \frac{1}{l} \int_0^l \sigma_z \, dz \tag{9-13}$$

is the average stress in the fiber of length l at failure. The average fiber stress may be written

$$\bar{\sigma} = \frac{1}{l} [\sigma_f (l - l_c) + \beta \sigma_f l_c] = \sigma_f \left(1 - \frac{1 - \beta}{l/l_c}\right) \tag{9-14}$$

where $\beta \sigma_f$ gives the average stress within a distance $l_c/2$ of the fiber end and $\beta = \frac{1}{2}$ for the case shown in Fig. 9-25. Therefore the strength of a discontinuous fiber composite is given by

$$\sigma_c = \sigma_f V_f \left(1 - \frac{1 - \beta}{l/l_c}\right) + \sigma'_m(1 - V_f) \tag{9-15}$$

If $l/l_c = 10$ and $\beta = \frac{1}{2}$, then the strength of the discontinuous fiber composite will be 95% of the continuous fiber system. For most fiber composites, l_c is about five times the fiber diameter, thus requiring a length-to-diameter ratio of about 50 to ensure maximum strengthening in the composite. Figure 9-26 shows how the strength of a discontinuous W wire-Cu matrix composite varies with volume fraction of W.

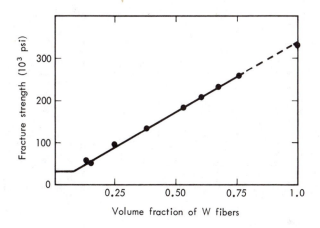

FIG. 9-26   Variation of fracture strength with volume fraction of discontinuous W fibers (0.005 in. in diameter) in a copper matrix

Fiber composites can even maintain high strengths at elevated temperatures where the matrix strength decreases appreciably—provided three conditions are met. First, the fibers must be long enough to ensure that they are appreciably loaded. For example, if the matrix strength decreases by a factor of ten with increasing temperature, then l_c increases by a factor of ten. Secondly, the fracture strength of the fiber must not change appreciably in the temperature range of interest. This condition is easily fulfilled by many high melting temperature fibers such as W, Al_2O_3, B, etc. And lastly, the fiber-matrix system must be chemically stable at elevated temperatures. That is, the fiber must not be degraded by chemical attack, diffusion, alloy formation, or any other interaction with the matrix.

Other types of composite materials, such as concrete, are structurally too complicated to lend themselves to simple mathematical analysis as do fiber composites. However, the general principles that govern the strength and fracture resistance of these more complicated composites are the same (strong, hard particles dispersed in a matrix, discontinuous crack paths, etc.)

An important process used in the production of composite materials is *sintering*. In this process, particles of the individual phases are pressed into a die that has the shape of the desired part; then the compact is heated to a sufficiently high temperature that the particles can bond together. The sintering processs can occur by liquid phase or solid phase diffusion. The driving force for sintering is the surface energy of the particles, since the free surface area is reduced during sintering. Some products, such as tungsten carbide-cobalt cutting tool bits or dies, are made by mixing fine powders of WC and Co together and heating to a temperature above the melting point of the metallic cobalt. The liquid cobalt wets the WC particles and fills in the empty spaces between them, forming a continuous matrix. The

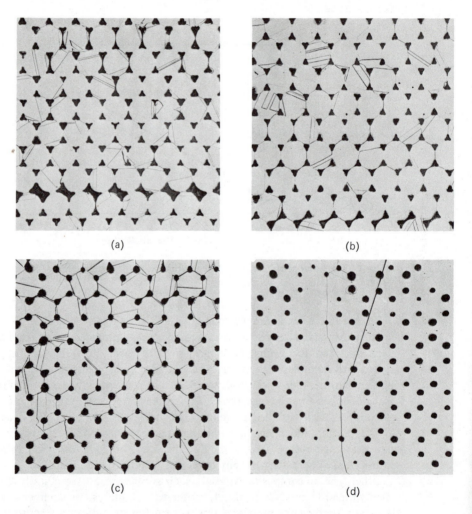

(a)   (b)

(c)   (d)

FIG. 9-27 Structure of 0.005 in. in diameter copper wires stacked in a close packed array and then sintered at 1075°C for: (a) 4 hr, (b) 8 hr, (c) 32 hr, and (d) 408 hr. Photos courtesy American Society for Metals from *Powder Metallurgy in Nuclear Engineering* ("Factors Affecting Sintering" by L. L. Seigle and A. L. Pranatis) (1958) Used with permission.

bond is maintained after the cobalt has solidified upon cooling, and a strong ceramic-metal composite (*cermet*) is formed.

Certain parts, such as refractories for furnace liners, must operate at temperatures above the melting point of inexpensive metallic binders. Therefore the part cannot contain any metallic additive, and a single-phase ceramic (e.g., MgO) rather than a composite is required. These parts are also made by sintering powders together. However, since there is only one phase present, the particles themselves must change shape and continuously bond together during sintering; if this is not done cracks and porosity will exist and the parts will have low strength. An example of solid state sintering is shown in Fig. 9-27. To bring the grains in continuous contact along the boundary that separates them, atoms diffuse from the points of contact to the free surfaces, while there is a flux of vacancies in the reverse direction. The annihilation of vacancies at the boundary separating the two grains results in a net mass transfer from that region, causing the centers of the two grains to move closer together. Ultimately the grains impinge along their entire surface, the porosity disappears, and the sintering process is complete. Since the sintering process is diffusion-controlled (thermally activated) the rate of sintering increases exponentially with temperature and most sintering operations are carried out just below the melting temperature.

PROBLEMS

9-1 Suppose that the brass described by Example 9-1 is being hot rolled and 50% recrystallization must occur in 10 sec in order that the alloy remain soft enough not to crack. What is the minimum safe rolling temperature?

9-2 Heavily cold worked brass has a stored energy of 2×10^9 ergs/cm^3. The grain boundary energy γ is 600 ergs/cm^2. Calculate the critical size and critical free energy for a nucleus of recrystallized material.

9-3 Why is lead a soft material at ambient temperature?

9-4 How could you salvage a batch of aluminum–4% copper rivets that had been inadvertently age-hardened prior to being driven?

9-5 Consider a precipitation-hardened system with precipitates of α in a β matrix. If the average spacing between the centers of precipitates on a slip plane is λ and the precipitates are spherical with radius r, compute the stress necessary to move a dislocation along the slip plane, through the precipitates, in terms of increasing the precipitate-matrix surface area (surface energy γ).

9-6 During studies of age hardening in aluminum alloys it was noted that maximum hardness could be achieved by aging for 10 hr at 327°C or 280 hr at 227°C. How long would it take at 257°C?

9-7 Why is it preferable to precipitation harden alloys (such as Al-4.5% Cu) by first quenching to a low temperature and then reheating rather than by quenching directly to the precipitation-hardening temperature?

9-8 The grain size of a pure copper sample can be refined (made smaller) by heating it after it has been cold worked. The grain size of a large-grained iron sample can be refined by heating it without any prior cold working. Explain this difference.

9-9 Consider an iron-0.3% carbon alloy.
    (a) Draw the microstructure of the alloy at 850, 720, and 25°C, assuming furnace (equilibrium) cooling from 900°C.
    (b) Draw the microstructures observed at these temperatures if the alloy is quenched from 1000°C to room temperature. $M_S = 100$°C; $M_F = -100$°C.
    (c) Show on the same graph the effect of aging time at 300°C on the strength of this alloy after it has been (1) furnace cooled to room temperature from 900°C; (2) quenched to room temperature from 900°C; (3) air cooled to room temperature from 900°C.

9-10 Draw a schematic T-T-T curve for a plain carbon steel containing 1.2% carbon. Label all phase regions.

9-11 Several samples of a 1080 steel are heated to 900°C for one hour. Then different samples are subjected to the following thermal treatments:
    (a) three samples are quenched to 650°C, held for 1, 10 and 1000 sec. respectively, and quenched to room temperature.
    (b) three samples are quenched to 300°C, held for 100, 1000 and 10,000 sec., respectively, and quenched to room temperature.
    (c) one sample is quenched to room temperature, re-heated to 650°C and held for 10 hr. and then quenched to room temperature.
Draw the microstructure after each of these thermal treatments.

9-12 Why is carburizing treatment of steel carried out when the steel is austenitic rather than ferritic?

9-13 The percent of pearlite formed during isothermal decomposition of austenite for a time t at the temperature corresponding to the nose of the T-T-T curve (423°C) is

$$\frac{\ln t - \ln t_{P_S}}{\ln t_{P_F} - \ln t_{P_S}}$$

where t_{P_S} is the time at which pearlite starts to form and t_{P_F} is the time at which the structure is 100% pearlite. Suppose, for simplicity, that the shape and position of the curve under continuous cooling are the same as for isothermal transformation. Experiments indicate that a cooling rate of 50°C/sec is required to form a completely martensitic structure in an alloy cooled from 723°C to room temperature. The UTS of the martensite is 300 ksi. It has also been shown that when an alloy is quenched from 723°C to 423°C, held for a time of 60 sec, and then quenched to room temperature, the $UTS = 190$ ksi. What is the UTS of an alloy that is quenched from 723°C to 423°C, held for 200 sec, and then quenched to room temperature?

9-14 It is observed that the hardenability of a eutectoid carbon steel (0.8% C) depends on the prior austenite grain size, with the hardenability increasing for larger grain sizes. Give an explanation for this effect.

9-15 Consider a composite material containing 40% high-strength steel wires (fracture strength 400,000 psi) in aluminum that is being strained longitudinally. This composite fractures when the fracture strength of the fibers is reached. If the flow curve of the aluminum is given by $\sigma = 22,000\epsilon^{0.2}$, and if isostrain deformation prevails, what is the fracture strength of the composite?

9-16 Determine an expression for the minimum volume fraction of discontinuous fibers for which Eq. (9-15) is applicable.

9-17 Show that if $l < l_c$, Eq. (9-15) is incorrect for discontinous Al_2O_3 fibers. Derive the appropriate expression.

9-18 Suppose discontinuous Al_2O_3 fibers 0.0005 in. in diameter are embedded in a silver (Ag) matrix. If the UTS of the Ag is 18,000 psi at room temperature and only 2500 psi at 600°C, calculate the length of the Al_2O_3 fibers at each temperature necessary to assure the composite having 95% of the strength of a continuous fiber composite. Assume perfect bonding between fiber and matrix.

9-19 Can you devise a technique employing eutectic solidification to produce a fiber composite?

9-20 In a sintering process it is observed that voids, or pores, disappear more rapidly when in intimate contact with a grain boundary. Suggest a reason why this is the case.

9-21 Dispersion-hardened materials are similar to precipitation-hardened materials, with the exception that the dispersed second phase is insoluble in the matrix in the solid state at all temperatures. Suggest a possible materials-processing technique to produce a dispersion-hardened material (e.g., ThO_2 particles dispersed in Ni, MgO dispersed in Ag, or Al_2O_3 dispersed in Al).

9-22 In this chapter we have considered precipitation hardening in conjunction with metallic systems. Discuss the importance of precipitation hardening in conjunction with ceramic systems (see Fig. 4-20).

BIBLIOGRAPHY

R. M. Brick, R. B. Gordon, and A. Phillips, *Structure and Properties of Alloys*, Ch. 4–9. New York: McGraw-Hill, 1965.

J. Burke, *Kinetics of Phase Transformations in Metals*. New York: Pergamon Press, 1965.

J. E. Byrne, *Recovery, Recrystallization and Grain Growth*. New York: Macmillan, 1965.

D. K. Felbeck, *Introduction to Strengthening Mechanisms*. Englewood Cliffs, N.J.: Prentice-Hall, 1968.

M. E. Fine, *Phase Transformations in Condensed Systems*. New York: Macmillan, 1964.

A. G. Guy, *Elements of Physical Metallurgy*, Ch. 12–14. Reading, Mass.: Addison-Wesley, 1959.

H. W. Hayden, W. G. Moffatt, and J. Wulff, *The Structure and Properties of Materials*, Vol. III, Ch. 8–9. New York: John Wiley, 1965.

Metals Handbook, Vol. 2, *Heat Treating, Cleaning and Finishing*. Cleveland: American Society for Metals, 1964.

R. E. Reed-Hill, *Physical Metallurgy Principles*, Ch. 7, 9, 17. New York: Van Nostrand Reinhold, 1964.

C. O. Smith, *The Science of Engineering Materials*, Ch. 9–12. Englewood Cliffs, N.J.: Prentice-Hall, 1969.

10
Deformation of Amorphous Materials

10-1 INTRODUCTION

Just as there are great differences in the mechanical behavior of amorphous and crystalline materials, we shall see in this chapter that there are equally striking differences between the properties of different classes of amorphous materials. To aid in our discussion we shall once again divide amorphous materials into the two general classes—network solids and solids composed of long chainlike molecules (polymers). Recall that ordinary window glass is an example of a network solid, while natural rubber is representative of the long-chain polymer solids. The room temperature response of these two materials to an applied load is perhaps indicative of the vast spectrum of behavior exhibited by amorphous materials. Glass exhibits only small ($< 1\%$) elastic strains, while natural rubber can be stretched elastically to strains of several hundred percent. This difference in behavior is related to differences in the internal structure of the solids. The glass deforms elastically by stretching and bending strong covalent bonds, while the rubber elongates by straightening out tangled long-chained molecules.

A second difference between these two classes of materials involves their response to static loads. Network solids behave like viscous liquids when subjected to a static load. The time-dependent strain is permanent and not recoverable when the load is removed. Most polymers, on the other hand, exhibit time-dependent strain that is at least partially recoverable when the load is released. Again, this difference in behavior can be interpreted in light of the internal structure. In network solids time-dependent flow involves the breaking of strong covalent bonds, while in polymers the time-dependent component of the strain is associated with the breaking of weak interchain bonds, the strong covalent bonds remaining unbroken. The polymer chains simply straighten slowly when subjected to a load, and when the load is released they slowly curl back into the original, tangled array.

In this chapter we shall relate the elastic and plastic deformation of these two classes of amorphous materials to their internal structure. Although our attention will be focused on the microscopic aspects of the deformation process, we will also concentrate on the relationships between deformation mechanisms and observed properties.

10-2 NETWORK AMORPHOUS SOLIDS DISTINGUISHED FROM SUPERCOOLED LIQUIDS

In Chapter 2 it was pointed out that the internal structure of a network amorphous solid is qualitatively similar to the structure of a liquid. In this section we examine this comparison in some detail and discuss some of the subtle differences between supercooled liquids and amorphous solids. This discussion will provide a basis for understanding the mechanisms of deformation in network amorphous solids and will also provide some insight into the mechanical behavior of polymer materials.

There are two completely distinct mechanisms by which a liquid can solidify upon cooling: it can *crystallize* and form a crystalline solid, or it can *vitrify* and form a glass.* This behavior is illustrated in Fig. 10-1, where we show the volume-temperature behavior of a glass-forming material (e.g., SiO_2) as it cools. At high temperatures the material behaves as a normal liquid. As the temperature is reduced, the volume contracts continuously until the equilibrium melting temperature T_m is reached; at this point isothermal crystallization may occur as heat is extracted from the material. Crystallization is accompanied by a sharp volume change as the molecules rearrange themselves into a periodic array. When crystallization is complete, the temperature can continue to fall, resulting in a further volume decrease due to thermal contraction.

For some materials no crystallization takes place at T_m if the material is cooled fairly rapidly. This is because the crystallization process requires the motion and rearrangement of the molecules in order to form a stable nucleus and have the nucleus grow. The ease with which nuclei form and grow therefore depends critically on the ease with which molecules can move past one another in the liquid. Liquids that are highly viscous at the melting temperature therefore have a great deal of difficulty crystallizing, while low-viscosity liquids crystallize easily. High viscosities result from large attractive forces between molecules and/or large, irregularly shaped molecules that have difficulty sliding past one another. Liquid metals and salts, in which the atoms or molecules are essentially spherical, have

---

*In Section 10.5 we shall see that polymeric materials behave in the same fashion.

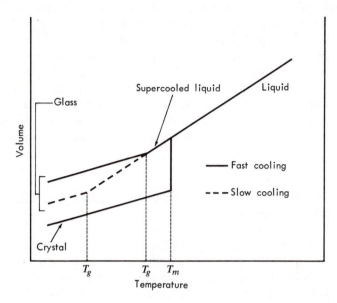

FIG. 10-1 Volume-temperature relationship for the solidification of a liquid into either a crystal or a glass

low viscosities (≈ 0.01 poise) at their melting temperature and crystallize quite easily. Oxides, silicates, borates, phosphates, and other glass-forming materials have much higher viscosities at the melting temperature (≈ 1000 poises).

If crystallization does not occur at T_m, then the liquid can simply continue to cool as a supercooled liquid. The volume of the liquid continues to decrease in a smooth fashion as the temperature is lowered below T_m. The viscosity increases as the liquid is cooled further. Eventually a temperature range is reached at which the viscosity of the liquid increases very sharply (for example, several orders of magnitude within a few degrees Celsius) to a value so high ($\approx 10^{13}$ poises) that viscous flow or molecular rearrangement in the glass is negligible. Below this temperature, called *glass temperature*, T_g, the supercooled liquid acts like a solid and is referred to as a glass even though the internal structure is still completely amorphous and liquidlike. Essentially, the molecules are frozen into position and can no longer move past one another.* Once the molecules are effectively frozen into position, the thermal contraction (volume change) of the glass is about the same as that of the crystalline solid, since now upon further cooling there is no molecular rearrangement in the glass similar to that taking place in the liquid. The value of T_g depends on the rate of cooling. *Liquids that are cooled slowly have a lower T_g and more closely approach the crystalline state than liquids that are cooled rapidly.*

All glasses or liquids that are held at temperatures slightly below T_m, where the

---

*While it is not strictly true that the molecules cannot move past one another, they move so slowly that we can consider them to be stationary.

FIG. 10-2 Microstructure of glass held several days just below melting temperature, T_m, showing formation of crystals in amorphous matrix. Photo courtesy R. J. Charles, the General Electric Research and Development Center. Used with permission.

viscosity is fairly low, will tend toward the equilibrium crystalline structure. In fact, it is possible to produce *crystalline glass* by this very technique. Figure 10-2 shows an ordinary soda-lime-silica glass that has partially crystallized after being held just below T_m for several days. At room temperature the viscosity of most glasses is so high that the amorphous structure remains stable over a long period of time. For example, the structure of many glass vases manufactured by the Egyptians over 3500 years ago remains nearly 100% amorphous.

10-3 DEFORMATION OF NETWORK AMORPHOUS SOLIDS

Deformation of network amorphous solids occurs predominantly as either elastic or plastic (irrecoverable) strain. The elastic strains are normally quite small due to strong covalent bonds in the network solids. Most glasses, for example, have an elastic modulus of $\approx 10^7$ psi; and, correspondingly, a stress of 100,000 psi will produce an elastic strain of only $\approx 1\%$.

Plastic deformation in glasses occurs by the local rearrangement of atoms or molecules. It occurs in response to imposed stresses, whether they be internal stresses introduced during fabrication or stresses due to externally applied loads. This type of deformation is illustrated in Fig. 10-3. The molecules are forced to positions of metastable equilibrium by the stress. Some bonds are highly strained, with the consequence that the overall energy of the system is lowered if these bonds

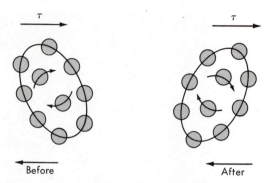

FIG. 10-3 Schematic illustration of atom motion during viscous deformation of a glass

are broken and new ones formed. However, to break the existing bonds requires a momentary increase in the energy, just as an atom moving in a crystalline solid must surpass an energy barrier in moving from one atomic site to another. This energy barrier can be overcome with the aid of thermal activation; that is, thermal energy assists the molecule in moving to a more stable position. Permanent plastic deformation accompanies this molecular movement.

The rate of plastic deformation in most glasses is directly proportional to the stress (or forces acting on the atomic bonds), and we can write

$$\dot{\epsilon} = \frac{\sigma}{\eta} \tag{10-1}$$

where the proportionality constant η is the *viscosity*. Materials that obey this relationship are called *Newtonian liquids*. Since the deformation process is thermally activated, the viscosity must be a function of temperature of the form

$$\eta = \eta_0 \exp\left(\frac{\Delta E_V}{RT}\right) \tag{10-2}$$

where η_0 is a constant and ΔE_V is the activation energy for the viscous motion of molecules. Some typical values for ΔE_V are given in Table 10-1. To reiterate, the high ΔE_V values for the glasses are the result of large, irregularly shaped molecules and strong covalent bonds. ΔE_V is generally fairly constant over a wide range in temperature but can vary appreciably in some instances. For example, if molecules

TABLE 10-1

Typical Activation Energies for Viscous Flow of Liquids near the Melting Temperature

Material (Bond Type)	ΔE_V (kcal/mole)
Nonpolar liquids	0.5–1
Hydrogen bond	2–10
Metals	0.5–2
Ionics	3–10
Covalent (glasses)	20–150

dissociate at high temperatures, then the individual atoms can move about more easily than large molecules and ΔE_V decreases, resulting in a viscosity that is less strongly temperature dependent. In addition, at low temperatures most glasses do not behave as Newtonian liquids (the strain rate is no longer linearly proportional to the stress), and the effective viscosity tends toward a limiting value (about 10^{20} poises for soda-lime-silica glass). A typical example of the range of viscosity with temperature is shown in Fig. 10-4.

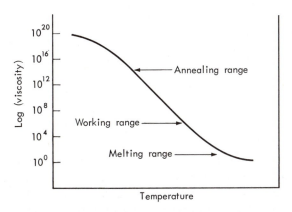

FIG. 10-4   Typical viscosity-temperature relationship for a silica glass

EXAMPLE 10-1   The activation energy for viscous flow of a soda-lime-silica glass is 85 kcal/mole. If the strain rate at 1000°C under a stress of 1000 psi is 10^{-6} per min, what would be the strain rate at 1150°C under the same stress?

Solution: We know $\dot{\epsilon} = \eta_0^{-1} \exp(-\Delta E_V/RT)\sigma$ and at constant stress $\dot{\epsilon} = K \exp(-\Delta E_V/RT)$, where K is a constant. If we know the strain rate at any one temperature, we can find it at any other from the simple ratio

$$\frac{\dot{\epsilon}(T_1)}{\dot{\epsilon}(T_2)} = \frac{\exp(\Delta E_V/RT_2)}{\exp(\Delta E_V/RT_1)}$$

and if we take $T_1 = 1150°C$, $T_2 = 1000°C$, and $\dot{\epsilon}(T_2) = 10^{-6}$ per min,

$$\dot{\epsilon}(1150°C) = 10^{-6}\frac{\exp\left[\dfrac{85,000}{1.98(1273)}\right]}{\exp\left[\dfrac{85,000}{1.98(1423)}\right]} = 3.2 \times 10^{-5} \text{ per min}$$

A simplified mechanical analog to the deformation of glass is shown in Fig. 10-5(a). The spring represents pure elastic deformation, while the dashpot represents pure viscous flow. The response of the spring to an applied load is instantaneous, while elongation of the dashpot obeys Eq. (10-1). Figure 10-5(b) illustrates the deformation–time curve for the model under constant stress. After an initial elastic elongation there is a region of constant strain rate represented by the motion of the piston in the dashpot, the rate of flow being proportional to the viscosity

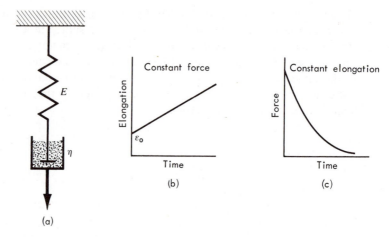

FIG. 10-5 (a) Mechanical analogy to deformation of glass; (b) elongation-time relationship; (c) force-time relationship

of the liquid in the dashpot. Note that the deformation due to the extension of the spring is recoverable when the stress is removed but that the extension of the dashpot represents permanent plastic flow. This simple model also predicts the time response of a glass subjected to a fixed strain [Fig. 10-5(c)]. At $t = 0$ the stress necessary to maintain the strain is proportional to the elastic constant of the spring (i.e., pure elastic strain). However, as the dashpot slowly elongates, the stress required to maintain the strain decreases until it eventually approaches zero.

EXAMPLE 10-2

A stress of 45,000 psi is applied to a piece of glass straining it 0.2%. After 1 yr the stress relaxes to 30,000 psi, maintaining the same strain on the glass of 0.2%. If the glass is maintained in this stretched position for 2 yr, to what value will the stress have fallen?

Solution:

The plastic strain rate $\dot{\epsilon}_p$ (viscous deformation) is given by

$$\dot{\epsilon}_p = -\dot{\epsilon}_e = -\frac{1}{E}\frac{d\sigma}{dt}$$

where $\dot{\epsilon}_e$ is the elastic strain rate. Using Eq. (10.1), we can write

$$\frac{\sigma}{\eta} = -\frac{1}{E}\frac{d\sigma}{dt}$$

or upon integration,

$$\frac{\sigma}{\sigma_0} = \exp\left(-\frac{t}{\tau}\right)$$

where σ_0 (equal to 45,000 psi) and τ are constants. Evaluating τ from

$$\frac{30,000}{45,000} = \exp\left(-\frac{1}{\tau}\right), \qquad \tau = 2.5 \text{ yr}$$

we have for $t = 2$ yr

$$\sigma = 45,000 \exp\left(\frac{-2}{2.5}\right) = 20,250 \text{ psi}$$

There are three important temperature ranges (or viscosity ranges) for most glasses. These are illustrated in Fig. 10-4. In the *melting range* (viscosity ≈ 100 poises) the glass (or, more correctly, the liquid) is so fluid that it cannot support its own weight and behaves much like a syrup, readily assuming the shape of any container into which it is placed. In the *working range* (viscosity $\approx 10^4$–10^8 poises) the glass can be easily formed to any shape by processes such as drawing, blowing, and rolling, but it has sufficient rigidity to support its own weight for a short period of time. In the *annealing range* the viscosity is about 10^{13} poises. This is the range to which glasses are heated to remove any internal stresses introduced during fabrication or during the cooling cycle following fabrication. The viscosity in the annealing range is such that large internal stresses can be relieved by localized rearrangement of molecules in a short period of time (≈ 15 min) with no overall shape change during the annealing treatment.

10-4 DEFORMATION OF LONG-CHAIN POLYMERIC SOLIDS

At high temperatures a linear polymer is an amorphous, rubbery liquid. The chains are interpenetrating random coils undergoing constant vibration and re-arrangement. At low temperatures the same polymer is a hard, rigid solid, often behaving mechanically much like the glasses described in the previous section. This strong temperature dependence of the strength of the polymer is the result of a change in structure. Therefore, as a prerequisite to our discussion of the mechanical properties of polymers we must examine in some detail the internal structure of the polymer as a function of temperature.

The volume-temperature behavior of an organic polymer is shown in Fig. 10-6. This curve is very similar to that shown for network solids in Fig. 10-1, with a few exceptions. As in the case of network solids, the polymer can either solidify as an amorphous solid or as a crystalline solid. Generally, polymers with highly irregular structures (e.g., heavily branched polymers or copolymers with several different kinds of monomers) find it difficult to fit into a crystalline array and are completely amorphous as solids, while symmetric molecular chains lead to crystalline structures. If the polymer does not crystallize, the volume-temperature behavior will follow curve $ABCD$ in Fig. 10-6. In region C (supercooled liquid) the molecules are constantly vibrating and rearranging their positions; as a consequence, the polymer is flexible. Upon further cooling below the glass transition temperature T_g, the vibrations lessen, the chain mobility becomes negligible, and the polymer behaves as a hard, glassy solid.

The properties of a crystalline polymer are also considerably different above and below T_g. Below T_g the polymer consists of a mass of crystallites imbedded in a glassy amorphous matrix* and responds to an applied stress in a brittle fashion. Between T_g and T_m the structure is composed of crystallites imbedded in a flexible amorphous matrix and the material tends to be tough rather than brittle. In the

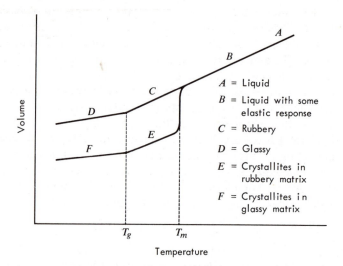

FIG. 10-6   Volume-temperature relationship for long-chain polymer

following paragraphs we shall consider the mechanical properties of polymers in further detail, focusing our attention on amorphous polymers.

The variation in mechanical properties with temperature for a polymeric solid can be conveniently described with reference to the variation in elastic properties with temperature. In Fig. 10-7 we have plotted the variation of elastic modulus with temperature for an *amorphous* polymer. Because of the *viscoelastic* response of polymers (time-dependent elasticity) in some temperature ranges, the value of

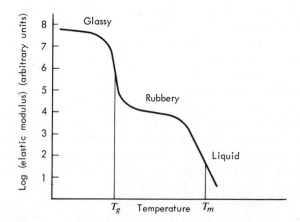

FIG. 10-7   Schematic illustration of the variation in elastic modulus with temperature for a long-chain polymer

---

*It is very difficult to produce polymers with 100% crystallinity.

elastic modulus is time-dependent. The value of the modulus shown in Fig. 10-7 is merely the proportionality constant between stress and strain for a fixed time of testing.

At temperatures below T_g every polymer chain is essentially frozen into one of its many possible geometrical arrangements, and the material behaves as a glassy elastic solid. When subjected to a small tensile stress the sample responds with a small, time-independent, reversible elastic strain. This strain is produced by stretching and bending of both the primary covalent interatomic bonds and the weaker intermolecular bonds. The value of Young's modulus in this temperature range is $\approx 10^6$ psi.

At temperatures above the glass transition temperature the strain is both more extensive and more time-dependent. In this temperature range the chainlike molecules are in a state of constant motion, continually changing their shape as well as their position. The application of a stress to the sample biases the random arrangement of these molecules in favor of certain configurations that allow the applied stress to do work. For example, a tensile stress favors arrangements in which the molecules are extended and oriented parallel to the stress, a configuration which produces an overall elongation of the sample. This elongation is time-dependent because the motion of the chains (breaking intermolecular bonds, etc.) is a thermally activated process dependent upon thermal vibrations. At temperatures only slightly above T_g the strain is almost entirely reversible and is caused by rearrangements of short segments within the chain. At higher temperatures (near T_m) permanent sliding of entire polymer chains past one another causes unrecoverable viscous flow to occur.

The response of a polymer to an applied stress can be qualitatively represented by the mechanical analog shown in Fig. 10-8. The spring in the top section of the model represents pure instantaneous elastic deformation (bond stretching) and is the dominant deformation mode at low temperatures. The spring and dashpot are parallel in the middle section, representing time-dependent elastic deformation

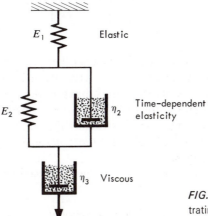

FIG. 10-8 Mechanical analog illustrating response of polymer to applied load

(chain straightening), indicative of deformation at temperatures between T_g and T_m. The single dashpot in the lower section approximates the viscous, irrecoverable flow (relative chain motion) that takes place at temperatures near and above T_m.

EXAMPLE 10-3

Imagine that a polymeric material can be represented by the mechanical analog shown in the accompanying diagram. This type of model is known as a *Voigt model*. Find the equation relating stress, strain, and time. *Hint:* the strain in the spring and dashpot are always equal.

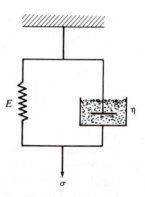

Solution:

Since the total stress σ must be taken up by the elastic spring and the viscous dashpot, we can write

$$\sigma = \sigma_{\text{elas}} + \sigma_{\text{visc}}$$

or

$$\sigma = E\epsilon + \eta \frac{d\epsilon}{dt}$$

which, upon integrating and using the initial condition $\epsilon = 0$ at $t = 0$, yields

$$\epsilon = \frac{\sigma}{E}\left[1 - \exp\left(-\frac{E}{\eta}t\right)\right]$$

The variation of elastic modulus or any other mechanical property with temperature is very dependent on the internal structure of the polymer. That is, any structural feature of the polymer that limits the ease with which molecules can uncoil or slide past one another will greatly influence the mechanical strength of the polymer. Those characteristics of polymeric structure that are important in determining the strength fall into four general categories: *molecular weight, crystallinity, crosslinking*, and *chain stiffening*.

The effect of molecular weight* on the tensile strength of a rubber is illustrated

---

*Molecular weight refers here to the total weight of the long-chain molecule; therefore, it is a direct measure of the length of the molecule.

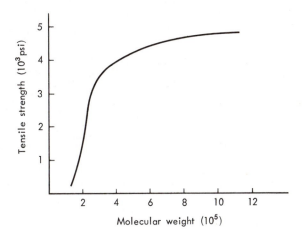

FIG. 10-9 Tensile strength of butyl rubber as a function of molecular weight

in Fig. 10-9. The shape of this curve can be explained in the following fashion. As the molecular weight of a polymer decreases, the density of chain ends per unit volume increases. These chain ends are ineffective in supporting a stress because they may readily slide past other chains. Consequently, as the density of chain ends increases, the strength decreases. This decrease in strength occurs as a loss in load-carrying capacity in the temperature range between T_g and T_m, where polymers deform by the straightening and sliding of chains, and as a decrease in liquid viscosity at temperatures above T_m.

A polymer that is partially crystalline generally has higher strength than the same polymer with an amorphous structure. The higher strength (or higher modulus) is a result of the large number and regular spacing of the interchain bonds in a crystalline structure. In the amorphous polymer the number of interchain bonds is less and their spacing is somewhat erratic, allowing sections of the chain to extend rather freely when a stress is applied. A typical example of the variation in elastic modulus for amorphous and crystalline polymers is illustrated in Fig. 10-10. The modulus of the crystalline polystyrene is higher than that of the amorphous form at all temperatures. In addition, the crystalline polymer does not exhibit the large elastic strains characteristic of amorphous polymers, for in a crystalline structure the chains are already aligned and cannot easily extend. Some typical examples of highly crystalline polymers include nylon, polystyrene, Dacron, and Mylar.

Noting that strength increases with both molecular weight and degree of crystallinity, it is possible to sketch out a map showing the expected properties of polymeric materials on the basis of these two variables. This is shown in Fig. 10-11 for the case of polyethylene. Other polymers of high molecular weight and low crystallinity may be either hard and brittle or soft and flexible, depending on the degree of interchain bonding (cross-links) and on the inherent stiffness of the polymeric chains.

As discussed in Chapter 3, cross-linking involves the formation of strong cova-

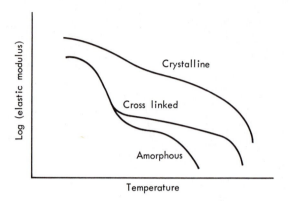

FIG. 10-10 Variation in elastic modulus with temperature for amorphous and crystalline polystyrene

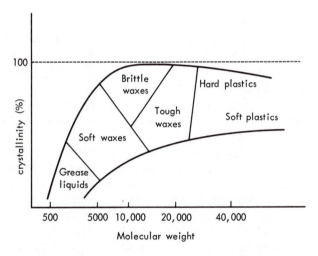

FIG. 10-11 Physical properties of polyethylene as a function of molecular weight and crystallinity

lent bonds between individual polymer chains. The vulcanization of natural rubber provides a good illustration of the effects of cross-linking on mechanical strength. Table 10-2 lists various properties of natural rubber (no cross-links) and vulcanized rubber (cross-linked with sulfur). It is apparent that cross-linking increases the mechanical strength and significantly changes other physical and mechanical properties. In fact, fully cross-linked rubber, known as ebonite, is very hard and brittle at room temperature. In addition to cross-linking, branching also raises the strength of polymers by making it more difficult for the individual chains to slide past one another. Typical materials strengthened by cross-linking include the hard rubbers, thermosetting resins, polyesters, polyurethanes, and phenol-formaldehyde (Bakelite).

TABLE 10-2

Properties Typical of Raw and Vulcanized Natural Rubber

Property	Raw Rubber	Vulcanized Rubber
Tensile strength, psi	300	3000
Elongation at break (%)	1200	800
Permanent set	large	small
Rapidity of retraction (snap)	good	very good
Water absorption	large	small
Solvent resistance (hydrocarbons)	soluble	swells only

EXAMPLE 10-4   Radiation (gamma-ray) cross-linking of partially crystalline polyethylene has a beneficial influence on its mechanical behavior and resistance to deformation at high temperatures. How, then, can you account for the observation that the elastic modulus of polyethylene initially decreases and then progressively rises with increasing amounts of radiation? This effect also appears when the density is plotted as a function of gamma-ray dose.

Solution: Two structural changes occur simultaneously in irradiated polyethylene. One is the destruction of crystalline regions with a resultant drop in density. The other change is the formation of cross-links that bind the macromolecules together and lead to tighter packing of the chains. The initial decreases in elastic modulus and density are due to the loss of crystallinity, while at higher gamma-ray doses, increased cross-linking more than offsets the drop in crystallinity and both the modulus and density increase.

In addition to cross-linking individual polymer chains with single atoms (as in the vulcanization of rubber) or molecules, polymers can be effectively strengthened by the use of *fillers*. These fillers effectively bond a large number of polymeric chains together and restrict their motion relative to one another. Carbon black is an outstanding reinforcing filler for both natural and synthetic rubbers. The added effects of reinforcement with carbon black on the properties of rubber over vulcanization alone are illustrated in Table 10-3.

TABLE 10-3

Effects of Vulcanization and Reinforcement on Properties of Typical Elastomers

Property	Natural Rubber	Butyl	Styrene-Butadiene Based Rubber
Tensile strength, psi			
Vulcanized	3000	400	3000
Reinforced	4500	3000	3000
Elongation at break, %			
Vulcanized	800	800	1000
Reinforced	600	500	400
Tensile stress at 400% elongation, psi			
Vulcanized	400	200	300
Reinforced	2500	2000	1000

If a polymer is fully strengthened by crystallization or cross-linking or some combination of both, then the strength of the polymer will depend to a large extent on the inherent strength of the chain. Hence, a polymer with a particularly stiff chain has the potentiality to be much stronger than a polymer with a flexible chain. There are two ways to stiffen a polymeric chain. One is to hang bulky side groups of atoms on the chain to restrict bending, and the other is to produce a chain with an inherently stiff backbone. An example of the former case is polymethyl methacrylate, which has bulky methyl (CH_3) and methacrylate ($COOCH_3$) side groups attached to carbon atoms along the chain. Polymers with inherently stiff backbones are generally composed of ring-shaped molecules. One example is the group of polymers known as the *ladder molecules*. The backbone of these polymers is composed of benzene-type rings, with each carbon atom bonded to two others along the chain [see Fig. 10-12(a)]. As a result of the arrangement the chain remains

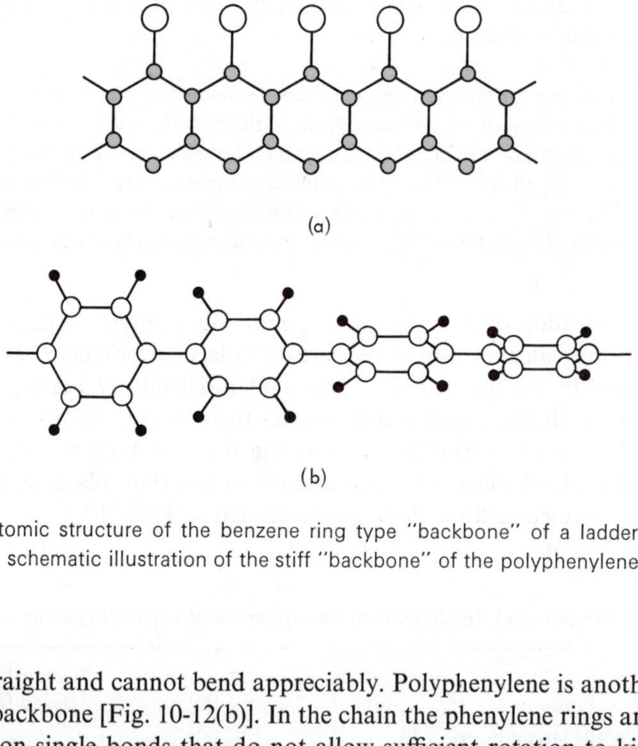

(a)

(b)

FIG. 10-12 (a) Typical atomic structure of the benzene ring type "backbone" of a ladder molecule; (b) schematic illustration of the stiff "backbone" of the polyphenylene molecule

relatively straight and cannot bend appreciably. Polyphenylene is another polymer with a stiff backbone [Fig. 10-12(b)]. In the chain the phenylene rings are linked by carbon-carbon single bonds that do not allow sufficient rotation to kink or bend the chain. Other polymers with stiff chains are polystyrene, polycarbonates, polyesters, polyethers, and cellulose.

One disadvantage of stiffening chains with the aid of bulky side groups is that these polymers are generally amorphous and have rather open structures. Consequently, they are relatively easy to dissolve and are subject to swelling or penetration by certain solvents. We can state, in general, that the higher the degree of crystallinity and the greater the number of cross-links,* the higher the resistance to solution and swelling. Therefore, those polymers with inflexible chains are generally superior to those with bulky side groups when put into service in corrosive environments.

Combining the various strengthening mechanisms described above makes it possible to impart various properties to polymers. Table 10-4 lists the various possibilities and gives some examples and uses.

TABLE 10-4
Characteristics and Examples of Polymers with Different Internal Structures

Polymer characteristics	Examples	Uses
Flexible and crystallizable chains	Polyethylene Polypropylene Polyvinyl chloride Nylon	Pails, pipes, thin films Steering wheels Plastic pipes Stockings, other clothing
Cross-linked amorphous networks of flexible chains	Phenol-formaldehyde Vulcanized rubber Styrenated polyester	Appliance casings Tires, transport belts, hoses Finish on automobiles and appliances
Rigid chains	Polyamides Ladder molecules	High-temperature insulation Heat shields
Crystalline domains in a viscous network	Dacron Cellulose acetate	Fibers and films Fibers and films
Moderate cross-linking with some crystallinity	Neoprene Polyisoprene	Oil-resistant rubber goods Resilient rubber goods
Rigid chains, partly cross-linked	Heat-resistant materials	Jet and rocket engines and plasma technology
Crystalline domains with rigid chains between them and cross-linking between chains	Materials of high strength and temperature resistance	Buildings and vehicles

10-5 ELASTOMERS

To conclude our discussion of the mechanical behavior of polymeric materials, let us examine the amorphous, rubberlike polymers that are known as *elastomers*. These materials have two unique elastic properties. First, they can extend elastically for strains up to several hundred percent. This is illustrated in the stress-strain curve for several vulcanized rubbers in Fig. 10-13(a). Secondly, in the temperature range where large elastic elongations are observed, the elastic stiffness or elastic modulus actually increases with increasing temperature, contrary to the behavior of other amorphous and crystalline materials [see Fig. 10-13(b)]. As

---

*Note that the data in Table 10-2 indicate that bonding the polymeric chains together with strong cross-links also increases the resistance to swelling and solution by organic solvents.

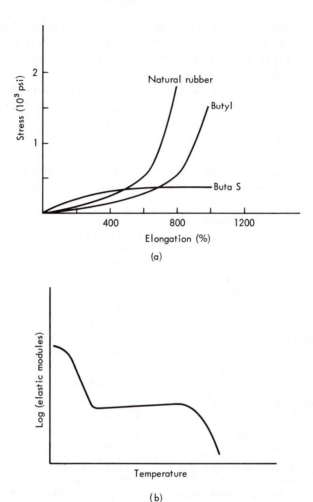

FIG. 10-13 (a) Typical stress-strain curve for a series of vulcanized rubbers; (b) schematic illustration of the increase in modulus with increasing temperature for an elastomer

discussed previously, the elastic properties of these elastomeric materials come from the fact that the polymer chains are kinked by the random orientation of the side groups along the chains. While this molecular feature is necessary for the elastic response, it does not by itself explain the unique properties of elastomers. As we shall demonstrate below, a thermodynamic property—namely, the concept of *entropy* (or *randomness*)—plays a central role in the explanation of these physical properties.

Consider the polymer chain that extends between points *A* and *B* in Fig. 10-14. Because of the kinks in the polymer chain, the chain does not extend directly from *A* to *B*. Two possible paths or configurations are shown in the figure. Because there are many paths that the chain may take, the configurational entropy of the system

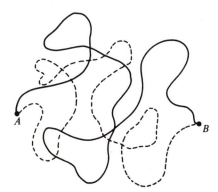

FIG. 10-14 Two possible paths for a long-chain molecule extending between points A and B

is high. As described in Chapter 4, we can compute the configurational entropy of each chain by

$$S_{\text{config}} = k \ln \Omega \qquad (10\text{-}3)$$

where Ω represents the number of ways the molecule can be arranged. Now when a tensile force is applied to the molecule, the points A and B move further apart. When the points A and B become further apart, the number of ways the molecule can be arranged is reduced (in number of paths) and the entropy is correspondingly decreased.

We now must ask, "What is the force required to produce an elongation?" To answer this question, let a tensile force f act on the elastomer and produce an elongation dx. The work done or energy change of the system is $f(dx)$. If the elongation takes place at constant temperature and pressure, the force f can be related to the change in free energy of the system by the relation

$$f = \left(\frac{\partial G}{\partial x}\right)_{T,p} = \left(\frac{\partial E}{\partial x}\right)_{T,p} - T\left(\frac{\partial S}{\partial x}\right)_{T,p} \qquad (10\text{-}4)$$

Now if the extension of the elastomer occurs merely by the individual chains reorienting themselves and extending slightly (producing no overall change in the bonding), then the internal energy is not a function of x and

$$\left(\frac{\partial E}{\partial x}\right)_{T,p} = 0$$

Consequently,

$$f = -T\left(\frac{\partial S}{\partial x}\right)_{T,p} = -Tk \ln \frac{\Omega_f}{\Omega_i} \qquad (10\text{-}5)$$

where the subscripts of Ω signify final (f) and initial (i) states. Since Ω is reduced as the chain is extended, the free energy of the system increases as the length increases, and a positive tensile force is required to straighten out the chain. When the force is released the chains will again coil up, because this configuration offers the greatest entropy and consequently the lowest free energy.

Equation (10-5) predicts that the force required to produce a given elongation is proportional to the absolute temperature. Thus, the force increases as the temperature increases, and consequently the elastic modulus increases. This is consistent with the intermediate temperature behavior shown in Fig. 10-13(b).

The statement that the internal energy does not change during the elongation, i.e., $(\partial E/\partial x)_{T,p} = 0$, does not hold for all elongations. When the elongations are very large, and the chains are all stretched out in a parallel fashion, a phenomenon known as *stress-induced crystallization* occurs. The elongation actually produces partial crystallization in the polymer by mechanically lining up the chains. Accompanying this partial crystallization is a rapid decrease in internal energy along with the expected decrease in entropy. The net force required to produce further elongation will then no longer be simply proportional to the entropy change, but will be some function of the change in both internal energy and entropy.

EXAMPLE 10-5 Suggest an experiment to determine if stress-induced crystallization occurs when a rubber band is stretched.

Solution: There are two obvious choices. First, crystallization could be detected by comparing X-ray diffraction patterns of the stretched and unstretched rubber. Perhaps a simpler technique is to measure any temperature change associated with stretching. If partial crystallization accompanies stretching, then there should be a temperature increase accompanying stretching (energy released when interchain bonds are formed) and a corresponding temperature decrease when the applied stress is removed and the bonds are broken. These temperature changes can actually be detected by using your lips as the temperature sensor with a large rubber band.

PROBLEMS

10-1 A sample of methylmethacrylate

$$CH_2{=}C\begin{array}{l}\diagup CH_3 \\ \diagdown COOCH_3\end{array}$$

has been polymerized. The resulting long-chain polymer has 10,000 molecules along its length. Draw the structure for the repeating unit and calculate the polymer molecular weight.

10-2 Specify the shapes and bonding requirements of polymer molecules to match the following property requirements:
(a) A very high melting crystalline polymer ($T_m > 400°C$)
(b) A crystalline polymer with a low melting point ($T_m < 0°C$)
(c) A permanently amorphous polymer with a very high T_g
(d) A permanently amorphous polymer with a very low T_g

10-3 The activation energy for viscous flow of a glass is 60 kcal/mole. If the strain rate at 1000°C under a stress σ is 10^{-4} per min, what would be the strain rate at 1200°C under the same stress?

10-4 A piece of glass is strained 0.05%, requiring a stress of 10,000 psi. If the strain is maintained constant for 10 days the required stress drops to 3000 psi. What would be the required stress after 30 days?

10-5 If a piece of rubber is strained 100%, requiring a stress of 3000 psi, and then maintained at that strain, it is observed that the time for the stress to drop to $\exp(-1)$ of its initial value is 25 days. What is the stress after 5 days, 100 days, and 1 yr? Assume the rate of stress relaxation is proportional to the stress.

10-6 Write an expression relating strain, stress, and time for the mechanical analog shown in Fig. 10-8. *Hint:* see Example 10.3.

10-7 Two models for representing cross-linked polymers are shown in the accompanying diagram. Write an expression relating stress, strain, and time for each of these models. By appropriate rearrangement of the constants in these equations, show how the parameters of one model could be obtained from those of the other. What is the equilibrium $(t = \infty)$ and instantaneous $(t = 0)$ modulus for each of these models?

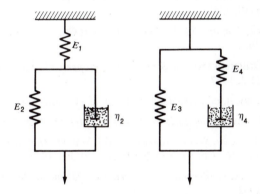

10-8 A polymeric material behaves like the mechanical analog shown in Example 10-3. If $E = 10^7$ dynes/cm² and $\eta = 10^7$ poises and the material is subjected to a uniaxial stress $\sigma(t)$ that varies with time according to

$$\sigma(t) = 10^4 \sin(2\pi\nu t) \text{ dynes/cm}^2$$

where $\nu = 10^{-1}$ sec⁻¹, calculate $\epsilon(t)$.

10-9 The material in Problem 10-8 is subjected to a constant strain rate tensile test where the strain rate is 10^{-2} sec⁻¹. Calculate $\sigma(t)$ and plot the stress-strain curve. Do the same for a strain rate of 10^{-5} sec⁻¹.

10-10 Show by means of simple sketches how the creep curve for the model shown in Fig. 10-8 would be affected by the following changes in parameters:
(a) Change in E_1
(b) Change in η_3
(c) Change in E_2 and η_2

10-11 It is observed that the fracture stress of rubber increases as the strain rate increases. Discuss this observation in light of the molecular structure of rubber.

10-12 What factors do you feel are important in considering the oxidation resistance of a polymer?

10-13 A rubber band is stretched at room temperature and then immersed in a bath maintained at $-20°C$. At this temperature the rubber band does not contract when let go. Why?

10-14 Discuss why the mechanical strength of a polymer may increase after long exposure to air (oxygen).

10-15 Describe the differences in the stress-strain curves that you would expect to find for an elastomer, an amorphous polymer below T_g, and a crystalline polymer above T_g.

BIBLIOGRAPHY

T. Alfrey and E. F. Gurnee, *Organic Polymers*, Ch. 4. Englewood Cliffs, N.J.: Prentice-Hall, 1967.

E. Baer, *Engineering Design for Plastics*, Ch. 3-4. New York: Van Nostrand Reinhold, 1964.

A. T. DiBenedetto, *The Structure and Property of Materials*, Ch. 12. New York: McGraw-Hill, 1967.

F. R. Eirich, ed., *Rheology*, Vol. I, Ch. 11. New York: Academic Press, 1956.

J. D. Ferry, *Viscoelastic Properties of Polymers*. New York: John Wiley, 1961.

H. W. Hayden, W. G. Moffatt, and J. Wulff, *The Structure and Properties of Materials*, Vol. III, Ch. 10. New York: John Wiley, 1965.

L. E. Nielsen, *Mechanical Properties of Polymers*. New York: Van Nostrand Reinhold, 1962.

A. V. Tobolsky, *Properties and Structure of Polymers*. New York: John Wiley, 1960.

11
Electrons in Solids

11-1 INTRODUCTION

In this part of the book we shall begin to study the electronic properties of engineering materials. We shall describe how materials behave when they are subjected to externally applied electric and magnetic fields and to electromagnetic radiation. Because in most cases the response of a material to an electric or magnetic field is determined by the behavior of the electrons, these properties are termed *electronic properties*. The electronic properties of materials that are of engineering importance are the properties due mainly to the behavior of the outermost electrons of the atom. For this reason our attention will be focused on these outer electrons. We shall find that when electric and magnetic fields are applied to a material, the electrons within the interior of the atoms are largely shielded from the fields by the outer electrons and hence do not contribute significantly to the overall electronic response of the material.

Even though our attention is focused only on the outer electrons, we shall find that there can be great variations in the electronic behavior of a material, depending on how tightly the outer electrons are bound to the atoms. In the one extreme the outer electrons may be rather tightly bound to their parent atoms and not free to wander about in the crystal. In this case the electrons are usually called *valence electrons* (for example, 3p and 3s electrons in silicon). In the other extreme, outer electrons can be so loosely bound to the atoms that they move about in the crystal with almost no regard to the positions of the atoms. In this case the electrons are referred to as *free electrons*. The outermost electrons in alkali metals are free electrons and are responsible for the high thermal and electrical conductivities of these metals.

Because electronic properties are defined by the way in which electrons respond to electric and magnetic fields and to electromagnetic radiation, it is instructive to examine the nature of these responses and to classify the electronic properties of materials on the basis of these interactions. The electronic properties of engineering materials may be categorized as shown in Table 11-1.

TABLE 11-1

Categorization of Electronic Properties and the Responsible Electronic Interactions

Class of Electronic Properties	Applied Field	Responsible Property of the Electron (or Ion)	Electronic Response
I. Electrical conduction	Electric field	Coulomb charge	Migration of electrons and ions (electric current)
II. Electric polarization	Electric field	Coulomb charge	Static displacement of electrons and ions
III. Magnetization	Magnetic field	Electron spin and orbital motion	Spins and orbital motion align parallel (or antiparallel) to the magnetic field
IV. Electromagnetic radiation and absorption	Electromagnetic radiation (X rays, visual light, infrared, etc.)	Discrete or quantized electronic energy levels	Electron is promoted to higher or lower energy state by absorption or emission of electromagnetic radiation

When an electric field is applied to a material, the component electrons and ions experience coulombic forces and tend to move in the direction of these forces. If the charges are not free to move large distances, as in the case of ionic materials at low temperatures, the material will simply become *electrically polarized*. The positive ions will be displaced from their mean positions in the direction of the applied field, while the negative charges will be displaced in the opposite direction. When the electric field is removed, the ions will return to their equilibrium positions. There is usually a mechanical strain associated with electric polarization. In direct correspondence, when such a crystal is mechanically strained, an electric field is induced, and because of this, crystals that can be polarized are capable of acting as electromechanical energy converters (transducers). Quartz, Rochelle salt, and tourmaline are examples of such ionic crystals; these are used in microphones and phonographs because they become electrically polarized when they change shape.

When the electrons or ions in a material are free to move large distances in response to an applied electric field (voltage gradient), the material is called a *conductor*. Pure metals, especially copper, silver, gold, and aluminum, are excellent conductors of electricity because they contain many electrons that are free to move under an electric field. In the solid state circuits that are now common in electronic technology, the conductors that connect one element of a circuit to another are usually very thin films of metal, such as aluminum, that are placed on the surface of the solid state device by vapor deposition. Because they are so thin and can dissipate heat to the substrate, these films can carry current densities as high as 10,000 A/cm^2 without overheating. This is about 100 times higher than typical current densities in house wiring.

When an electric current flows in a circular path an axial magnetic field is produced. At the atomic level this means that when an electron moves in a circular orbit or when it spins on its axis, a tiny magnetic dipole is produced. Because of this, the atoms in a material may behave as if they were tiny magnets, free to rotate about their equilibrium positions. When a magnetic field is applied, the magnetic dipoles of the atoms tend to become aligned parallel to the applied magnetic field and a magnetic polarization results. The ability of a material to be become magnetically polarized allows us to produce permanent magnets, electromagnets, electric motors, high-power transformers, communication transformers, and many other electromagnetic devices.

11-2 ELECTRONIC ENERGY LEVELS AND BANDS

We now begin a study of the properties and behavior of the outermost electrons of atoms in solids. In Chapter 2 it was shown that, according to the Heisenberg uncertainty principle, the exact positions of electrons in materials are not known. In fact, a detailed knowledge of the positions of the electrons in a material is not even required for an adequate description of electronic properties. What is important is a knowledge of the energy levels that are available to the electrons. In this chapter we shall pay very little attention to the locations of electrons, but we shall study very carefully the possible energy levels that the electrons can occupy.

There are two extreme and limiting states for the outer electrons in materials that may be described in a relatively simple way. In one limit we may say that the electron "sees" mainly the other electrons and nucleus of the atom in which it resides (*tightly bound or valence electrons*); in the other limit the electron may be viewed as completely separated from the atom and free to wander about in the material (*free electrons*). Intermediate between these two extremes we find electrons that are itinerant and move from the region of one atom to another, but not without regard to the potential fields of the atomic nuclei (*quasi-free electrons*). In this chapter we shall show that one may describe the quasi-free electrons either from the point of view of tightly-bound electrons with appropriate modifications or from the point of view of free electrons.

In order that we may understand the factors that control the energies of the outermost electrons in materials, let us consider what happens when two hydrogen atoms are brought into contact. When the atoms are widely separated, the $1s$ electrons in each atom see only the nucleus of the atom in which they reside, and

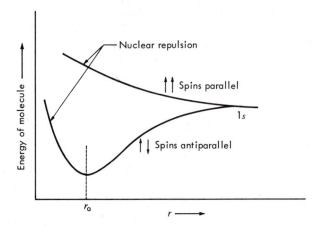

FIG. 11-1   Schematic representation of energy levels for the H_2 molecule as a function of atomic separation r. r_0 is the equilibrium separation distance.

the allowed energy levels for the electron are simply the atomic energy levels. The atomic energy level diagram for hydrogen is shown in Fig. 11-1. When the atoms are widely separated, the spins of the electrons in each atom may be chosen arbitrarily. The energies of the two atoms which make up the molecule are the same, regardless of which spin state the electrons happen to adopt. As the atoms are allowed to come into close proximity, the electron clouds overlap and the electrons begin to see the potential fields of both nuclei.

The wave mechanical treatment of this state is complicated, but the results of the analysis may be simply stated. When the electron clouds overlap, the $1s$ atomic energy level subdivides into one of two separate levels, depending on the relative spins of the two electrons. This means that the energy of the molecule may have one of two values, depending on the spins of the electrons. As shown in Fig. 11-1, when the spins of the two electrons are antiparallel the energy of the molecule is lower than the energies of the two isolated atoms. The physical meaning of this lower energy state is that when the atoms are close together the electrons may spend much of their time between the two positively charged nuclei. Thus, the coulomb energy of this state is lower than the total energy of the isolated atoms. The wave mechanical treatment also shows that when the spins of the electrons are parallel, the corresponding energy of the molecule is higher than the energies of the two isolated atoms. The total energy of the molecule increases very sharply when the two nuclei begin to repel each other, regardless of the relative spins of the electrons. As was discussed in Chapter 2, the equilibrium separation between the atoms corresponds to the separation distance that gives the minimum energy. As shown in Fig. 11-1, the minimum energy is achieved by having antiparallel spins and by having the atoms separated by a distance r_0.

The purpose of studying the atomic and molecular energy levels in hydrogen is to show that when atoms are brought into close proximity the outer electrons begin to interact and the electronic energy levels—and hence the atomic and molecular energy levels—become subdivided. In the case of two hydrogen atoms, the 1s atomic energy level subdivides into two separate and distinct energy levels.

We shall now turn our attention to the electronic energy levels of the outermost electrons of sodium atoms which are brought together to form a crystalline solid. Let us consider first an infinite array of sodium atoms situated on a BCC lattice but spaced so far apart that even the outer electrons of the atoms do not interact. In this case the outer electrons see only the other electrons and the nuclei of the atoms in which they reside; their energies are the same as if they were associated with isolated atoms (see Fig. 11-2). For sodium there is essentially one outer electron per

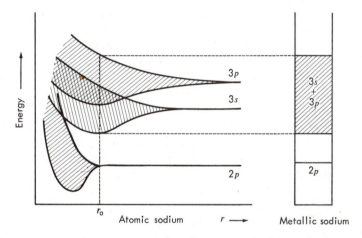

FIG. 11-2   Schematic representation of electronic energy levels for sodium as a function of interatomic separation distance r. r_0 is the equilibrium separation distance.

atom, and it resides in the 3s state in the free atom. We now imagine that the spacing between the atoms in this lattice is decreased to the point where the outer electrons of the atoms begin to overlap and interact. As was the case of the hydrogen molecule, the overlapping may be expected to cause the electronic energy level (in this case, the 3s level) to subdivide into more than one level. One way to see that this must occur is to recognize that the *Pauli exclusion principle* requires that no more than two electrons in a given system can reside in the same energy state at the same time and that those two electrons must have opposite spins. It is clear, then, that when the outer electrons in the atom begin to overlap, the electrons can no longer be treated as belonging to their parent atoms. Rather, we must consider that the outer electrons belong to the crystal as a whole. The Pauli principle requires that these outer electrons adopt different energy levels. As a consequence, the single energy level that describes all the 3s electrons in the separated atoms subdivides into a very large number of closely spaced energy levels as shown in Fig. 11-2. This array of closely spaced electronic energy levels is usually called an *energy band*.

In addition to the fact that the outer electronic energy levels are subdivided into bands, Fig. 11-2 also shows that when the separation distance between the atoms is very small, the average energy of a given band begins to increase sharply. This rise in electronic energy is due to the fact that the space between the atoms is so small that the kinetic energies of the electrons begin to rise (see Section 11-3).

It should be noted that the equilibrium spacing between the atoms, as denoted by r_0 in Fig. 11-2, is determined by the condition of minimum *total energy*. The fact that the equilibrium spacing occurs near the energy minimum in the 3s band simply means that most of the electrons in the band have minimum energies near that separation distance and that the ionic repulsion (which is not included in Fig. 11-2) does not have a large influence on the separation distance between the atoms. The 2p level in metallic sodium is not subdivided into a band because the 2p electrons do not overlap and interact at the equilibrium separation distance.

We have shown that when atoms are brought into close proximity the energy level of the outer electrons subdivides into an energy band. Since the formation of the energy band is due to the overlap and interaction of the outer electrons, we call these *quasi-free* electrons. That is, the electrons are not tightly bound to their parent atoms, but neither do they disregard the presence of the atoms. We have shown that electronic energy bands can be explained by starting with the electronic energy levels within the atom. Now we shall show that the energy bands can also be described by starting with perfectly *free electrons* which do not interact with the ion cores at all. In this case we are essentially assuming that the outer electrons are sufficiently free from their parent atoms that they may be treated as if they were moving about in an otherwise empty box. To relate the free-electron approach to the energy bands shown in Fig. 11-2 it might be noted in advance that the least energetic electron in the free-electron treatment will have a total energy that corresponds to the lowest point in the 3s energy band at the equilibrium spacing between atoms, r_0.

11-3 FREE-ELECTRON THEORY OF METALS

Before treating free electrons, we should mention the *Tolman effect*, an experimental observation that strongly supports this approach. When a metal is made to experience a very high acceleration or deceleration, a voltage gradient in the metal can be detected. Simply stated, this means that when a metal sample is accelerated, the electrons are forced toward one end of the sample, just as passengers standing in a bus are forced toward the rear of the bus when the driver rapidly accelerates from a standstill. One may imagine that the electrons are relatively free and darting about in an otherwise empty box. The inertia of the electron gas causes a voltage gradient to be established when the metal is quickly accelerated.

The idea of free electrons was first proposed to explain the very high electrical and thermal conductivities exhibited by metals. It was first thought that the free electrons could be treated as if they were particles of an ideal gas and that each electron in the so-called electron gas could be treated as a classical particle having a thermal energy of $\frac{3}{2}kT$. Although this view is qualitatively accurate, it predicts a very large electronic contribution to the specific heat of metals that is not observed. The wave mechanical description of free electrons was given by Sommerfeld in the late 1920s, and the electronic specific heat anomaly was removed. The description of free electrons which follows in this section is a simplified version of Sommerfeld's theory. It exhibits all of the important aspects of the wave mechanical description of electrons in solids.

We shall consider first a one-dimensional model for free electrons because it is easy to treat and because the important wave mechanical aspects of the problem are easy to see in this case. Suppose we have a linear array of atoms as shown in Fig. 11-3. Each atom is supposed to supply one or more electrons to the free-elec-

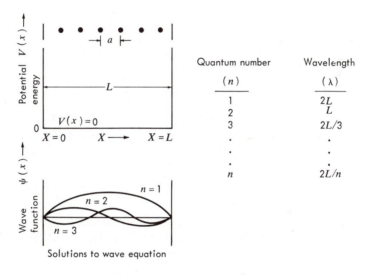

FIG. 11-3   Potential energy and wave functions for a one-dimensional metal

tron gas. If, for example, the atoms are alkali metal atoms, the valence is one and each atom gives one free electron to the system. The potential energy of the electrons is also illustrated in Fig. 11-3. We shall suppose that an electron has an arbitrarily chosen zero potential energy inside the metal. As mentioned before, the potential energy inside the metal actually coincides with the bottom edge of the $3s$ energy band shown in Fig. 11-2. We shall also suppose that the electrons have an infinite potential energy outside the metal. This simply means that the electrons are required to stay inside the metal. One may assume that the electrons are held inside the metal by their coulomb attraction to the positive ions of the lattice. We now wish to determine the kinetic energies that the electrons inside the metal may have. We can do this by solving the *Schrödinger equation* for one dimension.

In Chapter 2 it was pointed out that the motion of electrons is described by the Schrödinger equation. This basic equation is written as

$$\frac{d^2\psi}{dx^2} + \frac{8\pi^2 m}{h^2}(E - V)\psi = 0 \qquad (11\text{-}1)$$

where $\psi(x)$ is the *wave function* for the electron, V is the potential energy for the electron, and E is the total energy. As discussed in Chapter 2, the probability of finding an electron in the region from x to $x + dx$ is $|\psi|^2 dx$, so that the square of the wave function may be thought of as the density of free electrons. The quantity $(E - V)$ is simply the kinetic energy of the free electron and is written as

$$E - V = \frac{1}{2}mv^2 = \frac{p^2}{2m} \qquad (11\text{-}2)$$

where p is the momentum. Since electrons may be treated as waves we may use the deBroglie relation to express the momentum of an electron in terms of its corresponding wave length:

$$p = \frac{h}{\lambda} \qquad (11\text{-}3)$$

where h is Planck's constant. With Eqs. (11-2) and (11-3), the Schrödinger equation becomes

$$\frac{d^2\psi}{dx^2} + \left(\frac{2\pi}{\lambda}\right)^2 \psi = 0 \qquad (11\text{-}4)$$

This equation is identical to the equation for a vibrating string if ψ is interpreted as the amplitude of the string's vibration and λ corresponds to the wavelengths of the standing waves that are possible for the string. The general solution to Eq. (11-4) is

$$\psi(x) = A \cos\left(\frac{2\pi x}{\lambda}\right) + B \sin\left(\frac{2\pi x}{\lambda}\right) \qquad (11\text{-}5)$$

where the constants A and B must be selected according to the boundary conditions for our problem. Before finding the values of A and B, let us verify that Eq. (11-5) is a solution to Schrödinger's equation. Differentiating Eq. (11-5) once gives

$$\frac{d\psi}{dx} = -A\left(\frac{2\pi}{\lambda}\right) \sin\left(\frac{2\pi \lambda}{\lambda}\right) + B\left(\frac{2\pi}{\lambda}\right) \cos\left(\frac{2\pi x}{\lambda}\right) \qquad (11\text{-}6)$$

and a second differentiation yields

$$\frac{d^2\psi}{dx^2} = -\left(\frac{2\pi}{\lambda}\right)^2 \left[A \cos\left(\frac{2\pi x}{\lambda}\right) + B \sin\left(\frac{2\pi x}{\lambda}\right) \right] \qquad (11\text{-}7)$$

Using Eq. (11-5) with Eq. (11-7) gives

$$\frac{d^2\psi}{dx^2} = -\left(\frac{2\pi}{\lambda}\right)^2 \psi \qquad (11\text{-}8)$$

which is Schrödinger's equation for this problem. This exercise demonstrates that Eq. (11-5) is a general solution for the wave function for free electrons in a one-dimensional metal.

Now we may solve for the constants A and B using the boundary conditions for our problem. The boundary conditions may be stated as follows:

1. The wave function (or the probability for finding the electron) goes to zero at $x = 0$ and $x = L$. This is because the potential energy barrier at the end of

the one-dimensional solid is infinitely high and there is no probability that the electron may be found outside the metal.

2. The probability of finding the electron *somewhere* inside the metal is unity. This boundary condition also comes from the fact that with infinite potential energy barriers at the ends, the electrons must be found inside the metal.

The boundary conditions may also be stated mathematically using the wave function:

$$1. \quad \psi(0) = 0, \quad \psi(L) = 0 \tag{11-9}$$

$$2. \quad \int_0^L \psi^2(x)\, dx = 1 \tag{11-10}$$

Application of the first boundary condition to our solution [Eq. (11-5)] gives

$$\psi(0) = A \cos\left(\frac{2\pi(0)}{\lambda}\right) + B \sin\left(\frac{2\pi(0)}{\lambda}\right) = 0$$

or

$$A = 0 \tag{11-11}$$

With this boundary condition satisfied, the wave function is written as

$$\psi(x) = B \sin\frac{2\pi x}{\lambda} \tag{11-12}$$

Now we require that the wave function be zero at the opposite end of the metal. This is expressed as

$$\psi(L) = 0 = B \sin\frac{2\pi L}{\lambda} \tag{11-13}$$

Equation (11-13) is satisfied either by requiring that $2L/\lambda$ be an *integer* or by having $B = 0$. Since making $B = 0$ would lead to the trivial result that $\psi(x) = 0$ everywhere inside the metal (that is, one possible solution is that there is no electron at all in the wire), we consider only the cases in which $2L/\lambda$ is an integer. We shall use the fact that $2L/\lambda$ is an integer as soon as we determine the value of B that will satisfy the boundary condition (11-10). Applying Eq. (11-10) to Eq. (11-12) gives

$$B^2 \int_0^L \sin^2\left(\frac{2\pi x}{\lambda}\right) dx = 1 \tag{11-14}$$

Using this relation and the fact that $2L/\lambda$ is an integer it can be shown that

$$B = \sqrt{\frac{2}{L}} \tag{11-15}$$

Thus, the solution we have for the wave function for the free electron in the metal is

$$\psi(x) = \sqrt{\frac{2}{L}} \sin\left(\frac{2\pi x}{\lambda}\right) \tag{11-16}$$

Now let us return to the boundary condition that forces the wave function to be zero at $x = L$. We showed that

$$\frac{2L}{\lambda} = n \qquad n = 1, 2, 3, 4 \ldots$$

This means that the wavelengths for the free electrons in the metal may *not* have arbitrary values. Rather, they are given by

$$\lambda_n = \frac{2L}{n} \qquad n = 1, 2, 3, \ldots \qquad (11\text{-}17)$$

where n is the quantum number associated with the discrete wavelengths. Putting Eq. (11-17) into (11-16) gives

$$\psi_n(x) = \sqrt{\frac{2}{L}} \sin\left(\frac{2\pi x}{\lambda_n}\right) \qquad n = 1, 2, 3, \ldots \qquad (11\text{-}18)$$

or

$$\psi_n(x) = \sqrt{\frac{2}{L}} \sin\left(\frac{n\pi x}{L}\right) \qquad n = 1, 2, 3, \ldots \qquad (11\text{-}19)$$

The wave functions for $n = 1$, 2, and 3 are illustrated in Fig. 11-3. The solutions correspond to the solutions for the standing waves of a vibrating string.

Since the kinetic energies of free electrons can be expressed in terms of the deBroglie wavelengths, and the wavelengths are quantized according to Eq. (11-17), it follows that the kinetic energies are also quantized. The kinetic energy is expressed as

$$\text{K.E.} = \frac{1}{2}mv^2 = \frac{p^2}{2m} = \frac{h^2}{2m\lambda^2} \qquad (11\text{-}20)$$

Using Eq. (11-17), we see that the quantized kinetic energies become

$$E_n = \frac{h^2 n^2}{8mL^2} \qquad n = 1, 2, 3, \ldots \qquad (11\text{-}21)$$

The physical interpretation of Eq. (11-21) is that the energies of free electrons in a one-dimensional metal of length L may not be chosen arbitrarily. Rather, the energies must be one of a set of discrete values.*

EXAMPLE 11-1

Suppose we have a one-dimensional monovalent metal in which the atoms are spaced 2 A apart (see Fig. 11-3). Calculate the kinetic energy of the most energetic electron if the free electrons occupy the lowest possible energy levels.

Solution:

If there are N atoms, and hence N free electrons, in a wire of length L, the quantum number n associated with the most energetic electron is $n = N/2$. The factor of $\frac{1}{2}$ is applied because there can be two electrons in each quantum state. The energy of the most energetic electron is then

$$E_{\max} = \frac{h^2}{8mL^2}\left(\frac{N}{2}\right)^2 = \frac{h^2}{32ma^2}$$

where $a = L/N$ is the spacing between the atoms. Thus, the maximum energy is

$$E_{\max} = \frac{(6.63 \times 10^{-27})^2}{32(9.11 \times 10^{-28})(2 \times 10^{-8})^2} = 3.76 \times 10^{-12} \text{ erg}$$

$$E_{\max} = \frac{(3.76 \times 10^{-12})(\text{erg})}{(1.6 \times 10^{-12})(\text{erg/eV})} = 2.35 \text{ eV}$$

---

*As we have taken the potential energy $V = 0$, the kinetic energy is equal to total energy E of the electron. Throughout the remainder of this chapter we shall use the symbols K.E. and E interchangeably.

If we had treated the problem of electrons moving in a three-dimensional box with sides of length L (instead of a one-dimensional solid of length L), the allowed values of kinetic energies would have to be expressed as

$$E_{n_x, n_y, n_y} = \frac{h^2}{8mL^2}(n_x^2 + n_y^2 + n_z^2) \tag{11-22}$$

where n_x, n_y, and n_z are the quantum numbers associated with the discrete wavelengths that the electron must have as it moves parallel to each of the three edges of the cube. Equation (11-22) is simply a three-dimensional generalization of Eq. (11-21). A given set of integers n_x, n_y, n_z is called a *quantum state* and represents a state in which not more than two electrons may reside.

We may now show that the description of free electrons we have just given leads to the concepts of an energy band. In this respect the free-electron model is an alternate way of describing the energy bands that are formed from discrete electronic energy levels when atoms are brought close together. Equation (11-22) predicts that the free electrons in a metal may not all have the same energy. It follows, therefore, that the energies of the free electrons fall into a band of closely spaced energy levels.

11-4 QUANTUM STATES AND ENERGY LEVELS

We now direct our attention to the energy levels for the free electrons and the *quantum states* that occupy each energy level. The *quantum states* and *energy levels* for free electrons can be described with the use of Eq. (11-22). Let us first consider the lowest kinetic energy that a free electron may have. According to Eq. (11-22), the lowest kinetic energy is described by the condition that the quantum numbers n_x, n_y, and n_z are all unity. In this case the kinetic energy is $3h^2/8mL^2$. This particular set of quantum numbers describes a *quantum state*, and the energy of the state is called an *energy level*. The Pauli exclusion principle allows for two electrons of opposite spins to reside in a particular quantum state. This means that a total of two electrons can exist in the state corresponding to $n_x = n_y = n_z = 1$.

Now let us examine the next highest allowed energy level for free electrons. Letting $n_x = 2$, $n_y = n_z = 1$, the kinetic energy becomes $6h^2/8mL^2$. Thus, we see that the kinetic energy in the second energy level is exactly twice the kinetic energy of the lowest energy level. Again using the Pauli exclusion principle, two electrons (with opposite spins) can exist in the state $n_x = 2$, $n_y = n_z = 1$. It is evident that there are two other quantum states (sets of quantum numbers) that give the same energy level. They are $n_x = n_y = 1$, $n_z = 2$ and $n_x = n_z = 1$, $n_y = 2$. Therefore, there are three quantum states associated with the energy level $6h^2/8mL^2$ and a total of six electrons that can reside at that level. The energy levels and the quantum states for some of the lowest energy-free electrons are described in Table 11-2. Note that since there can be up to two electrons in each state the total number of quantum states may be thought of as half the number of acceptable homes for free electrons.

TABLE 11-2

Quantum States and Energy Levels for Free Electrons

Quantum States			Energy Level $(h^2/8mL^2)$	Number of Quantum States	Number of Electron Homes
n_x	n_y	n_z			
1	1	1	3	1	2
2	1	1	6		
1	2	1	6	3	6
1	1	2	6		
2	2	1	9		
2	1	2	9	3	6
1	2	2	9		
3	1	1			
1	3	1	11	3	6
1	1	3			
2	2	2	12	1	2
1	2	3	14		
1	3	2	14		
2	1	3	14	6	12
2	3	1	14		
3	1	2	14		
3	2	1	14		

EXAMPLE 11-2

Show that a free electron may not exist in a state for which $n_x = 1$, $n_y = 1$, $n_z = 0$.

Solution:

According to Eq. (11-17), the wavelength associated with the z component of the motion of the electron is expressed as

$$\lambda_{n_z} = \frac{2L}{n_z}$$

It is clear that if $n_z = 0$, the wavelength of the electron becomes infinitely large and the electron may not necessarily be found inside the metal. Since one of our boundary conditions is that electrons must be found somewhere inside the metal, we may not allow any of the quantum numbers to be zero.

An energy level diagram for free electrons is shown in Fig. 11-4. Note that the allowed values of energies match the discrete energies given in Table 11-2. The number of states at each energy level is indicated in the figure.

It also should be noted that the quantum states and energy levels described in Table 11-2 and Fig. 11-4 contain only a very few free electrons. For a monovalent metal there are about 10^{22} free electrons per cubic centimeter. Therefore, the energy levels we have described hardly begin to account for the free electrons in the system. It is evident that we need a huge number of quantum states and energy levels to account for 10^{22} electrons.

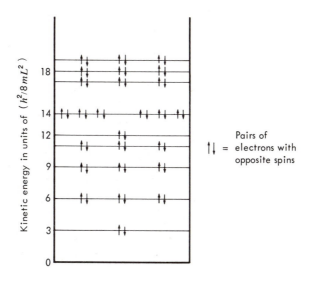

FIG. 11-4 Energy levels for free electrons in a metal

EXAMPLE 11-3

Determine the number of free electron quantum states and free electrons that are included in the following energy range in a metallic cube with edge length L:

$$50\frac{h^2}{8mL^2} \leq E \leq 51\frac{h^2}{8mL^2}$$

Solution:

The kinetic energies of free electrons in a metallic cube are given by Eq. (11-22) as

$$E_{n_x n_y n_z} = \frac{h^2}{8mL^2}(n_x^2 + n_y^2 + n_z^2)$$

Note that an energy of $51h^2/8mL^2$ is achieved by letting the quantum numbers have the values 7, 1, 1 and 5, 5, 1, while an energy of $50h^2/8mL^2$ is achieved with the integers 3, 4, 5. Since there are three different ways of arranging the integers 7, 1, 1 and 5, 5, 1 and six ways of mixing the integers 3, 4, 5 there are a total of 12 quantum states in the energy range specified. Since there may be two electrons in each quantum state, the energy range given can contain as many as 24 electrons.

The energies of large numbers of free electrons can best be described with use of the concept of *density of states*. The quantity is designated $N(E)$ and is defined in such a way that $N(E)\,dE$ is *the number of quantum states per unit crystal volume between the energy levels E and $E + dE$*. $N(E)$ is expressed as a function of the kinetic energy simply because the number of quantum states at a given energy level increases as the energy increases. There are so many levels available to free electrons, and energy levels are so close together (about 10^{-12} eV apart), that discrete available energies may be considered as a quasi-continuous band of allowed energies. We may also treat the density of states as a smooth, continuous variable.

The density of states, which is used to define the distribution of energies for the free electrons, can be computed with the use of the construction in Fig. 11-5, where

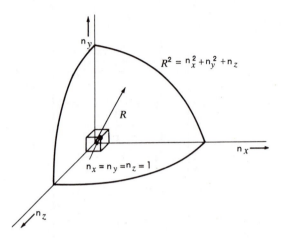

$$R^2 = n_x^2 + n_y^2 + n_z$$

$$n_x = n_y = n_z = 1$$

FIG. 11-5   Graphical representation of electronic quantum states in a cube of metal

the allowed quantum states for free electrons are identified as lattice points on a cubic lattice. It is evident that the lattice points that are most remote from the origin correspond to the highest quantum numbers $(n_x n_y n_z)$ and hence describe the highest energy levels. In fact, the square of the radial distance R is equal to the sum of the squares of the quantum numbers:

$$R^2 = n_x^2 + n_y^2 + n_z^2 \tag{11-23}$$

Since it is precisely this quantity which enters the expression for the energies of the free electrons, the energy level for a given electron can be written as

$$E = \frac{h^2}{8mL^2} R^2 \tag{11-24}$$

In the quantum space system in Fig. 11-5, each quantum state occupies a unit volume. Therefore, the number of quantum states, within a radius R, is simply equal to $\frac{1}{8}(\frac{4}{3}\pi R^3)$. The factor $\frac{1}{8}$ comes from the fact that we are considering only positive values of the quantum numbers [see Eq. (11-17)]; hence we are concerned with only one octant in quantum space. Combining the above expressions gives the number of quantum states per unit crystal volume with energy E or less, N_{qs}, as

$$N_{qs} = \frac{\pi}{6}\left(\frac{8m}{h^2}\right)^{3/2} E^{3/2} \tag{11-25}$$

so that the density of states may be expressed as

$$N(E) = \frac{dN_{qs}}{dE} = \frac{\pi}{4}\left(\frac{8m}{h^2}\right)^{3/2} E^{1/2} \tag{11-26}$$

The physical interpretation of Eq. (11-26) is that the density of states increases parabolically with the energy. It means that more quantum states (or electron homes) are available at higher energy levels. A graphical representation of the density of states for free electrons is given in Fig. 11-6.

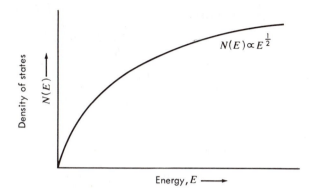

FIG. 11-6   Density of states $N(E)$ as a function of energy level

11-5 FERMI-DIRAC STATISTICS

Thus far our discusion has been concerned only with the energy states and levels that are available to the electrons. We have not yet considered how the energy levels are actually occupied at a given temperature. We now turn our attention to the occupancy of the energy levels.

In Chapter 5 we made extensive use of the Boltzmann statistical formula for the distribution of the energies of atoms. This expression is written as

$$P(E) = \exp\left(-\frac{E}{kT}\right) \tag{11-27}$$

where $P(E)$ is the probability that a given atom has a thermal energy E or greater. According to the Boltzmann formula, at low temperatures all atoms in a material tend to adopt the lowest possible energy level. There is no restriction on how many atoms may reside in the lowest energy level, and at 0°K all atoms have the same lowest energy. The Boltzmann formula does not predict the distribution of electronic energies because it does not obey the Pauli exclusion principle. The proper distribution function for electrons which is consistent with the Pauli principle is called the *Fermi function*. We now turn our attention to that distribution function.

Suppose that a metal sample is in the form of a cube with sides of length L and contains a total of N free electrons. Now at 0°K the electrons will occupy the lowest possible energy levels consistent with the Pauli exclusion principle. When all of the N electrons have occupied the lowest possible energy levels, we call the energy of the most energetic electron the *Fermi energy*, E_F. All electrons will have energies E_F or less. According to Eq. (11-25), the total number of quantum states with energy E_F or less is

$$N_{qs}L^3 = \frac{\pi}{6}\left(\frac{8mL^2E_F}{h^2}\right)^{2/3} \tag{11-28}$$

Since each quantum state will accept two electrons, we have $N_{qs}L^3 = N/2$, and the

Fermi energy can be calculated from

$$\frac{N}{2} = \frac{\pi}{6}\left(\frac{8mL^2 E_F}{h^2}\right)^{3/2} \qquad (11\text{-}29)$$

Rearrangement of Eq. (11-29) gives

$$E_F = \frac{h^2}{8m}\left(\frac{3N}{\pi L^3}\right)^{2/3} \qquad (11\text{-}30)$$

Since $N/L^3 = \eta$ is the number of free electrons per unit crystal volume, the Fermi energy may finally be expressed as

$$E_F = \frac{h^2}{8m}\left(\frac{3\eta}{\pi}\right)^{2/3} \qquad (11\text{-}31)$$

The meaning of Eq. (11-31) is that the energy of the most energetic free electron depends only on physical constants and the number of free electrons per unit volume. It is evident from this expression that the Fermi energy depends on the valence of the metal in question (number of free electrons contributed by each atom) and the crystal structure (number of atoms per unit volume). The Fermi energies for several monovalent metals are given in Table 11-3.

EXAMPLE 11-4

Compute the kinetic energy and the velocity of the most energetic free electrons in sodium.

Solution:

Using Eq. (11-31) and the information in Table 11-3, the Fermi energy for sodium is expressed as

$$E_F = \frac{h^2}{8m}\left(\frac{3\eta}{\pi}\right)^{2/3}$$

and has the value

$$E_F = \frac{(6.63 \times 10^{-27})^2}{8(9.11 \times 10^{-28})}\left(\frac{3}{\pi}\frac{2}{(4.28 \times 10^{-8})^3}\right)^{2/3}$$

$$= 5.05 \times 10^{-12}\ \text{erg} = 3.16\ \text{eV}$$

The velocity of the electron may be expressed by

$$v = \left(\frac{2E_F}{m}\right)^{1/2}$$

$$= \left(\frac{(2)(5.05 \times 10^{-12})}{9.11 \times 10^{-28}}\right)^{1/2}$$

$$= 1.05 \times 10^8\ \text{cm/sec}$$

We now turn our attention to the exact distribution of the energies of the electrons in solids. The basic question we are trying to answer is, "What fraction of the electrons have energies in a particular energy range?" We shall see that, in general, the electrons tend to adopt the lowest energies, as discussed above, but that this picture is not completely accurate. In fact, because the electrons can be thermally excited to higher energy levels, some electrons will always be found in energy levels

TABLE 11-3

Fermi Energies for Several Monovalent Metals

Metal	Crystal Structure	Lattice Parameter a (A)	Number of free electrons per unit volume (cm^{-3})	E_F (eV) (Measured from the Bottom of the Free Electron Energy Band)
Cs	BCC	6.05	$2/a^3 = 9.03 \times 10^{21}$	1.53
Rb	BCC	5.62	$2/a^3 = 1.13 \times 10^{22}$	1.82
K	BCC	5.33	$2/a^3 = 1.32 \times 10^{22}$	2.14
Na	BCC	4.28	$2/a^3 = 2.55 \times 10^{22}$	3.16
Li	BCC	3.50	$2/a^3 = 4.66 \times 10^{22}$	4.72
Ag	FCC	4.07	$4/a^3 = 5.96 \times 10^{22}$	5.51
Cu	FCC	3.60	$4/a^3 = 8.52 \times 10^{22}$	7.04

that do not represent the condition of minimum electronic energy. Therefore, we shall see that although the great majority of the electrons adopt low energy levels, a very important small fraction are thermally excited to higher energy levels. The distribution of the energies of the free electrons is described by the *Fermi function*, which is written as:

$$P(E) = \frac{1}{\exp[(E - E_F)/kT] + 1} \qquad (11\text{-}32)$$

and is graphically represented in Fig. 11-7. In this expression $P(E)$ is the probability that an energy level E is fully occupied and E_F is the Fermi energy. Notice that when

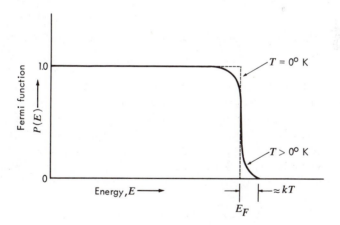

FIG. 11-7 Fermi distribution function. $P(E)$ is the probability that the energy level E is occupied. E_F = Fermi energy.

$E > E_F$, $P(E) \to 0$ as $T \to 0$, and that when $E < E_F$, $P(E) \to 1$ as $T \to 0$. The physical interpretation of the Fermi function is that at $0°K$, all energy levels below E_F are filled and all levels above E_F are empty. When the temperature is finite, the occupation of the energy levels near the Fermi energy varies gradually with energy. The probability that an energy level is fully occupied varies smoothly from unity below E_F to a value near zero above E_F.

EXAMPLE 11-5

The Fermi energy of silver is 5.51 eV. Determine the energy level for which the probability of occupancy is 0.1 at 300°K. Find the energy level for which the probability of occupancy is 0.9, and find the energy range for which the probability varies from 0.9 to 0.1.

Solution:

From Eq. (11-32) we can write

$$0.1 = \frac{1}{\exp\left[(E(0.1) - E_F)/kT\right] + 1}$$

or

$$\exp\left[(E(0.1) - E_F)/kT\right] = 9$$
$$E(0.1) = E_F + 2.2kT$$

At 300°K, $kT = 0.0259$ eV and

$$E(0.1) = 5.51 + 0.057 = 5.57 \text{ eV}$$

At $P(E) = 0.9$ we have:

$$0.9 = \frac{1}{\exp\left[(E(0.9) - E_F)/kT\right] + 1}$$

or

$$\exp\left[(E(0.9) - E_F)/kT\right] = 0.11$$
$$E(0.9) = E_F - 2.21\,kT$$

At 300°K, $kT = 0.0259$ eV and

$$E(0.9) = 5.51 - 0.057 = 5.45 \text{ eV}$$

Thus, the probability of occupancy drops from 0.9 to 0.1 in the energy range 5.45 eV to 5.57 eV.

The example above illustrates that the energy range over which the occupancy probability drops from 0.9 to 0.1 is about $4.4kT$. At room temperature this transition range is about 0.1 eV and is a very small fraction of the Fermi energy. One may think of the perturbed part of the Fermi function as a region inside which some electrons below E_F receive thermal energy kT and are excited to higher energy levels (above E_F). Since electrons are continually being thermally excited to higher energy levels, the distribution of occupancy given by the Fermi function is an equilibrium distribution.

The product of the density of states $N(E)$ and the Fermi function $P(E)$ is proportional to the *electron distribution function* $F(E)$, which is defined in such a way that $F(E)\,dE$ is the number of free electrons per unit crystal volume that have energies between E and $E + dE$. With this definition, $F(E)$ is written as

$$F(E) = 2N(E)P(E) \tag{11-33}$$

and is as illustrated in Fig. 11-8. Notice that according to the electron distribution function the energy level that contains the most electrons lies just below the Fermi energy. While there are some free electrons above E_F, the number of free electrons

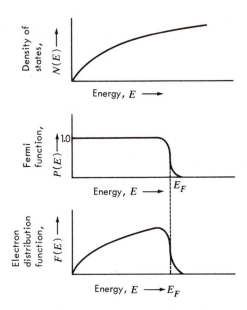

FIG. 11-8 Electron distribution function as a function of the energy level. $F(E) = 2N(E)P(E)$.

having energies well above E_F is very small indeed. Figure 11-8 also indicates that only a very few electrons have the lowest energies.

The free-electron model we have just developed predicts that the energy levels for the free electrons form a quasi-continuous band of allowed energies. As mentioned before, the free-electron model represents an alternate way to describe the energy bands that were illustrated in Fig. 11-2. The difference between the results of the free-electron treatment and the representation in Fig. 11-2 is that the free electron band is quasi-continuous and does not contain energy gaps. The absence of gaps in the energy band comes from our assumption that the electrons are perfectly free.

In the next section we begin to study the interactions between the free electrons and the ions that are arranged periodically in a crystal. We are interested mainly in the effects of the periodic potentials of the ions on the energy band structure of the electrons. We shall show that the free electron energy band develops gaps and becomes discontinuous when the effects of the ion cores are taken into account.

11-6 QUASI-FREE ELECTRONS IN A PERIODIC POTENTIAL FIELD

We now wish to examine the interaction between moving free electrons and the ion cores. For a perfectly free electron, the kinetic energy may be expressed as a function of the deBroglie wavelength, λ:

$$\text{K.E.} = \frac{1}{2}mv^2 = \frac{p^2}{2m} = \frac{h^2}{2m\lambda^2} \qquad (11\text{-}20)$$

It is often convenient when dealing with waves to use the *wave number* or *wave vector* instead of the wavelength itself. This quantity is defined as

$$|\mathbf{k}| = \frac{2\pi}{\lambda} \tag{11-34}$$

and is related to the momentum of the electron by

$$|\mathbf{k}| = \left(\frac{2\pi}{h}\right)p \tag{11-35}$$

The wave vector $\mathbf{k}$ is convenient because it expresses the direction of the electron motion as well as its wavelength. In terms of the wave vector, the kinetic energy of a free electron may be expressed as

$$\text{K.E.} = \frac{h^2}{8\pi^2 m}|\mathbf{k}|^2 \tag{11-36}$$

Thus, for free electrons there is a parabolic relation between the wave vector and kinetic energy. This relationship is illustrated in Fig. 11-9. It should be noted that

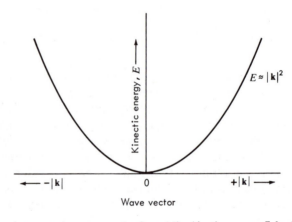

FIG. 11-9   Parabolic relation between the wave vector $\mathbf{k}$ and the kinetic energy E for free electrons

the wave vector in Fig. 11-9 is a quasi-continuous variable. Actually, since the wavelengths are quantized, the wave vectors are also quantized. However, the allowed wave vectors are so close together that $|\mathbf{k}|$ may be treated as an almost continuous variable. We now may examine the effect of the periodic array of ions in the solid.

Since electrons can be treated as waves, they are diffracted by the periodic arrays of ions that they encounter. Suppose that an electron with a wavelength λ moves in a *one-dimensional* metal along a linear array of ions spaced a distance a apart (see Fig. 11-10). According to Bragg's law, diffraction will take place when

$$n\lambda = 2d \sin \theta \tag{11-37}$$

where n is an integer indicative of the order of reflection and not to be confused with the quantum number n. In the one-dimensional problem here, the electron

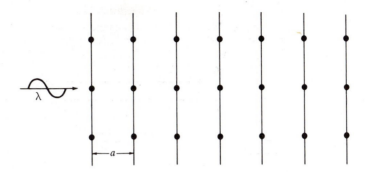

FIG. 11-10   Electron waves of wavelength λ incident on a set of atomic planes of spacing a.

may be considered to be moving normal to a set of diffracting planes with spacing a. Thus, we may set $\theta = \pi/2$ and $d = a$ so that the critical wavelengths for diffraction are

$$\lambda_c = \frac{2a}{n} \qquad n = \pm 1 \pm 2 \pm 3 \dots \tag{11-38}$$

The physical interpretation of Eq. (11-38) is that the free electron will be diffracted whenever the equation is satisfied. This also means that there are critical wave vectors, $\mathbf{k}_c$, for which diffraction of the free electrons takes place. From the definition of the wave vector, they are

$$|\mathbf{k}_c| = \frac{n\pi}{a} \qquad n = \pm 1 \pm 2 \dots \tag{11-39}$$

It should be noted here that since we are considering only a one-dimensional model, the electrons that satisfy the diffraction condition [either Eq. (11-38) or Eq. (11-39)] are simply reflected back (diffracted) in the opposite direction.

The energy-wave vector curve for completely free electrons is perfectly parabolic (see Fig. 11-9). When the electrons are allowed to interact with the ions in the lattice (diffraction), perturbations in the E-k curve appear at the critical values of $\mathbf{k}$ as shown in Fig. 11-11. It is beyond the scope of this book to explain why there is an energy discontinuity in the energy bands at the critical values of the wave vector. One may simply imagine that the energy discontinuity comes from internal diffraction of the free electrons, preventing electrons with specific energies from moving in certain directions.

The important feature of Fig. 11-11 is that the energy band for the free electrons becomes discontinuous when the electrons interact with the lattice. The regions over which the E-k curve is quasi-continuous are called *Brillouin zones*, after the famous French scientist, Leon Brillouin. The electrons with the lowest kinetic energies are contained in the first Brillouin zone, while the more energetic electrons reside in second-, third-, or higher-order zones. The vertical bar graph shown in Fig. 11-11 is a more common representation of the band structure for the electrons. In this representation the Brillouin zones are separated by energy gaps or discontinuities that arise out of the interaction between the moving electrons and the peri-

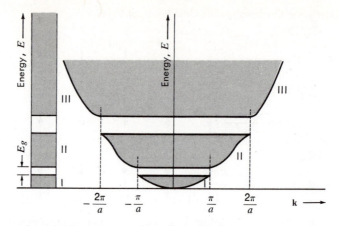

FIG. 11-11 Brillouin zones for quasi-free electrons (one-dimensional)

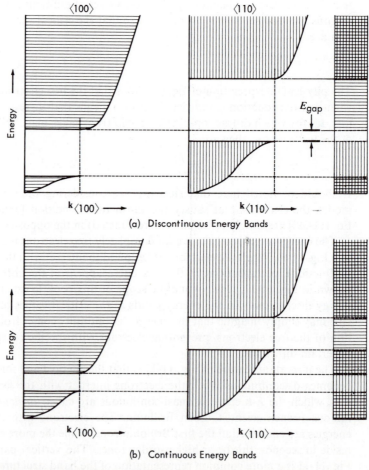

(a) Discontinuous Energy Bands

(b) Continuous Energy Bands

FIG. 11-12 Illustration of the summation of Brillouin zones for the different crystallographic directions

odic array of ions. Again, it must be emphasized that our discussion thus far has been somewhat hypothetical, in the sense that the electron is assumed to be moving along a one-dimensional array of ions. We shall now consider briefly the nature of Brillouin zones in three-dimensional solids.

To use the form of the E-$\mathbf{k}$ curve shown in Fig. 11-11 for a three-dimensional crystal we must consider the summation of a set of E-$\mathbf{k}$ curves for electrons moving in all directions. For example, the E-$\mathbf{k}$ curves for $\langle 100 \rangle$ and $\langle 110 \rangle$ directions are illustrated in Fig. 11-12. Notice that there are two possible cases to consider. The Brillouin zones for these two different directions can either overlap or maintain an energy discontinuity. For the case of overlapping zones, the first Brillouin zone cannot be completely filled up with electrons before the second zone begins to fill; that is, the electrons in the bottom of the second zone are less energetic than electrons in the top of the first zone.

After summing the E-$\mathbf{k}$ curves for all crystallographic directions, the three-dimensional energy band structure *may* or *may not* have energy gaps in the Brillouin zone structure. Those materials with energy gaps are generally those with strong bonding of the outer electrons (appreciable component of covalent or ionic bonding) and hence appreciable perturbation from the simple free-electron theory. We shall see in the next section that the presence or absence of these energy gaps is of considerable importance in the description of conductors, semiconductors, and insulators.

11-7 A PHYSICAL BASIS FOR DESCRIBING METALS, SEMICONDUCTORS, AND INSULATORS

The central idea in distinguishing between conductors, semiconductors, and insulators is that electronic conduction depends on the extent to which the outer electrons in a material may be excited to higher energy states and move in the direction of the applied field. In general, *to achieve higher momentum states (with higher velocities), there must be an empty energy state into which the electron may be excited.*

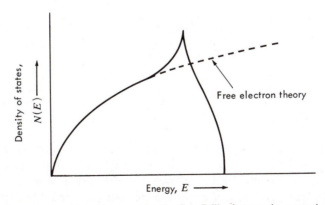

FIG. 11-13 Density of states as a function of energy for the first Brillouin zone in a metal

In other words, to achieve a net flow of electrons in one direction, some electrons must change their wave vectors, thereby increasing their energy. Because of this, the distinction between metals, semiconductors, and insulators is based on the electronic band structures that control the ease with which electrons can achieve higher kinetic energies and velocities.

Because the energy bands for free electrons are subdivided into zones, the density-of-states curve for a given zone deviates considerably from a parabola at the top of the zone (highest energies in the zone). A typical $N(E)$ curve for a Brillouin zone is shown in Fig. 11-13. Notice that the density of states falls to zero at the very top of the zone. Now we may formulate a physical basis for describing metals, semiconductors, and insulators with the use of $N(E)$ curves of this form.

Metals

A metal may be represented by the band structures shown in Fig. 11-14(a) and (b). In Fig. 11-14(a) the two zones overlap, but in Fig. 11-14(b) the zones are sepa-

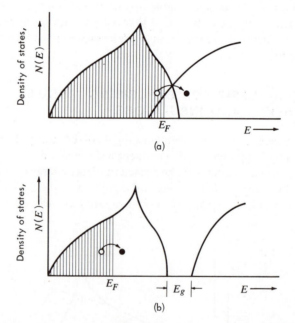

FIG. 11-14 Density of states curves for metals showing cases (a) where Brillouin zones overlap, and (b) where an energy gap is present

rated by an energy gap or discontinuity. In both cases, however, the Fermi energy is positioned in such a way that there are many empty quantum states and energy levels just above the Fermi level. This means that when an electric field is applied the free electrons can easily assume states having higher kinetic energies and can therefore easily conduct electricity. There is almost no energy barrier for the electrons to overcome as they adopt higher velocities and higher kinetic energies. We see from this argument that if the density of states is high at the Fermi level, the material in question will be a metal and it will have metallic electrical properties.

EXAMPLE 11-6

Use Eq. (11-24) and refer to Fig. 11-5 to show that the energy levels for free electrons in metals are separated by less than one-billionth of an electron volt.

Solution:

According to Eq. (11-24) and Fig. 11-5, energy levels for free electrons are separated by an energy difference of not more than

$$\Delta E \approx \frac{h^2}{8mL^2}$$

Taking $L \approx 0.1$ cm,

$$\Delta E \approx \frac{(6.63 \times 10^{-27})^2}{8(9.11) \times 10^{-28})(10^{-2})} = 0.6 \times 10^{-24} \text{ erg}$$

$$\Delta E \approx 0.38 \times 10^{-12} \text{ eV (less than one-billionth of an electron volt)}$$

Semiconductors

The band structure for a semiconductor is illustrated in Fig. 11-15. In this case at 0°K the first Brillouin zone is completely filled with electrons and the second zone

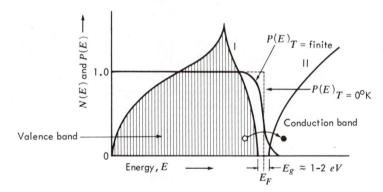

FIG. 11-15 Density of states curves for a semiconductor

is completely empty. The first zone (the last zone that is completely filled with electrons at 0°K) is called the *valence band*. The next zone into which electrons are excited is called the *conduction band*. The energy gap between the two bands is about 1–2 eV for most semiconductors (see Chapter 12). At 0°K, electronic conduction can occur only if the electrons at the top of the first zone are given sufficient energy by the applied field to be excited into the second zone, where higher energy states are available to allow a net electron flux in one direction. Since the Fermi energy is the energy below which all levels are occupied at 0°K and above which all levels are empty, it follows that the Fermi energy for the semiconductor that is described in Fig. 11-15 is in the gap between the two bands. When the temperature is raised to a finite value, the Fermi function is distorted as shown in Fig. 11-15, and some of the electrons are thermally excited from the first zone to the second. The "tails" of the Fermi function indicate that, at a temperature T, some electrons are found in the second zone and an equal number of electrons are missing from the first zone. The number of electrons that are thermally excited into the second zone depends sensitively on the temperature, as we shall see in the next chapter.

EXAMPLE 11-7

The band gap for pure germanium is 0.7 eV. Assume that the Fermi energy lies in the middle of the gap and compute the probability of occupancy of the lowest energy level, E_c, in the conduction band at room temperature. Compute the temperature at which the probability will be double the room temperature value. At what temperature will the probability be greater by an order of magnitude?

Solution:

At room temperature $kT = \frac{1}{40}$ eV, so that

$$E_c - E_F = 0.35 \text{ eV} \gg kT$$

and thus the Fermi function [Eq. (11-32)] becomes

$$P(E_c) \approx \exp - \frac{(E_c - E_F)}{kT} = \exp\left(-\frac{0.35}{\frac{1}{40}}\right)$$
$$= \exp(-14) = 8.1 \times 10^{-7}$$

The temperature at which the occupancy probability is doubled is

$$P(E_c) = 16.2 \times 10^{-7}$$
$$\frac{E_c - E_F}{kT} = 12.4$$
$$T = \frac{(0.35)(1.6 \times 10^{-12})}{(1.38 \times 10^{-16})(12.4)} = 328°\text{K} = 55°\text{C}$$

Thus, if room temperature is 20°C, the occupancy probability at the bottom of the conduction band doubles when the temperature is raised 35°C.

The temperature at which the occupancy probability is increased by an order of magnitude is

$$P(E_c) = 8.1 \times 10^{-6}$$
$$\frac{E_c - E_F}{kT} = 11.0$$
$$T = 369°\text{K} = 96°\text{C}$$

From this discussion we see that a semiconductor is a material having energy bands that are separated by a small energy gap. Because of this energy gap, only a limited number of electrons are available to contribute to the conduction process. That is, at any finite temperature, only those electrons excited to the conduction band contribute to current flow. As shown in Example 11-7, the number of electrons excited to the conduction band depends strongly on the temperature.

Insulators

The band structure of an insulator is one in which two zones, one of which is empty and the other of which is filled, are separated by a large energy gap (several electron volts). This band structure is illustrated in Fig. 11-16. As in the case of

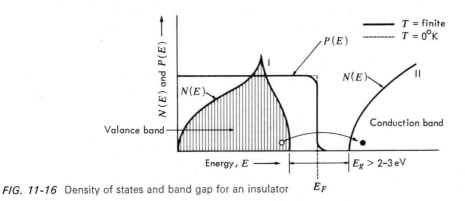

FIG. 11-16 Density of states and band gap for an insulator

semiconductors, the Fermi energy falls inside the energy gap. In the case of insulators, however, the energy gap is so large that it may not be surmounted either by thermal energy or by the application of a field. In terms of the Fermi function, which is also shown in Fig. 11-16, the number of electrons that exists several electron volts above the Fermi level is negligibly small. We may suppose that essentially no electrons are excited from the first zone into the second in an insulator. Since it is not possible for electrons to be excited to states corresponding to higher kinetic energies, and thus higher velocities, it follows that the material will not conduct electricity.

11-8 STATIC ELECTRONIC PROPERTIES OF METALS

One of the assumptions made in our simplified treatment of the free-electron theory of metals was that free electrons do not escape from the metal in which they reside. We supposed that the electrons see an infinite potential barrier at the metal surface (see Fig. 11-3). In fact, the potential barrier at the surface is finite and some of the more energetic electrons can escape. The operation of a vacuum tube, for example, depends on the ability of the material in the cathode to emit electrons when it is heated. Free electrons are also emitted when metals are subjected to high electric fields or when electromagnetic radiation falls on a metal surface. All of these properties can be explained in terms of the distribution of electronic energies in the metal and the magnitude of the potential barrier at the surface. They are called *static* electronic properties because they do not depend on the rate at which charge can be moved about within the metal.

Figure 11-17 is a representation of the potential energy of the free electrons in a one-dimensional metal. Notice that a finite potential energy barrier, W, exists at each end of the solid. In order for an electron to escape from the metal it must have a kinetic energy equal to or greater than W. Since the free electrons in the metal have kinetic energies ranging from nearly zero at the bottom of the potential energy well to E_F at the Fermi level, the electrons must absorb energies between W and $(W - E_F)$ to be ejected from the solid. The critical work that must be sup-

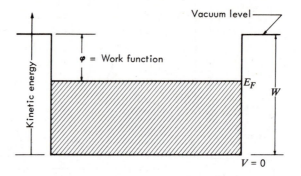

FIG. 11-17 Relation between the Fermi energy and the work function of a metal. The vacuum level is the energy of a free electron in vacuum.

plied to eject a free electron from the solid is called the *work function* and is denoted by $\varphi = W - E_F$. It is the most important parameter in our description of the static electronic properties of metals.

Contact Potential

Consider the energy level diagram for two different metals, A and B, shown in Fig. 11-18. In general, the metals have different potential energy barriers, W_A and W_B, at the surface and different Fermi energies, $E_F(A)$ and $E_F(B)$, and hence different work functions, φ_A and φ_B. It is evident that the free electrons at the Fermi energy in metal A are more energetic than the electrons at the Fermi level in metal

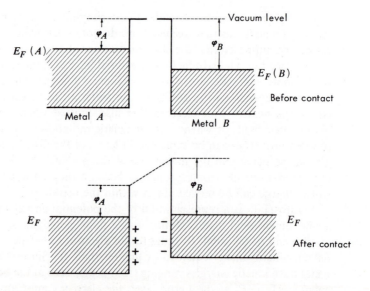

FIG. 11-18 Contact potential between two metals

B. As a consequence, when the two metals are brought into contact, electrons flow from metal *A* to metal *B* and reduce their kinetic energies. As electrons flow from one metal to the other, an electrostatic potential (voltage) tending to reverse the flow of electrons begins to build up. In terms of the potential energy diagram in Fig. 11-18, the potential energy of the electrons in metal *B* is raised relative to potential energy in metal *A*. Electron flow stops when the electrostatic potential energy difference between the two metals is equal to the difference in the work functions. Figure 11-18 illustrates the equilibrium state for the electrons after the metals have made contact. At equilibrium, the energy required to move an electron from metal *A* to metal *B* is equal to zero. It is zero because the energy gained by the electron when it passes from metal *A* to metal *B*, $(\varphi_B - \varphi_A)$, is exactly equal to the energy lost as the electron moves through the contact potential (voltage), V_c. At equilibrium the contact voltage is expressed as

$$V_c = \frac{\varphi_B - \varphi_A}{e} \qquad (11\text{-}40)$$

where *e* is the charge of the electron. We shall see in Chapter 13 that the contact potential produced when a metal makes contact with a semiconductor, or when two semiconductors are brought into contact, is of considerable importance in explaining the electrical properties of junctions between dissimilar materials. The work functions and contact potentials for several metals are given in Table 11-4.

TABLE 11-4
Work Functions and Contact Potentials for Several Metals

Metal	Work Function (eV)	Contact Potential* (volts)	Metal	Work Function (eV)	Contact Potential* (volts)
Ag	4.73	+0.59	K	2.24	+3.08
Al	4.08	+1.24	Li	2.49	+2.83
Au	4.82	+0.50	Mg	3.68	+1.64
Ba	2.48	+2.84	Mo	4.20	+1.12
Be	3.92	+1.40	Na	2.28	+3.04
Bi	4.25	+1.07	Ni	5.01	+0.31
Cd	4.07	+1.25	Pd	4.97	+0.35
Co	4.40	+0.92	Pt	5.32	0.00
Cr	4.60	+0.72	Rb	2.09	+3.23
Cs	1.81	+3.51	Ta	4.19	+1.13
Cu	4.46	+0.86	Th	3.35	+1.97
Hg	4.53	+0.79	W	4.50	+0.82

*Contact potential of metal in question with respect to platinum. (Metals with work functions smaller than Pt will contribute electrons to Pt and will become positively charged.)

Photoelectric Phenomena in Metals

Since the potential energy barrier at the surface of a metal is finite, the free electrons may be ejected from the metal if they receive sufficient energy from an external source. Alkali metals, for example, will emit free electrons when they are exposed to visible light. In this case the photons must be sufficiently energetic to remove the electrons from the metal. Figure 11-19 illustrates the events that take place when

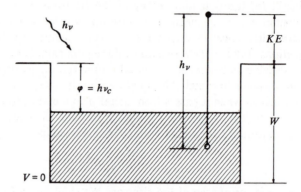

FIG. 11-19 Conservation of energy of an electron ejected from a metal by a photon with an energy $h\upsilon < h\upsilon_c$. K.E. is the kinetic energy of the ejected electron.

a free electron is ejected from a metal by a photon of energy $h\nu$. It is evident that there is a critical threshold energy, $h\nu_c$, that must be supplied to remove an electron from the surface. The threshold energy is equal to the work function for the metal in question. Figure 11-20 indicates the critical photon wavelengths for ejecting free electrons from several metals. Note that when the wavelength of the incident

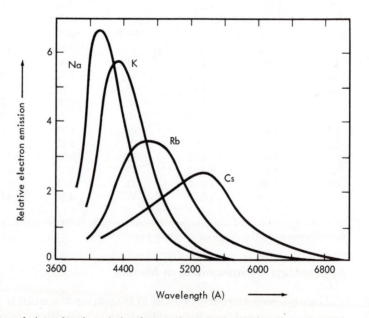

FIG. 11-20 Illustration of photoelectric emission from various alkali metals

radiation is greater than the critical wavelength, electrons will not be ejected ($hv < W - E_F$). As the incident photons become more energetic, $hv > W - E_F$, they have sufficient energy to eject more of the free electrons. The decrease in the number of emitted electrons, which is shown in Fig. 11-20 at short wavelengths, is due to a decrease in the efficiency of the electronic excitation.

The photoelectric effect may be used to determine the work function of a metal. As is evident from Fig. 11-20, one simply measures the longest photon wavelength λ_c that will cause electrons to be emitted from the surface. The work function is then simply calculated as

$$\varphi = \frac{hc}{\lambda_c} \tag{11-41}$$

The photoelectric effect is the basis for the operation for a *photocathode*, a device, somewhat like a vacuum tube, which will conduct electricity when it is exposed to light radiation of the proper wavelength. Photocathodes are sometimes constructed of alkali metals such as lithium, sodium, potassium, or cesium, although specially designed alloys are usually more efficient emitters of electrons.

Thermionic Emission

One of the most important static electronic properties of metals is their ability to emit electrons when they are heated. Figure 11-21 illustrates the thermal excitation of electrons from the surface. According to the electron distribution function, a small fraction of the free electrons will have kinetic energies that are sufficient

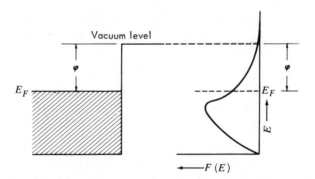

FIG. 11-21 Schematic representation of thermionic emission

for them to escape from the surface. In order to predict the electron currents produced by thermionic emission we must concern ourselves with the momenta of the electrons as well as their energies. When an electron is moving normal to the surface all of its kinetic energy may be applied to the potential energy barrier. Thus, for electrons that move normal to the surface the critical kinetic energy needed to escape from the surface must be W. However, when the electron strikes the surface at an angle from the inside, only the normal component of its momentum can be used in crossing the barrier. As a result, not all electrons that have kinetic energies greater than W can escape from the metal.

EXAMPLE 11-8

Suppose that an electron approaches the free surface of a metal along a direction that makes an angle of 60° with the surface normal. Calculate the kinetic energy of the electron when it is sufficiently energetic to escape the metal if $W = 4$ eV (see Fig. 11-19).

Solution:

Let p_x represent the component of momentum perpendicular to the surface. Then the electron will escape when

$$\frac{p_x^2}{2m} \geq 4 \text{ eV}$$

The total momentum of the electron is

$$p = \frac{p_x}{\cos \theta} \qquad \theta = \frac{\pi}{3}$$

and the kinetic energy is

$$\frac{p^2}{2m} = \frac{p_x^2}{2m \cos^2 \theta}$$

$$\text{K.E.} = \frac{4 \text{ eV}}{(0.5)^2} = 16 \text{ eV}$$

Because the momenta of the electrons as well as their kinetic energies must be taken into account, calculation of the electron current emitted from a metal surface is not straightforward, and we shall simply give the result of such an analysis. The *Richardson-Dushman equation* is used to compute the thermionic current density from a metal surface. It is derived from the free-electron theory of metals and is expressed as

$$J = AT^2 \exp\left(-\varphi/kT\right) \tag{11.42}$$

where A is a constant that involves the physical constants and has the value of $120 \ A/\text{cm}^2 \cdot {}^\circ K^2$, T is the absolute temperature in $^\circ K$, and φ is the work function of the metal.

EXAMPLE 11-9

It is proposed to make a filament with a tungsten wire, 0.01 cm in diameter, that is heated to 2500°K. Calculate the length of the wire needed to produce an electron current of 100 mA. $\varphi = 4.5$ eV.

Solution:

From Eq. (11-42)

$$J = (120)(2500)^2 \exp\left[\frac{-(4.5)(1.6 \times 10^{-12})}{(1.38 \times 10^{-16})(2500)}\right]$$

$$J = 0.67 \text{ A/cm}^2$$

$$\text{Area} = \pi(0.01)l = \frac{10^{-1}}{0.67} \frac{\text{current}}{\text{current density}}$$

$$l = \frac{15}{\pi} = 4.8 \text{ cm}$$

PROBLEMS

11-1 Why can we make the general statement that the kinetic energy of an electron increases when it is crowded into a small space?

11-2 Refer to Fig. 11-2 and explain what would happen to the $2p$ electronic energy level if metallic sodium were subjected to a large hydrostatic pressure.

11-3 A one-dimensional solid (a straight wire) 1 cm in length is composed of monovalent atoms that are spaced exactly 2 A along the length of the wire. Calculate the velocity of the free electron with the lowest kinetic energy.

11-4 Show that Eq. (11-16) satisfies the boundary condition given by Eq. (11-10).

11-5 Consider two long, straight metal wires of length L. If the wires are placed end to end and brought into contact, new electronic energy levels and states are permitted, since the free electrons may have wavelengths that are integral fractions of the length $2L$. Why, then, is not a great deal of heat given up when two wires are joined and when some of the electrons assume lower kinetic energies?

11-6 Suppose a metal were to undergo a phase transformation from one crystal structure to another. How might the average energy of the free electrons be changed? Suppose that lithium were to experience a phase transformation in which the mass density was increased by 10%. In what way, and to what extent, would the average energy of the free electrons change?

11-7 Explain why the Fermi energy (or the average kinetic energy of the free electrons) is involved in the following physical processes:
(a) The formation of vacancies in metals. (Does the energy needed to form a vacancy depend on E_F?)
(b) The hydrostatic compression of metals. (Does the bulk elastic modulus depend on E_F?)

11-8 Using what you know about the density of states for free electrons, write an expression for the average kinetic energy for the free electrons in a metal with a Fermi energy E_F.

11-9 Using the Fermi function, evaluate the temperature at which there is a 1% probability that an electron in a solid will have an energy 0.5 eV above the Fermi energy of 5 eV.

11-10 Consider a one-dimensional wire of length L in which there are N atoms in a row spaced a distance a apart. Calculate the total number of electrons that would be required to fill the first Brillouin zone.

11-11 Show that the Fermi function is symmetric in the sense that $P(E_F + \Delta E) \equiv 1 - P(E_F - \Delta E)$.

11-12 Explain why the density-of-states curve shown in Fig. 11-13 exceeds the density-of-states relationship for free electrons. [*Hint:* the peak in the $N(E)$ curve occurs when the Fermi surface touches the Brillouin zone.] Construction of a three-dimensional E-k_x-k_y surface should show how the density of states can exceed the free electron value.

11-13 The energy band gap for an insulator is 4.3 eV. What is the probability that the lowest energy level in the conduction band is filled with electrons at 1500°C?

11-14 The work functions for Na, Mg, and Cu are

Na   2.28 eV
Mg   3.70 eV
Cu   4.48 eV

Make a schematic plot of the rate of photoemission for each metal as a function of the wavelength of the incident radiation. Put the curves for each metal on the same graph so that their relative positions may be compared. Rate these metals in terms of their potential for producing electrons by thermionic emission at a given temperature. Assuming that each metal is operated at its melting temperature, which of the metals has the highest potential for thermionic emission of electrons?

$$T_M(\text{Na}) = 97.8°\text{C}$$
$$T_M(\text{Mg}) = 650°\text{C}$$
$$T_M(\text{Cu}) = 1083°\text{C}$$

11-15 Determine the magnitude and sign of the contact voltage that would arise if aluminum were brought into contact with magnesium.

11-16 Calculate the wavelength of the least energetic photon that would be required to cause photoemission from Cs. What wavelength would be needed to eject electrons from W?

11-17 A thin film of Ta 10 μ thick is to be used as a planar source for thermionic emission of electrons. The film is to be heated to 2500°K. Calculate the current that would emanate from each square centimeter of surface.

BIBLIOGRAPHY

L. V. AZAROFF and J. J. BROPHY, *Electronic Processes in Materials*, Ch. 6–7. New York: McGraw-Hill, 1963.

A. H. COTTRELL, *Theoretical Structural Metallurgy*, Ch. 4–5. New York: St. Martin's Press, 1957.

N. CUSACK, *The Electrical and Magnetic Properties of Solids*, Ch. 1–4. London: Longmans, Green, 1958.

D. F. GIBBONS, *Physical Metallurgy*, Ch. 3, ed. R. W. Cahn. New York: John Wiley, 1965.

T. S. HUTCHISON and D. C. BAIRD, *The Physics of Engineering Solids*, Ch. 8–11. New York: John Wiley, 1968.

R. M. ROSE, L. A. SHEPARD, and J. WULFF, *The Structure and Properties of Materials*, Vol. IV, Ch. 1–2. New York: John Wiley, 1966.

R. L. SPROULL, *Modern Physics: A Textbook for Engineers*, Ch. 8. New York: John Wiley, 1956.

C. A. WERT and R. M. THOMPSON, *Physics of Solids*, Ch. 9–10. New York: McGraw-Hill, 1964.

12

Electronic Transport

12-1 INTRODUCTION

In this chapter our attention is focused on the relation between applied electric fields and the resulting electric currents in materials. We shall describe the events that take place when electric fields are applied to metals, semiconductors, and insulators, and we shall pay particular attention to the material variables that control the resistance to the flow of electric current. We shall see that an understanding of the factors that control the electrical resistance of a metal is important because the resistance ultimately determines the amount of heat generated in an electric conductor and thus limits the current that a conductor of a given size may carry. We shall also see that the unique properties of superconducting materials, whose electrical resistance becomes vanishingly small at very low temperatures, may be used to transmit electric power with almost 100% efficiency or may be used in electromagnets to produce very high magnetic fields. As mentioned before, the principles of electronic transport in semiconductors are extremely important because they form the basis for the design and utilization of transistors and other solid state devices.

12-2 A PHYSICAL BASIS FOR CONDUCTION IN MATERIALS

Before treating the problem of electronic conduction in materials in a quantitative way, it is instructive to examine the relation between bonding and the current-carrying abilities of materials. The electrical resistivities of several different types of materials are given in Table 12-1. The predominant type of bonding for each material is also indicated in the table. We immediately see a correlation between the bonding type and the electrical resistivity. Materials with metallic bonding, whether crystalline or liquid, have electrical resistivities that range between 10^{-8} and 10^{-7} ohm·m. Materials with electrical resistivities of this magnitude are commonly called *conductors*. The table also shows that materials with covalent bonds have electrical resistivities that range from about 1 ohm·m (germanium) to 10^{17} ohm·m (polystyrene). Materials with resistivities of the order of 1 ohm·m are termed *semiconductors*, while *materials with* resistivities greater than 10^{12} ohm·m are called *insulators*. Ionic materials have very high electrical resistivities, and thus they are considered to be insulators.

TABLE 12-1

The Relation between Bonding and the Electrical Resistivities of Materials*

Material*	Bonding Type	Electrical Resistivity (approximate) (ohm·m)
Silver	Metallic	1.6×10^{-8}
Aluminum	Metallic	2.8×10^{-8}
Copper	Metallic	1.7×10^{-8}
Gold	Metallic	2.4×10^{-8}
Iron	Metallic	8.8×10^{-8}
Nickel	Metallic	7.2×10^{-8}
Sodium	Metallic	4.3×10^{-8}
Mercury (liquid)	Metallic	9.6×10^{-7}
Lithium (liquid)	Metallic	4.5×10^{-7}
Indium-antimonide	Covalent	2.0×10^{-4}
Germanium	Covalent	0.5
Silicon	Covalent	3.0×10^{3}
Silver-bromide	Ionic	1×10^{6}
Gallium-arsenide	Covalent	2.0×10^{6}
Strontium-Titanate	Ionic	2×10^{9}
Window glass	Ionic	5×10^{9}
Nickel-oxide	Ionic	1×10^{11}
Gallium-phosphide	Covalent	6.0×10^{13}
Diamond	Covalent	1×10^{14}
Cadmium-sulfide	Ionic-covalent	2×10^{14}
Quartz	Ionic	3×10^{14}
Mica	Ionic	9×10^{14}
Polyethylene	Covalent†	10^{16}
Zinc-selenide	Ionic-covalent	2×10^{16}
Polystyrene	Covalent†	10^{17}

*The resistivities given are for pure materials. The compounds have stoichiometric compositions. All resistivity values were taken at room temperature except for lithium (230°C)

†Only the intramolecular bonding is covalent in nature.

Now we may introduce a very simple model for describing conductors, semiconductors, and insulators. This model, schematically illustrated in Fig. 12-1, is

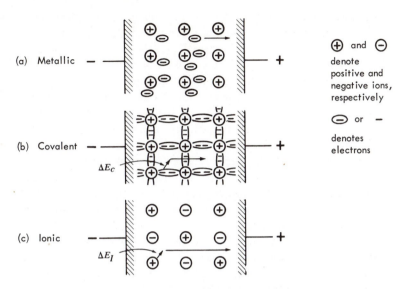

FIG. 12-1 Schematic representation of the mechanisms of electronic conduction in metallic, covalent, and ionic materials. ΔE_c and ΔE_I represent the critical energies needed to excite electrons into a conducting state in covalent and ionic materials, respectively.

based on the idea that if electrons are not free to migrate in response to an applied electric field, they must be excited from the bound states in which they reside by a kind of internal ionization process.

In the preceding chapter we showed that the outer electrons in metallic atoms become free when the electron clouds from neighboring atoms begin to overlap. We saw that materials with metallic bonding should be good conductors of electricity because the free electrons can easily respond to the application of an electric field and can provide high electric currents [see Fig. 12-1(a)].

In Chapter 2 it was pointed out that covalent bonds are formed when electronic orbitals or clouds from neighboring atoms overlap. The bonding differs from metallic bonding in the sense that in covalent bonding the shared electrons are localized and remain in the vicinity of the atoms that they bond together. Because of this, the electrons are unable to move freely through the material, and the covalent materials are clearly not good conductors. It is possible to have some electronic conduction take place in covalent materials, however. One may view this conduction process as being dependent on a kind of ionization phenomenon in which electrons are robbed from the covalent bond [see Fig. 12-1(b)]. In crystalline covalent materials such as Si, Ge, InSb, and GaAs, the energy required to remove an electron from the covalent bond is of the order of 1 eV. The thermal energy that an electron may receive is of the order of kT and has the value 0.025 eV at room temperature. Since kT at room temperature is much smaller than the energy needed to remove an electron from a covalent bond, it follows that only a very few electrons are free to conduct electricity. This explains why the electrical resistivities of these materials are several orders of magnitude higher than those for metals. As the temperature is raised, the number of electrons that are thermally excited from the covalent bond is increased, and the electrical resistivity decreases accordingly. Also, in organic materials with covalent bonds electronic conduction occurs only when electrons can be removed from the covalent bonds. Generally speaking, the energies needed to excite electrons from covalent bonds in polymers are of the order of serveral electron volts; thus, most polymer materials are insulators. As shown in Table 12-1, both polyethylene and polystyrene have electrical resistivities of the order of 10^{16}–10^{17} ohm·m. In the case of some crystalline polymers, the energies needed to free the electrons from covalent bonds are of the order of 1 eV; here we have what are known as *organic semiconductors*. Anthracene ($C_{10}H_8$) and napthalene ($C_{14}H_{10}$) are examples of organic semiconductors. The critical energies needed to free electrons for conduction in these materials are 1.35 eV and 1.85 eV, respectively.

Ionic materials are insulators because the energy required to excite an electron from one of the ions is very large, usually of the order of several electron volts. In Fig. 12-1(c) the term ΔE_I corresponds to the internal ionization energy required to free an electron for the conduction process.

12-3 A CLASSICAL TREATMENT OF CONDUCTION IN METALS

In order to develop a physical picture of the process of conduction, we shall analyze the physics of conduction in metals on the assumption that the electrons are entirely free and may be treated as classical particles. While this treatment is clearly inaccurate because it does not account for the quantized kinetic energies of the free electrons, it nevertheless allows us to introduce several important parameters that maintain their significance when the quantum mechanical treatment of conduction is carried out.

Consider a wire of diameter d and length L. According to *Ohm's law*, the electric current I passing through any homogeneous material is proportional to the applied voltage V through the relation

$$I = \frac{V}{R} \tag{12-1}$$

where R is the electrical resistance in ohms when the current is expressed in amperes and the voltage is expressed in volts. The electrical resistance is proportional to the length of the wire and inversely proportional to the cross-sectional area and thus is not an intrinsic property of a material. Therefore, it is appropriate to focus our attention on the resistance of a sample of unit length and unit cross-sectional area. This intrinsic quantity is known as the *electrical resistivity* and is expressed as

$$\rho = \frac{RA}{L} \tag{12-2}$$

where $A =$ area. If we define the *current density* J as $J = I/A$ and the *electric field strength* (or *voltage gradient*) as $E = V/L$, then Ohm's law can be expressed as

$$E = \rho J \qquad (12\text{-}3)$$

or

$$J = \sigma E \qquad (12\text{-}4)$$

where σ, the *electrical conductivity*, is the reciprocal of the *electrical resistivity* ρ and is expressed in mhos per meter in the MKS system of units.

EXAMPLE 12-1

Consider a solid state circuit of the kind shown in Fig. 1-8. The conductors in such a circuit are often produced by vapor-depositing thin strips of aluminum onto the surface. Assume that the circuit is formed on a silicon wafer 1mm square and 0.2mm in thickness. The conductors are approximately 5×10^{-3} mm in thickness and 0.025 mm in width. The conductors cover about 10% of one side of the wafer and carry a current density as high as $10^4 A/cm^2$. Estimate the time that would be required for the conductors to heat to their melting temperature if they did not lose their heat to the silicon wafer and the surrounding atmosphere. Take the heat capacity C_p of aluminum to be 0.25 cal/g·°C and the density to be 2.69 g/cc.

Solution:

The power loss due to resistive loss in a conductor is $I^2 R$, so that the rate of heat production per unit volume is expressed as J^2/σ and is given in W/cm^3. For the aluminum conductors, the rate of heat production is

$$\dot{Q} = \frac{1}{\sigma} J^2 = (2.8 \times 10^{-8})(100)10^8 = 2.8 \times 10^2 \; W/cm^3$$

$$(\text{ohm} \cdot m)(cm/m)(A^2/cm^4)$$

or

$$\dot{Q} = (2.8 \times 10^2)(14.33) = 4.0 \times 10^3 \; cal/cm^3 \cdot min$$

where 14.33 is a unit conversion factor. The time τ required for the aluminum strips to reach their melting temperature if they were not in thermal contact with the silicon wafer is obtained by letting

$$\tau \dot{Q} = (\rho C_p)_{Al}(T_M - 300) \qquad (T_M = 933°K)$$

$$\tau = \frac{(2.69)(0.25)(633)}{4.0 \times 10^3} = 0.106 \; min$$

Thus, the aluminum conductors would melt very quickly if they were not in thermal contact with the silicon wafer.

Equations (12-3) and (12-4) represent a phenomenological description of the behavior of homogeneous materials. As discussed in the previous section, the electrical conductivity or resistivity of a material greatly depends on the type of bonding involved and depends to a less extent on the atom arrangements in that material. We shall now present a microscopic theory of conduction that conforms to the phenomenological description just given. This theory, which is usually called the *free-electron theory of conduction*, provides a basis for understanding the conduction phenomenon in metals as well as in nonmetallic solids.

Consider an isolated free electron with charge $-|e|$ and mass m within a metallic sample. Let us consider the events that take place when an electric field (voltage gradient) E is applied to the sample. The electric field exerts a force $|e| E$ on the electron so that its subsequent motion is governed by

$$|e|E = ma \qquad (12\text{-}5)$$

or

$$\frac{dv}{dt} = a = \frac{|e|E}{m} \qquad (12\text{-}6)$$

where v is the velocity. This equation indicates that if the electric field is constant and the electron is perfectly free (i.e., no other forces on the electron), the velocity of the electron increases linearly with time from the instant that the electric field is applied. If we think of the meaning of this result in terms of large numbers of free electrons, we see immediately that Eq. (12-6) predicts that the electric current in a metal increases linearly without limit when a voltage is applied. This is clearly not consistent with experience. In order to modify our picture so that it may be consistent with Ohm's law (constant current for a given applied voltage), it must be recognized that electrons periodically collide with the atoms in the lattice and lose their kinetic energy. The process of collision is sometimes called *scattering* and is a central concept in the free-electron theory of metallic resistivity. After such a collision the electrons are free to accelerate in the applied field as before. Under these conditions the velocity of a given electron will vary in a "saw-tooth" fashion with time, as shown in Fig. 12-2. It is conventional to define the time elapsed be-

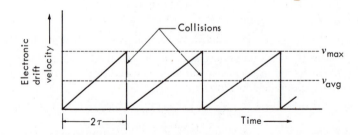

FIG. 12-2 Schematic representation of the acceleration and deceleration (collisions) associated with electronic migration in response to an applied voltage

tween collisions as 2τ, where τ is called the *lifetime* or *relaxation time* for the free electrons. The average drift velocity of the electron may be expressed as

$$\bar{v} = \frac{|e|E}{m}\tau \qquad (12\text{-}7)$$

If it is assumed that all free electrons behave in the manner described by Fig. 12-2, then Eq. (12-7) may be taken as an accurate representation of the average drift velocity of all of the electrons participating in the conduction process. In this approximation the current density may be expressed directly as

$$J = \eta|e|\bar{v} \qquad (12\text{-}8)$$

where η is the number of conduction electrons per unit volume. Using Eq. (12-7) and comparing this equation with Ohm's law [Eq. (12-4)], we find that the electrical conductivity may be expressed as

$$\sigma = \frac{\eta |e|^2 \tau}{m} \tag{12-9}$$

Before we discuss the physical meaning of Eq. (12-9), it will be worthwhile to examine the validity of the free-electron model that we have used to derive this equation. In the first place, it is erroneous to suppose that the free electrons are at rest before the electric field is applied. We saw in the previous chapter that the wave nature of electrons demands that all free electrons move with finite speed and that the velocities they adopt are not chosen arbitrarily, but rather are fixed by the dimensions of the sample in which they reside. This fact suggests that our analysis of the migration of free electrons in an electric field must be interpreted to refer to the "drift" motion of the electrons which is superimposed on the inherent or thermal free-electron motion.

A second inaccurate feature of the free-electron picture of conduction that we have just developed relates to the assumption that all the electrons are free to migrate in response to the applied field. The quantum mechanical view of free electrons developed in the previous chapter holds that the kinetic energies of the free electrons are quantized, and thus that their kinetic energies may not be changed arbitrarily. The allowed values of momentum for free electrons in a two-dimensional metal are illustrated in Fig. 12-3. When there is no applied electric field, the

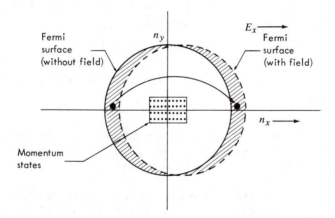

FIG. 12-3   Representation of the redistribution of the occupancy of electronic states when a voltage is applied to a metal

net momentum of all the free electrons is zero. There are as many electrons moving in the $+x$ direction as are moving in the $-x$ direction. Each point in Fig. 12-3 corresponds to an allowed value of momentum for the free electrons. When an electric field is applied, the electrons are not free to accelerate in the classical sense. The effect of the applied field is to change the occupancy pattern of the allowed energies. An electric field in the $+x$ direction causes the occupied momentum states to be shifted in the $+x$ direction as shown. One may consider that all the electrons have increased their momentum in the $+x$ direction by the same amount. An alternate way to view the conduction process is to imagine that some of the electrons at the Fermi level are simply transported from one side of the momentum diagram to the other.

This discussion makes it clear that the free-electron model of conduction just presented is not consistent with quantum mechanical principles. A model of free-electron conduction which is consistent with the quantum mechanical restrictions was first presented by Sommerfeld in the late 1920s. Interestingly, the result of his analysis is an equation similar in form to Eq. (12-9):

$$\sigma = \frac{\eta |e|^2 \tau_F}{m} \tag{12-10}$$

where τ_F is the relaxation time for electrons at the Fermi level.

The relaxation time for electrons may be related to the mean free path between collisions, $\bar{l}$, by

$$\tau_F = \frac{\bar{l}}{v_F} \tag{12-11}$$

where v_F is the thermal velocity of electrons at the Fermi energy E_F. It should be noted that Eq. (12-11) is accurate only if the drift velocity caused by the applied field is small compared to the thermal or absolute velocity of the electrons at E_F. Since v_F can be calculated from a knowledge of the Fermi energy, substitution of Eq. (12-11) into Eq. (12-10) yields

$$\sigma = \frac{\eta |e|^2 \bar{l}}{m v_F} \tag{12-12}$$

where $\bar{l}$, the mean free path between collisions, is the principal unknown for a given model.

EXAMPLE 12-2

A sodium wire 1 mm in diameter carries a current of 1 mA at room temperature. Calculate the drift velocity $\bar{v}$ of the free electrons and compare it with the thermal velocity v_F of electrons at the Fermi energy.

Solution:

Sodium is monovalent and has a BCC crystal structure with a lattice parameter of 4.28 A. Using Eq. (12-8), we find that the drift velocity becomes

$$\bar{v} = \frac{J}{\eta |e|} = \frac{(10^{-3})/\pi(0.05)^2}{[2/(4.28 \times 10^{-8})^3](1.6 \times 10^{-19})}$$

$$= 3.13 \times 10^5 \text{ cm/sec}$$

The thermal velocity is expressed as

$$v_F = \left(\frac{2E_F}{m}\right)^{1/2}$$

and, using Eq. (11-31), as

$$v_F = \frac{h}{2m}\left(\frac{3\eta}{\pi}\right)^{1/3}$$

$$= \frac{6.63 \times 10^{-27}}{2(9.11 \times 10^{-28})}\left(\frac{3}{\pi}\frac{2}{(4.28 \times 10^{-8})^3}\right)^{1/3}$$

$$= 1.05 \times 10^8 \text{ cm/sec}$$

This example demonstrates that the assumption that $v_F \gg \bar{v}$ is very good indeed.

12-4 CONTRIBUTIONS TO THE ELECTRICAL RESISTIVITY OF METALS

From Eq. (12-12) we may express the electrical resistivity of a metal as

$$\rho = \frac{mv_F}{\eta|e|^2\bar{l}} \tag{12-13}$$

The parameter on which we now focus our attention is $\bar{l}$, the mean free path between collisions, since this is the most important factor determining the electrical resistivity of a metal. Thus far in our discussion we have assumed that the process of scattering simply involves the collision of an electron with an ion core, and we have not paid attention to the factors that contribute to the probability of a collision. We shall see shortly that atomic vibrations, impurity atoms, and structural imperfections all make a contribution to the probability of collision and thus all contribute to the electrical resistivity of a metal.

Suppose the mean free path due to thermal vibrations is $\bar{l}_{TH}$ and that due to impurities is $\bar{l}_I$. It is clear that when an electron drifts a distance L in a metal, it will have suffered $L/\bar{l}_{TH}$ collisions due to lattice vibrations and $L/\bar{l}_I$ collisions due to impurities. Since the total number of scattering events may be expressed as $L/\bar{l}_{TH} + L/\bar{l}_I$, it follows that the mean free path $\bar{l}$ due to both thermal vibrations and impurity atoms may be expressed by

$$\frac{1}{\bar{l}} = \frac{L/\bar{l}_{TH} + L/\bar{l}_I}{L} = \frac{1}{\bar{l}_{TH}} + \frac{1}{\bar{l}_I} \tag{12-14}$$

Comparison of Eq. (12-14) with Eq. (12-13) immediately suggests that the electrical resistivity caused by scattering from lattice vibrations and impurity atoms may be written as the sum of the resistivity due to each factor separately:

$$\rho = \rho_{TH} + \rho_I \tag{12-15}$$

Equations (12-14) and (12-15) may be generalized to include any factor that contributes to the electrical resistivity of metals. This is a law of additive resistivities and is known as *Mathiessen's rule*. We now focus our attention on each of the factors that contribute to the resistivity of a metal.

Thermal Component of Resistivity

One may think of the thermal component of electrical resistivity as arising from the interaction between moving electrons and atomic vibrations. Since the amplitude of atomic vibration increases with temperature and since the probability that an electron collides with a given atom depends directly on the area swept out by the vibratory motion, it can be argued that the mean free path between collisions decreases as the temperature is raised. It can be shown that the mean free path between collisions is proportional to the reciprocal of the absolute temperature:

$$\bar{l} = \frac{\alpha}{T} \tag{12-16}$$

where α is a proportionality constant. Using Eq. (12-13), we find that the thermal contribution to the electrical resistivity of a metal is proportional to the absolute temperature:

$$\rho_{TH} = \frac{mv_F}{\eta |e|^2} \frac{T}{\alpha} \tag{12-17}$$

The temperature dependence of the electrical resistivity as expressed in this equation is confirmed by the resistivity data shown in Fig. 12-4. It is generally true that

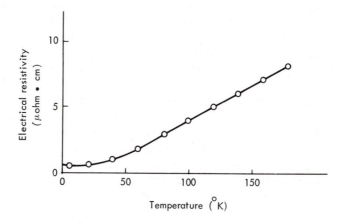

FIG. 12-4   Temperature dependence of the electrical resistivity of tantalum

when the electrical resistivity of a metal or alloy is known at two different temperatures, the electrical resistivity at a third temperature may be obtained by linear extrapolation.

EXAMPLE 12-3

The electrical resistivity of pure tantalum wire is $35\mu \cdot \text{ohm} \cdot \text{cm}$ at $100°K$ and $65\mu \cdot \text{ohm} \cdot \text{cm}$ at $160°K$. Calculate the electrical resistivity at room temperature ($20°C$).

Solution:

Because the electrical resistivity varies linearly with the absolute temperature, we may write

$$\frac{\rho_{160} - \rho_{100}}{160 - 100} = \frac{\rho_{293} - \rho_{100}}{293 - 100}$$

$$\rho_{293} = 35 + \tfrac{193}{60}(30) = 131\mu \cdot \text{ohm} \cdot \text{cm} \text{ (at room temperature)}$$

Impurity Component of Resistivity

In general, the electrical resistivity of a metal is increased by the addition of impurities. The increase is caused by the increased scattering due to the presence of the impurity atoms. Each impurity atom, by virtue of its different size and valence, may be viewed as a lattice defect, and hence causes electrons to be scattered. If we imagine that an electron is traveling along a row of atoms in a very dilute solid solution, the mean free path between impurity atoms $\bar{l}_I$ is related to the atomic fraction of impurities x_I through the relation

$$\frac{1}{l_I} = \frac{x_I}{a} \tag{12-18}$$

where a is the distance between adjacent atoms in the row. Using Eq. (12-13), we find that the impurity component of the electrical resistivity is proportional to the concentration of impurities:

$$\rho_I = \frac{mv_F}{\eta |e|^2}\left(\frac{x_I}{a}\right) \tag{12-19}$$

The data in Fig. 12-5 demonstrate that the electrical resistivity of dilute alloys is

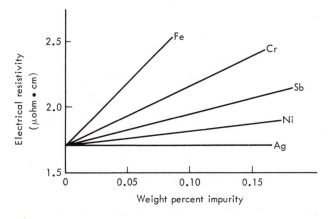

FIG. 12-5   The effect of small additions of various elements to room temperature electrical resistivity of pure copper

indeed proportional to the impurity atom concentration, a relation that is known as *Nordheim's rule*.

EXAMPLE 12-4   Estimate the change in electrical resistivity caused by the addition of 1 atomic percent Ni to copper.

Solution: We use Eq. (12-19) to determine the component of resistivity due to the presence of impurity atoms. In that expression, v_F is evaluated as follows:

$$\frac{1}{2} m v_F^2 = E_F = \frac{h^2}{8m} \left(\frac{3\eta}{\pi}\right)^{2/3}$$

$$v_F = \frac{h}{2m} \left(\frac{3\eta}{\pi}\right)^{1/3}$$

$$v_F = \frac{(6.63 \times 10^{-27})}{2(9.11 \times 10^{-28})} \left(\frac{3}{\pi} \frac{4}{(3.60 \times 10^{-8})^3}\right)^{1/3} = 1.57 \times 10^8 \text{ cm/sec}$$

$$\rho_I = \frac{(9.11 \times 10^{-31})(1.57 \times 10^6)}{[4/(3.60 \times 10^{-10})^3](1.6 \times 10^{-19})^2} \frac{0.01}{2.5 \times 10^{-10}}$$

$$= 2.63 \times 10^{-8} \text{ ohm} \cdot \text{m}$$

Note that this value is a good estimate of the measured change in resistivity when 1 atomic percent Ni is added to pure copper (see Figs. 12-5 and 12-6).

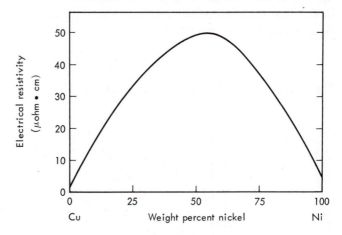

FIG. 12-6 Composition dependence of the electrical resistivity of the Ni-Cu alloy system at 0°C.

Two features of the computations in Example 12-4 should be considered carefully. For computing the thermal velocity of electrons at the Fermi energy, it was convenient to use the CGS system of units. However, the computation of the resistivity involves electrical units and it is more convenient in that case to work the problem in the MKS system of units. In the latter calculation the electronic mass is expressed in kilograms, the velocity in meters per second, and the lattice parameter in meters.

The dependence of the electrical resistivity on the solute concentration in a solid solution is not linear if the solute concentration is large. The reason for this comes from the fact that when the concentration of solute atoms becomes very large ($x_{\text{solute}} \approx 1$), it is the *solvent* atoms that may be viewed as the imperfections in a solute matrix. To take this fact into account, we may express the mean free path between collisions as

$$\frac{1}{\bar{l}} = \frac{x_I(1 - x_I)}{a} \qquad (12\text{-}20)$$

where $x_I(1 - x_I)$ is the probability that a given pair of adjacent atoms are dissimilar. Notice that when $x_I \ll 1$, this equation reduces to Eq. (12-18); hence, Nordheim's rule for dilute solutions is predicted. Equation (12-20) predicts that the electrical resistivity of a solid solution should be proportional to $x_I(1 - x_I)$. The data shown in Fig. 12-6 illustrate that this composition dependence is exhibited by the Cu-Ni alloy system at 0°C. It should be remembered that the Cu-Ni alloy system is isomorphous and that at 0°C the two metals are completely soluble in each other. Equation (12-20) is, of course, not applicable when the alloy is not a single phase.

Imperfection Component of Resistivity

The electrical resistivity of a given metal at a particular temperature generally increases as the density of imperfections increases. The reason for this increase is due to the fact that electrons are preferentially scattered at lattice imperfections. It is well established that vacancies and self-interstitials, dislocations, grain boundaries, and phase boundaries all contribute significantly to the electrical resistivity of metals and alloys.

Resistivity of Multiphase Alloys

When an alloy consists of more than one metallic phase, the electrical resistivity depends on the resistivities of the component phases and their volume fractions. Consider an alloy having three phases, α, β, and γ, with resistivities ρ_α, ρ_β, and ρ_γ. Unless the phases are in the form of plates or rods exactly parallel to the flow of current, it is reasonable to suppose that the current passes through each of the

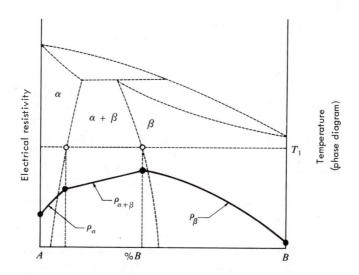

FIG. 12-7 Schematic representation of the variation of electrical resistivity with composition in a two-phase alloy

phases in series. In this case the electrical resistivity for the multiphase alloy may be expressed as

$$\rho = \rho_\alpha x_\alpha + \rho_\beta x_\beta + \rho_\gamma x_\gamma \tag{12-21}$$

where x_α, x_β, and x_γ are the volume fractions of the component phases. The prediction of Eq. (12-21) is that the electrical resistivity of a binary alloy varies linearly with composition in a two-phase region. This behavior is illustrated in Fig. 12-7.

12-5 SUPERCONDUCTIVITY

We have shown that the electrical resistivity of a metal decreases as the temperature decreases and hence reaches a very low value near $0°K$. The resistivity of metals with especially high conductivity, such as Cu, Al, Ag, or Au, reaches a plateau as the temperature approaches $0°K$. These metals, therefore, appear to exhibit a residual resistivity at absolute zero. In 1911 Kamerlingh Onnes of Leiden University discovered that Hg behaves in a dramatically different way. He found that below about $4°K$ the electrical resistivity of Hg becomes immeasurably small. The name that has been given to the phenomenon he discovered is *superconductivity*. Since this discovery, many metals and hundreds of alloys and compounds have been found to exhibit superconductivity at low temperatures.

In the past few years superconducting materials have received a great deal of attention because of their potential technological significance. One of the more obvious potential applications of superconducting materials would be as conductors in power transmission lines. The absence of resistivity offers the possibility of the transmission of power at 100% efficiency. Another important application of superconductivity involves the use of superconductors to make very high field electromagnets. Generally speaking, the factor that limits the intensity of the magnetic field that can be produced by a current-carrying coil is the rate of removal of the heat produced by the resistive losses in the wire. Coils made from superconducting wires would obviously not be limited by heating since joule heating does not occur in superconductors. For these reasons the phenomenon of superconductivity may be expected to become even more technologically significant in the future. One of the main factors that prevents the widespread application of superconductivity at the present time is the excessive cost of operating devices at the low temperature needed to obtain the superconducting phenomenon.

A semi-schematic representation of the low-temperature resistivity of a normal metal and two superconductors is shown in Fig. 12-8. For the superconductors there is a *critical temperature* or *transition temperature*, T_c, below which the resistivity appears to go to zero. A material held at temperatures lower than T_c is said to be in the *superconducting state*, while a material whose temperature is above T_c is said to be in the *normal state*. It should be noted that the transition between the superconducting state and the normal state may be very sharp and may occur within a few millidegrees, or it may be so broad as to occur over several degrees. We shall see that the structure and purity of the superconductor are the most important factors that determine the sharpness of the transition.

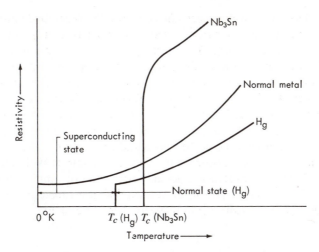

FIG. 12-8   Electrical resistivity of normal and superconducting materials as a function of temperature

The Meissner Effect

The magnetic properties of superconductors are as remarkable as their electronic properties. In addition, an explanation of superconductivity involves both magnetic and electronic phenomena in an inseparable way. The relation between these phenomena is apparent in the *Meissner effect*, to which we now direct our attention.

If a material is a perfect conductor and exhibits no electrical resistance, it will not be penetrated by magnetic flux when it is brought into a magnetic field. The reason for the flux exclusion is that as the flux begins to enter the sample, electric currents are induced on the surface of the superconductor and an opposing magnetic field is produced. Since the electric current remains undiminished in a superconductor, the reverse field persists and the applied magnetic field does not penetrate the sample. This behavior is referred to as perfect diamagnetism and is discussed in Chapter 14. The exclusion of the magnetic flux is illustrated in Fig. 12-9.

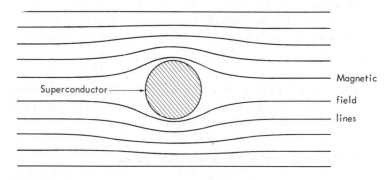

FIG. 12-9   Exclusion of magnetic field lines from a superconductor

The fact that flux does not penetrate the superconductor could be interpreted to mean that the superconductor simply has a very low conductivity and that given sufficient time the magnetic field would eventually permeate the sample. This possibility can be disproved with a different experiment. When a superconductor is heated above the transition temperature, it becomes normal and is easily penetrated by a magnetic field. When such a sample is cooled below T_c in the presence of a magnetic field, it will spontaneously expel the magnetic field and return to exactly the same state that was produced by applying the magnetic field to the sample while in the superconducting state. This experiment demonstrates that the superconducting state is an equilibrium phase in the thermodynamic sense and that the resistance is actually zero and not simply a low value.

The effects we have described imply that a superconductor will exclude a magnetic field, regardless of its intensity. The curves in Fig. 12-10, however, indicate that

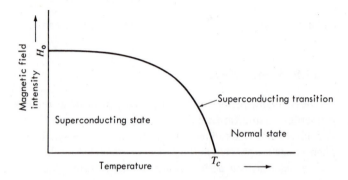

FIG. 12-10 Dependence of the superconducting transition on magnetic field intensity and temperature

this is not true. Superconductivity can be destroyed either by raising the temperature above T_c or by increasing the magnetic field above a critical value H_c. We shall see later that because there is a magnetic field associated with an electric current, superconductivity can also be destroyed if the current density exceeds a critical value J_c.

The boundary between the superconducting state and the normal state as illustrated in Fig. 12-10 is parabolic in shape and can be represented by

$$H_c = H_0 \left(1 - \frac{T^2}{T_c^2}\right) \tag{12-22}$$

where H_0 is the critical field at $0°K$ and T is the absolute temperature. The details of how the magnetic field penetrates the sample and causes the transition to the normal state will be examined shortly.

A Classical Model for Superconduction

Quantum mechanical arguments can be offered to explain how a state with zero resistivity may be produced, but they are well beyond the scope of our discussion.

Instead, we shall consider a simple classical picture of the events that cause super-conductivity.

The main cause of superconductivity is due to the strong coupling between pairs of moving electrons and the vibrating atoms. The important characteristic of the coupling is that pairs of electrons develop an attractive interaction through their simultaneous interaction with the vibrating atoms. While the details of these inter-actions are not important for our purposes, it is instructive to develop an intuitive model for the pairing. One may imagine that both of the electrons in a given elec-tron pair are attracted to the positive ion cores and that because of this mutual attraction the electrons in the pair are held together. As electron pairs migrate in response to an applied electric field, the individual electrons can suffer collisions with the lattice only if the thermal energy available in a given vibrating atom is sufficient to break the coupling between the electrons. If the temperature is below the critical temperature, the thermal energy is not sufficient to destroy the coupling, and the paired electrons move without suffering collisions. Since electrical resistance comes only from electron collisions, it is clear that the collisionless electron motion is the mechanistic definition of superconductivity.

The idea that the interaction between moving electrons and the atom vibrations is important in superconductivity is supported by the fact that most of the super-conducting metals do not have particularly high conductivities at normal tempera-tures. Such superconducting metals as Hg and Sn are relatively poor metallic conductors in their normal state. The poor conductivity of these metals is an indi-cation that there is a strong interaction between the free electrons and the vibrating atoms. In fact, it is this strong interaction which causes the superconducting state to exist at low temperatures. Metals such as Ag, Au, and Cu are excellent conductors at room temperature because the free electrons do not interact strongly with the vibrating atoms. For this same reason these metals should not be, and indeed are not, superconductors at low temperatures. Thus, the general rule is that metals with poor conductivities at room temperature are likely to be superconductors at low temperatures, whereas excellent metallic conductors do not have the capacity to become superconductors. But this rule does not always hold, since Al, which has a relatively high room temperature conductivity, becomes superconducting at very low temperatures.

An extremely simple account of the mechanism of superconductivity is contain-ed in the following analogy. Electrons moving in superconductors may be viewed as a group of soldiers marching along in formation with arms locked together. If one of the soldiers stumbles, he is helped along by the others until he regains his balance. In this way the electrons are allowed to move in the crystal without losing any of their momentum to the lattice.

Hard and Soft Superconductors

The classification of superconductors is based on the way in which the transition between the superconducting state and the normal state occurs. However, it may be said at the outset that *"soft" superconductors* are generally ones in which super-conductivity can be destroyed by a small magnetic field or by a low current density.

"*Hard*" *superconductors*, on the other hand, are characterized by a high critical magnetic field and a high critical current density.

The electric current induced in a superconductor by the application of a magnetic field gives rise to an internal magnetic moment, $-M$, which opposes the applied field and causes the flux exclusion. As shown in Fig. 12-11, the reverse magnetic

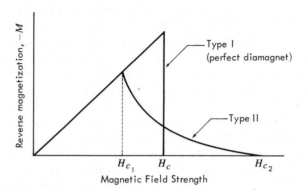

FIG. 12-11 Reverse magnetic moment $(-M)$ in type I and type II superconductors as a function of the applied magnetic field

moment varies directly with the applied field in the superconducting state. The transition to the normal state may take place in one of two ways. For type I superconductors, the transition occurs sharply at the critical field H_c. In this case the magnetic field is excluded completely until the critical field is reached. Above the critical field, the magnetic field penetrates the sample completely and the superconductor becomes normal. If the penetration of the magnetic field takes place gradually and the transition occurs over a range of values of the magnetic field, the superconductor is referred to as type II. Type I superconductors are soft, while type II superconductors may be either hard or soft, depending on the internal structure of the material.

The flux penetration that occurs in type II superconductors is illustrated in Fig. 12-12. We may imagine that the supercurrents that flow in both type I and type II superconductors follow the paths shown in the figure. In the case of type I superconductors the internal components of the supercurrents cancel out exactly and the current is confined to flow around the surface. It is the surface current that provides the reverse magnetic moment $-M$ in this class of superconductors. The internal supercurrents in type II superconductors do not cancel exactly when the superconductor has been partially penetrated by the flux. Figure 12-12 illustrates the partial cancellation of the internal supercurrents that occurs in type II superconductors and shows how the residual supercurrents provide internal magnetic *fluxoids*. The fluxoids may be considered normal regions in a superconducting matrix. In a similar way we can view the superconducting phase as those regions that carry supercurrents. The fluxoids are thus surrounded by superconducting toroids.

The existence of fluxoids in type II superconductors provides a means by which

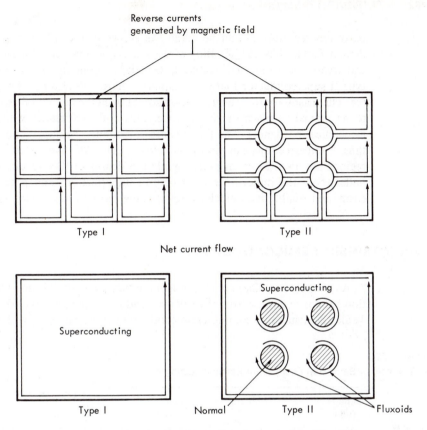

Reverse currents
generated by magnetic field

Type I

Type II

Net current flow

Superconducting

Type I

Superconducting

Normal   Type II   Fluxoids

FIG. 12-12 Illustration of supercurrents in type I and type II superconductors. Fluxoids are illustrated for type II superconductors.

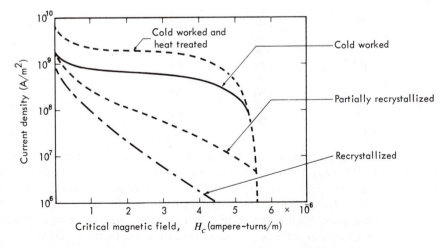

FIG. 12-13 Critical current density as a function of critical field for Nb$_3$Zr with different thermal and mechanical treatments

hard superconductors can be made. As shown in Fig. 12-13, the critical current density J_c of an Nb–25% Zr alloy wire is a sensitive function of thermal and mechanical treatment given that material. Generally speaking, it may be said that a type II superconductor that is mechanically hard will also be a hard superconductor. The reason for this generalization comes from the fact that fluxoids are believed to be pinned in place by the imperfections and precipitates which also provide mechanical hardening. The fluxoids are surrounded by superconducting torroids; thus, if imperfections can pin the fluxoids in place, the associated superconducting regions are also frozen in place. While the details of fluxoid pinning are not yet understood, the importance of structural imperfections and precipitates is quite clear, and qualitative guidelines for making hard superconductors are available.

12-6 INTRINSIC SEMICONDUCTORS

As discussed in Chapter 11, semiconductors are materials in which the conduction band is separated from the valence band by an energy gap of the order of 1 eV. Table 12-2 lists some common semiconducting materials along with their respective

TABLE 12-2
A Summary of Band-Gap Energies for Semiconducting Materials

General Type	Material	Band Gap (eV)
IV	Si	1.1
	Ge	0.68
III-V	AlSb	2.4
	GaAs	1.4
	GaSb	0.67
	InP	1.25
	InAs	0.33
	InSb	0.18
II-VI	ZnTe	2.1
	CdSe	1.7
	CdTe	1.5
	CdS	2.4
	HgSe	0.6

band-gap energies. As shown in Fig. 12-14, the valence band in semiconductors is almost fully occupied with electrons, while the conduction band is nearly empty. For the special case of an intrinsic semiconductor (pure elements such as Si or Ge or stoichiometric compounds such as InSb, GaAs, or AlP), the valence band is completely filled at 0°K, while the conduction band is completely empty. At finite temperatures some of the valence electrons are thermally excited across the energy gap into the conduction band, as illustrated in the figure. As we shall see shortly,

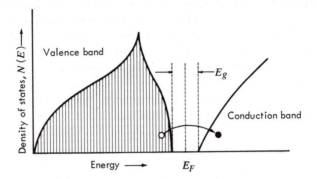

FIG. 12-14 Valence and conduction bands in semiconductors

the number of electrons that are excited to the conduction band is a strong function of the band-gap energy and the absolute temperature.

Before describing intrinsic semiconductors in a quantitative way, it will be worthwhile to develop a simple physical picture of the valence and conduction electrons in a covalent material. Consider the schematic illustration of covalently bonded atoms in Fig. 12-15. The circles in this figure represent the ion cores, and the short lines signify the electrons in the valence band.

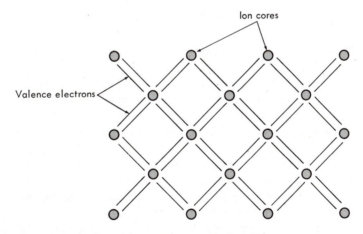

FIG. 12-15 Pictorial representation of valence electrons in a covalent crystal

Since there are eight electrons surrounding each ion and these electrons are shared with the four nearest neighbors, each atom is imagined to contribute four electrons to the valence band. The model can be used to represent such elemental semiconductors as Si or Ge since these atoms contribute four valence electrons each. The model may also be used to represent compound semiconductors in which the average contribution of valence electrons by each atom is four. This is true for III-V compounds such as GaAs or InSb and II-VI compounds such as CdSe or ZnS. In all these compounds the valence band is completely filled at 0°K when there are equal numbers of each kind of atom. For example, Ga atoms contribute only three valence electrons each, but since As atoms have five valence electrons, the stoichiometric compound GaAs has an average of four valence electrons for each atom.

The process of exciting an electron from the valence band to the conduction band may be viewed in either of the two ways shown in Fig. 12-16. The excitation requires

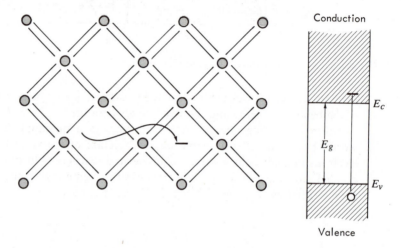

FIG. 12-16 Representations of the formation of electrons and holes in semiconductors

that the valence electron be removed from the valence band and that it be physically separated from the electron hole that is produced. When the electron is moved sufficiently far from the hole, it is free to participate in conduction. The electron is then said to be in the conduction band.

The center or hole that is left in the valence electron structure represents a net positive charge. For this reason alone it is clear that there is a binding energy between the hole and the electron and that a critical amount of energy must be supplied to excite the electron to the conduction band. Actually the energy required to separate the electron from the hole involves considerations other than coulombic forces. Nevertheless, the idea of electrostatic attraction gives an intuitive picture of the band-gap energy.

The energy-band picture of the excitation deals only with the energy required for the process and does not take into account the locations of the electrons involved. In this representation the excitation of an electron from the valence band to the conduction band simply requires that one of the valence electrons receives sufficient energy to be elevated to the conduction band. An electron-hole pair is always created when an electron is excited across the energy gap.

Let us now consider the process of conduction in a semiconductor. It is clear

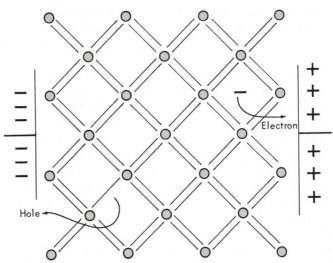

FIG. 12-17   Conduction by the migration of electrons and holes in response to applied voltage

from Fig. 12-17 that both electrons and holes can respond to an applied field. The conduction electrons are obviously attracted to the positive terminal, and the positively charged holes are attracted to the negative terminal. The figure illustrates the fact that the motion of the hole is a virtual motion that actually involves a coordinated movement of the valence electrons. The valence electrons exchange positions with the hole in such a manner that the electrons themselves migrate to the positive terminal. Therefore, each time the electron hole moves from one site to the next, one of the valence electrons shifts in the opposite direction.

In considering the conductivity of a semiconductor, we must consider the current that is carried by both the electrons in the conduction band and the migrating holes in the valence band. Using Eq. (12-8), we may express the current density J in a semiconductor as

$$J = \eta_e |e| v_e + \eta_h |e| v_h \tag{12.23}$$

where η_e and η_h are the numbers of conduction electrons and holes per unit volume, $|e|$ is the electronic charge (electron holes have a charge of $+|e|$), and v_e and v_h are the drift velocities for the electrons and holes, respectively. Dividing each side of the equation by the voltage gradient E and using Eq. (12-4), we may write the conductivity as

$$\sigma = \eta_e |e| \left(\frac{v_e}{E}\right) + \eta_h |e| \left(\frac{v_h}{E}\right) \tag{12-24}$$

The first term in this equation is referred to as the *electronic conductivity*, and the second term is known as the *hole conductivity*. The quantities v_e/E and v_h/E are called the *electronic* and *hole mobilities* because they are measures of how fast the electrons and holes will drift for a unit applied field. It is common to use the symbols μ_e and μ_h to express the mobilities of electrons and holes, respectively, so that the conductivity of a semiconductor may be expressed in final form as

$$\sigma = \eta_e |e| \mu_e + \eta_h |e| \mu_h \qquad (12\text{-}25)$$

We may refer to Eq. (12-25) to describe how the factors that control the conductivity of semiconductors are different from those that control the conductivity of metallic conductors. For metals, the number of free electrons that participate in conduction is essentially fixed by the valence of the metal and cannot be changed appreciably. In contrast, the mobility of the free electrons in metals can be significantly changed through structural control. In the case of semiconductors the importance of the two variables is just reversed. The mobility of conduction electrons and holes changes as the structural perfection changes, but these variations are overshadowed by the changes that can be made in the densities of electrons and holes. In Section 12-4 our attention was focused mainly on factors that affect the mobilities of the free electrons in metals. In the following pages we shall be concerned mainly with the variables that affect the densities of electrons and holes in semiconductors.

12-7 DENSITY OF CONDUCTION ELECTRONS AND HOLES IN INTRINSIC SEMICONDUCTORS

The computation of the density of electrons in the conduction band and of holes in the valence band of a semiconductor is illustrated in Fig. 12-18. The top

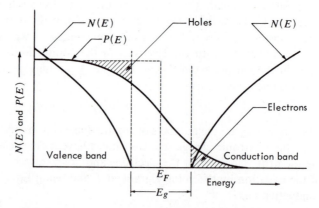

FIG. 12-18 Illustration of computation of density of electrons and holes in semiconductors. $N(E)$ and $P(E)$ are density of states and probability of occupancy, respectively.

of the valence band is taken as a reference of zero energy ($E_v = 0$). The energy at the bottom of the conduction band is denoted as E_c, and the band-gap energy is E_g. The Fermi level resides somewhere within the band gap and is denoted by E_F. As indicated in the figure, the number of electrons in the conduction band and the number of holes in the valence band depend on where the Fermi level is positioned in the gap, the density of states in the valence band and conduction band, and the absolute temperature. We shall assume that the density of states in the conduction band follows the form of Eq. (11-26), varying as $(E - E_c)^{1/2}$, and that the density of states in the valence band varies parabolically with decreasing energy according to $(E_v - E)^{1/2}$.

Using Eq. (11-33), we may now express the total number of electrons in the conduction band as

$$\eta_e = \int_{E_c}^{\infty} 2N(E)P(E) \, dE \qquad (12\text{-}26)$$

The integral of this equation represents an integration over the entire conduction band. The fact that the upper limit of the integral is ∞ and not the maximum energy for electrons in the conduction band (approximately the same as the work function) can be justified by noting that most of the electrons in the conduction band are near the bottom of the band and that only a very few electrons are energetic enough to escape from the material. Thus, the conduction band may be imagined to extend without bound above the band gap. Taking

$$N(E) = \frac{\pi}{4} \left(\frac{8m}{h^2}\right)^{3/2} (E - E_c)^{1/2} \qquad (12\text{-}27)$$

and the exponential approximation of the Fermi function

$$P(E) = \frac{1}{1 + \exp\left[(E - E_F)/kT\right]} \approx \exp\left(-\frac{E - E_F}{kT}\right) \qquad (12\text{-}28)$$

and noting that $E_c = E_g$, we may write

$$\eta_e = \frac{\pi}{2} \left(\frac{8m}{h^2}\right)^{3/2} \exp\left(-\frac{E_g - E_F}{kT}\right)(kT)^{3/2} \int_0^{\infty} x^{1/2} \exp\left(-x\right) dx \qquad (12\text{-}29)$$

where $x = (E - E_g)$. In final form,

$$\eta_e = 2\left(\frac{2\pi mk}{h^2}\right)^{3/2} T^{3/2} \exp\left(-\frac{E_g - E_F}{kT}\right) \qquad (12\text{-}30)$$

Using numerical values for the physical constants, the density of conduction electrons may be given as

$$\eta_e = 4.8 \times 10^{21} \, T^{3/2} \exp\left(-\frac{E_g - E_F}{kT}\right) \text{ electrons/m}^3 \qquad (12\text{-}31)$$

where T is expressed in degrees Kelvin. A similar integration would yield the hole concentration:

$$\eta_h = 4.8 \times 10^{21} \, T^{3/2} \exp\left(-\frac{E_F}{kT}\right) \text{ holes/m}^3 \qquad (12\text{-}32)$$

Equations (12-31) and (12-32) are general equations for the concentrations of conduction electrons and holes and they apply to both intrinsic and extrinsic*

---

*We shall see in Section 12.8 that *extrinsic* semiconductors are semiconductors in which excess conduction electrons or holes are produced by impurities or imperfections.

semiconductors. It should be noted that the important parameters in these equations are the Fermi energy, the gap energy, and the absolute temperature. Of these quantities, the gap energy is a material parameter and the temperature is fixed by the environment; the Fermi energy, as we shall see, depends in turn on the temperature and the density of states for the semiconductor in question. To understand how the Fermi energy may depend on the purity of a semiconductor and on the temperature, it is necessary to recall that the Fermi energy as defined in the Fermi function is that energy level for which the probability of occupancy is 0.5.

Let us first consider the case of an intrinsic semiconductor. As indicated before, at 0°K the valence band of an intrinsic semiconductor is filled and the conduction band is empty. At finite temperatures some of the valence electrons are excited into the conduction band, leaving holes in the valence band. Thus, the number of electrons in the conduction band is equal to the number of holes in the valence band for an intrinsic semiconductor. This information, coupled with the assumption that the density of states in the valence band mirrors the density of states in the conduction band, leads directly to the conclusion that the Fermi energy lies midway between the valence and conduction bands:

$$E_F = \frac{E_g}{2} \tag{12-33}$$

This result is qualitatively evident in Fig. 12-18 and is required mathematically if one sets $\eta_e = \eta_h$ with Eqs. (12-31) and (12-32). Therefore, for the case of intrinsic semiconductors, the conductivity may be expressed—using Eqs. (12-25), (12-31), and (12-32)—as

$$\sigma = |e|(\mu_e + \mu_h)\, 4.8 \times 10^{21}\, T^{3/2} \exp\left(-\frac{E_g}{2kT}\right) \tag{12-34}$$

Since the exponential term in Eq. (12-34) is much more strongly temperature dependent than $T^{3/2}$, the temperature dependence of the electrical conductivity follows an exponential law. Taking natural logarithms of the terms in Eq. (12-34), we have

$$\ln\left(\frac{\sigma}{\sigma_0}\right) = -\frac{E_g}{2kT} \tag{12-35}$$

where σ_0 depends on the mobilities of the electrons and holes and is slightly temperature-dependent through the $T^{3/2}$ term. Equation (12-35) implies that if $\ln \sigma$ is measured and plotted against $1/T$, the resulting slope will be $-E_g/2k$. The conductivity of pure Ge is shown in Fig. 12-19 as a function of the absolute temperature. From the slope of that curve the band-gap energy for germanium ($E_g = 0.68$ eV) may be deduced.

The significance of Eq. (12-35) is that the conductivity of an intrinsic semiconductor increases as the temperature increases. This is just the opposite of what was found for metals. Another important feature of Eq. (12-35) and Fig. 12-19 is that the band-gap energy for an intrinsic semiconductor can easily be determined by measuring the temperature dependence of the electrical conductivity.

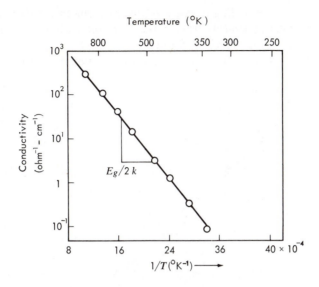

FIG. 12-19 Temperature dependence of the conductivity of intrinsic germanium

EXAMPLE 12-5

The electrical resistivity of pure silicon is 3.0×10^3 ohm·m at room temperature (300°K). Determine the conductivity at 250°C (523°K).

Solution:

According to Table 12-2, the band gap of silicon is 1.1 eV. Using Eq. (12-35), we may write

$$\sigma = \sigma_0 \exp\left(-\frac{E_g}{2kT}\right)$$

or

$$\sigma_{RT} = \sigma_0 \exp\left(-\frac{E_g}{2kT_{300}}\right)$$

$$\sigma_{523} = \sigma_0 \exp\left(-\frac{E_g}{2kT_{523}}\right)$$

so that

$$\sigma_{523} = \sigma_{RT} \exp\left(-\frac{E_g}{2kT_{523}} + \frac{E_g}{2kT_{300}}\right)$$

$$\sigma_{523} = \frac{1}{3.0 \times 10^3} \exp\left[-\frac{(1.1)(1.6 \times 10^{-12})}{2(1.38 \times 10^{-16})}\left(\frac{1}{523} - \frac{1}{300}\right)\right]$$

$$= 2.67 \text{ mho/m}$$

(The electrical conductivity of silicon increases by a factor of 8000 when the temperature is raised from room temperature to 250°C.)

12-8 EXTRINSIC SEMICONDUCTORS

Extrinsic semiconductors are impure or imperfect semiconductors that contain energy levels within their band gap. These energy levels produce an excess of either conduction electrons or holes. The most common extrinsic semiconductors are produced by adding trivalent (elements from group III on the periodic chart) or pentavalent (elements from group V on the periodic chart) impurity atoms to an intrinsic semiconducting crystal.

As shown in Fig. 12-20, substitutional impurity atoms from group V provide

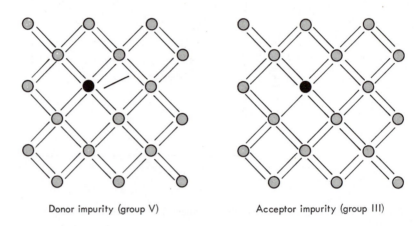

Donor impurity (group V)   Acceptor impurity (group III)

FIG. 12-20 Donor and acceptor impurity atoms in semiconductors

five valence electrons to the crystal. Since only four of these are needed for covalent bonding, each pentavalent impurity atom donates one excess electron, which can be used in conduction. Because these impurity atoms donate extra electrons to the crystal, they are called *donor* impurity atoms. These semiconductors are called *n-type* because the carriers are negatively charged. Elements such as P, As, and Sb are commonly used as donor impurities in silicon and germanium.

Acceptor impurity atoms are elements from group III; they contribute only three valence electrons to the crystal. In this case there is an electron deficiency in the valence band near the impurity atom. The impurity atom can complete the covalent bonding to its neighboring atoms only if another valence electron is accepted by the impurity. The result of this is the production of a hole in the valence band which is free to move and provide for conduction. These semiconductors are called *p-type* because the holes that carry the current are positively charged. Elements such as B, Al, Ga, and Zn are common acceptor impurity atoms in silicon and germanium.

As in the case of intrinsic semiconductors, extrinsic semiconductors may be described with an energy-level diagram. Figure 12-21 is a schematic illustration of the energy levels that are provided by impurity atoms and donor atoms in a semiconductor. It should be noted at the outset that because the impurity atoms are imperfections and interrupt the perfect periodicity of the lattice, energy levels within the band gap that were previously forbidden are no longer disallowed. Whether the impurity energy level actually resides in the band gap depends on the following factors.

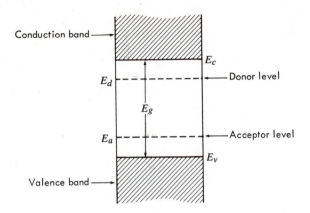

FIG. 12-21   Donor and acceptor energy levels in semiconductors

Consider first the donor atoms. There are two factors that are involved in finding the *donor level* E_d. On the one hand, there is clearly a positive electrostatic energy required to separate the donor electron from the donor impurity atom. This factor alone would necessarily place E_d below the bottom of the conduction band. On the other hand, it should be recognized that if the tendency for the impurity atom to have the proper number of valence electrons for covalent bonding is stronger than that for the solvent atoms, the donor electron would be spontaneously given to the conduction band and the donor energy level would reside within the conduction band. It is the competition between these two factors that determines where the energy level for the donor electron resides. Generally speaking, the coulombic or ionic considerations are usually dominant, and the donor level resides in the band gap just below the bottom of the conduction band.

A similar argument may be advanced to arrive at the position of the acceptor level produced by acceptor impurity atoms. Again, it is usually the case that the ionic or coulombic factors are dominant and that the acceptor level resides within the band gap. In this case a positive energy $E_a - E_v$ is needed to excite an electron from the valence band to the impurity atom. The donor and acceptor levels for several impurity atoms in silicon and germanium are shown in Tables 12-3 and 12-4.

TABLE 12-3
Donor and Acceptor Impurity Levels in Silicon (E_g = 1.1 eV)

Donor Impurities		Acceptor Impurities	
Impurity	$E_c - E_d$ (eV)	Impurity	$E_a - E_v$ (eV)
P	0.044	B	0.045
As	0.049	Al	0.057
Sb	0.039	Ga	0.067
Li	0.033	In	0.16

TABLE 12-4

Donor and Acceptor Impurity Levels in Germanium (E_g = 0.68 eV)

Donor Impurities		Acceptor Impurities	
Impurity	$E_c - E_d$ (eV)	Impurity	$E_a - E_v$ (eV)
Li	0.0093	Zn	0.029
P	0.012	B	0.0104
As	0.0127	Al	0.0102
Sb	0.0097	Ga	0.0108
Bi	0.012	In	0.0112
		Cu	0.040

12-9 TEMPERATURE DEPENDENCE OF CARRIER CONCENTRATIONS IN EXTRINSIC SEMICONDUCTORS

The temperature dependence of the carrier concentrations in extrinsic semiconductors can be described by considering a specific example. Let us suppose that we have an *n*-type semiconductor composed of 2×10^{21} As atoms per cubic meter in a silicon crystal. Figure 12-22 illustrates the relevant energy levels for this system

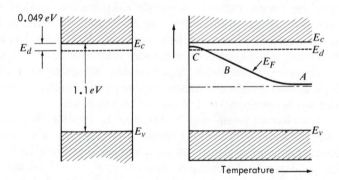

FIG. 12-22 Energy levels in an *n*-type extrinsic semiconductor and dependence of the Fermi energy E_F on temperature

(taken from Tables 12-2 and 12-3). A schematic illustration of the variation of the Fermi energy with temperature is also shown in the figure. At 0°K all the donor electrons reside within the donor level and the conduction band is entirely empty. Because of the shape of the Fermi function (see Fig. 11-8) the Fermi energy must lie between E_d and the bottom of the conduction band at 0°K. As the temperature is raised, some of the donor electrons are thermally excited into the conduction band and the Fermi energy is correspondingly lowered. This region is denoted by *C* in Fig. 12-22. When about half the donor electrons are excited to the conduction band, the Fermi energy must coincide with E_d. This occurs because E_F is that energy level for which the probability of occupancy is 0.5. As the temperature is raised even higher, more of the donor electrons are excited into the conduction band and the donor level becomes less than half filled. At an intermediate temperature, the donor level is almost completely ionized, and further increases in temperature cause the Fermi energy to decrease almost linearly with temperature (region *B*). When the temperature reaches a very high value, the number of conduction electrons begins to increase rapidly with temperature. This is due to the excitation of electrons from the valence band into the conduction band. At temperatures of this magnitude the semiconductor behaves as an intrinsic semiconductor because the number of intrinsic electrons greatly exceeds the extrinsic electron concentration. Under these conditions the Fermi level resides near the center of the band gap as in the case of intrinsic semiconductors (region *A*).

Let us now estimate the temperatures at which the donor level becomes ionized and the valence band begins to contribute significant numbers of electrons to the conduction band. For the case of As in Si the energy needed to excite an electron from the donor level to the conduction band is $E_c - E_d = 0.05$ eV. At very low temperatures we may estimate the density of conduction electrons as

$$\eta_e = \eta_d \exp\left(-\frac{E_c - E_d}{kT}\right) \tag{12-36}$$

where η_d is the concentration of donor impurity atoms. It is clear that this relation expresses the fact that when $T \rightarrow 0$, $\eta_e \rightarrow 0$ and when $kT \gg (E_c - E_d)$, $\eta_e \rightarrow \eta_d$.

The variation of η_e with temperature for silicon crystals with various concentrations of As atoms is shown in Fig. 12-23. In the low temperature range (called the *impurity range*) the variation of η_e with temperature follows Eq. (12-36). Accordingly, the slope of the curve in the low-temperature domain is $-(E_c - E_d)/k$. At higher temperatures, in the *exhaustion range*, the donor atoms are completely ionized and the concentration of conduction electrons is independent of temperature.

We may estimate the temperature at which the intrinsic semiconduction begins to overshadow the extrinsic behavior by finding that temperature for which η_e due to intrinsic excitation is equal to η_d. Setting

$$\eta_d = 4.8 \times 10^{21} T^{3/2} \exp\left(-\frac{E_g}{2kT}\right) \tag{12-37}$$

and solving for the temperature gives

$$T = \frac{E_g}{2k \ln\left(4.8 \times 10^{21} T^{3/2}/\eta_d\right)} \tag{12-38}$$

Solving this equation by trial and error for 2×10^{21} As atoms in a cubic meter of silicon gives $T = 628°$K. As shown in Fig. 12-23, above this temperature the density of electrons in the conduction band follows a curve that applies for intrinsic semiconductors. This is denoted as the *intrinsic* region.

The variation of the concentration of carriers with temperature in extrinsic semiconductors usually follows the trends shown in Fig. 12-23. This result applies to *p*-type as well as to *n*-type semiconductors. In general, extrinsic semiconductors are operated in the exhaustion range. In this case the density of carriers, and hence

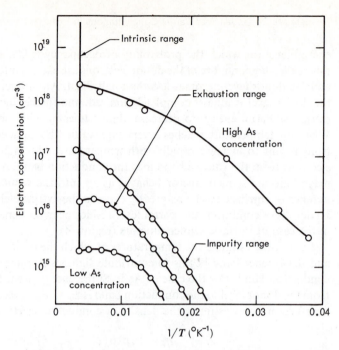

FIG. 12-23 Carrier concentration vs. temperature for a series of *n*-type silicon samples with different As concentrations

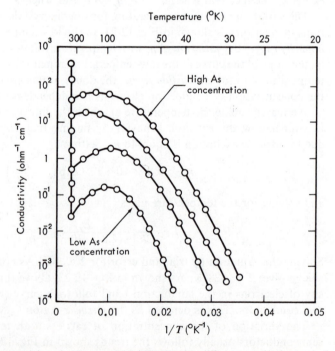

FIG. 12-24 Electrical conductivity of silicon crystals doped with different concentrations of As impurity atoms

the electrical conductivity, is essentially constant and fixed by the impurity concentration. It is this fact that allows the conductivity of extrinsic semiconductors to be precisely controlled by controlling the concentration of the impurity atoms. The electrical conductivity of As-doped silicon crystals is shown in Fig. 12-24. Generally speaking, the conductivity follows the variation of the electron concentration with temperature, as shown in Fig. 12-23. The principal difference between the curves in Fig. 12-23 and Fig. 12-24 is found at temperatures just below the intrinsic range. The decrease in the electrical conductivity with increasing temperature in this range is caused by a decrease in the mobilities of the electrons.

EXAMPLE 12-6

An n-type germanium semiconductor has a hole concentration of 10^{22} holes/m³ at 500°K. Calculate the concentration of electrons in the conduction band. Determine the concentration of As atoms that would be required to produce this concentration of electrons and holes at this temperature.

Solution:

The Fermi energy may be determined by using Eq. (12-32):

$$\eta_h = 4.8 \times 10^{21} T^{3/2} \exp\left(-\frac{E_F}{kT}\right)$$

$$E_F = kT \ln\left(\frac{4.8 \times 10^{21} T^{3/2}}{\eta_h}\right)$$

$$= (1.38 \times 10^{-16})(500) \ln\left(\frac{4.8 \times 10^{21}(500)^{3/2}}{10^{22}}\right)$$

$$= 5.92 \times 10^{-13} \text{ erg} = 0.37 \text{ eV}$$

The concentration of electrons in the conduction band can be obtained directly from Eq. (12-31):

$$\eta_e = (4.8 \times 10^{21})(500)^{3/2} \exp\left[-\frac{(0.68 - 0.37)(1.6 \times 10^{-12})}{(1.38 \times 10^{-16})(500)}\right]$$

$$= 4.08 \times 10^{22} \text{ electrons/m}^3$$

The difference between η_e and η_h represents the concentration of extrinsic electrons and hence is equal to the concentration of donor atoms (As):

$$\eta_{As} = \eta_e - \eta_h = 3.08 \times 10^{22} \text{ atoms/m}^3$$

The impurity concentration may be expressed as

$$X_{As} = \frac{\eta_{As}}{\eta_{Ge}} = \frac{3.08 \times 10^{22}}{8/(5.66 \times 10^{-10})^3} = 7 \times 10^{-7}$$

$$= 7 \times 10^{-5}\% \text{ As, or 0.7 parts per million As}$$

12-10 HALL EFFECT

In the previous sections our attention has been focused mainly on the densities of carriers in semiconductors. Thus far in our discussion the carrier density has been obtained by computation. We shall now see that the density of carriers can be measured through the use of the *Hall effect*.

Consider the crystal shown in Fig. 12-25, which is assumed to contain conduc-

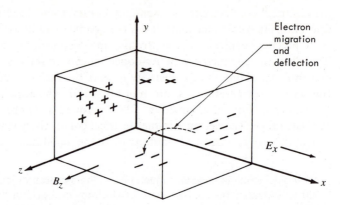

FIG. 12-25 Illustration of Hall effect geometry. Electric field is applied in the x direction, and magnetic field is applied in the z direction.

tion electrons as carriers. An electric field $\mathbf{E}$ is applied in the positive x direction and a magnetic field $\mathbf{B}$ is applied in the positive z direction. Electrons migrate in the negative x direction in response to the applied electric field, and the drift velocity of the electrons may be expressed as

$$\mathbf{v}_e = -\mu \mathbf{E} \tag{12-39}$$

where μ is the mobility of the conduction electrons. As the electrons move through the magnetic field, they experience a force that tends to deflect them in the negative y direction. This force is expressed as

$$\mathbf{F} = -|e|\mathbf{v}_e \times \mathbf{B} \tag{12-40}$$

where $|e|$ is the magnitude of the charge on the carrier. Combining Eqs. (12-39) and (12-40) gives the y component of the force exerted on the moving electrons in Fig. 12-25:

$$F_y = -|e|\mu B_z E_x \tag{12-41}$$

Now, as electrons collect at the bottom face of the crystal, a voltage gradient is established in the crystal in the y direction. This voltage is called the *Hall voltage*, and it provides a y component of force on the electrons which tends to counteract the force described by Eq. (12-41). The restoring force that comes from the Hall voltage is expressed as

$$F_y = -|e|E_y \tag{12-42}$$

where E_y is the Hall voltage gradient, which is negative for the crystal shown in Fig. 12-25.

At equilibrium the force imposed by the magnetic field on the moving electrons is just balanced by the force due to the Hall voltage. This equilibrium condition is expressed as

$$F_y^{(B)} + F_y^{(\text{Hall})} = 0 \tag{12-43}$$

$$-|e|B_z E_x \mu - |e|E_y = 0 \tag{12-44}$$

or

$$E_y = -B_z E_x \mu \qquad (12\text{-}45)$$

It is convenient to define a *Hall coefficient*, R, as

$$R = \frac{E_y}{B_z J_x} \qquad (12\text{-}46)$$

where J_x is the positive current density in the x direction. Using Eq. (12-45) and recalling that $J_x = \sigma E_x$, we can express the Hall coefficient as

$$R = -\frac{\mu E_x}{J_x} = -\frac{\mu}{\sigma} \qquad (12\text{-}47)$$

Since $\sigma = \eta_e |e| \mu$, it follows that

$$R = -\frac{1}{\eta_e |e|} \qquad (12\text{-}48)$$

Equation (12-48) indicates that if the Hall coefficient R is measured directly by measuring the Hall voltage E_y for a given magnetic field B_z and imposed current density J_x [see Eq. (12-46)], the density of carriers η_e can be computed directly. The sign of the Hall coefficient indicates the sign of the charge on the carriers. For the case of negatively charged conduction electrons, R is negative, as shown in Eq. (12-48). Had we considered the case of conduction by positively charged holes, the Hall coefficients would have been positive. Thus, we see that the type (n or p) and density of carriers in semiconductors can be easily measured by simply measuring the sign and magnitude of the Hall voltage in a Hall effect experiment. By measuring both the conductivity and the carrier concentration the mobility of the carriers can be found directly using $\sigma = \eta_e |e| \mu$.

EXAMPLE 12-7   Consider an extrinsic semiconducting crystal oriented as shown in Fig. 12-25. The application of an electric field $E_x = +62.5$ V/m and a magnetic field $B_z = +1$ Wb/m² causes a positive current density $J_x = 3.0 \times 10^4$ A/m² and a positive Hall voltage $E_y = +18.7$ V/m. (The positive y face of the crystal becomes negatively charged.) Determine the sign and density of the carriers in this semiconductor, and compute the mobility of these carriers.

Solution: Using Eq. (12-46), we find that the Hall coefficient is

$$R = \frac{E_y}{B_z J_x} = \frac{18.7}{(1)(3.0 \times 10^4)} = +6.25 \times 10^{-4} \text{ m}^3/\text{C}$$

From the sign of the Hall coefficient we find that the carriers are *holes*. The hole concentration can be obtained from

$$R = \frac{1}{\eta_h |e|}; \qquad \eta_h = \frac{1}{(6.25 \times 10^{-4})(1.6 \times 10^{-19})}$$
$$= 10^{22} \text{ holes/m}^3$$

The hole mobility can be determined from

$$\mu_h = \frac{\sigma}{\eta_h |e|} = \frac{J_x}{\eta_h |e| E_x} = \frac{(3.0 \times 10^4)}{(10^{22})(1.6 \times 10^{-19})(62.5)}$$
$$= 0.3 \text{ m}^2/\text{V} \cdot \text{sec}$$

12-11 EFFECTS OF STRUCTURAL IMPERFECTIONS AND IMPURITIES ON SEMICONDUCTORS

Our discussion of semiconductors thus far has been based on the idea that the crystals in question are structurally perfect and that the only impurities are those that are intentionally added. Neither of these conditions actually exists in practice, and a great deal of work is necessary even to approximate them. In this section we shall describe the deleterious effects that are caused by structural imperfections and unwanted impurities. These considerations will provide the motive for establishing processes to control the purity and perfection of single crystals of semiconductors.

Single Crystals vs. Polycrystals

There are two important reasons why useful semiconductors must be in the form of single crystals. In the first place, grain boundaries represent imperfect regions of the material where electrons may be preferentially trapped. Specifically, it is expected that grain boundaries would provide extraneous energy levels within the band gap and that the electrical conductivity could not easily be controlled by controlling the donor or acceptor concentrations. The other reason that polycrystalline semiconductors would not be practical relates to the physical distribution of the impurity atoms within the material. Impurity atoms usually tend to segregate preferentially to grain boundary regions, and hence cause chemical heterogeneities within polycrystalline materials. In the case of semiconductors this would cause the electrical properties to be inhomogeneous, and the overall electrical behavior of the material would be unpredictable. For these reasons it is essential that semiconductor devices be made by using single crystals.

Effects of Dislocations

Dislocations in semiconductors are generally deleterious because they serve as traps for electrons or holes. It has been shown that in germanium the covalent bond structure is such that dislocations tend to accept electrons. The acceptor level provided by the dislocations is usually in the upper half of the band gap and has an effect on the concentrations of carriers only in the case of n-type semiconductors at low temperatures. Figure 12-26 is an illustration of the acceptor level provided by dislocations. For n-type semiconductors at low temperatures, the donor electrons tend to be trapped in the dislocation levels and are not so easily excited to the conduction band. At higher temperatures, in the exhaustion range and above, the dislocation energy levels are ionized and the density of conduction electrons is not affected by the presence of the dislocations. In the case of p-type semiconductors the dislocations do not have an effect on the density of holes because the acceptor energy level is high relative to the impurity acceptor level.

Dislocations in semiconductors also have an effect on the mobilities of the carri-

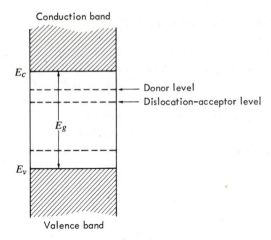

FIG. 12-26 Dislocation-acceptor energy levels in a semiconductor

ers, just as in the case of metals. This effect is generally small and not of great significance.

Unwanted Impurities in Semiconductors

Since the properties of semiconductors are based on the unique electrical properties that are provided by the donor and acceptor impurity atoms, it is clear that impurity atoms other than those intentionally added are undesirable. They are unwanted because they provide either donor or acceptor levels within the band gap and hence degrade the precision with which the electrical conductivity can be controlled.

One of the problems encountered early in the development of semiconductors was associated with the presence of unwanted impurity atoms. At one stage in the processing of germanium crystals they were heated in water in a copper tank. After such a treatment the carriers in the crystals had such short lifetimes that they were essentially useless. The infliction was termed "Deathnium" until it was discovered that heating in a copper tank caused copper impurities to be absorbed by the crystals. As indicated in Table 12-4, copper impurities in germanium act as acceptors and hence provide a trap within the band gap where recombination of electrons and holes can occur. Reference to Fig. 12-23 indicates that the impurities that are intentionally added to semiconductors are usually present in concentrations of between 0.01 and 10 parts per million. Thus, it is clear that very small impurity concentrations can drastically affect the properties of semiconductors.

12-12 APPLICATIONS OF STRUCTURAL CONTROL

In the previous sections of this chapter it has been emphasized that the electronic transport properties of materials depend on both purity and structural perfection. In this section we shall see how such factors can be controlled.

Control of Purity—Zone Refining

One of the prerequisites for making useful semiconductors and for achieving high conductivities in metals is the availability of materials with extremely high purity (with impurities of less than a few parts per billion in the case of semiconductors). One of the methods for obtaining such high-purity materials is through the use of a process known as *zone refining*.

To illustrate how zone refining operates, let us consider the phase diagram shown in Fig. 12-27. For this hypothetical system, the composition of the liquidus at any

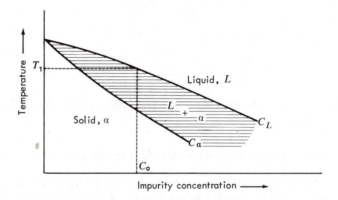

FIG. 12-27 Partial phase diagram representing the distribution of impurities in solid and liquid phases

given temperature is greater than the composition of the solidus, and hence the distribution coefficient $K = C_s/C_l$ is less than one ($K < 1$). The zone refining process we shall describe can be modified and applied to alloys for which $K > 1$, but the process is more complicated and less effective than for the case of $K < 1$.

Suppose we start with a sample that has the impurity concentration C_0. Notice that if the alloy is heated above T_1 it is entirely liquid and that if that liquid is slowly cooled below T_1 the first solid that forms has the composition KC_0. If we would be satisfied with the production of just a small amount of material, we could freeze a small part of the sample and discard the less pure liquid. The small sample with composition KC_0 could then be remelted and partially frozen to produce a still smaller quantity of material with composition K^2C_0. This process of partial freezing could be continued, and after n cycles the composition of the remaining bit of solid would be K^nC_0. It is clear that unlimited purity could be achieved with this method but that the amount of material purified would be so small that the method would be impractical.

We now turn our attention to *zone refining*, a process by which very high purity can be achieved in a large fraction of a given sample. We shall see that the fundamental principles that underlie the impractical scheme mentioned in the previous paragraph also apply to zone refining. In zone refining a liquid zone is produced in a bar sample by heating a portion of the bar with induction heating or electron-

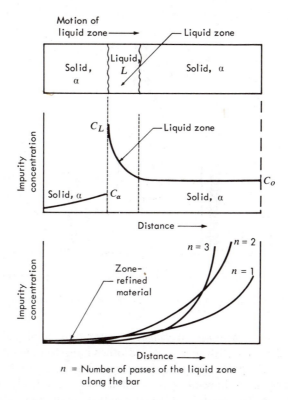

FIG. 12-28 Illustration of the process of zone refining. Each pass pushes impurity atoms toward the right end of the bar.

beam heating methods. Such a zone is shown in Fig. 12-28 along with the composition profiles produced by various passes of the liquid zone down the bar. The basic idea of the technique is that because $K < 1$, the solid produced is purer than the liquid zone with which it is in equilibrium. This means that when the zone is made to pass through the bar by moving the bar relative to the heating source, impurities are preferentially segregated to the liquid zone and passed to the end of the bar. Since the liquid zone finally freezes, the impurities are eventually frozen in the bar, but only at the end. By making several passes with the liquid zone, the impurity content of the major portion of the bar can be made very low. After many such passes the impure end of the sample is removed and the remaining portion is highly purified. It should be pointed out that the zone refining process is one of the techniques of structural control that had to be developed before the technology of semiconductors could be achieved.

Preparation of Single Crystals

As discussed in the previous section, highly perfect single crystals are needed for the manufacture of semiconductor devices. Of the many methods that can be

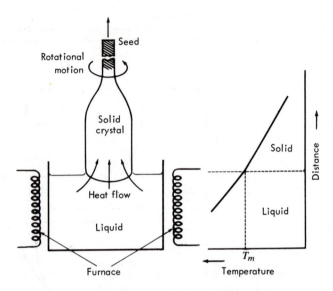

FIG. 12-29 Schematic illustration of crystal growing with the Czochralski technique

used, the *Czochralski technique* is commonly employed to produce crystals with low densities of dislocations and other imperfections. A schematic representation of a crystal-growing furnace is shown in Fig. 12-29. In the operation of the furnace a seed crystal is dipped into the molten bath and is slowly pulled out with the rotating pull rod. The crystal cools as it is pulled out of the molten bath, so that the solid-liquid interface remains near the top surface of the liquid region. Beginning with the seed, which partially remelts when it enters the bath, atoms from the liquid phase are attached to the growing seed as it is removed from the bath. The atoms that attach to the seed extend the crystal structure of the seed. In this way the crystal that is finally produced has the orientation of the seed. Crystals of silicon or germanium grown by this method are usually about 1–3 in. in diameter and approximately 12–24 in. in length. Semiconducting devices are produced from single-crystal wafers that are cut from the melt-grown crystal.

It should be noted that the perfection of the melt-grown crystal depends on the perfection of the seed crystal and the velocity of the liquid-solid interface during crystal growth. If dislocations are in the seed, they will extend into the crystal during crystal growth. Hence, a highly perfect seed is needed to produce a crystal with a low dislocation density. Even if the seed is relatively perfect, imperfections can be induced into the growing crystal at the liquid-solid interface by local stresses. Generally speaking, a crystal must be grown slowly (growth velocity ≈ 1 cm/hr) in order to prevent the formation of dislocations and other imperfections. This is understandable since the atoms that attach to the crystal must have enough time to find an equilibrium position on the crystal. Crystals that are grown too fast entrap defects because the attaching atoms are trapped in nonequilibrium positions by the other atoms that are simultaneously being adopted by the crystal.

It is also necessary to point out that the composition of the crystal can be controlled by controlling the composition of the bath from which the crystal is drawn. Usually it is necessary to make additions to the bath during crystal growth to maintain the composition of the crystal, because the growing crystal usually has a composition different from that of the bath.

PROBLEMS

12-1 Solid state microelectronic circuits and devices are often connected to other electrical circuits with fine gold wires. In certain cases consideration must be given to the resistance of the lead wires themselves. Calculate the resistance of a 1-mil (0.001 in. in diameter) gold wire 5 cm in length.

12-2 Use your knowledge of bonding to explain why long-chain polymers are generally poor conductors. Explain why we can use the "large energy gap" concept even though the polymers are often amorphous and therefore do not conform to the crystalline model on which the band-gap picture was based.

12-3 Explain why the electrical resistivity of a metal that is quenched from a high temperature is greater than the resistivity for a slowly cooled sample.

12-4 Explain why the electrical resistivity of fine metal wires should become temperature independent at very low temperatures (see Fig. 12-4, for example). Explain why the critical temperature below which the resistivity is constant with temperature increases as the diameter of the wire decreases. How should this temperature-independent region depend on the state of perfection of the metal?

12-5 Write all the equations and explain how you could estimate the electrical resistivity of a well-annealed alloy of silver +1 atomic percent Pd at 4°K. Be sure to tell how to compute all quantities not found in tables of physical constants. The crystal structure of silver is FCC with a lattice parameter of 4.06 A. Silver is monovalent.

12-6 Use your knowledge of the factors that control the electronic conductivity of metals and group IV semiconductors to suggest some ways by which the conductivity of polymers might be improved.

12-7 Explain why there is a correlation between mechanical strength and hardness and the electrical resistivity of metallic conductors.

12-8 The electrical resistance of a pure metal can be increased by plastic deformation or by the addition of impurity atoms. Estimate which of the following changes would increase the resistance most:
(a) The addition of impurity atoms to a concentration of 1 ppm (one impurity atom for each million solvent atoms).
(b) The production of a dislocation density of 10^8 cm^{-2} by plastic deformation.
(c) The development of a grain size 0.01 mm in diameter.
One may assume that the atomic density of this metal is about 10^{23} atoms/cm^3.

12-9 Using a sketch, show the difference in electron energy-band structure among an insulator, a semiconductor, and a metallic conductor. Explain why an insulator is not a good conductor.

12-10 Compute the band-gap energy for germanium from the data given in Fig. 12-19.

12-11 Explain why nonstoichiometric III-V compounds are expected to be extrinsic semiconductors.

12-12 Explain why the electrical conductivity of silicon crystals doped with As atoms can decrease with increasing temperature at 200°K. (See Fig. 12-24.)

12-13 A p-type germanium semiconductor has a hole concentration of 10^{22} holes/m^3 at 500°K. Calculate the concentration of electrons in the conduction band. The band-gap energy for germanium is 0.68 eV.

12-14 Show that when an n-type semiconductor is held in a temperature range where all of the donor levels are ionized and none of the electrons are excited from the valence band, the Fermi energy decreases approximately linearly as the temperature increases.

12-15 Explain why the addition of Sb (which has a valence of $5+$) to Si results in an n-type semiconductor at room temperature. Explain the effect of doping the above with boron, which has a valence of 3.

12-16 Explain why interstitial Li atoms in Si should act as donors.

12-17 Wafers of p-type semiconductors are usually cut from a large single crystal of the p-type material. The electrical conductivity is proportional to the concentration of p-type impurity atoms. Suppose we decide to grow the large crystals by first placing the desired alloy in a refractory boat, melting the entire alloy, and then freezing the alloy from one end. After this procedure we find that only the wafers cut from somewhere near the center of the crystal will have the desired electrical conductivity. Wafers cut from the head of the crystal (first to freeze) have a low conductivity, while wafers taken from the butt end have a high conductivity. Explain this result and indicate how the process could be changed to produce a large number of wafers with the proper electrical conductivity.

12-18 It is known that lattice vacancies in silicon can act as *acceptors* of electrons and can seriously affect the electrical properties of extrinsic semiconductors. Suppose that a transistor (semiconductor device) is being made with thin silicon wafers that are produced by vapor deposition onto an inert substrate. The process is successful in producing single crystals of silicon, but the concentration of lattice vacancies is always very high ($\approx 10^{13}$ vacancies/cm^3) just after vapor deposition. Attempts to remove the vacancies by heating the wafers to just below the melting temperature have failed because the films cool so quickly after the anneal that the vacancy concentration is still too high. It is now suggested that the wafers be heated to just below the melting temperature of silicon and that they be slowly cooled to a temperature where the equilibrium vacancy concentration falls below a certain critical value. (We will require the vacancy concentration to be equal to or less than 10^{12} vacancies/cm^3.) Calculate the temperature to which the wafers should be slowly cooled. The formation energy for vacancies in silicon is 2.3 eV per vacancy. Explain why it is all right to rapidly cool the wafers once they have reached the critical temperature.

12-19 Suppose we want to determine the number of free electrons per unit volume that participate in conduction in a monovalent metal at room temperature. We can use two different approaches:

(a) *Theoretical:* Assume that the electrons are perfectly free and calculate the number of conduction electrons in a unit volume at room temperature. We know the metal to have an FCC structure with a lattice parameter of 3.6×10^{-10} m.

(b) *Experimental:* A Hall effect experiment is carried out and the following results are obtained:

$$V_x = 1 \text{ V}$$
$$V_y = 10^{-6} \text{ V}$$
$$B_z = 1 \text{ W/m}^2$$
$$J_x = 3.5 \times 10^6 \text{ A/m}^2$$

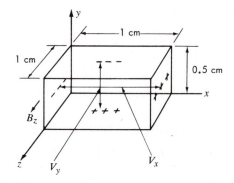

(1) Calculate the Hall constant.
(2) Determine the density of carriers (free electrons that participate in conduction).
(3) Calculate the Hall mobility.
(4) Calculate the resistivity.

(c) How does the theoretical estimate of the number of conduction electrons compare with the experiment?

12-20 As an engineer working for an electronics company, you and your associates have been asked to design some equipment that must operate at liquid nitrogen temperature ($78°\text{K}$). One part of the equipment involves the use of an *n*-type semiconductor material. The electrical design requires the electrical conductivity to be $2.75 \text{ A/V} \cdot \text{m}$. Your associates have decided to use antimony-doped silicon for the device. They conclude that the number of conduction electrons needed can be calculated in the following way:

$$\eta_e = \frac{\sigma}{e \mu}$$

since

$$\sigma = 2.75 \text{ A/V} \cdot \text{m}$$
$$e = 1.602 \times 10^{-19} \text{ C}$$
$$\mu = 0.17 \text{ m}^2/\text{V} \cdot \text{sec}$$
$$\eta_e = \frac{2.75}{(1.602 \times 10^{-19})(0.17)} \approx 10^{22} \text{ electrons/m}^3$$

Your co-workers have assumed that you need only specify that there be 10^{22} Sb atoms/m^3 to get the desired electrical conductivity. You insist, however, that because of the low temperature it cannot simply be assumed that all of the donors are ionized. Therefore, you may need to specify a higher donor concentration than 10^{22} atoms/m^3. The band gap for silicon at 78°K is 1.15 eV and the donor level lies 0.039 eV below the conduction band.

(a) Calculate the Fermi energy in the material your associates have suggested.

(b) Calculate the number of electrons in the conduction band.

(c) What recommendations would you make to your boss in this matter (i.e., will 10^{22} Sb atoms/m^3 be enough to obtain the desired conductivity)?

BIBLIOGRAPHY

J. C. ANDERSON and K. D. LEAVER, *Materials Science*, Ch. 12-13. New York: Van Nostrand Reinhold, 1969.

L. V. AZAROFF, *Introduction to Solids*, Ch. 12. New York: McGraw-Hill, 1960.

L. V. AZAROFF and J. J. BROPHY, *Electronic Processes in Materials*, Ch. 8. New York: McGraw-Hill, 1963.

R. H. BUBE, *Photoconductivity of Solids*, Ch. 2. New York: John Wiley, 1960.

R. W. HANKS, *Materials Engineering Science, An Introduction*, Ch. 10. New York: Harcourt Brace Jovanovich, 1970.

N. B. HANNAY, ed., *Semiconductors*, Ch. 1. New York: Van Nostrand Reinhold, 1959.

T. S. HUTCHISON and D. C. BAIRD, *The Physics of Engineering Solids*, Ch. 12, 14. New York: John Wiley, 1968.

A. NUSSBAUM, *Electronic and Magnetic Behavior of Materials*, Ch. 2. Englewood Cliffs, N.J.: Prentice-Hall, 1967.

W. G. PFANN, *Zone Melting*. New York: John Wiley, 1958.

R. M. ROSE, L. A. SHEPARD, and J. WULFF, *The Structure and Properties of Materials*, Vol. IV, Ch. 4–5. New York: John Wiley, 1966.

R. L. SPROULL, *Modern Physics, A Textbook for Engineers*, Ch. 11. New York: John Wiley, 1956.

L. H. VAN VLACK, *Materials Science for Engineers*, Ch. 14–15. Reading, Mass.: Addison-Wesley, 1970.

C. A. WERT and R. M. THOMSON, *Physics of Solids*, Ch. 11–13. New York: McGraw-Hill, 1964.

13
Electrical Properties of Junctions

13-1 INTRODUCTION

The electronic properties described in the previous chapters are called *bulk* or *homogeneous* properties because they depend mainly on the behavior of electrons in the interior of a material. The electrical conductivity of metals, for example, depends on the number of free electrons per unit volume (which in turn depends on the crystal structure and valence) and the mean free path between collisions (which in turn depends mainly of the state of perfection within the interior of the crystal). In this case the free surfaces and other interfaces play only a minor role because they affect only a small fraction of the atoms and electrons in the material. To illustrate the relatively small contribution by surfaces to bulk properties, we may note that the fraction of atoms located at the surface of a sphere of material with radius R in which the atoms are spaced a distance r apart is about $3r/R$. Evidently even when a crystal is as small as 1000 A in diameter less than 1 % of the atoms lie on the surface. It is clear that the fraction of atoms located at the surface of a crystal of macroscopic size is extremely small and hence, aside from their effects on electronic mean free paths, *surfaces* and *internal boundaries* are expected to make negligible contributions to bulk electronic properties.

We now turn our attention to the electronic properties that are uniquely related to *free surfaces* and *junctions* (*interfaces*) between dissimilar materials. These properties are called *heterogeneous* because they involve *surfaces* and *internal boundaries*. The fact that only a small fraction of the electrons in a solid reside near surfaces and junctions does not diminish the importance of these properties because it is *only* the interface electrons that contribute to these electronic properties.

The electrical properties of junctions between dissimilar materials are of great interest because they give rise to phenomena that are used in making such devices as transistors, solar batteries, and solid state refrigerators. It is because there are so many practical applications of the properties of junctions that the subject deserves treatment as a separate chapter in this book. Emphasis is placed on semiconductors and metal-semiconductor junctions since these materials provide the basis for most solid state devices. The properties of junction diodes, transistors, and thermoelectric devices are described in terms of the electronic structures of junctions.

13-2 SURFACE STATES

The peculiar electronic band structures that are formed at interfaces may be introduced by considering a free surface. We shall study the effects of a free surface on the electronic structure of an *n*-type semiconductor. Consider the energy band diagrams in Fig. 13-1. Figure 13-1(a) corresponds to the bulk or interior of an *n*-type semiconductor. As shown in Fig. 13-1(b), the effect of a free surface is to provide extra energy levels within the band gap (as well as within the valence and conduction bands). These new levels arise, first, because of the defect nature of the free surface and, second, because impurity atoms and other phases such as oxides are likely to be present at the surface.

Because the atoms at the surface share valence electrons with fewer neighbors than atoms in the interior, the surface atoms do not have sufficient electrons in their vicinity to satisfy the conditions for covalent bonding. Consequently, surfaces may provide acceptor states that serve as traps for conduction electrons. In addition, impurities that reside at the surface can provide new energy levels within the band gap since the adsorbed atoms can simply act as either donor or acceptor impurities, as discussed in the previous chapter. The energy levels, or states, that are available only at the surface are called *surface states*, whether the states arise from the defect nature of the surface or from adsorbed impurity atoms.

The effect of surface states on the band structure of the *n*-type semiconductor is shown in Fig. 13-1(b) and 13-1(c). The availability of extra energy levels within the band gap at the surface allows some of the electrons, which would normally lie either in the donor level or in the conduction band, to reside in the low-energy surface states. That is, the conduction electrons adjacent to the surface (at *A*) lower their energy by dropping into the surface states that lie in the band gap. When this happens, the Fermi energy at the surface is decreased as shown in Fig. 13-1(b). Now, conduction electrons from the interior of the crystal flow toward the surface in response to the gradient in the Fermi energy (i.e., there is an electronic potential energy gradient). As electrons collect at the surface, the region becomes negatively charged and the electronic energy levels near the surface are raised.

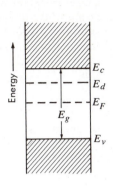

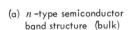

(a) *n*–type semiconductor
band structure (bulk)

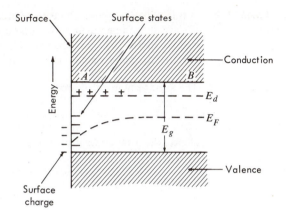

(b) *n*–type semiconductor band
structure near a free surface

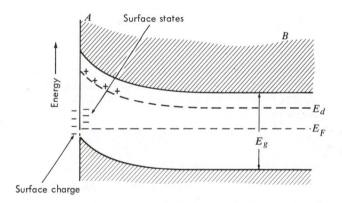

(c) Distortion of the band structure of an
n–type semiconductor near a free surface

FIG. 13-1   Illustrations of the band structure of an *n*-type semiconductor near a free surface
at equilibrium. The Fermi energy is the same everywhere in the semiconductor.

The distortion of the band structure shown in Fig. 13-1(c) is caused by the electrostatic charge that accumulates on the surface. Electrons continue to migrate to
the surface until the Fermi energy has the same value everywhere in the material.
This condition of equilibrium (constant Fermi energy) is achieved when the energy
gained by bringing an electron from the interior to the surface is equal to the work
required to move the electron against the electrostatic potential (surface charge).
In our subsequent discussions of the elctronic properties of junctions we shall
always require that the Fermi energy be constant and independent of position at
equilibrium.

EXAMPLE 13-1

A thin film of n-type silicon is to be prepared in such a manner that the surface states are exactly compensated by the donor electrons. The plane of the thin film is {111}, and the film is 1000 A in thickness. Calculate the concentration of As that would be required to compensate for the surface states. The lattice parameter of Si is 5.42 A.

Solution:

The average number of atoms on a {111} surface of Si is

$$\rho_s = \frac{1}{(a/\sqrt{2})(\sqrt{3}\,a/2\sqrt{2})} = \frac{4}{a^2\sqrt{3}}\ \text{atoms/cm}^2$$

Hence, the total number of surface atoms for a film of unit area is

$$N_s = 2\ \text{(two surfaces)}\ \rho_s = \frac{8}{a^2\sqrt{3}}\ \text{atoms/cm}^2$$

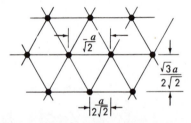

Each atom at the surface has on the average one-half the four neighboring atoms it needs to satisfy the covalent bonding requirements. Therefore, two electrons must be supplied for each surface atom to fill the surface states. Consequently, the donor concentration must be

$$\eta_d = \frac{2 \cdot N_s}{\text{Vol}} = \frac{16}{a^2\sqrt{3}\,t}$$

where t is the film thickness. For $t = 1000$ A and $a = 5.42$ A,

$$N_d = 3.14 \times 10^{20}\ \text{As atoms/cm}^3$$

or a fractional concentration of

$$\frac{3.14 \times 10^{20}}{5.0 \times 10^{22}} = 6.3 \times 10^{-3}$$

13-3 PROPERTIES OF JUNCTIONS

Having described the special electronic features of free surfaces, we are now prepared to consider the electronic structures of junctions between dissimilar materials. We begin this discussion by considering the electronic aspects of the junction between two dissimilar metals.

Metal-Metal Junction

As discussed in Chapter 11, when metals A and B with work functions $\varphi_A > \varphi_B$ (see Fig. 13-2) are brought into close contact, electrons flow from metal B to metal

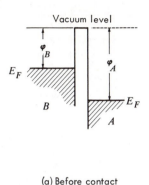

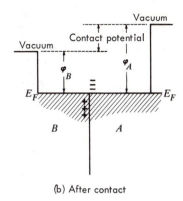

(a) Before contact (b) After contact

FIG. 13-2   Illustrations of the contact potential formed when metals with different work functions are brought into contact

A. Electrons pass the metal-metal junction until a reverse voltage (contact potential) is established to counteract the driving force for electron flow ($\varphi_A - \varphi_B$). Equilibrium in this case is obtained when the energy gained by moving an electron from metal *B* to metal *A* ($\varphi_A - \varphi_B$) is exactly equal to the work required to move the electron through the electric field established by the contact potential. When this condition is reached, the net flux of electrons across the junction is zero.

It should be noted that the contact potential between two metals is not a potential difference in the sense of a battery. For example, if the opposite ends of a metal-metal junction are connected with a wire, the wire will not carry current because there is no net potential difference between the free electrons in the two metals. Consequently, the contact potential may not be measured directly with a volt meter. We may illustrate this with Fig. 13-3. Consider the junction between metals *A* and *B* and the contacts with the lead wire *C*. After the metals are joined and the lead wire has been attached at each end, the Fermi level of all three metals will be the same. The voltage drop in moving an electron (from left to right) from the lead wire *C* to metal *A* is $\varphi_C - \varphi_A$, and the voltage drop in going from *A* to *B* is $\varphi_A - \varphi_B$. As the electron is moved from metal *B* to the lead wire *C* on the right, there is voltage drop (actually an increment) in the amount $\varphi_B - \varphi_C$. It is clear that since $(\varphi_C - \varphi_A) + (\varphi_A - \varphi_B) + (\varphi_B - \varphi_C) = 0$, the net voltage drop across the junction is zero and a voltmeter will not register a potential difference.

Measurement of the contact voltage basically involves a measurement of the amount of charge that passes from one metal to the other. For example, if two metals are contacted and subsequently separated, they will act as two oppositely charged plates. A voltage may be induced in an external circuit by moving the charged plates normal to each other. In this way the contact voltage can be indirectly measured. Another method for measuring the contact voltage is illustrated in Fig. 13-4. When two metals are contacted and then separated, they are separately charged and the electric field in the gap between the two metals can be used to deflect charged particles. For example, an electron that passes between the two

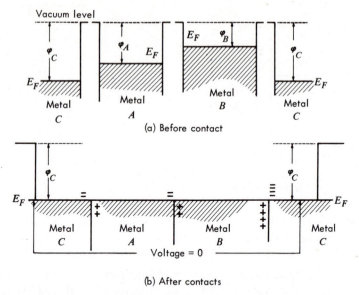

(a) Before contact

(b) After contacts

FIG. 13-3 Contact voltages produced by contacting metals *A* and *B* with lead wire *C*. The voltage between the two ends of the lead wire is zero.

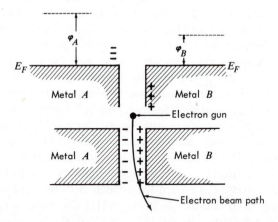

FIG. 13-4 Illustration of a technique for measuring the contact voltage between two metals

metals will be deflected toward the metal with the smallest work function. The magnitude of the deflection gives a measure of the charges that were exchanged when the metals were brought into contact. This information in turn leads to a determination of the contact voltage.

When two metals are joined, a certain amount of charge is transported across the junction, and electric currents necessarily flow for a short period of time. This current flow is important only if we are concerned with voltages and currents the instant after the contact is made. After the Fermi levels of the two metals are equilibrated, any net current that flows across the junction must be equal to the current flow in the two metals in response to an applied voltage. Since the junction will allow currrent to pass in either direction with the same facility, and since the steady state current that flows across the junction is proportional to the applied voltage (Ohm's law), the junction is said to be *ohmic*. We shall see later that the junctions between metals and semiconductors and between different semiconductors are very often not ohmic in their behavior. That is, current will pass more easily in one direction than the other.

Metal-Semiconductor Junction

A metal-semiconductor junction may be *ohmic* or *rectifying*, depending on the nature of the semiconductor (*p* or *n*) and the relative work functions of the two materials. Let us consider first an example of an ohmic contact. Figure 13-5 illustrates the energy-band picture of the junction between a metal and a *p*-type semiconductor. The band structures of the two materials before contact is made are shown in Fig. 13-5(a). Since in this example the work function of the metal φ_m is

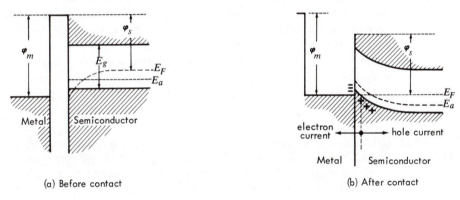

(a) Before contact   (b) After contact

FIG. 13-5   Illustration of the band structure and contact voltage for an ohmic contact between a metal and a *p*-type semiconductor

larger than the work function of the semiconductor φ_s, electrons flow from the semiconductor to the metal when contact is made. This means that there will be an electron deficiency on the semiconductor side of the junction and an excess electron concentration on the metal side. Electrons continue to flow across the junction until the reduction of kinetic energy on moving from the semiconductor to the metal, $\varphi_m - \varphi_s$, is equal to the increase in potential energy caused by the contact potential. When equilibrium is achieved, the Fermi energy, which includes the effects of the contact potential, is the same on each side of the junction. It is conventional to represent the Fermi energy by a horizontal line in energy-band diagrams. Use of this representation makes it necessary to distort the valence and conduction bands as shown in Fig. 13-5(b) in order to properly reflect the occupancy of the levels and bands at all points in the semiconductor. According to Fig. 13-5(b), since E_F in the semiconductor near the junction is far below the conduction band, the conduction electron concentration in that region is lower than in the bulk of the semiconductor.

This is to be expected, since some of the electrons in the semiconductor passed into the metal when contact was made.

Having described the band structures associated with the metal-semiconductor junction, we are now prepared to consider the electrical properties of the junction. Consider first the case in which the metal side of the junction is made more positive than the semiconductor side. In this case, electrons flow from the valence band of the semiconductor into the metal. Since the valence band is not completely filled near the metal junction, the semiconductor acts as a metal at that point. Near the point where the top of the valence band drops below the Fermi energy, the conduction mechanism in the semiconductor changes from electron current (near the metal) to hole current (far away from the interface). In the transition region, electron-hole pairs in the valence band are formed by exciting valence electrons to higher available energy levels within the valence band. Because the electrons and holes in the valence band can be so easily created, the junction does not provide a barrier to current flow. When the voltage bias on the junction is reversed, electrons flow from the metal into the semiconductor and combine with holes near the place where the Fermi energy coincides with the top of the valence band. Thus, we see that electric current easily flows in either direction. This junction is said to be ohmic because the currents that flow depend only on the resistivities of the component materials and not on the features of the interface.

An analysis similar to the one presented here would show that when a junction is made between a metal and an n-type semiconductor, the work function of the metal φ_m must be smaller than that for the semiconductor in order for the junction to be ohmic.

EXAMPLE 13-2

To utilize p-n junction devices, it is necessary to make ohmic (nonrectifying) electrical contacts to both n and p sides of the junction, as shown in the accompanying diagram. Determine which of the following metals should be used to make the contacts to a Si device. The work function of Si is 4.2 eV.

Metal	Work Function (eV)
Ba	2.39
Ca	2.75
As	5.2
Mg	3.46
Tl	3.7
Ni	4.90

Solution:

To make an ohmic contact with the p-type material, the contact metal must have a work function larger than that for Si. Thus, for contact 1, we could use either Ni or As. Contact 2 will be ohmic if $\varphi_2 < \varphi_s$; hence, any of the metals Ba, Ca, Mg, or Tl would form ohmic contacts on the n-type side.

We now consider the case of *nonohmic* or *rectifying contacts* between metals and semiconductors. It may be said at the outset that rectifying contacts occur when $\varphi_m > \varphi_s$ for n-type semiconductors and when $\varphi_m < \varphi_s$ for p-type semiconductors. Let us consider the case of metal-n-type semiconductor junctions for which φ_m

$> \varphi_s$, and whose energy band diagrams are shown in Fig. 13-6. When contact between the two materials is made, electrons flow from the semiconductor to the

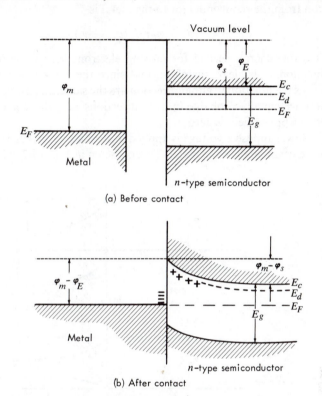

(a) Before contact

(b) After contact

FIG. 13-6   Energy band structures for rectifying contact between a metal and n-type semiconductor (a) before and (b) after contact is made.

metal (because $\varphi_m > \varphi_s$) and the energy bands in the semiconductor become distorted as shown. The energy-level configuration shown in Fig. 13-6(b) represents an equilibrium state for which the reduction in the kinetic energy of an electron going from the semiconductor to the metal is exactly balanced by the increase in potential energy caused by the contact potential. When there is no voltage applied to the junction, the net electric current that crosses the interface is zero. This can be seen easily by considering the energy change associated with electrons moving across the junction in either direction. For an electron at the Fermi energy to move from the metal to the semiconductor, it must receive an energy $\varphi_m - \varphi_E$* so that it can be excited to the conduction band in the semiconductor. For an electron to move from the semiconductor to the metal, it must first be excited to the conduction band, a process that requires an energy

$$E_c - E_F = (\varphi_m - \varphi_E) - (\varphi_m - \varphi_s) = \varphi_s - \varphi_E$$

---

*φ_E is the energy needed to excite an electron from the bottom of the conduction band to the vacuum level (see Fig. 13-6).

and then moved to the metal-semiconductor junction, a process which requires an additional energy $(\varphi_m - \varphi_s)$. Thus, we see that the energy needed to move an electron from the semiconductor to the metal is

$$(\varphi_s - \varphi_E) + (\varphi_m - \varphi_s) = \varphi_m - \varphi_E$$

the same energy needed for moving electrons across the junction in the reverse direction. It is clear, therefore, that since the energy barriers to electron motion across the junction in either direction are the same, electronic jumps across the junction are equally probable in both directions and the net current in the absence of an applied field is zero.

Now, suppose a voltage is applied in such a way that the semiconductor is made more negative than the metal. This is shown in Fig. 13-7, where the potential ener-

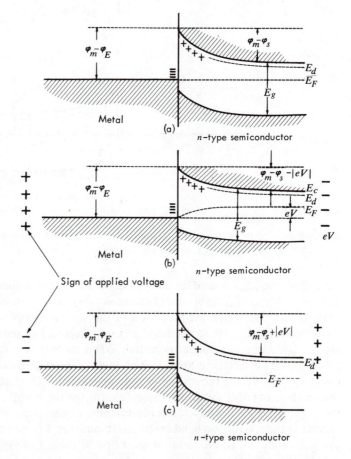

FIG. 13-7   Illustration of the band structures for a metal-*n*-type semiconductor junction. The band structure for zero bias is shown in (a), while (b) and (c) show the effects of applied voltages. The effect of a forward bias is shown in (b); the effect of a reverse bias is shown in (c).

gies of electrons in the semiconductor are greater than corresponding energies in the metal in the amount eV, where V is the applied voltage. The voltage does not change the energy barrier, which must be overcome by electrons that pass from the metal. Also, it is apparent that the electron concentration in the conduction band of the semiconductor is unchanged (since both E_c and E_F in the semiconductor are raised equally by the applied voltage). The only feature of the band structure that is changed is the energy barrier preventing electrons from flowing from the semiconductor to the metal. This energy barrier is decreased from $\varphi_m - \varphi_s$ to $\varphi_m - \varphi_s - eV$. It follows, therefore, that the applied voltage increases the electronic current flowing from the semiconductor to the metal. It is easy to see that if eV becomes very large, the energy barrier facing the conduction electrons in the semiconductor can become arbitrarily small and the current can become very large. When the voltage is applied in this manner, the junction is said to be operating under *forward bias*.

When the sign of the voltage is reversed, the potential energy of the electrons in the semiconductor is reduced relative to the potential energy of electrons in the metal, as shown in Fig. 13-7(c). In this circumstance the electronic current flow from the metal to the semiconductor is still controlled by the energy barrier $\varphi_m - \varphi_E$ and hence is not affected by the applied voltage. Again, it is only the current flowing from the semiconductor which is changed by the applied voltage. In this case, which is called *reverse bias*, the energy barrier discouraging electrons from passing from the conduction band of the semiconductor to the metal is increased to $\varphi_m - \varphi_s + eV$. Therefore, as the voltage across the junction is increased, the current flowing from the semiconductor to the metal is reduced and a net electronic flow from the metal to the semiconductor occurs. It is clear in this case that there is a limit to the current that will flow across the junction. The limiting current is reached when the forward current (from semiconductor to metal) is negligible compared to the reverse current (from metal to semiconductor). The maximum reverse current is usually rather small, of the order of a few milliamperes per square centimeter of the junction area, even though the reverse voltage can be as large as several hundred volts. In the forward direction the current density can reach many amperes per square centimeter with a forward bias of just a few volts.

The important feature of the rectifying metal-semiconductor junction is that current passes easily when the junction is subjected to forward bias, whereas the current that passes the junction is very small and essentially voltage independent under reverse bias. The current-voltage relation for a rectifying junction is illustrated in Fig. 13-8.

Before leaving the subject of metal-semiconductor junctions, it is worthwhile to point out an important engineering problem that relates to the type of junction formed between a semiconductor and a metal. All semiconductor devices, such as diodes or transistors, must be connected to other parts of electronic circuits, usually with metallic lead wires. It is generally desirable for the lead wires to make ohmic junctions with the semiconductor so that no rectification will occur because of the dissimilarity between the semiconductor and the lead wire. In the simplest case this means that the metal for the lead wire for *p*-type semiconductors must be chosen so

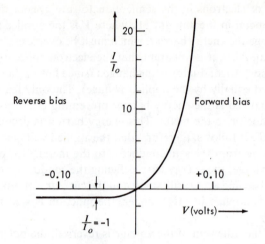

FIG. 13-8 Current voltage characteristics of a rectifying junction. I_0 is the limiting reverse current.

that $\varphi_m > \varphi_s$ and for ohmic junctions with n-type semiconductors, φ_m must be less than φ_s. The work functions of several metals and semiconductors are shown in Table 13-1.

TABLE 13-1
Work Functions for Several Metals and Semiconductors

Metals	φ (eV)	Semiconductors	φ (eV)
W	4.5	Si	4.4
Sb	4.1	Ge	4.7
Al	4.2	InSb	5.6
Mg	3.6	GaAs	4.8
Ba	2.5	Bi_2Te_3	5.3
Cs	1.9	GaP	4.8
Sn	4.4	CdSe	6.5
In	3.8	CdS	6.0

p-n Semiconductor Junction

By far the most important type of junction is the so-called *p-n junction*, which is formed at the interface between *p*-type material and *n*-type material in a semiconductor. As will be discussed later, such a junction is usually formed by diffusing one type of impurity atom (say, *n*-type) into an already extrinsic semiconductor (say, *p*-type). Consequently, the *p-n* junction is a diffuse interface that separates the *n*- and *p*-type regions of a single semiconductor crystal. In this section we shall focus our attention on the rectifying properties of *p-n* junctions.

Consider the energy-band diagram for a *p-n* junction shown in Fig. 13-9. From

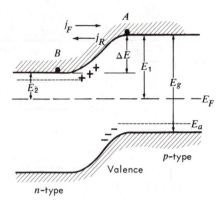

FIG. 13-9 Band structure of a *p-n* junction

our discussions in the previous chapter, we know that the Fermi energy in *n*-type material is greater than in the *p*-type region. This is, of course, a direct consequence of the donor atoms that reside in *n*-type material and the acceptor atoms found in the *p*-type region. If we imagine that the two regions are brought into close contact, electrons will flow from the *n*-type region to the *p*-type region and the energy bands will be distorted as shown in Fig. 13-9.

We now turn our attention to the rectifying properties of the junction. When no voltage is applied across the junction, the forward current (electron flow from *n*-type to *p*-type) and reverse current (electron flow from *p* to *n*) are equal, and the net current is zero. This can be shown by considering the processes that control the forward and reverse currents. The forward current j_F is proportional to the concentration of electrons in the conduction band of the *n*-type material (say, at point *B* in Fig. 13-9), $N_c \exp(-E_2/kT)$, and the probability that an electron at *B* receives sufficient energy to move into the *p*-type region in the conduction band, exp $(-\Delta E/kT)$. Consequently, the forward current can be expressed as

$$j_F = AN_c \exp\left(-\frac{E_2}{kT}\right) \exp\left(-\frac{\Delta E}{kT}\right) = AN_c \exp\left(-\frac{E_1}{kT}\right) \qquad (13\text{-}1)$$

where A is a constant, N_c is the density of states in the conduction band, and $E_1 = E_2 + \Delta E$. Similarly, the reverse current is simply proportional to the density of electrons that exist in the conduction band in the *p*-type material (point *A* in Fig. 13-9). Since there is no energy barrier to move from *A* to *B*, the reverse current is given by

$$j_R = AN_c \exp\left(-\frac{E_1}{kT}\right) \qquad (13\text{-}2)$$

It is clear from this analysis that $j_F = j_R$ and that no current flows across the *p-n* junction when there is no applied voltage.

Now, let us consider how a *p-n* junction responds to the application of a voltage *V*. Suppose that a forward bias voltage is applied as shown in Fig. 13-10(a). The imposed voltage reduces the potential energies of the electrons in the *p*-type material by $-(eV)$ and, consequently, causes a reduction in the energy barrier ΔE which must be overcome by conduction electrons moving across the junction into the

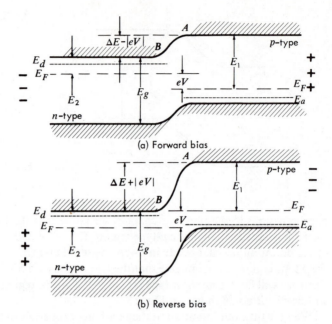

FIG. 13-10 Band structure of a *p-n* junction showing forward and reverse bias

p-type material. In this case the forward current j_F is expressed as

$$j_F = AN_c \exp\left(-\frac{E_2}{kT}\right)\exp\left(-\frac{\Delta E - |eV|}{kT}\right) \tag{13-3}$$

while the reverse current is unchanged by the applied voltage

$$j_R = AN_c \exp\left(-\frac{E_1}{kT}\right) \tag{13-4}$$

Now, the net current may be found from

$$j = j_F - j_R$$

$$= AN_c \exp\left(-\frac{E_1}{kT}\right)\exp\left[+\frac{(eV)}{kT}\right] - AN_c \exp\left(-\frac{E_1}{kT}\right)$$

$$= j_R\left\{\exp\left[+\frac{(eV)}{kT}\right] - 1\right\} \tag{13-5}$$

It is evident from Eq. (13-5) that the net current that flows across a *p-n* junction in response to a forward voltage increases linearly with voltage when the voltage is small

$$\exp\left(+\frac{|eV|}{kT}\right) - 1 \approx 1 + \frac{|eV|}{kT} - 1 = \frac{|eV|}{kT}$$

and exponentially with voltage when the voltage is large

$$\exp\left(+\frac{|eV|}{kT}\right) - 1 \approx \exp\left(+\frac{|eV|}{kT}\right)$$

These features are evident in Fig. 13-8, which shows the current-voltage relation for a rectifying junction.

Consider the application of a reverse bias voltage. As shown in Fig. 13-10 (b), the reverse bias increases the potential energy of the electrons in the p-type material and hence increases the barrier to forward current. In this circumstance the forward current is

$$j_F = AN_c \exp\left(-\frac{E_2}{kT}\right) \exp\left(-\frac{\Delta E + |eV|}{kT}\right) \tag{13-6}$$

while the reverse current again remains unchanged:

$$j_R = AN_c \exp\left(-\frac{E_1}{kT}\right) \tag{13-7}$$

Combining Eqs. (13-6) and (13-7), we find that the net forward current is

$$j = j_F - j_R$$
$$= j_R\left[\exp\left(-\frac{|eV|}{kT}\right) - 1\right] \tag{13-8}$$

Equation (13-8) indicates that when $V = 0$, $j = 0$ as expected. It also shows the limiting current for large reverse bias voltages. As $|V| \rightarrow \infty$, $\exp\left(-|eV|/kT\right) \rightarrow 0$ and $j \rightarrow -j_R$. Thus, we see that, regardless of the magnitude of the voltage, the current in reverse bias is limited by j_R, which is in itself unaffected by the applied voltage. It should be noted that Eq. (13-8) is also in agreement with the features of the voltage-current relation shown in Fig. 13-8. In addition, it should be pointed out that the junction behavior represented by Eqs. (13-5) and (13-8) can be represented by a single equation:

$$j = j_R\left[\exp\left(\frac{eV}{kT}\right) - 1\right] \tag{13-9}$$

where V is taken positive for forward bias and negative for reverse bias.

Equation (13-9), called the *junction equation*, is the single most important equation in this chapter, as it embodies the essential features of a rectifying junction. The equation clearly expresses the fact that current flows more easily across the junction in the forward direction than it does in the reverse direction.

EXAMPLE 13-3

A p-n junction at 300°K carries a current of 1 A/cm² when a bias of 0.1 V is applied in the forward direction. Calculate the current if the polarity of the voltage is reversed. At 300°K, $kT = \frac{1}{40}$ eV.

Solution:

In the forward direction we have from Eq. (13-9)

$$j = 1 \text{ A/cm}^2 = j_R\left[\exp\left(\frac{eV}{kT}\right) - 1\right]$$

so

$$j_R = \frac{1}{\exp\left(eV/kT\right) - 1} \text{ A/cm}^2$$

In reverse bias

$$j = j_R \left[\exp\left(-\frac{eV}{kT}\right) - 1 \right] = \frac{\exp\left(-\frac{eV}{kT}\right) - 1}{\exp\left(+\frac{eV}{kT}\right) - 1}$$

Since $eV/kT = 4$,

$$j = \frac{\exp(-4) - 1}{\exp(+4) - 1} = 1.85 \times 10^{-2} \text{ A/cm}^2$$

We have just shown that current flows across a p-n junction more easily in one direction (forward bias) than in the other (reverse bias). Although we have treated this on the basis of electron flow, the same results would be obtained if hole current were treated. We may now consider a more intuitive way of understanding these properties. Consider the p-n junction depicted in Fig. 13-11(a). The holes in the

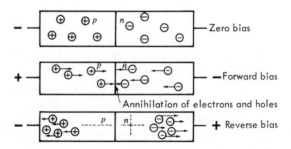

FIG. 13-11 Pictorial representation of forward and reverse bias for a p-n junction

p-type material are denoted by $\oplus$, and the donor electrons are indicated by $\ominus$. When a forward bias is applied, the holes and electrons are driven toward the p-n junction, and current crosses the junction as the holes and electrons annihilate each other [Fig. 13-11(b)]. In reverse bias both electrons and holes migrate away from the junction, leaving a space charge in the vicinity of the p-n junction (because of the absence of the electrons and holes). Since the carrier concentration near the p-n junction is greatly diminished, it follows that the current flowing across the junction will be correspondingly small. Thus, we can see intuitively that a forward bias will allow current to flow naturally across the junction, whereas a reverse voltage simply leads to the development of a space charge across the junction and very low current.

EXAMPLE 13-4

A rectifying contact is made between a metal and an n-type semiconductor (see Fig. 13-7). How will the maximum current in reverse bias depend on temperature?

Solution:

When the junction is in reverse bias, the current is controlled by the thermal excitation of electrons from the metal to the conduction band of the semiconductor. The reverse current is proportional to $\exp\{-[\varphi_m - \varphi_s + (E_c - E_F)]/kT\}$; consequently, the slope of a plot of $\ln j_R$ vs. $1/T$ should be equal to $-[\varphi_m - \varphi_s + (E_c - E_F)]/kT$. (See Fig. 13-7.)

13-4 SEMICONDUCTOR DEVICES BASED ON THE ELECTRICAL CHARACTERISTICS OF *p-n* JUNCTIONS

The basic rectifying properties of *p-n* junctions were discussed in the previous section and are quantitatively expressed by the junction equation (13-9). In this section we shall briefly consider how the current-voltage characteristics of *p-n* junctions can be put to use.

Diodes

The simplest and most obvious application of the *p-n* junction is as a diode or current rectifier. Because of the asymmetry of the current-voltage relation for a *p-n* junction, an a-c signal can be converted to a d-c signal. As shown in Fig. 13-12,

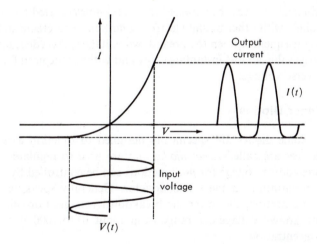

FIG. 13-12 Illustration of the rectifying properties of a *p-n* junction. Use of the *p-n* junction as a diode

when an a-c voltage is imposed across a *p-n* junction, the resulting current is primarily unidirectional. The resulting output current signal can be supplemented and smoothed with other electronic devices and circuits to produce a steady d-c signal.

The current-voltage relation, illustrated in Fig. 13-8 and described by Eq. (13-9), does not hold when the applied reverse voltage exceeds a critical value called the *breakdown voltage*. As shown in Fig. 13-13, the reverse current flowing across a *p-n* junction rises sharply when the breakdown voltage is exceeded. Although a detailed explanation of the breakdown phenomenon is beyond the scope of this book, certain features of the effect can be simply stated. At a sufficiently large reverse voltage, electrons and holes in the junction are accelerated to such a high velocity that they collide with bound electrons and holes and excite them to the conduction and valence band, respectively. The process is sometimes called an

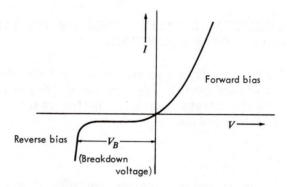

FIG. 13-13 Illustration of the current-voltage characteristics for a diode showing breakdown phenomena

avalanche process because electrons and holes created by the first collisions in turn collide with other bound electrons and holes and create still more free carriers. As a consequence, when the breakdown voltage is exceeded, the current rises to a very large value almost instantaneously and without the need for further increases in the reverse voltage.

Zener Diodes

Some diodes are specifically designed to function as *breakdown diodes*. Such devices are called *Zener diodes* and are used to regulate voltages in circuits. The breakdown voltage for *p-n* junctions can be controlled by controlling the impurity concentration in the semiconductor. Generally speaking, the higher the carrier concentration, the lower the breakdown voltage. For silicon semiconductors, the breakdown voltage can range from about 10 to 1000 V, depending on the carrier concentration.

Tunnel Diodes

Tunnel diodes, also called *Esaki diodes* after the discoverer, are *p-n* junctions that can be made to exhibit negative resistance. They are important because of their use as oscillators and their application in amplifier circuits. The characteristic property of the tunnel diode is illustrated in Fig. 13-14. It is evident that as the forward voltage increases, the current first increases, then beyond V_p it decreases, and finally increases again when the voltage exceeds V_c. It is clear that the junction exhibits negative resistance in the voltage range $V_p < V < V_c$.

The behavior of tunnel diodes can best be described by referring to the energy band picture of the *p-n* junction. Suppose, as shown in Fig. 13-15, that the junction is doped with a sufficiently high concentration of carriers that the Fermi level resides within the valence band in the *p*-type material and in the conduction band in the *n*-type material. When a forward bias is applied to the junction, electrons are

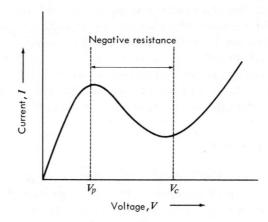

FIG. 13-14 Illustration of the voltage-current relation for a tunnel diode showing negative resistance

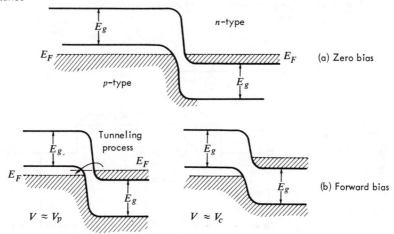

FIG. 13-15 Energy band structures for a *p-n* junction showing how negative resistance can be produced by forward bias

able to "jump" across the forbidden gap that exists at the *p-n* junction. This "jumping" process is called *tunneling* and is a wave mechanical process whose explanation is beyond the scope of this book. By way of analogy, one might consider the tunneling process here to be similar to the process by which an electron can penetrate or surmount a potential-energy barrier even when its average kinetic energy is less than the potential energy required. It is sufficient for our purposes here to state that tunneling can occur only if there are unoccupied energy levels on the opposite side of the junction. If the forward bias is sufficiently small [as in the first illustration in Fig. 13-15(b)], electrons can tunnel through the junction because there are empty energy levels into which the electrons may move. When forward voltage is increased to V_p, the energy bands are shifted in such a way that the Fermi energy in the n-type material is adjacent to the forbidden energy gap in the p-type material. Consequently, there are no empty energy levels to accept the tunneling electrons, and the tunneling current becomes small. This causes the current to decrease with voltage as shown in Fig. 13-14. At sufficiently large forward voltages the tunneling probability becomes exceedingly small and the diode behaves in a normal fashion. The increase in current that occurs when $V > V_c$ is explained by the argument that led to the junction equation (13-9).

The Transistor

By far the most important application of p-n junctions has been in transistors. These devices are analogous to triodes in conventional electronic circuits. They are important because of the amplifying characteristics and the resultant revolutionary changes they have caused in communications.

Some terminology associated with the n-p-n junction transistor is given in Fig. 13-16. The transistor is composed of two p-n junctions, which are placed back

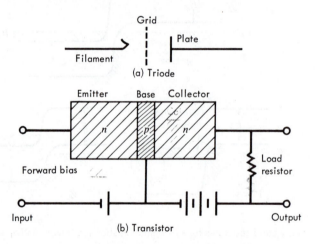

FIG. 13-16 Illustration of the elements of a conventional triode and a transistor

to back as shown. The components of the transistor are the *emitter*, *base*, and *collector*, which are analogous to the filament, grid, and plate, respectively, in a conventional triode.

The p-n junction between the emitter and base is operated in forward bias, while a reverse bias is imposed on the other junction. Consequently, electrons flow easily from the emitter to the base. Because of the reverse bias on the junction between the base and the collector, electrons would not ordinarily be expected to flow in the collector part of the circuit. However, a special effect occurs in the case of the junction transistor. If the base is sufficiently thin, usually of the order of 10^{-3} cm in thickness, then most of the electrons injected into the base by the emitter pass through the base and into the collector. Essentially, we can imagine the width of the base to be less than the mean free path of the electrons. Once an electron reaches the collector, it is driven through the load resistor by the reverse voltage. It is this feature which allows the transistor to be used to amplify electrical signals. A small input signal is impressed across the *p-n* junction with a forward bias. This causes a base current flow, most of which passes through the base into the collector. As the collector current increases, there is an increased voltage drop across the load resistor. Now, because a small increase in voltage across the emitter junction produces a large increase in collector current, the *n-p-n* transistor acts as a voltage amplifier. Increasing the load resistance and applied voltage increases the amplification, since the reverse current in the base-collector *n-p* junction is essentially independent of applied voltage. A simlar explanation using holes as carriers instead of electrons could be used to rationalize the operation of a *p-n-p* junction transistor.

13-5 TECHNIQUES FOR MAKING *p-n* JUNCTIONS

The addition of electrically active impurity atoms to semiconductors is referred to as *doping*. As shown in Fig.13-17, there are a variety of techniques for doping semiconductors and for making *p-n* junctions. Central to all of these methods is the concept of *compensation*. If a material is *n*-type by virtue of a given concentration of donor impurity atoms, it can be made *p*-type only after each of the donor atoms has been compensated with an acceptor. This means that an acceptor atom with its associated hole is needed to tie up or compensate for the extra electron supplied by each donor. For singly ionized impurities, it takes one acceptor atom to compensate for the effects of one donor atom, and vice versa. When an *n*-type material is doped with acceptors to the point at which the acceptor concentration exceeds the donor concentration, the material becomes *p*-type. In all the methods discussed below, the *p-n* junction is created by adding compensating impurity atoms at a concentration sufficient to reverse the sign of the majority carrier.

Crystal-Growth Technique

From a conceptual point of view, one of the simplest methods for making a *p-n* juction involves changing the composition of the liquid solution from which a semiconductor crystal is being grown. This process is illustrated in Fig. 13-17(a). Suppose, for example, that an *n*-type crystal is being grown by a Czochralski technique in which the liquid bath is doped with a certain concentration, C_L, of donor atoms. From the principles discussed in Chapter 4, we can predict that the donor concentration in the crystal in equilibrium with the liquid and being grown from it is KC_L, where K is the solute distribution coefficient defined as the ratio of the solute concentration in the solid to the solute concentration in the liquid. It is clear that if the composition of the liquid bath from which the crystal is growing is suddenly changed (say, by the addition of a high concentration of acceptor atoms), the composition of that portion of the crystal which is subsequently grown will also

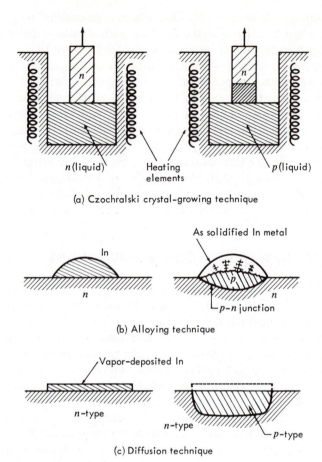

(a) Czochralski crystal-growing technique

(b) Alloying technique

(c) Diffusion technique

FIG. 13-17 Techniques for making *p-n* junctions

change. In this way, it is possible to produce a *p-n* juction within a semiconductor crystal.

Alloying Technique

The alloying method for producing a *p-n* junction is illustrated in Fig. 13-17(b). This technique involves melting a pellet of the desired impurity material and allowing crystal growth and solidification to take place on cooling. When the pellet is molten, part of the semiconductor dissolves into the liquid, thereby making the liquid zone an alloy of the impurity material (say, In) and the host an extrinsic semiconductor (say, *n*-type Ge). On cooling, the molten zone begins to freeze by forming a layer of doped semiconductor crystal onto the semiconductor substrate. In a later stage of cooling, the remaining liquid zone freezes, thereby forming a polycrystalline as-cast structure. The *p-n* junction formed by this technique is found at the point at which the alloy began its growth on the semiconductor substrate. The last part of the pellet to be frozen is mainly metallic In, and it acts as the contact for one side of the *p-n* junction.

Diffusion Technique

The most common technique for making a *p-n* junction, and the one which is most easily controlled, involves solid state diffusion of the appropriate impurity atoms (say, acceptors) into an already extrinsic *n*-type semiconductor. Usually a thin layer of the dopant is vapor-deposited onto the surface of the semiconductor and subsequently introduced into the crystal by diffusion. The point in the crystal at which the concentration of dopant exceeds the initial extrinsic impurity concentration is the exact center of the *p-n* junction. Since diffusion phenomena are usually "well behaved" and can be accurately predicted by the diffusion laws described in Chapter 5, the diffusion technique represents a precise method for producing a *p-n* junction with well-defined electrical characteristics.

The precision with which doping can be carried out by the diffusion technique has allowed the development of solid state microelectronic circuits such as the one shown in Fig. 1-8(c). These devices are unique because they involve the construction of entire circuits within very tiny semiconductor crystals. The techniques by which *p-n* junctions and other circuit elements are introduced into semiconductor crystals involves *masking* portions of a semiconductor crystal in order to allow doping to occur only on selected places. By sequentially masking and doping by vapor deposition and diffusion, multiple *n* and *p* layers can be produced and arranged in such a way as to perform ordinary circuit functions. Suppose, for example, that we wish to build the amplifier circuit illustrated in Fig. 13-18(a). Figure 13-18(b) shows a schematic cutaway view of an integrated circuit comparable to the one in Fig. 13-18(a). To build this circuit, we might start with an essentially pure intrinsic semiconductor crystal such as Si. The first step might involve the formation of the *p*-type regions that are associated with the two diodes and the transistor in the circuit. These regions might be formed by placing a mask onto the surface of the Si crystal and vapor-depositing a layer of acceptor impurity atoms (such as In) onto the surface. Subsequent annealing at high temperatures would allow the In atoms to diffuse into the crystal and produce the *p*-type regions shown. A second step in the process might involve the formation of the resistors (*R*) by masking and doping to the appropriate impurity concentration (remember that since the resistivity is directly related to the carrier concentration, resistors in semiconductors can easily be formed by controlled doping). Subsequent steps in the formation of the device involve additional masking and diffusion of appropriate impurities in the right places. In order to insure that the elements of the circuit are electrically isolated and that they cannot be shorted at the surface, it is common practice to develop an insulating oxide layer (SiO_2) on the surface as shown. Finally, the circuit elements are connected to each other and to external leads with thin metallic strips, which are vapor-deposited onto the SiO_2 surface. Electrical connections are made to the circuit elements by etching away the oxide coating at appropriate places prior to depositing the metallic strips.

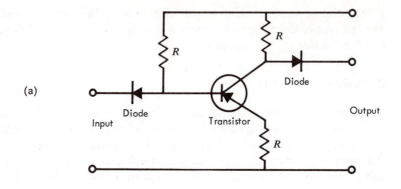

(a)

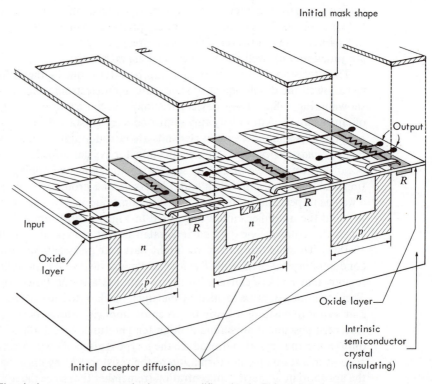

FIG. 13-18(a) Electrical components constituting an amplifier circuit. This entire circuit can be produced in a single crystal as shown in (b).   (b) Solid state microelectronic amplifier circuit. This circuit is equivalent to the schematic circuit shown in (a).

13-6 THERMOELECTRIC PROPERTIES OF MATERIALS

Not all the thermoelectric properties to be discussed here are junction properties. For example, the *Seebeck effect* and the *Thomson effect* are found in homogeneous materials. We shall see that the *Peltier effect*, on the other hand, arises only at the junction between two dissimilar materials.

Seebeck Effect

As shown in Fig. 13-19, when a metal is subjected to a temperature gradient, electrons at the hotter end of the bar may have higher kinetic energies than the electrons in the colder region. As a consequence, electrons tend to flow from the hot region to the cold in order to reduce their average kinetic energy. However, as electrons diffuse toward the cold end, a *Seebeck* voltage is produce in the bar. This

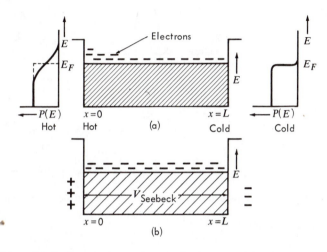

FIG. 13-19 (a) Nonequilibrium distribution of electronic energies in a metal bar in a temperature gradient. (b) Equilibrium distribution of electronic energies giving rise to a Seebeck voltage in the bar.

voltage tends to drive the electrons back toward the hot end of the bar. Equilibrium is achieved when the Seebeck potential exactly balances the thermal driving force for electron flow. Although this description of the Seebeck effect has been phrased in terms that apply only to metals, the effect is found in semiconductors, too. In fact, as shown in Table 13-2, the magnitude of the Seebeck voltage is usually larger for semiconductors than it is for metals.

TABLE 13-2
Seebeck Voltage (for ΔT of 1°K)

Material	Temperature (°C)	$V_{\text{Seebeck}} (V_{\text{hot}} - V_{\text{cold}})$ (μV)
Al	100	-0.20
Cu	100	$+3.98$
Ag	100	$+3.68$
W	100	$+5.0$
$(Bi, Sb)_2 Te_3$	25	$+195$
$Bi_2 (Te, Se)_3$	100	-210
ZnSb	200	$+220$
InSb	500	-130
Ge	700	-210
TiO_2	725	-200

Let us now consider how the Seebeck potential for a given material might be measured. Suppose, as shown in Fig. 13-20, that material A is subjected to temperatures T_H (hot) and T_C (cold). As mentioned before, the temperature difference will induce a Seebeck voltage that is proportional to $\Delta T = T_H - T_C$. If we try to meas-

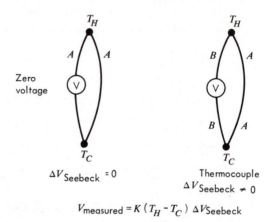

$$V_{measured} = K\,(T_H - T_C)\,\Delta V_{Seebeck}$$

FIG. 13-20 Illustration of voltage measurements on wires in a temperature gradient. Voltages are produced when the wires are different and have different Seebeck potentials (thermocouple).

ure the Seebeck voltage by connecting a voltmeter in the circuit using lead wires of material A, the Seebeck potential produced in the lead wire will exactly nullify the potential to be measured, and the voltmeter will not record a net potential difference. A potential difference will arise only if the lead wires are dissimilar to the material being studied. In this case the measured voltage is proportional to the difference in Seebeck voltages for the two materials.

The effects we have described form the basis for the operation of a *thermocouple*. The circuit shown in Fig. 13-20 may be used to describe a thermocouple. The voltage recorded by the voltmeter depends approximately linearly on the temperature difference between the junctions and on the difference in Seebeck potentials for the two materials. Some common materials in use today as thermocouple wires are copper-constantan (60% Cu–40% Ni) for temperatures up to 300°C; chromel (90% Ni–10% Cr)–alumel (94% Ni–20 Al–3% Mn–1% Si) up to 1200°C; platinum–platinum-rhodium alloys up to 1500°C; and tungsten-rhenium alloys above 1500°C.

Thomson Effect

Consider again the case of a rod of material subjected to a temperature gradient. If a voltage is applied in such a way that electrons are made to flow from the cold end of the rod to the hot end, the migrating electrons will spontaneously absorb heat from the bar, thereby producing a *thermoelectric refrigeration* effect. If the voltage is reversed and the electrons drift from the hot end to the cold end in response to the applied voltage, the moving electrons will be required to give up heat to the rod. This effect of heat production and/or absorption which occurs when an electric current flows in a temperature gradient is called the *Thomson effect*. It is easy to see that the effect can, in principle, be used as a thermoelectric refrigerator, although such thermoelectric devices are usually based on the *Peltier effect*.

Peltier Effect

The Peltier effect involves the evolution or absorption of heat that is created when an electric current flows across a junction between two dissimilar materials. The effect is reversible in the sense that if heat is evolved when the current flows in one direction, heat will be absorbed at an equivalent rate if the current is reversed. Heat production or absorption at a junction is to be added to the normal joule heating effects that accompany the flow of current in normal conductors.

We may develop some understanding of the *Peltier effect* by considering what happens when current flows across an ohmic contact between a metal and an *n*-type semiconductor. Consider the junction shown in Fig. 13-21. As electrons flow

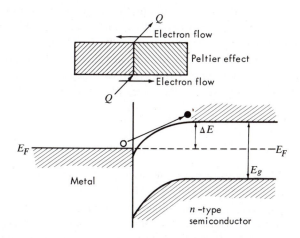

FIG. 13-21 Ohmic contact between a metal and an *n*-type semiconductor (illustration of the Peltier effect)

from the metal to the semiconductor, they must move from the Fermi level in the metal to the conduction band in the semiconductor. Consequently, the conduction electrons must increase their average kinetic energy by ΔE as they move from the metal to the semiconductor. This change in kinetic energy is accompanied by the *absorption* of heat; that is, heat or thermal energy is used to increase the average kinetic energy of the electrons. Now, suppose the current flow is reversed. In this case the electron kinetic energy decreases and there is an associated evolution of heat. Thus, we see that because the average kinetic energy of the electrons changes when the electrons cross the junction, heat is either absorbed or evolved, depending on the direction of the current.

A reversible Peltier effect occurs whenever current flows across an ohmic contact. It would be easy, for example, to extend the argument above to apply to the case of an ohmic contact between a metal and a *p*-type semiconductor.

The Peltier effect is important because it provides a basis for thermoelectric heating and refrigeration, and for thermoelectric generators of electric power. As current flows in a circuit that includes junctions between dissimilar materials, the Peltier effect causes heat to be transported from the cold junctions to the hot. If a sufficiently large number of such junctions are connected in series, it is possible to produce a workable refrigerator. Although such devices are relatively inefficient and cannot be operated at high currents because of joule heating, they are nonetheless silent, compact devices with important specialized applications. Compound semiconductors such as Bi_2Te_3 and $PbTe$ can be used to make thermoelectric refrigerators capable of producing a temperature differential of 80°C.

Another way to utilize the Peltier effect is as a thermoelectric generator. An electric current can be made to flow in a circuit that includes dissimilar materials if the junctions are held at different temperatures. Mechanistically this means that at the hot junction, for example, electrons are thermally pumped from one material to the other (from the metal to the semiconductor for the junction shown in Fig. 13-21), while at the opposite junction they are allowed to fall back in the reverse direction (by extracting heat). As a consequence, electric current flows in the circuit and an electric power is produced. As in the case of thermoelectric heaters and refrigerators, the thermoelectric power generator operates very inefficiently. Bi_2Te_3, Sb_2Se_3 (*n*- and *p*-types), and $GeTe$ (*p*-type) can be used to make thermoelectric generators with overall efficiencies of the order of 10%.

PROBLEMS

13-1 Illustrate the energy-band distortions that occur when an ohmic contact is made between a metal and an *n*-type semiconductor.

13-2 Illustrate the energy-band distortions that occur when a metal makes contact with a *p*-type semiconductor (after the Fermi energies are equilibrated). The Fermi energy of the metal is above the Fermi level in the *p*-type semiconductor before contact is made. Explain why this junction can act as a rectifier (i.e., will pass current in one direction and not the other). Indicate which polarity is *forward* and which is *reverse*.

13-3 Refer to Table 13-1 and explain why it might be difficult to make an ohmic contact with a *p*-type CdSe.

13-4 A 0.5-V potential is applied to a *p-n* junction by connecting the positive terminal of a battery to the *p* side of the junction and the negative terminal to the *n* side. The measured current density is 5 A/cm² at 20°C. Calculate the current density that would result if the polarity were reversed.

13-5 An *n*-type semiconductor material is to be manufactured by diffusing Sb atoms into a single crystal of silicon. It has been suggested that the appropriate composition can be achieved by coating a single crystal wafer of pure silicon with a thick layer of Sb and by performing a diffusion anneal. The required electrical properties demand that 0.01 atomic percent Sb be dissolved into the silicon wafers. Because the Sb atom is so much larger than the Si atom, there is an energy (strain energy) associated with the introduction of each solute, $\Delta E_s = 1$ eV.

   (a) Calculate the temperature at which the equilibrium composition will be 0.01 atomic percent Sb. (Assume there is plenty of Sb on the surface.)

   (b) If all of the 0.01 atomic percent Sb must remain in solution for the electrical properties to be achieved, how should the material be cooled after the diffusion anneal is complete?

   (c) What limitations must be put on the operating conditions for the device constructed with this material?

13-6 A wafer of germanium (0.5 cm in thickness) has been doped (during crystal growth) with phosphorus to the extent that it is *n*-type. In order to produce a *p-n* junction (and subsequently produce a transistor), we must diffuse gallium (a *p*-type dopant) into the surface of the crystal. The diffusion constants for gallium in germanium are:

$$D_0 = 5 \times 10^{-2} \text{ cm}^2/\text{sec}$$
$$\Delta E_D = 4.0 \times 10^{-12} \text{ erg/atom}$$
$$\text{Boltzmann constant } k = 1.38 \times 10^{16} \text{ erg/deg}$$

Calculate the time required for gallium to reach a depth of 0.1 mm if the diffusion is carried out at 800°C.

13-7 Sketch the band structure and describe the operation of a *p-n-p* transister.

13-8 Discuss how a transistor can be used as a current amplifier.

13-9 Explain why transistors cannot be made with polycrystalline semiconductors.

13-10 Explain why when growing a semiconductor crystal with a given donor concentration it is necessary to continually adjust the composition of the liquid bath during crystal growth.

13-11 How would the surface states shown in Fig. 13-1 affect the band structure of a *p*-type semiconductor?

13-12 Explain why the vacuum energy levels at each end of the metal-metal contact in Fig. 13-2 appear to be different.

13-13 Suppose the Fermi level of the metal in Fig. 13-5 had been even with the acceptor level in the semiconductor. Would the junction have been ohmic or rectifying? Why?

13-14 What do the positive charges in Fig. 13-6 denote?

13-15 Explain how the current must flow across a metal–*p*-type semiconductor junction to absorb heat.

BIBLIOGRAPHY

L. V. AZAROFF and J. J. BROPHY, *Electronic Processes in Materials*, Ch. 10–11. New York: McGraw-Hill, 1963.

A. J. DEKKER, *Solid State Physics*, Ch. 14. Englewood Cliffs, N.J.: Prentice-Hall, 1957.

T. S. HUTCHISON and D. C. BAIRD, *The Physics of Engineering Solids*, Ch. 14. New York: John Wiley, 1968.

C. KITTEL, *Introduction to Solid State Physics*, Ch. 13–14. New York: John Wiley, 1956.

R. M. ROSE, L. A. SHEPARD, and J. WULFF, *The Structure and Properties of Materials*, Vol. III, Ch. 5–8. New York: John Wiley, 1966.

R. A. SMITH, *Semiconductors*, Ch. 12. London: Cambridge University Press, 1959.

A. VAN DER ZIEL, *Solid State Physical Electronics*. Englewood Cliffs, N.J.: Prentice-Hall, 1957.

C. A. WERT and R. M. THOMSON, *Physics of Solids*, Ch. 13. New York: McGraw-Hill, 1964.

14
Magnetic Properties of Materials

14-1 INTRODUCTION

The phenomenon of magnetism in materials plays an important part in our everyday experience. It extends from the permanent magnets used to latch our refrigerator doors to the magnetic memory elements of our most sophisticated computers. Although we have known for centuries of the phenomenon of magnetism, it is only within the past few decades that our understanding of magnetism in materials has matured to the point where magnetic devices are becoming commonplace in technological applications This chapter deals with magnetic phenomena in materials, with particular emphasis on magnetic properties of engineering importance.

There are two general types of magnetism in materials. The first of these is a class we can call *induced magnetism*. In this class, the material is magnetized only when there is an applied magnetic field. *Diamagnetism* and *paramagnetism* are two forms of induced magnetism that will be discussed briefly in this chapter. The second general class of magnetic phenomena is referred to as *spontaneous magnetism*. As the name implies, spontaneous magnetism refers to the ability of a material to be magnetized and to retain its magnetic state even in the absence of an applied magnetic field. *Ferromagnetism* and *antiferromagnetism* are the common examples of this form of magnetic phenomena. Almost all magnetic materials that have important engineering applications belong to this class. Consequently, spontaneous magnetism will be emphasized in this chapter.

Spontaneously magnetized materials may be further classified with respect to the form of their use. Materials that can be magnetized and demagnetized easily are called *soft magnetic materials*. Materials such as silicon-iron and permalloy (iron-nickel alloy) are soft magnetic materials commonly used for transformer cores. Soft magnetic oxides such as ferrites are frequently used as memory cores in computers. By contrast, *hard magnetic materials* can be magnetized only by the application of strong magnetic fields and once magnetized are difficult to demagnetize. Hard magnetic materials such as alnico (Al-Ni-Co) and hard ferrites (barium ferrite) are used in the magnetized state as permanent magnets.

14-2 MAGNETIC FIELDS IN MATERIALS

Our discussion of magnetism in materials begins with a brief account of the basic properties of magnetic fields. Consider the long, thin bar magnet shown in Fig. 14-1(a). The magnet is said to have *magnetic poles* near each end which serve as *source* and *sink* for the associated magnetic field. It is evident that the poles arise as a consequence of the magnetic state of the material and that for each positive pole on one end there is a negative pole on the other. Although it has not yet been possible to isolate a single magnetic pole, it is nevertheless convenient to think of a unit pole or unit of "magnetic charge" that is so far removed from the other end of the dipole to which it belongs that it behaves as if it were truly isolated. This condition is illustrated in Fig. 14-1(b). Much of our understanding of magnetism in materials and many of the important magnetic relations can be derived from the concept of the *unit magnetic pole*.

By definition, when two unit magnetic poles of like sign are separated by a unit distance (1 cm), a repulsive force of one dyne is exerted between them. We imagine that a magnetic field emanates from each pole and acts on the other to produce the force. A related definition is that the magnetic field intensity 1 cm away from a unit pole is 1 *oersted*. Combining these two definitions, we may say that a force of 1 dyne is exerted on a unit magnetic pole when it resides in a magnetic field of 1 oersted. In our discussion of magnetic poles, it is convenient to imagine that magnetic *lines of force* emanate from each pole and constitute the magnetic field. It is conventional to let 4π lines of force come from each unit pole so that the line density 1 cm from the pole is 1 line/cm². With this related definition, 1 oersted is equal to 1 line/cm². Stated in another way, a force of 1 dyne is exerted on a unit pole when it is acted upon by a magnetic field of 1 line/cm².

A more general statement about the properties of magnetic poles is contained in *Coulomb's law*, which is: *The force exerted between two magnetic poles varies inversely with the square of the separation distance.* This is commonly expressed as

$$F = \frac{m_1 m_2}{\mu d^2} \tag{14-1}$$

where m_1 and m_2 are the pole strengths of the two poles in question, d is the separation distance, and μ is the *magnetic permeability* (to be defined later) for the medium

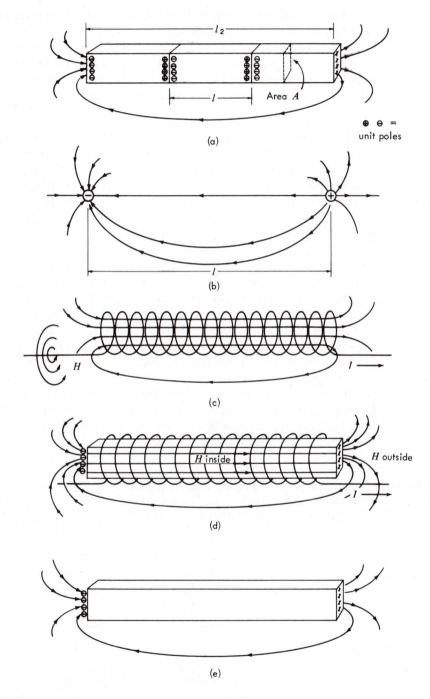

$\oplus$ $\ominus$ =
unit poles

(a)

(b)

(c)

(d)

(e)

FIG. 14-1   Schematic illustrations of magnets, magnetic fields, and magnetic poles

in which the poles reside. In the CGS system of units μ for a vacuum is one, and μ for air differs from unity by only one part in 10^8.

As was indicated earlier and will be made clearer later in this section, a unit magnetic pole does not exist by itself. It is always part of a *dipole*. A magnetic dipole simply consists of two magnetic poles of opposite signs separated by a given distance. The *magnetic dipole moment* $\bar{\mu}$ of any dipole is defined as

$$\bar{\mu} = ml \qquad (14\text{-}2)$$

where m is the pole strength of the dipole and l is the distance between the two poles. We shall see later that atoms can act as if they were magnetic dipoles in the sense that they behave as tiny permanent magnets. In this case the dipole itself becomes meaningful, and the single-pole concept does not apply. In the case of a macroscopic bar magnet, such as the one illustrated in Fig. 14-1(a), the dipole moment of the magnet is simply the product of the pole strength (i.e., the number of unit poles times the length) of the magnet.

Since magnetization in materials arises because of the collective behavior of all the individual atomic dipoles, the magnetic moment of any sample is simply the sum of the dipole moments in all the dipoles in the sample. Consider again the magnet in Fig. 14-1(a). In this hypothetical sample there are 12 dipoles, each having a dipole moment ml. The dipole moment of the entire sample is simply $12ml$, or the sum of the individual dipole moments. Note that the dipole moment can also be equivalently expressed in terms of the poles at the free ends of the sample (four dipoles with length l_2) since $4ml_2 = 12ml$.

A magnetic property of considerable interest is the *intensity of magnetization* or the *dipole moment per unit volume*. For the bar magnet of Fig. 14-1(a), the intensity of magnetization is given as $M = 4\,m/A$, where A is the cross-sectional area of the magnet. From this relation it is evident that the intensity of the magnetization is equivalent to the *density of the magnetic poles* on the end of the magnet.

EXAMPLE 14-1 Shown in the accompanying illustration are two bar magnets with pole strengths of 50 and 100, respectively. Calculate the force F_t needed to hold the magnets in the position shown. The magnets are in air.

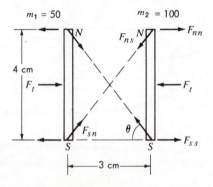

Solution: Using Coulomb's law, the repulsive forces F_{nn} and F_{ss} exerted between the two ends are

$$F_{nn} = F_{ss} = \frac{m_1 m_2}{d^2} = \frac{(50)(100)}{(3)^2} = 556 \text{ dynes}$$

The attractive forces between the poles of opposite signs are

$$F_{ns} = F_{sn} = \frac{m_1 m_2}{d^2} = \frac{(50)(100)}{(5)^2} = 200 \text{ dynes}$$

The angle θ is

$$\theta = \sin^{-1}\left(\tfrac{4}{5}\right) = 53.2°$$

so that the total force F_t is

$$F_t = F_{nn} + F_{ss} - 2F_{ns} \cos \theta = 2(556) - 2(200) \cos 53.2° = 1350 \text{ dynes}$$

We are now prepared to introduce the concept of the *magnetic induction field*. Consider the solenoid shown in Fig. 14-1(c), in which a direct current is flowing. Because of the magnetic field associated with electric current, there is an axial magnetic field produced within the solenoid as shown. The intensity of the field in that region is

$$H = \frac{0.4\pi n I}{l} \tag{14-3}$$

where H is expressed in oersteds, n is the number of turns, l is the length of the solenoid, and I is the current in amperes. The important aspect of Fig. 14-1(c) is that the magnetic field is *continuous*. In this particular case, since there is no material inside the solenoid to be magnetized, the field is referred to as either the *magnetic field strength* (H) or the *induction field* (B). We shall see that the induction field B is *always* continuous, whereas the magnetic field strength H may extend from one magnetic pole to another if magnetized materials are involved.

Now let us place a magnetic material within the solenoid and allow the material to become magnetized. This situation is depicted in Fig. 14-1(d). Because of the magnetization of the material within the solenoid, the field outside the solenoid is stronger than that in Fig. 14-1(c). The induction field (the magnetic field induced into the material) is simply given as the solenoid field plus the field that arises from the magnetization. To analyze this, let us suppose that the field from the solenoid is removed but that the magnetization of the sample is retained [see Fig. 14-1(e)]. The magnetic pole density on the end of the magnetized sample is m/A, so that the intensity of magnetism is also $M = m/A$. Now, since each pole emits 4π magnetic field lines, the field strength arising from the magnetization is simply $4\pi(m/A)$ $= 4\pi M$ lines/cm². When the solenoid produces a magnetic field H, the induction field is given as

$$B = H + 4\pi M \tag{14-4}$$

That is, the *induction field B* (which is the continuous field) is the sum of the *applied magnetic field H* and the field that arises from the *magnetization*, $4\pi M$. The fact that H is not continuous is depicted in Fig. 14-1(d), where the H field varies discontinuously at the end of the magnet. The H field outside the material (where $M = 0$) is stronger than inside because the B field is continuous:

$$H_{\text{in}} + 4\pi M = B = H_{\text{out}}$$

The unit of magnetic induction in the CGS system is the *gauss*. When a magnetic field of 1 oersted causes no magnetization (such as when it exists in a vacuum), it is equivalent to an induction of 1 gauss.

The energy associated with a magnetic field is proportional to the square of the magnetic field intensity H. This relation, which is particularly important to our understanding of magnetism in ferromagnetic materials, can be derived from our knowledge of magnetic poles. Suppose we have an infinitely long bar magnet with an intensity of magnetization M and a unit cross-sectional area. We can create a magnetic field by cutting the magnet in two, separating the two cut faces by a given distance—say, 1 cm. Now a magnetic field resides in the gap between the two cut faces of the magnet. Since there are no externally applied fields, and since the induction field is continuous, the field in the gap is $H = 4\pi M$.

Now let us consider how that magnetic field can be formed in another way. Imagine two plates of unit area separated by a unit distance (1 cm). Suppose that positive and negative poles are created in pairs between the plates and are separated and attached to the plates. The magnetic field in the gap associated with the poles becomes more intense as more and more poles are collected on the plates. When the number of unit poles on each plate is $m = M$, the magnetic field created will be the same as the field in the gap in the magnet. It is easy to calculate the work needed to form such a field. After some poles have already been transferred to the two plates, the magnetic field between the plates is $H = 4\pi m$. At this point, the work needed to bring dm unit poles to each plate is simply the product of the force on the poles ($H\,dm$) times the distance moved (1 cm): $dW = H\,dm$. Now, the process of bringing dm new poles to the two plates is equivalent to increasing the magnetic field by $dH = 4\pi\,dm$, so that $dW = (1/4\pi)\,H\,dH$. Therefore, the total work W required to create a magnetic field of intensity H in a unit volume is

$$W = \int_0^H \frac{1}{4\pi} H\,dH = \frac{1}{8\pi} H^2 \qquad (14\text{-}5)$$

We shall refer to this relation later when we consider the various energy factors that contribute to the structure and properties of magnetic domains.

Now we can define the quantities that describe the response of materials to magnetic fields. In the most general case, we may take the intensity of magnetization to be some function of the magnetizing field H. If we let the function be $M = f(H)$, we have from Eq. (14-4)

$$B = H + 4\pi f(H) \qquad (14\text{-}6)$$

For some materials such as diamagnetic and paramagnetic materials, the function $f(H)$ reduces to a linear function represented by

$$M = \chi H \qquad (14\text{-}7)$$

where χ is known as the *magnetic susceptibility*. We shall see that the susceptibility for paramagnetic materials is positive, whereas for diamagnetic materials it is negative. For materials that can exhibit spontaneous magnetization, the intensity of magnetization is usually a complex function of H, so that magnetic susceptibility has meaning only if we recognize that χ itself might depend on the strength of the applied field H.

The *magnetic permeability* μ is defined as the ratio of the induction field B to the magnetizing field H, which causes the induction

$$\mu = \frac{B}{H} \tag{14-8}$$

From Eq. (14-4) we see that

$$\frac{B}{H} = \mu = 1 + 4\pi \frac{M}{H} = 1 + 4\pi\chi \tag{14-9}$$

This relation shows that magnetic susceptibility and magnetic permeability are simply related and that when χ is a complex function of H, as in the case of soft ferromagnetic materials, μ also depends on H in a complicated way.

The response of various materials to the application of a magnetic field is indicated in Fig. 14-2. Here we see that diamagnetic bismuth and paramagnetic platinum both behave in a relatively simple fashion; the magnetization varies linearly

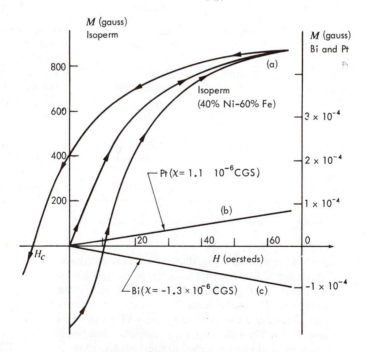

FIG. 14-2 Magnetization curves for (a) ferromagnetic Isoperm (40% Ni-60% Fe), (b) paramagnetic Pt, and (c) diamagnetic Bi

with the applied magnetic field. The magnetization induced in these materials is more than six orders of magnitude smaller than the magnetization developed in the ferromagnetic Ni-Fe alloy (Isoperm). In addition, the magnetization of diamagnetic and paramagnetic materials is reversible in the sense that the magnetic moment per unit volume depends only on the applied field and not on magnetic history. The ferromagnetic response is evidently more complicated. First, the intensity of magnetization for the ferromagnetic materials is extremely large even in relatively weak magnetic fields. Second, the magnetization curve is irreversible. It is evident that both the susceptibility and the permeability of the Isoperm vary with the applied magnetic field in a complex way. The magnetic hysteresis properties illustrated by this curve will be discussed at length in Section 14-7.

Some characteristics of the different types of magnetic behavior are summarized in Table 14-1. The table also indicates some typical materials for each of the different types of magnetic behavior.

TABLE 14-1

Characteristics of the Magnetic Susceptibility for Several Different Types of Magnetic Behavior

Type of Magnetic Behavior	Characteristics of Magnetic Susceptibility		Typical Materials
	Sign	Magnitude	
Dia-magnetism	Negative	Small* $\chi = $ constant	Organic materials, superconducting metals, and other metals (e.g., Bi)
Para-magnetism	Positive	Small $\chi = $ constant	Alkali and transition metals, rare earth elements
Ferro-magnetism	Positive	Large $\chi = f(H)$	Some transition metals (Fe, Ni, Co) and rare earth metals (Gd)
Antiferro-magnetism	Positive	Small $\chi = $ constant	Salts of transition elements (MnO)
Ferri-magnetism	Positive	Large $\chi = f(H)$	Ferrites ($MnFe_2O_4$, $ZnFe_2O_4$) and chromites

*Diamagnetic susceptibilities for superconducting metals are large.

EXAMPLE 14-2

Shown in the accompanying illustration are two magnets. One is a bar of Isoperm that has been permanently magnetized ($M = 400$ gauss), and the other is an electromagnet with Pt as the core. The bars have a cross-sectional area of 1 cm², and there are 1000 turns in the solenoid. Calculate the current, in amperes, that would be required to induce a magnetization of 400 gauss in the Pt rod. (i.e., what current would make the electromagnet equal in strength to the permanent magnet?)

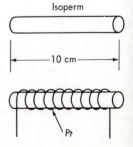

Isoperm

10 cm

Pt

Solution: From Fig. 14-2 the magnetic susceptibility of Pt is $\chi = 1.1 \times 10^{-6}$, so that to obtain $M = 400$ gauss a field of $H = M/\chi = 4 \times 10^8$ oersteds is required. From Eq. (14-3) we have

$$I = \frac{Hl}{0.4\pi n} = \frac{(4 \times 10^8)(10)}{0.4\pi(10^3)} = 3.18 \times 10^6 \text{ A}$$

This current is so large that joule heating would cause a normal solenoid wire to fail if the current flows for any significant period of time.

14-3 DIAMAGNETISM

All materials have the ability to display diamagnetic behavior. However, not all materials have a negative magnetic susceptibility. Some materials are also paramagnetic or ferromagnetic, and their diamagnetism becomes masked by these stronger effects. The statement that all materials, regardless of their state, may act diamagnetically is derived from the origin of diamagnetism, which is related to the properties of orbiting electrons. We shall now show how the intrinsic ability of a material to behave diamagnetically can be derived from a simple extension of Lenz's law. Consider a loop of electric current with its associated magnetic field. According to Lenz's law, when one attempts to change the magnetic field in the loop by the application of an external field, a current is induced into the loop in such a direction that the magnetic field tends to counteract the change imposed on the magnetic field by the external field. If the electrical resistance of the loop of wire were zero, the induced current would continue to flow as long as the external field were present. This is the case for electrons in orbits about nuclei. The effect of an imposed magnetic field is to induce into the atom a current that has a field that opposes the applied field. A very similar situation exists when a magnetic field is applied to a ring of superconducting material as noted in Chapter 12. Because of this diamagnetic interaction, all orbiting electrons will make a negative contribution to the susceptibility of the material in which they reside. A detailed analysis of the interaction between an externally applied magnetic field and the currents associated with orbiting electrons shows that the diamagnetic susceptibility for a material can be approximated by

$$\chi_{\text{dia}} = -4.7 \times 10^{-14} N Z \langle r^2 \rangle \tag{14-10}$$

where N is the number of atoms per unit volume, Z is the number of electrons in each atom, and $\langle r^2 \rangle$ is the mean square distance of the electrons to the nucleus. As expected, the negative susceptibility increases with both the atomic density and the number of orbiting electrons per atom. We may obtain an estimate of the order of magnitude for the diamagnetic susceptibility by noting that for most solids $N \approx 5 \times 10^{22}$ atoms/cm³ and $10 < Z < 100$. Taking $\langle r^2 \rangle \approx 10^{-16}$ cm², we have $\chi_{\text{dia}} \approx 10^{-5} - 10^{-7}$ in the CGS system of units. Table 14-2 shows that diamagnetic susceptibilities for most materials are indeed of the order of 10^{-6}.

It should be noted that the diamagnetic susceptibility is essentially temperature independent. This is in contrast to all of the other forms of magnetic behavior, which are characterized by temperature-dependent susceptibilities.

TABLE 14-2

Magnetic Susceptibilities of Some Diamagnetic and Paramagnetic Materials*

Material	$\chi \times 10^6$ (CGS units)	Material	$\chi \times 10^6$ (CGS units)
Al	+0.6	Ag	−0.20
Al_2O_3	−0.098	NaCl	−0.499
Sb	−0.87	Ta	+0.87
Be	−1.0	Sn	+0.025
Bi	−1.35	Ti	+1.25
Cd	−0.18	U	+2.6
Cu	−0.086	Zn	−0.157
Ga	−0.24	W	+0.28
Ge	−0.12	Zr	−0.45
Pt	+1.1	C_3H_6O (acetone)	−0.58
K	+0.52	C_6H_6 (benzene)	−0.712
Si	−0.13	C_2H_6O (Ethyl alcohol)	−0.744

*All values measured at room temperature

14-4 PARAMAGNETISM

Materials that exhibit paramagnetic behavior have a positive, reversible magnetic susceptibility that is independent of applied magnetic field and that varies reciprocally with temperature.* The comparison of paramagnetic and diamagnetic susceptibilities in Table 14-2 indicates that these forms of magnetic polarization are of comparable strength. Compared with ferromagnetism and ferrimagnetism, these susceptibilities are relatively weak.

Although a quantitatively accurate explanation of paramagnetism requires a quantum mechanical description, it is possible to give an intuitive account of paramagnetic behavior with a simple classical argument. In either case, the explanation of paramagnetism starts with the recognition that in some materials each atom can behave as if it were a tiny magnet. We shall see that the ability of an atom to exhibit a magnetic moment is related to unpaired orbital electronic motions and to the electronic spins of the outer electrons. In either case, we deal with a kind of current loop rather than a magnetic moment in the sense of Section 14-2. Before looking at the origin of atomic magnetism in detail, let us consider how a loop of current can behave as if it were a tiny magnet.

Consider the current loop and magnetic dipole shown in Fig 14-3. We imagine that the current loop and magnetic dipole are equivalent when they produce the same induction field. We treat the current loop as one turn taken from the long

---

*We shall see that some materials can be ferromagnetic or ferrimagnetic at low temperature and paramagnetic at high temperature. In these cases the paramagnetic susceptibility varies as $\chi \propto 1/(T - T_c)$, where T_c is the critical temperature above which the spontaneous magnetism disappears.

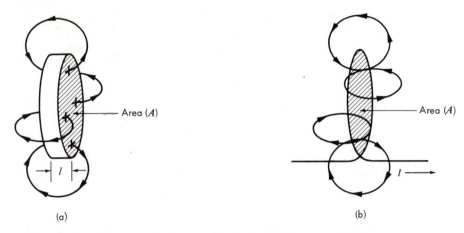

Area (A)

Area (A)

(a)

(b)

FIG. 14-3 Relation between (a) a permanent magnetic dipole and (b) a current loop

solenoid shown in Fig. 14-1(c), so that the magnetic field or induction field may be given by Eq. (14-3), or in this case

$$B = \frac{4\pi}{10}\frac{I}{l} \tag{14-11}$$

In the case of the magnet the induction field is

$$B = 4\pi M \tag{14-12}$$

where M is the intensity of magnetization or the pole density m/A. Remembering that the magnetic moment $\bar{\mu}$ is defined as

$$\bar{\mu} = MlA \tag{14-13}$$

with Eqs. (14-11) and (14-12), we have

$$\bar{\mu} = \frac{IA}{10} \tag{14-14}$$

We see, therefore, that the *loop of current* is equivalent to a *magnetic dipole moment*.

We have indicated that atomic dipole moments can arise out of orbital electronic motion and electron spin. Let us now consider the physical origin of these moments in more detail. First, it is clear that when an atom contains an uneven number of electrons, it will exhibit a net magnetic moment because of the magnetic moment associated with the orbital motion of the unpaired electron. This is illustrated in Fig. 14-4 for the case of lithium. In that illustration we see that the two $1s$ electrons move about the nucleus in opposite directions, resulting in cancellation of their associated magnetic fields. The single $2s$ electron is unpaired and hence provides a magnetic moment to the atom. The other, more important way in which atoms can develop magnetic moments is through unpaired electronic spins. Consider again the case of lithium as an example. According to the Pauli exclusion principle, the two electrons in the $1s$ state have opposite spins. This means that the magnetic moments associated with the spins of the $1s$ electrons are can-

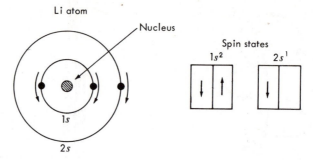

FIG. 14-4   Electronic orbital and spin states for electrons in Li

celled, and the spin of the unpaired $2s$ electron contributes to the magnetic moment of the atom. Because the alkali metals are characterized by their single s electron, it is not surprising that these metals are paramagnetic.

We may conclude this discussion by noting that materials have the capacity to exhibit paramagnetism if any of their component atoms exhibit a dipole moment. In the case of atoms of high atomic number, it is possible for diamagnetism to be stronger than paramagnetism and negative susceptibilities can be found even when there is an odd number of electrons per atom (bismuth, for example). In addition, as was indicated before, paramagnetism can be masked at low temperatures by the onset of spontaneous magnetization. Except for these cases, it is generally assumed that paramagnetism will be found in materials for which the atoms exhibit dipole moments.

Thus far our discussion has been confined to the atomic dipole aspects of paramagnetism. Let us now consider how an assemblage of atomic dipole moments can give rise to paramagnetic behavior. Consider the schematic illustration of atomic magnetic dipoles in Fig. 14-5. When a magnetic field is not applied to the material, the atomic dipoles point in random directions and undergo thermal fluctuations. The reason the dipoles assume random orientations is similar to the

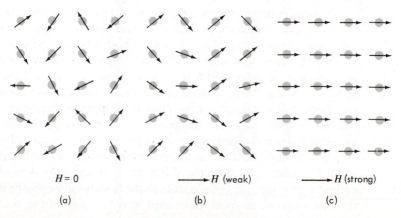

FIG. 14-5   Schematic representation of the atomic magnetic dipoles in a paramagnetic material as a function of the applied magnetic field H

reason that crystals tend naturally to become imperfect. That is, the entropy of the system is allowed to increase and hence the free energy decreases when the dipoles adopt random orientations. Now, when a weak magnetic field is applied [as in Fig. 14-5(b)], the atomic dipoles no longer adopt random orientations. The reason for this is that there is an interaction energy between the dipoles and the applied field that tends to favor alignment with the field. This energy is expressed as

$$E_{\text{dipole}} = -\bar{\mu} H \cos \theta \qquad (14\text{-}15)$$

where θ is the angle between the dipole axis and the applied field. It is clear from Eq. (14-15) that the total energy of this assemblage of dipoles would be minimum if they were to align perfectly with the field, but the entropy effects prevent that from happening by randomizing the orientations. One may imagine that the thermal fluctuations tend to make them randomly oriented. The effect of the field is to bring the dipoles into partial alignment. When a strong magnetic field is applied, the dipoles are forced to become more perfectly aligned, as shown in Fig. 14-5(c). In principle, perfect alignment would require infinitely large magnetic fields.

The paramagnetic susceptibility evidently depends on temperature. Simply stated, at low temperatures the randomizing effects of thermal agitation are small, and the dipoles can therefore be aligned more easily by the applied field. As a consequence, it can be shown that the susceptibility should vary as

$$\chi = \frac{C}{T} \qquad (14\text{-}16)$$

As mentioned before, when spontaneous magnetism exists below T_c, the temperature dependence of the paramagnetic susceptibility is expressed as

$$\chi = \frac{C}{T - T_c} \qquad (14\text{-}17)$$

where $T - T_c$ replaces the absolute temperature of Eq. (14-16). This relationship, called the *Curie-Weiss law*, is illustrated in Fig. 14-6 for nickel.

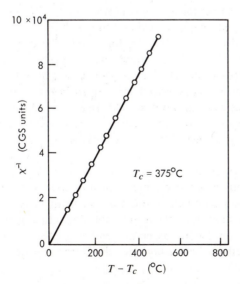

$T_c = 375°C$

FIG. 14-6 Experimental confirmation of the Curie-Weiss law for pure nickel

EXAMPLE 14-3

The magnetic susceptibility of Fe is about 2.5×10^{-4} at 900°C, and the Curie temperature [T_c in Eq. (14-17)] is 770°C. Calculate the susceptibility at 850°C and 950°C. The *measured* values for the susceptibility are $\chi_{850} = 4 \times 10^{-4}$ and $\chi_{950} = 2.5 \times 10^{-5}$. Explain why your estimate for 850°C is reasonably accurate while the calculated value for 950°C differs considerably from the measured value.

Solution:

Above the Curie temperature Fe is paramagnetic. Using Eq. (14-17), we have

$$\chi = \frac{C}{T - T_c}$$

At 900°C

$$\chi = 2.5 \times 10^{-4} = \frac{C}{900 - 770}; \qquad C = 3.25 \times 10^{-2}$$

Now, at 850°C

$$\chi_{850} = \frac{3.25 \times 10^{-2}}{850 - 770} = 4.06 \times 10^{-4}$$

and at 950°C

$$\chi_{950} = \frac{3.25 \times 10^{-2}}{950 - 770} = 1.8 \times 10^{-4}$$

The calculated value of χ_{850} compares favorably with the measured value of 4×10^{-4}. The difference between χ_{950} (calculated) and the measured value is caused by a phase change from BCC to FCC which occurs at 910°C. Thus, the Curie-Weiss law cannot be extrapolated to include new phases. The constant C corresponds to a given phase (BCC in this problem) and cannot be used for a different crystal structure.

14-5 SPONTANEOUS MAGNETIZATION

Thus far in this chapter we have been concerned only with types of magnetization induced by an applied magnetic field and in which the magnetization persists only as long as the field is maintained. In this section we turn our attention to *spontaneous* magnetism. This form of magnetism provides for most of the important engineering applications of magnetic phenomena. As the name implies, spontaneous magnetism is characterized by a polarization that is retained in a sample even in the absence of an applied field. In this section we explore briefly the physical basis for spontaneous magnetism or permanent magnetization, and in the next section we extend these ideas to account for the existence and properties of *magnetic domains*.

The mere existence of permanent magnets is the most direct and persuasive evidence for spontaneous magnetism. We shall now consider the basic cause of this important magnetic phenomenon. Since all forms of magnetism arise out of either orbital electronic motion or electronic spin (neglecting nuclear magnetism), an understanding of spontaneous or permanent magnetism must come from a study of atomic magnetism. Let us consider the ferromagnetic elements Fe, Ni, and Co of the first transition series. Since these materials exhibit permanent magnetism at sufficiently low temperature, it is evident that the atomic dipoles can be spontaneously aligned. Figure 14-7 is a schematic representation of the alignment

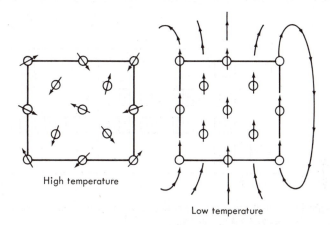

High temperature

Low temperature

FIG. 14-7   Schematic illustration of magnetic dipoles and external field for a material that is ferromagnetic at low temperatures and paramagnetic at high temperatures. $H_{\mathbf{applied}}$ is zero in both cases.

of atomic magnetic dipoles in a permanent magnet. The fundamental question that arises here is, "Why are the dipoles spontaneously aligned?" This basic question was posed in the early part of this century by Pierre Weiss, who postulated the existence of an *internal molecular magnetic field*, which causes the dipole alignment. It was not until after the development of quantum mechanics in the 1920s that an adequate explanation for this hypothetical internal field could be given. The basic cause of spontaneous magnetism is now understood, and it involves an electronic interaction called the *exchange interaction*. While this interaction is quantum mechanical in character and cannot be fully explained here, we can use classical terms to describe how it provides for the spontaneous dipole alignment. The exchange energy refers to that part of the electrostatic energy of a system of electrons which depends on the spin states of neighboring electrons. In the case of the ferromagnetic metals Fe, Ni, and Co, the exchange energy is lowest when the electronic spins of some of the $3d$ electrons in adjacent atoms are aligned parallel to each other. Consequently, ferromagnetic atoms in close proximity spontaneously develop parallel magnetic alignment. This condition is equivalent to local spontaneous magnetization.*

Our treatment of spontaneous magnetism thus far has been concerned only with the nature of the forces that cause the spontaneous dipole alignment. Let us now consider how the magnetic moment of a sample of, say, iron is derived from

---

*We shall see in the section on magnetic domains that it is possible for a material to possess spontaneous magnetism locally and yet not exhibit a macroscopic magnetic moment.

the magnetic moments of the spins of the 3d electrons. The electronic structure of iron is shown schematically in Fig. 14-8. We see that the spins of the electrons in the inner electron shells are all paired. That is, there are equal numbers of "up" spins and "down" spins. Because of the electron occupancy of the 3d shell there is

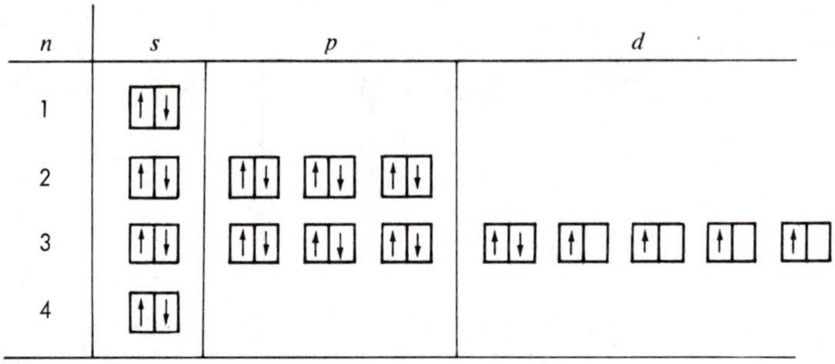

FIG. 14-8   Representation of the electronic structure of Fe. With spectral notation, the structure is $1s^2\ 2s^2\ 2p^6\ 3s^2\ 3p^6\ 4s^2\ 3d^6$.

a net number of four electrons with parallel spin. This means that the magnetic moment of an iron atom is equivalent to the magnetic moment of four spinning electrons. The magnetic moment associated with electron spin is 0.927×10^{-20} in the CGS system of units and is defined as 1 *Bohr magneton*. Thus, by this defini-tion the magnetic moment of an iron atom is 4 Bohr magnetons. Actually, in metal-lic iron the close proximity of the iron atoms not only causes the atomic magnetic moments to align spontaneously but also affects the electronic structure in such a way that the atomic magnetic moment is reduced to 2.2 Bohr magnetons. Similar effects cause the atomic magnetic moment in metallic Ni and Co to be smaller than the corresponding magnetic moment of the free atom. Now, if the atomic magnetic moments are perfectly aligned, as they are at very low temperatures, the intensity of magnetization or magnetic moment per unit volume can be computed directly from a knowledge of the crystal structure and the atomic magnetic moment.

EXAMPLE 14-4   The lattice parameter of BCC Fe at room temperature is 2.86 A.   Calculate the intensity of magnetization when all the atomic dipoles are in perfect alignment.

Solution: The intensity of magnetization is defined as the magnetic moment per unit volume. Since each atom in metallic iron has a magnetic moment of 2.2 Bohr magnetons,* and since there are two atoms per unit cell, the saturation magnetization is

$$M = \frac{2(2.2)(0.927 \times 10^{-20})}{(2.86 \times 10^{-8})^3} = 1750 \text{ gauss}$$

The measured value is 1714 gauss.

---

*Actually, the value of 2.2 Bohr magnetons comes from an experimental knowledge of the saturation magnetization.

The calculation of intensity of magnetization in Example 14-4 assumes that the dipoles are in perfect alignment. Actually, this is the case only at 0°K. At any finite temperature, thermal energy causes the magnetic dipoles to fluctuate about a particular direction and therefore to deviate from perfect alignment. The exchange energy that produces the dipole alignment is counteracted by the randomizing effects of thermal energy or entropy. The randomizing effects are more important at high temperature and eventually cause a significant reduction of the intensity of magnetization, shown schematically and graphically in Fig. 14-9. At the

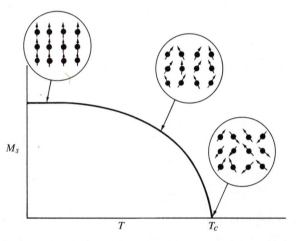

FIG. 14-9 Effect of temperature on the saturation magnetization M_s of a ferromagnetic material below the Curie temperature T_c

Curie temperature T_c, the thermal fluctuations of the magnetic dipoles become so significant that the effects of the exchange energy are no longer evident and ferromagnetism disappears altogether. Above the Curie temperature, the material behaves paramagnetically since the dipoles are brought into partial alignment only by the application of external fields. Since spontaneous magnetism does not exist above T_c, it follows that the interesting and useful ferromagnetic properties are not available at these temperatures.

For simplicity, we have chosen to discuss spontaneous magnetism by considering the special case of ferromagnetism. We shall now consider briefly the other forms of spontaneous magnetism: *antiferromagnetism* and *ferrimagnetism*.

The electronic structures of Cr and Mn are such that each atom exhibits a magnetic moment in much the same way as Fe (see Fig. 14-8). In these metals, however, the nature of the exchange interaction between electrons of neighboring atoms is such that the atomic magnetic dipoles of adjacent atoms are aligned antiparallel. As in the case of ferromagnetism, the perfection of the dipole alignment depends on temperature and there is a critical temperature, called the *Néel temperature*, above which the spontaneous alignment is completely destroyed. For Cr and Mn, the Néel temperatures are 1673° and 95°K, respectively. Antiferromagnetism is also found in nonmetallic substances such as MnO, NiO, and

MnS. In these compounds the magnetic moments of the transition metal ions are aligned in an antiparallel manner. Consider the structure of MnO shown in Fig. 14-10. MnO has the NaCl crystal structure, which means that the oxygen ions form

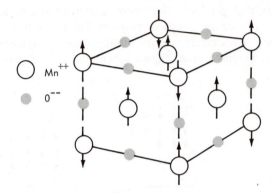

FIG. 14-10 Magnetic structure of MnO (only part of the magnetic unit cell is shown)

an FCC array and the manganese ions are fitted into the interstices as shown. If sufficient numbers of ions are considered, it is evident that half of the Mn dipoles point in the "up" direction while half point in the "down" direction. The antiparallel dipole arrangement in this material and in other antiferromagnetic materials comes from the nature of the exchange interaction. That is, in these materials the electrostatic energy of some of the outer electrons is lowest if the magnetic dipoles of adjacent cations are antiparallel.

Although antiferromagnets are interesting to the solid state scientist, they do not yet have important engineering applications. The primary reason for this is that these materials do not exhibit a net spontaneous magnetic moment. The antiparallel dipole arrangement causes perfect cancellation of the internal magnetic moments. For this reason we shall not discuss antiferromagnetic properties in further detail.

While antiferromagnetic materials are of limited practical utility, *ferrimagnetic materials* are of tremendous importance. *Ferrites* are an extremely important member of the ferrimagnetic family. The most important feature of these materials, and the feature that distinguishes them from ferromagnetic metals, is that they are insulating; that is, the changing magnetic fields that accompany magnetization do not induce significant electric currents. Consequently, eddy current losses ($I^2 R$) for these materials are extremely small compared to the losses for ferromagnetic metallic conductors. This feature allows ferrites to be used in very-high-frequency applications with very little power loss.

Ferrimagnetic materials or ferrites are oxides of transition metals that have the following chemical composition:

$$\text{MO} \cdot \text{Fe}_2\text{O}_3 \quad \text{or} \quad \text{M}^{2+} \text{Fe}_2^{3+} \text{O}_4^{2-}$$

where M represents a divalent cation. For example, nickel ferrite would be represented by $NiO \cdot Fe_2O_3$ and iron ferrite or magnetite by $FeO \cdot Fe_2O_3$ or Fe_3O_4. The most common magnetic ions in ferrites are Mn^{2+}, Fe^{2+}, Co^{2+}, Ni^{2+}, Cu^{2+}, and Zn^{2+} from the first transition series of the periodic table. The magnetic moments for these ions and for Fe^{3+} are shown schematically in Table 14-3. As we saw when we

TABLE 14-3

Electronic Structure and Magnetic Moments of Ions from the Transition Metal Series

Ion	Number of Electrons	Electronic Structure 3d Shell*					Ionic Magnetic moment (Bohr Magnetons)
Fe^{3+}	23	↑	↑	↑	↑	↑	5
Mn^{2+}	23	↑	↑	↑	↑	↑	5
Fe^{2+}	24	↑↓	↑	↑	↑	↑	4
Co^{2+}	25	↑↓	↑↓	↑	↑	↑	3
Ni^{2+}	26	↑↓	↑↓	↑↓	↑	↑	2
Cu^{2+}	27	↑↓	↑↓	↑↓	↑↓	↑	1
Zn^{2+}	28	↑↓	↑↓	↑↓	↑↓	↑↓	0

*All of these ions have an inner electronic structure of 18 electrons represented by $1s^2\ 2s^2\ 2p^6\ 3s^2\ 3p^6$, which does not exhibit a net magnetic moment.

studied the electronic structure of the iron atom, the electrons in the 3d shell assume parallel spin states whenever possible. In all the transition metal ions the 4s level is not occupied and the electrons partially fill the 3d shell. The magnetic moments for these ions vary from 5 Bohr magnetons for Fe^{3+} and Mn^{2+} to zero magnetic moment for Zn^{2+}. When these ions reside in ferrites, they retain their free ion magnetic moment, unlike the case of iron atoms in metallic iron or any of the other ferromagnetic metals. We may now consider how the properties of magnetic oxides are related to the moments of these individual ions.

The crystal structure of ferrites is dominated by the close-packed cubic arrangement of oxygen ions. The cations are located in the octahedral (O) and tetrahedral (T) interstitial sites. When the divalent cations (such as Ni^{2+}) reside in the T sites and the iron (Fe^{3+}) are in the O sites, the structure is called the *normal spinel* and it is not spontaneously magnetic. $ZnO \cdot Fe_2O_3$ is an example of such a nonmagnetic ferrite. The magnetic ferrites have the *inverted spinel* structure in which the divalent ions reside in the O sites, and the trivalent Fe ions are equally divided among the O and T interstitial positions. The magnetic coordination of the cations in both inverted and normal spinels is illustrated in Table 14-4, along with a similar illustration for magnetite (Fe ferrite).

In magnetic ferrites there is an antiferromagnetic interaction between the Fe^{3+} ions on the O sites and those on the T sites, causing the corresponding internal magnetic moments to cancel out. This cancellation allows the magnetic moment of the ferrite to be determined by the divalent cation. Thus, the magnetic moment

TABLE 14-4

Magnetic Coordinations of the Cations in Ferrites

Ferrite	Structure	Interstitial Site Occupancy		Net Magnetic Moment per Molecule (Bohr Magnetons)
		Tetrahedral (*T*)	Octahedral (*O*)	
$NiO \cdot Fe_2O_3$	Inverted spinel	$Fe^{3+}\downarrow\downarrow\downarrow\downarrow\downarrow$	$Ni^{2+}\uparrow\uparrow$, $Fe^{3+}\uparrow\uparrow\uparrow\uparrow\uparrow$	$2(\uparrow\uparrow)$
$ZnO \cdot Fe_2O_3$	Normal spinel	Zn^{2+}	$Fe^{3+}\uparrow\uparrow\uparrow\uparrow\uparrow$, $Fe^{3+}\downarrow\downarrow\downarrow\downarrow\downarrow$	0
$FeO \cdot Fe_2O_3$	Magnetite (inverted spinel)	$Fe^{3+}\downarrow\downarrow\downarrow\downarrow\downarrow$	$Fe^{2+}\uparrow\uparrow\uparrow\uparrow$, $Fe^{3+}\uparrow\uparrow\uparrow\uparrow\uparrow$	$4(\uparrow\uparrow\uparrow\uparrow)$
$CoO \cdot Fe_2O_3$	Inverted spinel	$Fe^{3+}\downarrow\downarrow\downarrow\downarrow\downarrow$	$Co^{2+}\uparrow\uparrow\uparrow$, $Fe^{3+}\uparrow\uparrow\uparrow\uparrow\uparrow$	$3(\uparrow\uparrow\uparrow)$

of each molecule of $CoO \cdot Fe_2O_3$ should be equal to the magnetic moment of Co^{2+}. Figure 14-11 shows that this is indeed the case. The magnetic moments of the ferrites can be accurately predicted from the ionic moments given in Table 14-3. The computation of the intensity of magnetization for these materials then reduces to a problem of calculating the volume associated with each molecule of the fer-

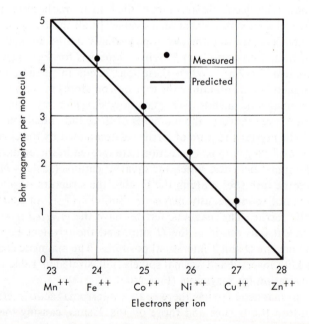

FIG. 14-11 Variation of the molecular magnetic moment of ferrites of the iron-group elements as a function of the electrons per ion

rite. Since there are four oxygen ions in the molecule and since these ions adopt an FCC configuration, the magnetic moment per unit volume is easily computed from information about the distance between oxygen ions or the lattice parameter.

EXAMPLE 14-5

The distance between the nearest neighbor oxygen ions in nickel ferrite is 2.95 A. Calculate the saturation intensity of magnetization.

Solution:

Each molecule of $NiFe_2O_4$ involves four oxygen ions. Therefore, we can consider a small "unit cell" cube with edge $a = \sqrt{2}\ (2.95) = 4.17$ A. Taking the molecular magnetic moment to be 2.0 Bohr magnetons (see Fig. 14-11), the saturation magnetization is

$$M_s = \frac{(2.0)(0.927 \times 10^{-20})}{(4.17 \times 10^{-8})^3} = 255 \text{ gauss}$$

To illustrate the extent to which the intensity of magnetization of a ferrite can be changed and precisely controlled, consider the magnetic properties of the ferrite alloy $Ni_{0.9}Co_{0.1}O \cdot Fe_2O_3$. This material can be produced by mixing nine parts $NiO \cdot Fe_2O_3$ with one part $CoO \cdot Fe_2O_3$. The mixed ferrite would probably be fired at a high temperature to allow the Ni^{2+} and Co^{2+} ions to interdiffuse into the two ferrites. This mixed ferrite has the inverted spinel structure, and the Ni^{2+} and Co^{2+} ions are located in the octahedral sites. The average magnetic moment for each molecule is $(0.1)(3) + (0.9)(2) = 2.1$ Bohr magnetons. The intensity of magnetization for this mixed oxide is computed by simply dividing the molecular magnetic moment (2.1 Bohr magnetons) by the volume associated with each molecule. It is clear from this discussion that the intensity of magnetization of this mixed oxide would fall between the values for the component oxides and would have a value in direct proportion to the atomic fractions of the two ferrites.

14-6 MAGNETIC DOMAINS

In the previous section we showed that the magnetic moments of neighboring atoms in ferromagnetic materials are spontaneously aligned below the Curie temperature. In this section we shall look at the consequences of this alignment when a large assemblage of atoms is considered. We begin by noting that ferromagnetic metals such as Fe and Ni can be demagnetized even below the Curie temperature. This leads directly to the question of how an assemblage of spontaneously magnetic atoms can be demagnetized. This problem was first recognized in 1907 by Weiss, who resolved the question by postulating that a bulk sample is subdivided into many different regions, called *domains*, which are magnetized in different directions. Figure 14-12 shows how the domain postulate permits a spontaneously magnetic material to be demagnetized. When the magnetic alignment is parallel over the entire sample, an external magnetic field is exhibited and the whole sample is said to be magnetized to saturation. On the other hand, when the sample is subdivided into domains with differently oriented magnetizations, the external field

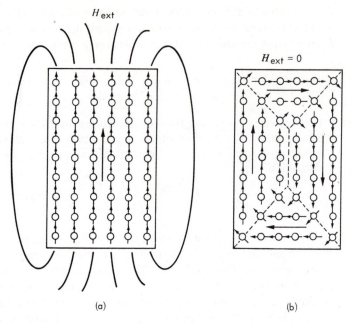

FIG. 14-12 Schematic representation of (a) the external magnetic field, H_{ext}, for a magnetized sample, and (b) magnetic domains

can be eliminated and the sample will behave as if it were not magnetized. The domain postulate, then, allows the spontaneous dipole alignment to persist over most of the sample and at the same time permits the sample to be demagnetized in a macroscopic sense. The domain postulate by Weiss was an extremely good guess, later proved correct when Francis Bitter and others first observed magnetic domains in ferromagnetic solids. The domain structure of a single crystal of Fe–3% Si is illustrated in Fig. 14-13.

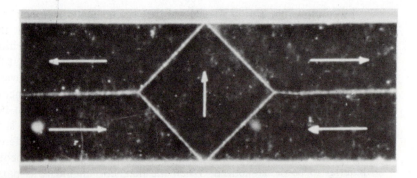

FIG. 14-13 Magnetic structure of a single crystal of Fe-3% Si as revealed with the Bitter technique. Photo courtesy R. W. DeBlois, The General Electric Research and Development Center, and C. D. Graham, the University of Pennsylvania. Used with permission

The properties of domains in spontaneously magnetic materials are usually very important in explaining magnetic properties, and so it is worthwhile to study their characteristics and properties before addressing the problems of ferromagnetic properties. It should be noted at the outset that the domain theory of spontaneously magnetized materials is general and can be applied equally well to both ferromagnetic metals and ferrimagnetic oxides.

We start this discussion of the physics of magnetic domains by discussing the various factors that contribute to the total magnetic energy of a given magnetic material. We shall see that some generalizations about these energy factors will enable us to explain many magnetic phenomena and will permit a rational discussion of magnetic hysteresis.

Exchange Energy

In the previous section we saw that the driving force for spontaneous magnetization is the *exchange energy*. It will be recalled that the exchange energy in a ferromagnetic material is lowest when the magnetic dipoles are aligned in a parallel fashion (antiparallel for the special case of antiferromagnetic materials). When a group of atoms is brought together to form a ferromagnetic crystal, the exchange energy (or total electrostatic energy) is lowest when all the atomic dipoles are aligned parallel. This means that if only the exchange energy is considered, each and every atomic dipole within the solid would be expected to adopt the same orientation. The effects of the exchange energy can be likened to strong springs connecting all the dipoles and forcing them into parallel alignment. Any perturbation in this perfect alignment would occur at the expense of the exchange energy or, in our analogy, would be resisted by the springs holding the dipoles in line.

It is clear from Fig. 14-12 that if the exchange energy in a magnetic solid were minimized absolutely, all of the dipoles would adopt precisely the same orientation and an external magnetic field H_{ext} would arise. We saw in Section 14-2 that the energy associated with a magnetic field is equal to $1/8\pi \int_V H^2 \, dV$, where H is the field and the integral is taken over the space occupied by the field. It is therefore evident that the total magnetic energy of a magnetic sample includes not only the exchange energy but the energy of the external field as well. In fact, we shall see that the lowest energy is achieved by reducing the energy of the external magnetic field at the expense of the exchange energy. We now turn our attention to the energy term that relates to the energy of the external magnetic field, the *magnetostatic energy*.

Magnetostatic Energy

As the exchange energy is the basic driving force for spontaneous magnetization, so the *magnetostatic energy* is the driving force for *domain formation*. We shall show in this section that the energy of the external magnetic field or the magnetostatic energy of a magnetized sample can be reduced through the formation of domains.

Consider the single domain sample shown in Fig. 14-14(a) with its associated

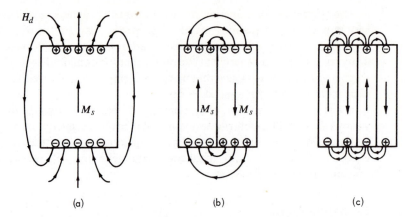

FIG. 14-14 Illustration of the external magnetic fields associated with various domain patterns

magnetic field. Since the external magnetic field extends from one end of the sample to the other, it is evident that the sample itself is sitting in its own magnetic field. The external field acts on the sample in such a way as to demagnetize it. For this reason it is common to call the external field the *demagnetizing field*, H_d. We now turn our attention to the energy of the demagnetizing field, the *magnetostatic energy*.

Let the intensity of magnetization in the sample be M_s. From our discussion of magnetic poles we know that the abrupt change in the normal component of the magnetization from M_s to zero at the top and bottom faces of the sample must be accompanied by an external (demagnetizing) field. As before, we imagine that the field lines arise from magnetic poles that exist at the top and bottom surfaces. The intensity of the demagnetizing field just outside the sample at the top and bottom surfaces is $H_d = 4\pi M_s$, and from Eq. (14-5) the energy associated with a tiny volume V in this region is

$$dE_{\text{mag}} = \frac{1}{8\pi} H_d^2 \, dV \tag{14-18}$$

or

$$dE_{\text{mag}} = 2\pi M_s^2 \, dV \tag{14-19}$$

Now the total magnetostatic energy of the sample shown in Fig. 14-14(a) is obtained by integration of the field energy over the entire space surrounding the sample;

$$E_{\text{mag}} = \frac{1}{8\pi} \int_V H_d^2 \, dV \tag{14-20}$$

Unfortunately, the magnetic field varies in a complicated way with position, and the magnetostatic energy cannot be expressed in more convenient form. Although we cannot compute the magnetostatic energy precisely, we can develop some insight into how the magnetostatic energy is changed when domains are formed.

Consider the sample shown in Fig. 14-14(b), which is subdivided into two domains. The intensity of magnetization in the two domains is still M_s, but the polarizations of the two domains are antiparallel. Since the intensity of polarization in the two regions is the same as in Fig. 14-14(a), the intensity and energy of the demagnetizing field just outside the sample are the same as before. However, because of the magnetization reversal on one side of the sample, the external or demagnetizing field can extend directly between the positive and negative poles on the same surface. This causes the external magnetic flux to be confined to the vicinity of the ends of the magnetic domains. Consequently, the magnetostatic energy is much lower than it was for the single domain since the volume containing the magnetic field is greatly reduced. If we permit more than two domains to be formed as shown in Fig. 14-14(c), the extent of the demagnetizing field is reduced even more and the total magnetostatic energy is even less.

EXAMPLE 14-6

A single crystal of cobalt shaped like a cube has the domain structure shown in the accompanying cutaway view. Assume that the demagnetizing fields are confined to semicylindrical regions on the ends of the sample, and assume that H_d is constant in those regions. Calculate the total magnetostatic energy for the sample. The intensity of magnetization of cobalt is 1420 gauss.

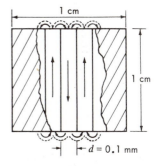

Solution:

The number of semicylindrical field regions for the sample is

$$N = 2\left(\frac{1}{d}\right) = 200$$

counting the fringes on both ends. The intensity of the demagnetizing field H_d is $4\pi M_s$. Therefore, the energy density in the demagnetization field is

$$E = \frac{1}{8\pi}(4\pi M_s)^2 = 2\pi M_s^2$$

The total magnetostatic energy is

$$E_{\text{mag}} = N\frac{1}{2}\left(\frac{\pi d^2}{4}\right)2\pi M_s^2 = \frac{\pi^2}{4}Nd^2 M_s^2$$

$$= \frac{\pi^2}{2}dM_s^2$$

$$= 0.99 \times 10^5 \text{ ergs}$$

It is clear from the above discussion and Example 14-6 that the reduction in the magnetostatic energy is in inverse proportion to the number of domains. As the domain size or width approaches zero, the residual magnetostatic energy also approaches zero. However, as more and more domains and domain boundaries are introduced into the crystal, the energy begins to rise again, as there is a *surface energy* associated with the domain boundaries. However, before considering the factors that contribute to the energy of domain boundaries, we must first consider the *magnetocrystalline anisotropy energy*.

Magnetocrystalline Anisotropy Energy

The magnetization in a ferromagnetic crystal tends to lie along certain crystallographic directions called "easy" directions of magnetization. In BCC iron the magnetization lies in the $\langle 100 \rangle$ direction, as shown in Fig. 14-15. The easy direc-

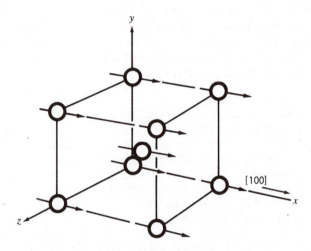

FIG. 14-15 Illustration of the easy magnetic axes for BCC Fe

tions of magnetizations for Ni (FCC) and Co (HCP) are $\langle 111 \rangle$ and $\langle 001 \rangle$, respectively. Other crystallographic directions are called "hard" directions, since it is more difficult to force the magnetization vectors to lie in these directions.

The tendency of the magnetization to lie in easy crystallographic directions is caused by the interactions between the magnetic dipoles and the crystal lattice. Evidently this interaction energy, called the *magnetocrystalline anisotropy energy*, is lowest when the polarization is in the easy crystallographic direction. Although it is possible for the dipoles to lie in hard directions in the crystal, such dipole orientations can be induced only by the application of strong magnetic fields. We may therefore consider the magnetocrystalline anisotropy energy to provide strong internal forces that hold the dipoles in the easy crystallograpahic directions.

The existence of crystal anisotropy forces and easy and hard directions of magnetization can be observed by comparing the magnetization curves for different crystallographic directions. Consider the two magnetization curves shown in Fig. 14-16 for a single crystal of cobalt. One of the curves corresponds to magnetization along the easy [001] direction, while the other is for magnetization along the hard [100] direction. It is evident that a much larger magnetic field is needed to fully magnetize the crystal in the hard direction. Also, it is clear from the shapes of the magnetization curves in the two directions that the energy needed to magnetize the crystal in the [100] direction is greater than the corresponding energy for magnetization in the [001] direction. Recalling the argument that led to Eq.

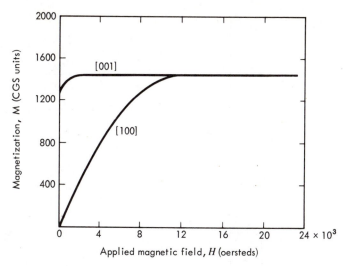

FIG. 14-16 Magnetization curves for cobalt single crystals in the easy [001] and hard [100] directions

(14-5), we can show that the energy needed to magnetize a unit volume of crystal is given as $\int_0^M H\, dM$. Consequently, the difference in energy of magnetization in the two directions, or the difference in magnetocrystalline anisotropy energy, is simply the area between the two curves in Fig. 14-16. This means that the energy density needed to rotate the magnetic dipoles from the easy [001] direction to the hard [100] direction is simply

$$\int_0^M (H_{[100]} - H_{[001]})\, dM$$

where $H_{[100]}$ and $H_{[001]}$ are the magnetic fields needed to produce magnetization in the [100] and [001] directions, respectively. Computing the area between the curves in Fig. 14-16, we find the difference in the crystal anisotropy energy per unit crystal volume to be 5.6×10^6 ergs.

Having discussed crystal anisotropy forces, we may now return to the problem of estimating the energy of a domain boundary.

Bloch Wall Energy

A *domain boundary* or *Bloch wall* is an interface that separates two regions of a crystal that are polarized in different directions. Although it is geometrically possible for the polarization on either side of a Bloch wall to be pointing in arbitrary directions, the energies of all but a few boundaries are so large that they almost never exist. The types of Bloch walls that commonly occur in iron are shown in Fig. 14-17. In general, there can be either *tilt* or *twist* domain boundaries. In iron, the twist boundaries can have misorientations of either 180° or 90°, as shown in the figure. These two misorientations arise because of the anisotropy forces that

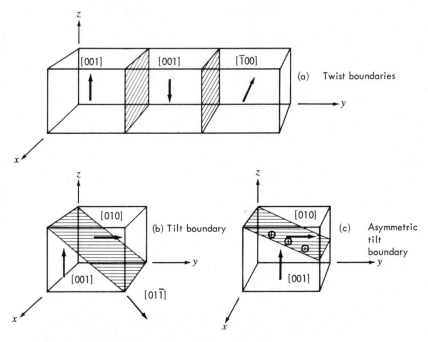

FIG. 14-17 Illustration of tilt and twist Bloch walls that occur in BCC Fe

hold the polarizations in the $\langle 100 \rangle$ directions. It is clear that a 180° wall is formed when the polarizations in adjacent domains are antiparallel. For example, a 180° Bloch wall might lie on a (010) plane separating domains polarized in the [001] and [00$\bar{1}$] directions. If the two domains were polarized in the [00$\bar{1}$] and [$\bar{1}$00] directions, the boundary between them would be a 90° twist boundary. Still another Bloch wall in cubic iron is the 90° tilt boundary. As shown in the figure, this type of boundary would lie on a {110} plane [(011) in the figure] and would separate domains polarized in the easy directions [[001] and [010] in the figure].

The Bloch walls that commonly occur are ones that exhibit no magnetic poles. Notice that for the Bloch walls shown in Fig. 14-17(a) and (b) the component of the magnetization normal to the boundary is the same in each domain. This means that the normal component of the magnetization does not change across the interface and no magnetic poles or demagnetizing fields arise. Inspection will show that if the Bloch walls are allowed to rotate out of the low index planes in which they lie (about the [100] axis), the normal component of the magnetization changes abruptly, magnetic poles arise, and a demagnetizing field is produced. This situation is illustrated in Fig. 14-17(c).

We now wish to consider the factors that contribute to the surface energy of a Bloch wall. The anatomy of a 180° Bloch wall is shown schematically in Fig. 14-18. There are essentially two forces to be considered: the *exchange energy* and the *magnetocrystalline anisotropy energy*. As discussed earlier, the exchange energy tends to force the dipoles into parallel alignment. However, in the region of a Bloch wall, the dipoles are forced to be misaligned. Therefore, part of the

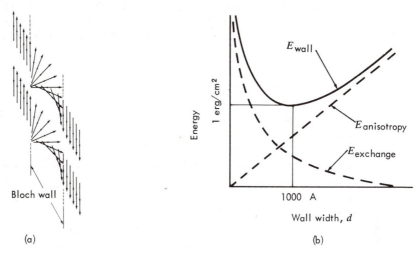

FIG. 14-18 Illustration of (a) magnetic dipole arrangments in a Bloch wall; and (b) relation between exchange energy, anisotropy energy and wall energy, and the wall width. The equilibrium wall width is about 1000 A, and the Bloch wall energy is about 1 erg/cm².

Bloch wall energy is related to the exchange energy associated with the forced misalignments of the dipoles in the wall. The magnetocrystalline anisotropy energy is lowest when the dipoles lie in an easy direction. It is evident from the figure that in a Bloch wall the dipoles are forced into hard directions. The magnetocrystalline anisotropy energy associated with having the dipole in hard directions represents the other part of the domain wall energy.

Let us now consider how these two contributions to the wall energy are affected by the width of the wall. If the wall were very wide, the dipoles could be nearly parallel, because the 180° misorientation would be apportioned to many dipole pairs and the exchange energy would consequently approach zero. However, a very wide Bloch wall would have an extremely high crystalline anisotropy energy, because all the dipoles in the wall would be forced to lie away from easy directions. Thus, we may consider that the exchange energy tends to widen the wall, while the anisotropy energy tends to narrow the wall. The competing effect of these two energies gives rise to an equilibrium wall width of about 1000 A at which the sum of these two energies is minimum. In iron, a Bloch wall of equilibrium width has a surface energy of about 1 erg/cm². The equilibrium width determination discussed here is shown schematically in Fig. 14-18.

It is now clear that while the introduction of Bloch walls into crystals does permit a reduction of the magnetostatic energy, it is also accompanied by an increase in the total surface energy associated with the walls.

EXAMPLE 14-7   Show why tilt boundaries in cobalt are unstable and almost never appear.

Solution: At room temperature cobalt has the HCP crystal structure with an easy axis of magnetization along the [001] direction (*c* axis). Since there is but one easy magnetic direction, it follows that polarizations in adjacent domains can only be antiparallel. Therefore, only 180° twist boundaries are possible, and tilt boundaries will not be formed.

Magnetostrictive Energy

When a ferromagnetic material is magnetized, it undergoes slight dimensional changes. To a first approximation the sample either extends or contracts in the direction of magnetization. We shall let λ, called the *magnetostriction constant*, represent the longitudinal strain induced by the magnetization, and λ_s will correspond to the strain that accompanies saturation in a given direction. Magnetization of a crystal of iron to saturation along the $\langle 100 \rangle$ direction causes a positive longitudinal strain of 19.5×10^{-6} to be induced: $\lambda_{\langle 100 \rangle} = 19.5 \times 10^{-6}$. Magnetization in the hard $\langle 111 \rangle$ direction produces a contraction in the magnetization direction: $\lambda_{\langle 111 \rangle} = -18.8 \times 10^{-6}$. The magnetostriction constant in polycrystalline iron is positive at low fields and negative at high fields. We shall see later that this behavior is related to the rotation of the magnetization from the easy to hard crystallographic directions when large fields are applied.

We may obtain a qualitative understanding of the forces that produce *magnetostriction* by considering the following simple model. As shown in Fig. 14-19,

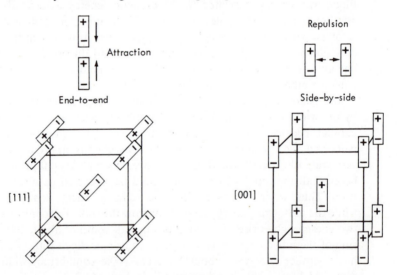

FIG. 14-19 Illustration of attraction and repulsion of differently oriented dipoles. Dipole arrangement in BCC Fe is magnetized in the $\langle 001 \rangle$ and $\langle 111 \rangle$ directions

the forces exerted between magnetic dipoles depend on their relative orientations. If the dipoles are end-to-end, they attract each other; if they are side-by-side, they repel one another. If we think of each atom as a tiny magnetic dipole, we see that the forces between the atoms depend on how the polarization is oriented in the crystal. For example, when BCC iron is magnetized along the $\langle 111 \rangle$ close-packed direction, the dipoles are arranged essentially in an end-to-end pattern. Thus, it is not surprising that we find a negative magnetostriction constant in this direction (the atoms in the close-packed row simply attract one another). On the other hand, when the magnetization of iron is in the $\langle 100 \rangle$ direction, the nearest neighbor dipoles are arranged in such a way that the side-by-side character of the dipoles is dominant and the atoms or dipoles repel each other. This explains why the magnetostriction constant in the $\langle 100 \rangle$ direction is positive. Although this model gives a simple way to think about magnetostriction, it cannot be extended too far or taken as quantitative, since it is a great oversimplification to imagine that the atoms can be treated as little magnets.

We now wish to consider how the magnetostrictive strain contributes to the energy and equilibrium configurations of domains. Consider the domain structure of a crystal of iron shown in Fig. 14-20. Notice that the triangularly shaped domains, called *closure domains*, at the ends of the sample provide an internal short circuit for the magnetic induction and that no external or internal poles are formed.

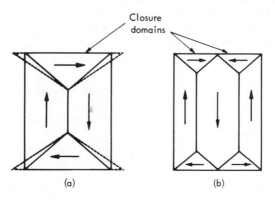

FIG. 14-20 Closure domains in BCC Fe

This configuration is particularly stable because there is no demagnetization field and hence no magnetostatic energy. This configuration is possible in iron because the cubic magnetic symmetry permits the magnetization in the closure domains to be parallel to the top and bottom surfaces. Since there is surface energy associated with the domain boundaries, it might appear that the lowest energy configuration would be the one shown in Fig. 14-20(a), since this configuration involves the least Bloch wall area. However, the fact that the closure domains are magnetized perpendicular to the primary domains introduces an elastic strain energy term into the total energy. This term is called the *magnetostrictive energy*. As shown in the figure, the magnetostriction in the closure domains would cause those regions to extend in the direction of magnetization if they were not constrained by the presence of the primary domains. To get a physical picture of the magnetostrictive energy we can imagine that the closure domains are first magnetized and allowed to extend and then compressed back into shape and filled into the triangular slot between the primary domains. The energy needed to squeeze the closure domain back into shape is called the magnetostrictive energy or *magnetoelastic energy*. This energy is directly proportional to the square of the magnetostriction constant, the elastic modulus, and the total volume of the closure domains. We see in Fig. 14-20(b) that the total closure domain volume can be reduced by making the primary domains smaller. The equilibrium domain configuration involves a balance between the effects of Bloch wall energy and magnetostrictive energy. The Bloch wall energy tends to make the domains large (with few domain boundaries), while the magnetoelastic energy tends to make the domains small (with small closure domains). Again, we find the equilibrium domain configuration to be the result of minimizing the sum of these two energy terms.

EXAMPLE 14-8

The accompanying figure shows the domain configuration in a sheet of Fe–3% Si. Calculate the equilibrium Bloch wall spacing S. For Fe–3% Si, γ_B (Bloch wall energy) $= 2$ ergs/cm^2, $\lambda_{\langle 100 \rangle} = 20 \times 10^{-6}$, and the elastic modulus in the $\langle 100 \rangle$ direction is $E_{\langle 100 \rangle} = 24.2 \times 10^{11}$ dynes/cm^2.

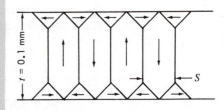

Solution:

To find the equilibrium value of S, we must minimize the total energy $E_{total} = E_w + E_{ms}$, where E_w is the total wall energy and E_{ms} is the total magnetostrictive (or magnetoelastic) energy. Consider a unit area of the sheet. The total wall energy in a unit area of the sheet is

$$E_w = \frac{1}{S}\gamma_B\left[(t-S) + 4\frac{S}{\sqrt{2}}\right] = \gamma_B\left(\frac{t}{S} + 1.83\right)$$

where $1/S$ is the number of unit sections of width S. The total magnetoelastic energy in a unit area is

$$E_{ms} = \frac{1}{S}\frac{1}{2}E_{\langle 100 \rangle}\lambda_{\langle 100 \rangle}^2\left(\frac{S}{\sqrt{2}}\right)^2$$

where $\frac{1}{2}E_{\langle 100 \rangle}\lambda_{\langle 100 \rangle}^2$ is the strain energy density in the closure domains and $(S/\sqrt{2})^2$ is the closure domain volume for a unit section. The total energy is

$$E_{total} = E_w + E_{ms} = \gamma_B\left(\frac{t}{S} + 1.83\right) + \frac{1}{4}E_{\langle 100 \rangle}\lambda_{\langle 100 \rangle}^2 S$$

Now, letting $dE_{total}/dS = 0$, we have

$$-\frac{\gamma_B t}{S^2} + \frac{1}{4}E_{\langle 100 \rangle}\lambda_{\langle 100 \rangle}^2 = 0$$

Hence,

$$S_{eq} = \sqrt{\frac{4\gamma_B t}{E_{\langle 100 \rangle}\lambda_{\langle 100 \rangle}^2}}$$

$$S_{eq} = \left(\frac{4(2)(0.1)}{(24.2 \times 10^{11})(20 \times 10^{-6})^2}\right)^{1/2} = 2.87 \times 10^{-2} \text{ cm}$$

We have seen how the domain structure in ferromagnetic materials is governed by the various contributions to the total magnetic energy. In every case we try to compute the total magnetic energy by including the exchange, magnetostatic, magnetocrystalline, Bloch wall, and magnetostrictive energies and we find the domain configuration for which the total energy is lowest. We shall see later that consideration of these energy terms will permit us to explain many important irreversible magnetic phenomena and will also allow us to establish guidelines for the design of soft and hard magnetic materials.

14-7 HYSTERESIS PROPERTIES OF FERROMAGNETIC AND FERRIMAGNETIC MATERIALS

Having described the properties of magnetic domains, we are now prepared to discuss the mechanisms responsible for magnetic hysteresis in spontaneously magnetic materials. We start this discussion with a phenomenological description of the hysteresis loop.

Determination of Hysteresis Loops

The hysteresis loop for a ferromagnetic or ferrimagnetic material can be obtained by measuring the intensity of magnetization M or the induction field B as a function of the applied field H. It is conventional to permit the applied field to cycle between positive and negative values and to record the entire cycle. Consider the circuit elements shown in Fig. 14-21 and an initially unmagnetized sample.

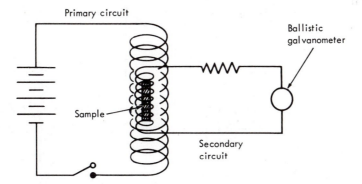

FIG. 14-21 Electrical circuit used for determining the B-H relationships for ferromagnetic and ferrimagnetic materials

When a direct current is allowed to flow in the primary coil, a uniform magnetic field is produced within the coil, and the sample responds by developing a magnetic moment. The magnetic flux lines associated with the magnetization of the sample induce a charge or current in the secondary coil that can be easily measured with a galvanometer. Each time the field is abruptly changed, the magnetization or magnetic flux in the sample changes and a fixed, measurable charge is induced into the secondary. A complete record of the magnetization of the sample as a function of the applied magnetic field can be obtained by keeping a running account of the increments in the magnetization or induction each time the field is changed. Using this technique, it is possible to determine the entire magnetic hysteresis loop by varying the applied magnetic field from positive to negative values. The general features of the hysteresis loop are shown in Fig. 14-22.

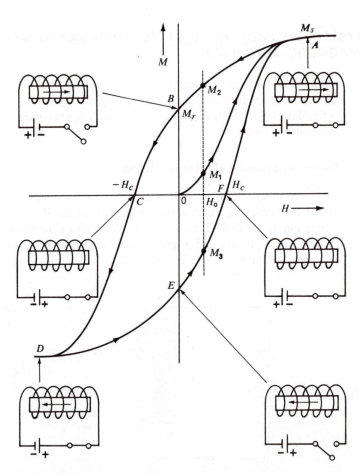

FIG. 14-22 Hysteresis loop for a ferromagnetic material

The hysteresis loop shown in Fig. 14-22 corresponds to the magnetization of a ferromagnetic sample that is initially unmagnetized. As the magnetic field is applied to the sample, the magnetization usually follows a smooth but irrational path such as OA. Usually the magnetization builds up by following an S-shaped curve. When the applied field is sufficiently strong, the sample begins to saturate; that is, the intensity of magnetization in the sample approaches the *saturation magnetization M_s*.

Now let us consider what happens when the magnetic field intensity begins to decrease. The hysteresis loop indicates that the magnetization does not follow the initial magnetization curve *AO*. Rather, it follows the hysteresis loop along *AB*. When the magnetic field is reduced to zero, the sample remains partially magnetized. At this point the residual magnetization in the sample is called the *remanent magnetization M_r*; B_r is the corresponding *remanent induction field*. Now, if the magnetic field is reversed, the sample will be forced to demagnetize along *BC*. As indicated by the hysteresis loop, a critical field, called the *coercive field H_c*, is required to reduce the magnetization to zero. If the magnetic field is increased negatively beyond the coercive field, the sample begins to develop a negative magnetization along *CD*. Eventually, when the reverse field is sufficiently strong, the sample begins to saturate in the reverse direction.

Now, if the reverse field is decreased, the magnetization changes again lag behind the changes in the applied magnetic field and the magnetization follows the path *DE*. The magnetization is $-M_r$ when the reverse field reduces to zero. When the magnetic field is reversed for the second time (becoming positive again), the magnetization follows *EF*. Again, a critical field H_c is needed to eliminate the remanent magnetization. If the magnetic field is continually increased, the sample eventually saturates again in the positive direction along *FA*. The sample has now been returned to the same state that was produced by the initial magnetization. Hereafter, each time the magnetic field oscillates between positive and negative values, the magnetization follows the hysteresis curve shown.

This discussion of the hysteresis curve indicates that the magnetization is history dependent. Consider the applied field H_0 shown in Fig. 14-22. It is clear that the magnetization may have one of three different values (M_1, M_2, or M_3), depending on the magnetic history. When the hysteresis loops are rectangular in shape, as in the case of some ferrites, there are two limiting magnetic states (i.e., $+M_s$ or $-M_s$) in which the material can exist. By simply applying a magnetic field of the proper sign, the material can be switched from one state to the other. In this way, these materials can be used as memory elements for computers.

Since the energy required to change the magnetization from M_1 to M_2 is $\int_{M_2}^{M_1} H \, dM$, it follows that the area of a magnetic hysteresis loop is equivalent to the energy absorbed per unit volume each time the hysteresis loop is traversed. Therefore, it is a relatively simple matter to determine the magnitude of the hysteresis losses by measuring the area of the *M-H* loop.

EXAMPLE 14-9

A rectangular hysteresis loop for a cobalt-iron ferrite ($Co_{0.1}Fe_{0.9}Fe_2O_4$) is shown in Fig. 14-27 (see p. 497). Calculate the energy absorbed in a unit volume of the sample when the magnetization varies through one entire cycle. Also, estimate the temperature rise that would occur if the hysteresis loss for one cycle were converted into heat adiabatically (with no heat losses to the surroundings). The heat capacity of the ferrite is about 0.2 cal/g·deg and the density is 5 g/cm³ (1 erg $= 2.39 \times 10^{-8}$ cal).

Solution:

The energy or the hysteresis loss is simply the area enclosed by the *M-H* hysteresis curve. Therefore, the hysteresis loss is

$$E_{hys} = \left(\frac{1}{4\pi}\right)4(2100)(8.5) = 5700 \text{ ergs/cm}^3$$
$$= (5700)(2.39 \times 10^{-8}) = 1.36 \times 10^{-5} \text{ cal/cm}^3$$

The temperature rise for one cycle is

$$\Delta T = \frac{E_{hys}}{\rho C_p} = \frac{(1.36 \times 10^{-5})}{(5)(0.2)} = 1.36 \times 10^{-5} \text{ °C}$$

It is clear that the magnetization would have to be reversed more than a million times for heating to become significant.

Domain Theory of Magnetization

We may now give a microscopic interpretation of the process of magnetization. In this treatment we shall be concerned mainly with the growth of some domains relative to others or, in other words, with the motion of domain boundaries.

Consider the domain structures of an iron crystal shown in Fig. 14-23. We see that when the magnetic field is zero, the size of the two domains magnetized in the positive direction is exactly the same as the size of the two domains with the opposite magnetization. Therefore, at zero field there is no net magnetic moment in the horizontal direction. When a positive magnetic field is applied, the domain boundaries respond by moving in the manner shown. The domain walls move in the direction that permits the most favorably oriented domain to grow. That is, the domains with magnetization parallel to the applied field grow, while the domains magnetized antiparallel to the field shrink. In this way the sample develops a net magnetic moment parallel to the field and becomes magnetized. It should be noted that the 90° walls in this sample also move when the field is applied. We see that the domain with magnetization normal to the field is less favorably oriented than the domain parallel to the field, but is more stable than the domain that is magnetized opposite to the applied field. As a consequence, the small domain with perpendicular magnetization grows at the expense of the antiparallel domains and is reduced in size by the growth of the domain parallel to the field. When the field is sufficiently strong, the 180° walls are driven to the edges of the sample and the perpendicular domain finally shrinks to nothing and disappears altogether. When this happens, the sample will be saturated, as a single domain with the saturation magnetization M_s will point in the direction of the applied field. The sample is then fully saturated and the magnetization process is complete.

We may now use the ideas expressed by Fig. 14-23 to explain the hysteresis properties of ferromagnetic materials. Let us consider first the initial stages of magnetization. As we saw in Fig. 14-23, the first step in the magnetization process involves the displacement of the domain boundaries by the applied field. When this displacement is sufficiently small, the walls move reversibly and return to their initial positions when the field is removed. This phenomenon arises because the domain boundaries are held in place in an unmagnetized sample through their interactions with other defects in the solid (inclusions, dislocations, and other heterogeneities). In this reversible region, the material does not display a hysteresis effect. This is indicated on the magnetization curve shown in Fig. 14-24.

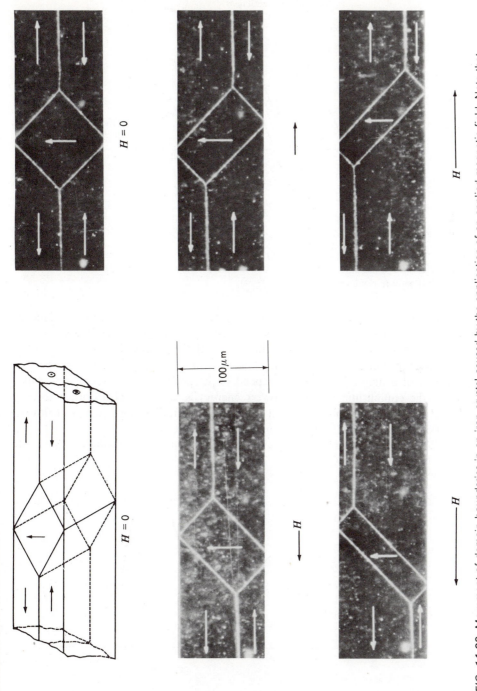

FIG. 14-23 Movement of domain boundaries in an iron crystal caused by the application of an applied magnetic field. Note that domains with magnetization aligned parallel to the applied field grow at the expense of the other domains. Photo courtesy R. W. DeBlois, The General Electric Research and Development Center, and C. D. Graham, the University of Pennsylvania. Used with permission.

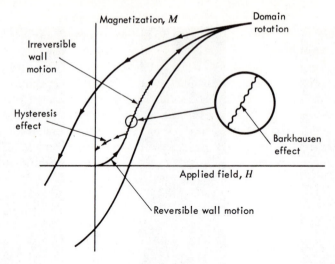

FIG. 14-24 Ferromagnetic hysteresis loop showing reversible and irreversible magnetization regions. The discontinuous magnetization behavior known as the *Barkhausen effect* is shown in the exploded view.

If the magnetization or magnetic field remains sufficiently small, the magnetization of the sample exactly retraces its values when the magnetic field is removed. The microscopic interpretation of this is that the domain walls return to their initial equilibrium positions when the field is removed. Figure 14-25 shows how the energy of a domain wall might depend on its position in a crystal. It is evident that the equilibrium position (lowest energy) is x_0 and that a critical force or field is needed to displace the wall from its position. If the field remains below the critical value, the wall displacements are reversible.

When the magnetic field is sufficiently large, the walls displace irreversibly;

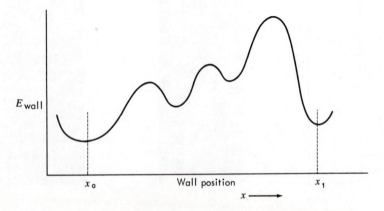

FIG. 14-25 Variation of the energy of a Bloch wall with position in an imperfect crystal. The energy minima and maxima are caused by the interaction of the Bloch wall with imperfections and inhomogeneities.

that is, they do not return to their initial positions when the field is removed. When this happens, the sample begins to show hysteresis effects. In this region, the walls cannot return to their initial positions when the field is removed because they are trapped into positions as in position x_1 in Fig. 14-25. Again, this trapping effect comes from the interaction of the walls with other defects in the material. The walls essentially hang on the other defects and are prevented from returning to their starting positions. As a consequence of this trapping, the material is left with a residual magnetization when the field is removed. This residual magnetic moment is produced because the forward movement of the walls in response to the applied field is not completely nullified by the reverse wall movement that occurs when the field is removed.

When the magnetization takes place irreversibly, it can occur discontinuously. The magnetization in this case occurs in steps called *Barkhausen jumps*. This discontinuous phenomenon is due to the rapid motion of groups of domain walls when they suddenly break free from the defects that hold them in place.

As noted earlier, sufficiently strong magnetic fields can drive all the domain boundaries from the material. In that case, each crystal in the sample becomes a single domain magnetized in the easy direction most closely aligned with the applied field. In the case of single crystals magnetized along the easy axis, the sample saturates as soon as the domain walls are driven out of the sample. However, if the easy axis of the single crystal is not parallel to the field, or if the sample is polycrystalline with randomly oriented grains, saturation is not achieved by removing all the domain walls. This is because some of the crystals are not yet magnetized parallel to the field. Further magnetization can occur only by the rotation of the dipoles from the direction of easy magnetization into the direction of the field. As this process occurs, the magnetization of the sample slowly approaches M_s as the field is increased. Because of the large magnetocrystalline anisotropy forces holding the magnetization in the easy directions, extremely large fields would be required to fully saturate a polycrystalline sample with randomly oriented grains.

Now, if the magnetic field is permitted to decrease, the dipoles begin to rotate back toward the easy direction of magnetization and the net magnetic moment decreases. In addition, the demagnetizing field produces reverse magnetic domains that allow the sample to be partially demagnetized. However, since the domain wall motion associated with demagnetization is driven by the demagnetizing field rather than by the applied field, and since the walls again encounter obstacles in the form of defects and defect clusters, the walls in general do not retrace their original path. Even when the applied field is reduced to zero, the domain walls will not have returned to their initial positions. As a result of the interference set up by the obstacles, the sample will exhibit a remanent magnetization M_r at zero field. If a reverse magnetic field is applied to assist the demagnetizing field, the walls can be forced past the obstacles that hold them back and the sample can be made to demagnetize. This process involves the irreversible motion of the domain walls past defects. The coercive field H_c is the additional field needed to eliminate the net magnetization.

We see from this discussion that the hysteresis properties come from the interaction of domain walls with inclusions and lattice defects. Generally speaking, materials with high defect and inclusion contents have large hysteresis loops. These materials are called *hard* magnetic materials and are characterized by large remanence and high coercive field. Such materials may be used as permanent magnet materials. When materials are to be used as soft magnetic materials, as in the case of transformer cores or electromagnets, it is important that the hysteresis loop be as small as possible. Generally, this is achieved by minimizing the inclusion content and the dislocation density. In the next section we shall see how the structures of specific magnetic materials can be tailored by using the general model of hysteresis given here.

Eddy Current Losses

Before leaving the subject of hysteresis loops, we shall discuss another important contribution to their shape and size. This topic relates to the energy losses that are due to the resistive losses associated with the electric currents that are induced by changing magnetic fields.

When the magnetization of a sample is changed at any finite rate, the induction of the magnetic flux causes transient voltage gradients to be set up within the material. If the magnetic material in question is a conductor, the induced voltages produce electric currents called *eddy currents*, and resistive energy losses occur. If the material is an insulator, as in the case of ferrimagnetic oxides, electric currents are not induced and there are no eddy current losses. Since the induced voltages and eddy current losses depend on the rate of change of the magnetic flux, the eddy current losses are frequency dependent. One can imagine that the hysteresis loop has a minimum size that depends on the defect concentration when it is determined at low frequencies. As the frequency increases, the loop area increases as the eddy current energy losses become significant. We shall discuss techniques for limiting the eddy current losses in metals in the next section.

14-8 MAGNETIC MATERIALS

In the previous sections of this chapter we have studied mainly the principles that govern the structure and properties of magnetic materials. We have seen how the domain structures that occur in magnetic materials can be explained in terms of the various contributions to the total magnetic energy, and we have shown that the response of magnetic domains to externally applied fields provides a basis for understanding hysteresis effects. In this section we shall show how these principles can be applied to real engineering materials. In particular, we shall discuss the magnetic properties of some commercially significant materials and show how they can be explained with the principles of magnetic domains. This discussion will illustrate how a knowledge of the properties of magnetic domains can be used to design materials with desirable magnetic properties.

Iron-Silicon Alloys

In the latter part of the nineteenth century, transformer cores for low-frequency power transformers were made of iron or steel. Because of the large hysteresis losses associated with these materials, the transformers of the day operated at very low efficiencies. At about the turn of the century it was discovered that the addition of a few percent silicon greatly increases the permeability of low-carbon steel and substantially reduces the hysteresis loss. It was found that the improvement in hysteresis properties came from the fact that silicon increases the electrical resistivity of steel and thereby reduces the eddy current losses. Consequently, silicon steels were, and still are, used for transformer core materials. Another early development in the design of transformer cores was also related to the minimization of eddy current losses. It was discovered that the circulating eddy currents that accompany magnetization of a core can be minimized if the core is constructed with laminated sheets of silicon steel separated by insulating spacers. The thin insulating spacers prevent long-range electric currents, and hence reduce eddy current losses. This laminated construction can be found in almost all the power transformers in use today.

In the early 1930s, the second major development in the field of transformer core materials occurred. It was discovered that silicon steels could be made to develop a *crystallographic texture* by appropriate rolling and heat treatment, and that this texture leads to increased permeability and smaller hysteresis losses. This process involves cold rolling and subsequent annealing and produces a sheet with a *grain-oriented* structure. Figure 14-26 shows the approximate orientation of the

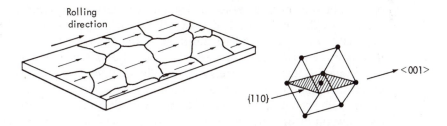

FIG. 14-26 Crystallographic orientation of grains in grain-oriented Fe-3% Si sheet

crystals in grain-oriented sheet. We see that the crystals become oriented in such a way that their easy axis of magnetization lies nearly parallel to the rolling direction. This means that grain-oriented sheet has magnetic properties similar to those of single crystals. When magnetized in the rolling direction, the sample becomes saturated as soon as the domain walls are driven from the sample. Almost no dipole rotation occurs, since the entire sample is oriented along the easy axis. Consequently, the hysteresis losses in grain-oriented sheet are particularly low, and silicon steels processed in this way make excellent transformer cores.

Still another significant development in silicon steels came in the 1940s, when it was learned that hysteresis losses in grain-oriented silicon steel can be decreased even further if the steel is annealed in a hydrogen atmosphere. Annealing in moist hydrogen at about 800°C followed by annealing at about 1200°C in dry hydrogen reduces the carbon content from about 0.1 to 0.01 weight percent. Since carbides in steels act as obstacles to the movement of domain walls and provide hysteresis loss effects, it seems likely that the beneficial effects of the hydrogen treatments are due to the removal of these inclusions from the steel.

In summary, the attractive magnetic properties of the silicon steels currently used for transformer cores are the result of several important developments in materials. The eddy currents are minimized by the addition of silicon and by laminated constructions, while the intrinsic permeability is maximized by developing grain-oriented sheet and by preparing the steel in such a way that the inclusion content is minimum.

Iron-Nickel Alloys (Permalloys)

The silicon steels discussed in the previous paragraphs are used mainly when the application involves transmission of large amounts of power. In contrast, the Fe-Ni alloys discussed here are used to detect or transmit the small signals used in communication. One of the more common *permalloy* compositions is 79% Ni-21% Fe. This particular alloy has a very high magnetic permeability. Consequently, it is ideally suited to applications that involve high gain. That is, small changes in the applied field produce very large changes in the magnetization or induction in this alloy.

To understand the high permeability of permalloys, let us consider the process of domain wall motion. When a domain boundary is displaced from its initial position by the applied field, the magnetic vectors in the vicinity of the wall are required to rotate through hard directions as the wall moves by. One of the factors that limit the permeability or ease with which the walls can be moved is the magnetocrystalline anisotropy energy. Generally speaking, a large anisotropy energy means that the Bloch walls are thin and have a large surface energy. These features in turn cause the walls to interact strongly with defects on the lattice. Conversely, for a low anisotropy energy, the Bloch walls are very thick and are not easily trapped by lattice imperfections. One of the reasons permalloy exhibits high permeability is that the crystal anisotropy is very small near the permalloy composition (79% Ni, 21% Fe). In fact, if the 79-21 permalloy is annealed at a high temperature and then quenched to room temperature, the anisotropy energy can be made vanishingly small.

A second related cause of the high permeability found in permalloy comes from the fact that the magnetostriction constant in the easy direction of magnetization is zero for the permalloy composition. This is important because one of the factors that limit domain wall movement is the local strain or distortion associated with the displacement of the wall. When the magnetostriction constant is zero, the strains associated with magnetization vanish and the wall moves very easily. Thus, we see that Bloch walls will move very easily if the anisotropy energy is small or zero and if the magnetostriction constant is zero. Both of these conditions are met in 79-21 permalloy. Consequently, this alloy exhibits very high magnetic permeabilities and is useful in high-gain communications applications.

Ferrites

Another class of soft magnetic materials of great technological importance are *ferrites* or *ferrimagnetic oxides*. We discussed the atomic aspects of the magnetic properties of these materials in Section 14-5. We now want to consider briefly their magnetic behavior and how they may be used.

Ferrites are important not only because they have useful magnetic properties but also because they are insulators and have very high electrical resistivities. This becomes important in magnetic applications that involve high frequencies. Because the eddy currents are the result of induced voltage gradients that increase with the rate of magnetization, it follows that the losses associated with these currents become very significant at high frequencies. Therefore, it is impractical if not impossible to use metallic magnetic materials in applications that involve high frequencies—it is necessary to use insulators. Ferrites or ferrimagnetic oxides are insulators and, therefore, can be used at high frequency. For example, the cores of very-high-frequency (kilocycles) transformers are fabricated from ferrites.

Another high-frequency application of ferrites involves their use as elements for information storage in computers. Ferrites with rectangular hysteresis loops of the kind shown in Fig. 14-27 can be used to store bits of information. Because of

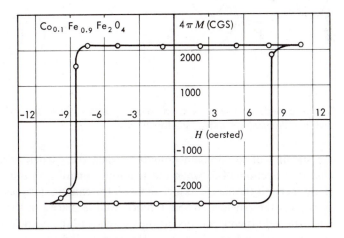

FIG. 14-27 Rectangular hysteresis loop for a cobalt-iron ferrite, $Co_{0.1}Fe_{0.9}Fe_2O_4$

the squareness of the loop, the material can be made to exist in one of two states, $M = +M_s$ or $M = -M_s$. By simply applying a magnetic field, the material can be switched from one state to the other. In any such device it is important that the switching occur in a very short period of time (this ultimately contributes to the speed of the computer). Ferrites are particularly attractive in this application because of their high resistivities. Because the eddy current losses are almost nonexistent, these materials can be switched from one state to the other in about 1 μ sec.

Fine-Particle Magnets

So far in this chapter we have been concerned with the bulk magnetic properties of materials. We now want to consider the special properties of *fine particles*. This information will permit us to describe the properties of one class of permanent magnet materials called *fine-particle magnets*. In addition, our discussion of the magnetic properties of fine particles will be of use in the next section, where we treat the properties of magnetic tapes.

Consider the magnetic particles shown in Fig. 14-28. These might be small particles of Fe or Co or a ferrite such as Fe_3O_4. We want to focus our attention on

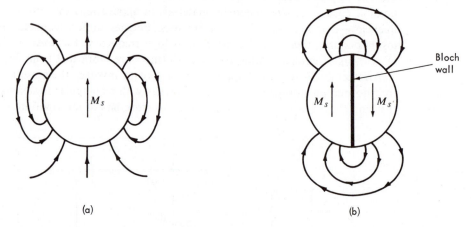

(a)   (b)

FIG. 14-28 Magnetic fields associated with fine particles with (a) a single domain and (b) with two domains separated by a Bloch wall

the stability of the domain boundary when the particles are very small. As discussed in Section 14-5, when the material contains no Bloch walls, the magnetostatic energy is the energy associated with the external magnetic field, which entirely surrounds the particle. When a Bloch wall is placed in the particle, the external magnetic field is concentrated at the two ends of the particle and the magnetostatic energy is reduced by about a factor of two. We see, therefore, that the driving force for the formation of the Bloch wall is the reduction of the magnetostatic energy that occurs when the wall is formed. To a first approximation, the reduction in magnetostatic energy is proportional to the volume of the particle. Therefore, the driving force for domain wall formation decreases as the particle size decreases, according to the relation: driving force $\propto r^3$.

Now let us consider the energy associated with the Bloch wall. Clearly, this energy must be less than the reduction in magnetostatic energy if the Bloch wall is to be stable. Since the Bloch wall is a surface area of πr^2, its energy is proportional to the square of the particle radius. Therefore, we find the reasonable result that the total Bloch wall energy also decreases when the particle size decreases. We see from this discussion that the driving force for wall formation varies as r^3, while the wall energy varies as r^2. This means that particles smaller than a critical size are too small to support a Bloch wall. That is, below a critical particle size, the energy of the particle is lowest if it remains as a single domain.

The idea that very small magnetic particles should be single domains can be seen in another way. We showed in Section 14-6 that the width of a Bloch wall depends on the balance between the effects of exchange energy and magnetocrystalline anistropy energy, and that the Bloch walls are typically 1000 A in thickness. Obviously, when the magnetic crystal in question is in the form of a particle that is smaller than the thickness of the Bloch wall, it becomes impossible for a Bloch wall to be contained in the particle. In this case the lowest energy is obtained by having all the dipoles reside in one of the easy directions. Again, we find a single domain is stable when the particle is small. Most single-domain particles that are used in fine-particle magnets or magnetic tapes have diameters that range from about 100 A to about 1000 A.

Now let us consider the magnetic properties of a particle that is too small to contain a Bloch wall. As shown in Fig. 14-29, the magnetization lies along an

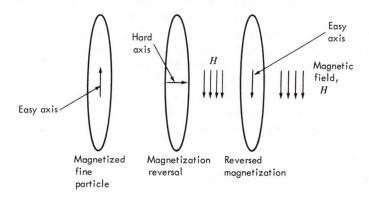

FIG. 14-29 Illustration of magnetization and magnetization reversal in fine magnetic particles

easy direction in the particle. When a strong reverse magnetic field is applied, the magnetization in the particle is reversed as shown. Since the particle is too small to contain a Bloch wall, the demagnetization cannot occur by domain wall motion. Instead, it is necessary for the magnetization to rotate from one easy direction through the hard direction into a reverse easy direction. Because of the strong anisotropy forces that hold the dipoles in the easy direction, the process of rigid dipole rotation can occur only if very large magnetic fields are applied. For this reason the coercive force of fine particles is usually quite large. Figure

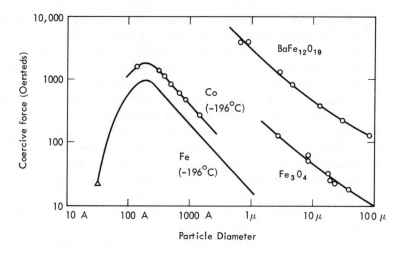

FIG. 14-30 Coercive force of some ferromagnetic and ferrimagnetic particles as a function of particle diameter

14-30 shows how the coercive force of some common fine particles increases as the particle diameter decreases.

Shape anisotropy is another important factor that contributes to the high coercive force of a small particle. If the shape of the particle is such that the easy axis is parallel to the major axis of the particle, then the magnetostatic energy of the particle also holds the polarization along the long axis of the particle. In other words, the magnetostatic energy increases as the dipoles are rotated from the long axis of the particle toward the short axis.

Fine particles of magnetic materials that are elongated in shape have very high coercive force and thus make excellent permanent magnets. Some of the properties of fine-particle magnet materials are given in Table 14-5. Generally speaking, good fine-particle magnets are made by packing elongated magnetic particles

TABLF 14-5

Properties of Fine-Particle Magnet Materials

Material	B_r* (gauss)	H_c* (oersteds)
Manganese-bismuth	4,800	6,000
Estimated upper limit	7,800	37,000
Barium ferrate	4,000	1,950
Estimated upper limit	4,650	17,000
Iron	9,000	720
Estimated upper limit	14,300	3,600
Iron-cobalt	9,050	1,050
Estimated upper limit	16,300	4,100

*The properties reported are maximum values. Generally, the high H_c values are found for magnets with a low particle-packing density, whereas the high values of B_r are associated with high particle-packing densities.

together in the presence of a magnetic field. The particles align parallel to the field and when packed together constitute a very hard magnetic material. The particles are usually packed tightly together so that the residual magnetization or induction is high. However, if the particles are packed together too tightly, they lose their single-domain characteristics and the coercive force is reduced. Consequently, there is an optimum packing density at which the product of the coercive force and remanent induction (called the *energy product*) is maximum.

The properties of alnico magnets may also be explained with the principles of fine-particle magnetism, even though these magnets are not really fine-particle magnets. Alnico is a cast alloy, containing Al, Ni, and Co as well as Fe, in which magnetic particles are formed through precipitation from solid solution in a magnetic field. The presence of the magnetic field is needed to force the tiny needle-shaped magnetic particles to form with their long magnetic axes in parallel alignment. Also, the rate of cooling must be carefully controlled so that the particles are large enough to exhibit a magnetic moment but not so large as to contain Bloch walls. When these precautions are taken, the precipitates have the right shape and orientation to act as fine particles, and they cause the coercive force of alnico to be very high.

Magnetic Tapes

Another important application of fine-particle magnetism is in magnetic recording tapes. Elongated, single-domain magnetic particles of such materials as $\gamma \, Fe_2O_3$ are imbedded into a plastic tape. Information is stored on the tape by passing the tape past a recording head, which produces a strong magnetizing field in direct proportion to the signal to be stored. The recording magnetic field magnetizes the tape by polarizing some of the fine particles in the tape as the tape passes through the recording head. The information is extracted from the tape by passing it through a pick-up device that is capable of sensing the magnetization in the tape. The particles in the tape must have sufficiently high coercive force that the fields from the tape itself (when it is wrapped up) do not destory the information on the tape by changing the residual magnetization in the tape.

PROBLEMS

14-1   Two bar magnets 5 cm in length with pole strengths $m = 50$ and $m = 75$ are placed end-to-end in air with a separation distance of 2 cm. Calculate the force required to hold the magnets in position.

14-2   A bar magnet 1 cm in length with a pole strength of 25 is placed in a magnetic field of 12 kilogauss (12,000 oersteds in air). Calculate the maximum torque that can be exerted on the magnet by the field.

14-3   An iron wire 1 mm in diameter is magnetized along its length. What is the maximum pole strength that can be produced at the end of the wire in the absence of an applied magnetic field?

14-4   Two very long bar magnets 1 cm in diameter with pole strengths $m = 100$ and $m = 200$ are placed end-to-end with the north pole of one magnet in contact with the south pole of the other. Estimate the work or energy required to separate the magnets by a distance of 1 cm.

14-5   Two metal bars are identical in color, weight, and shape. Both bars are ferromagnetic, but only one of the bars is permanently magnetized. Explain how it can be determined which bar is the permanent magnet without the use of auxiliary instruments.

14-6   One end of a soft iron rod 3 cm in diameter is placed in a solenoid 30 cm in length with 11,000 turns of copper wire 0.5 mm in diameter. A d-c power supply is used to produce a current of 2 A in the solenoid. Calculate the maximum force that can be exerted on the iron rod when the solenoid is energized. How does the force exerted on the rod depend on the length of the rod?

14-7   One end of a long, thin rod of tantalum is placed between the pole pieces of a very strong magnet that produces a field of 12 kilogauss. The rod is 0.5 cm in diameter. Calculate the force exerted on the rod by the field. How would measurement of this force be used to obtain an experimental determination of the magnetic susceptibility of Ta?

14-8   Tungsten is paramagnetic at room temperature with a susceptibility of 1.1×10^{-6} (CGS). Explain why tungsten might be expected to become diamagnetic at sufficiently high temperatures.

14-9   Zirconium undergoes a phase change from HCP to BCC at 862°C. How does the diamagnetic susceptibility change at this transition temperature?

14-10 The magnetic susceptibility of Fe at 900°C is about 2.5×10^{-4} (CGS) and the Curie temperature is 770°C. Estimate the magnetic susceptibility at 800°C.

14-11 The magnetic moment of each atom of cobalt in metallic cobalt is about 1.7 Bohr magnetons. Cobalt is HCP with lattice parameters $a = 2.50$ A and $c = 4.06$ A. Calculate the saturation magnetization.

14-12 A magnetic ferrite with the inverted spinel structure is to be prepared in such a way that the magnetic moment per molecule is exactly 2 Bohr magnetons. Refer to Fig. 14-11 and indicate how such a ferrite could be made. Estimate the composition of the ferrite having this particular magnetic moment.

14-13 Explain why the unit cell shown in Fig. 14-10 is a "chemical unit cell" but not a "magnetic unit cell."

14-14 The Curie temperatures of Ni and Fe are 735°C and 770°C, respectively. In which metal is the exchange energy larger?

14-15 Explain why the magnetization of a ferromagnetic solid tends to lie along a long axis of the body as opposed to the short axis (in the case of an elongated particle, for example).

14-16 Explain why Bloch walls are strongly pinned by voids or nonmagnetic inclusions.

14-17 Show why two 90° twist Bloch walls can annihilate under some circumstances and not under others.

14-18 Explain how and why the Bloch wall width should change with temperature just below the Curie temperature.

14-19 What is the sign of the magnetostriction coefficient in the rolling direction of grain-oriented Fe-Si sheet?

14-20 Why do Fe-Si transformer cores hum? Would you expect a permalloy to hum? Why or why not?

14-21 Assume that the energy of 180° Bloch walls in Co is about 1 erg/cm² and calculate the equilibrium domain size (width) for the configuration shown in Example 14-6.

14-22 Show that the rate of change of the intensity of magnetization in a ferromagnetic solid can be expressed as a product of the Bloch wall density and the average Bloch wall velocity along with appropriate magnetic constants.

14-23 Explain why the magnetostriction constant for polycrystalline Fe is positive at low fields and negative at high fields.

14-24 Suppose a tensile stress were exerted onto a sample of grain-oriented sheet of Fe-Si in the rolling direction. How would the magnetic domain structure be expected to change?

14-25 Estimate the hysteresis loss in ergs for each cycle for Isoperm using the M-H loop shown in Fig. 14-2.

BIBLIOGRAPHY

L. V. AZAROFF and J. J. BROPHY, *Electronic Processes in Materials*. New York: McGraw-Hill, 1963.

R. M. BOZORTH, *Ferromagnetism*. New York: Van Nostrand Reinhold, 1951.

S. CHIKAZUMI, *Physics of Magnetism*, Ch. 6–16. New York: John Wiley, 1964.

N. CUSACK, *The Electrical and Magnetic Properties of Solids*, Ch. 12–14. London: Longmans, Green, 1958.

A. J. DEKKER, *Solid State Physics*, Ch. 18–19. Englewood Cliffs, N.J.: Prentice-Hall, 1957.

J. E. GOLDMAN, ed., *The Science of Engineering Materials*, Ch. 12–13. New York: John Wiley, 1957.

C. KITTEL, *Introduction to Solid State Physics*, Ch. 9, 15. New York: John Wiley, 1955.

Magnetic Properties of Metals and Alloys. Metals Park, Ohio: American Society for Metals, 1959.

A. NUSSBAUM, *Electronic and Magnetic Behavior of Materials*, Ch. 1–4. Englewood Cliffs, N.J.: Prentice-Hall, 1967.

R. M. ROSE, L. A. SHEPARD, and J. WULFF, *The Structure and Properties of Materials*, Vol. IV, Ch. 9–10. New York: John Wiley, 1966.

L. H. VAN VLACK, *Materials Science for Engineers*, Ch. 16. Reading, Mass.: Addison-Wesley, 1970.

C. A. WERT and R. M. THOMSON, *Physics of Solids*, Ch. 18–20. New York: McGraw-Hill, 1964.

15
Optical Properties of Materials

15-1 INTRODUCTION

Thus far we have considered electronic properties in terms of the electrical and magnetic characteristics of materials. We now turn our attention to those electronic properties that we come in contact with every day, and oftentimes unconsciously—the optical properties. This broad category of properties includes the color of solids, the reason for the transparency of some materials and the opacity of others, and the physical understanding of such processes as the recording of a photographic image and the operation of a solid state laser. In every instance we shall see that there is a unifying concept that aids in our interpretation of the various phenomena, namely, that most of the optical properties result from the interaction of electromagnetic radiation with the electrons in a material. On this basis we can interpret optical properties in terms of what we know of electronic structure and how factors such as atomic bonding and impurity atoms influence the electronic structure.

We begin our discussion with a few brief descriptive definitions of important terms and then turn our attention to the examination of specific properties. Foremost in this discussion will be an emphasis of the physical principles involved rather than a detailed discussion of the interaction of electromagnetic radiation with electrons. This format is necessitated by the complex nature of the interaction. In most instances it is necessary to resort to quantum mechanics for a complete description of the important events and, as usual, these quantum mechanical calculations convey little physical meaning.

15-2 BASIC DEFINITIONS

Electromagnetic Spectrum

The electromagnetic spectrum is shown in Fig. 15-1. The range of wavelengths λ extends from high-energy gamma rays ($\lambda \approx 10^{-10}$ cm) to the long wavelength radio waves ($\lambda \approx 10^4$ cm). For most of our discussion we shall be concerned with the visible or nearly visible spectrum ($\lambda \approx 10^{-4}$–10^{-5} cm), although the conclusions we reach will apply to the interaction of materials with most electromagnetic radiation.

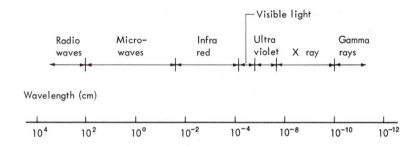

FIG. 15-1 The electromagnetic spectrum

Radiation can be obtained either as *monochromatic* radiation (single wavelength) or *polychromatic* radiation (continuous range of wavelength). Blue light with a wavelength equal to 4500 A is monochromatic, while white light is polychromatic. Ordinary, white light is composed of wave trains (photons) of different wavelengths with a random phase relationship. Sources of monochromatic radiation in which the phase relationship between individual photons is controlled are said to be *coherent* sources. Examples of coherent beams of electromagnetic radiation include radio waves emitted from an antenna and the more recently developed laser beams.

Absorption and Emission

It has long been known that individual atoms or molecules are capable of absorbing and emitting electromagnetic radiation of characteristic wavelengths. The exact wavelengths involved are solely a function of the electronic and nuclear structure of the atom or molecule. The origin of these characteristic absorption and emission spectra can be visualized with the aid of Fig. 15-2, which shows the electronic energy levels of a free sodium atom. We consider first the case of emission (absorption will turn out to be just the reverse). Suppose we somehow raise the energy of the valence electron ($3s$ electrons) to one of the higher allowed values shown in Fig. 15-2. The atom is then in an excited state and can lower its energy by

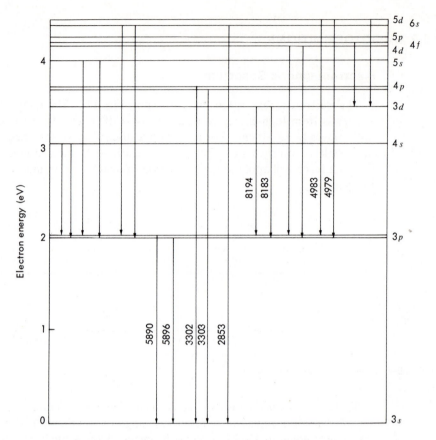

FIG. 15-2 Electron energy level diagram for a sodium atom. The photon wavelengths associated with some of the electronic transitions are given in units of 10^{-8} cm.

having the electron return to the 3s level, emitting a photon in the process. Some of the possible return paths are shown in the figure.

Because the energy levels are discrete, the photon emitted when the electron changes energy levels is characteristic of the sodium atom. Corresponding to each possible transition is a characteristic wavelength in the sodium emission spectrum. For example, the 3p to 3s transitions yield a wavelength of ≈5900 A, the characteristic yellow line in the sodium spectrum. Absorption is just the opposite of emission; i.e., the electron absorbs a photon of sufficient energy to enable the electron to move to a higher allowed energy level. It should be noted that absorption and emission need not take place along the same paths. For example, the 3s electron of the sodium atom could absorb a single photon and go from the 3s to the 5d level, whereas during the emission process one conceivable path is 5d → 3d → 3p → 3s, in which three different photons would be emitted. The actual sequence the transition takes depends on the probability of the transition occurring; determining this probability involves a quantum mechanical calculation.

EXAMPLE 15-1

The absorption edge of a material corresponds to the smallest energy required to break a bond and hence generate free carriers by absorption of photons. What is the lowest energy a photon can have in order to generate free carriers in the materials having the following absorption edges?

(a) ZnS: 3700 A
(b) CdS: 5150 A
(c) GaAs: 8900 A
(d) Si: 11,000 A
(e) InSb: 69,000 A

Solution:

The energy of a photon is given by

$$E = hv = \frac{hc}{\lambda} = \frac{1.24 \times 10^{-4}}{\lambda} \text{ eV}$$

for

(a) ZnS: $E = 3.35$ eV
(b) CdS: $E = 2.4$ eV
(c) GaAs: $E = 1.4$ eV
(d) Si: $E = 1.13$ eV
(e) InSb: $E = 0.18$ eV

When atoms or molecules come together to form liquids or solids, the outermost electronic energy levels become rather diffuse (see Section 11-2) and, correspondingly, some of the absorption and emission lines also become spread out over a range of wavelengths, oftentimes covering an appreciable portion of the visible spectrum. This condition is illustrated in Fig. 15-3. The spreading out of absorption

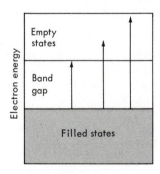
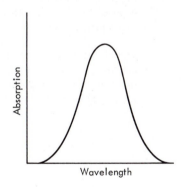

FIG. 15-3   Smearing out of absorption lines due to a continuous range of available energy levels

bands qualitatively explains such features as the color of glasses. Blue glass, for instance, has a strong absorption band in the red and yellow regions and is transparent in the blue region. Conversely, red glass suffers absorption in the blue and green regions, but the red wavelengths are highly transmitted.

Not all absorption or emission lines undergo appreciable alteration when atoms are bonded together; only the outermost electrons participate in the bonding process, and only these electrons have their energy levels changed appreciably. Electronic transitions involving the inner electrons are generally independent of the nature of the interatomic bonds and, correspondingly, any radiation emitted from transitions between these inner levels will be of a fixed wavelength, regardless of the surroundings of the atoms. Even when the atom exists as an impurity in a solid, it can still maintain characteristic emission and absorption spectra.

The processes of emission and absorption are extensively used in techniques of chemical analysis such as spectrographic analysis and atomic absorption spectroscopy. In addition, absorption has helped identify the chemical elements in the atmosphere of far-distant stars. Light from these stars must pass through the atmosphere of the star before it reaches the earth. This radiation suffers absorptions at the characteristic wavelengths corresponding to the elements composing the atmosphere. Thus, by determining the dark lines in the spectrum (absorption maxima), we can use this information to identify the elements in the solar atmospheres. This is how we have found that the chemical elements in the sun and stars are the same as those found on earth.

Thus far we have assumed that all electronic transitions are associated with either the emission or absorption of a photon. In practice this is not the case, and there exists a number of nonradiative processes by which electronic transitions can occur. For example, the impact of an electron with an atom or the collision between two atoms can cause electronic transitions to occur without the emission or absorption of a photon.

EXAMPLE 15-2

Suppose a material exhibits a strong absorption of blue light and appears red when exposed to white light. Schematically illustrate a series of electronic transitions that could produce these characteristics.

Solution:

The blue component is absorbed, thereby raising electrons to some high level. The electron energy then decreases by a non-radiative process to a metastable state and finally returns to the ground state, emitting a photon with a wavelength corresponding to red light (see accompanying illustration).

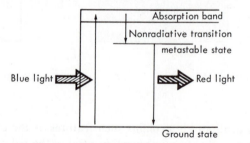

We have just considered what happens when the energy (frequency) of an incident photon corresponds closely to the difference between two allowed electronic energy levels. An electron at the lower energy can absorb the photon and be excited to the higher level (if it is empty), resulting in a characteristic absorption line. But what happens when the energy of the incident photon is appreciably different from that of an allowed electronic transition? We shall now consider this problem and show that the answer will lead to a qualitative understanding of the phenomena of *scattering*, *reflection*, and *refraction*.

Suppose that an electromagnetic wave is incident on a material. Even if the photon cannot produce a discrete electronic transition, there will still be an interaction with the atoms comprising the material. The alternating electric field of the incident wave will excite any charged particles (electrons, ions, dipole molecules) within the material into oscillation. Since the most important oscillations are those involving electrons, we shall focus our attention on their behavior. The incident radiation forces the electrons to oscillate at the frequency of the incoming wave while essentially not changing their energy. As the electrons oscillate, they produce radiation that is of the same frequency as the incident beam and that travels in all directions. The intensity of this scattered radiation is a function of the difference between the frequency of the incident radiation and the frequencies corresponding to the possible discrete electronic transitions. That is, for a given atom the scattered intensity is a function of the wavelength of the incident radiation. This simple fact explains such common features as the blue of the sky and red color of sunsets. The elements comprising our atmosphere scatter blue light to an appreciably greater extent than red light. Thus, the color of the sky, which is really just the scattered radiation from our atmosphere, appears blue, while the sunset, which is sunlight with most of the blue component scattered away, appears red.

The scattering of incident radiation, whether it results from absorption and emission or induced electron oscillation, results in the processes of reflection and refraction. The reflected beam is the backward-scattered radiation, while the refracted beam is the sum of the scattered and incident radiation in the forward direction.

The reflected beam does not originate solely from the atoms at the surface. The backward-scattered radiation comes from the interior of the sample as well, but the geometrical effect is equivalent to a reflection from the surface.

The refracted beam appears to travel at a different velocity in different materials. This is due to the fact that the interaction of the forward-scattered and incident beams produces a phase shift in the resultant refracted beam. This phase shift can be physically interpreted as a difference in velocity of the incident and refracted beams. This concept leads to the definition of the refractive index n, which is the ratio of the velocity of light in vacuum, c, to the velocity of light in the material, v, or

$$n = \frac{c}{v} \tag{15-1}$$

As might be expected, the refractive index is dependent on the strength of the scattered beam, which is in turn dependent on the number of scattering centers (electrons) per unit volume. Crystals with dense atomic packing or elements of high atomic number generally have the highest index of refraction. Also, because the scattered intensity is a function of wavelength, the index of refraction varies with wavelength for a given material. Some typical examples of refractive indices are shown in Table 15-1.

TABLE 15-1

Refractive Indices for Visible Radiation

Composition	Average Refractive Index
Silica glass, SiO_2	1.46
Soda-lime-silica glass	1.51
Borosilicate (Pyrex) glass	1.47
NaF	1.32
LiF	1.39
Quartz, SiO_2	1.55
MgO	1.74
Al_2O_3	1.76
PbO	2.61
SiC	2.68
TiO_2	2.71
PbS	3.91

The refractive index in crystals is not necessarily the same in all crystallographic directions (i.e., it is not always an isotropic property). Only in cubic materials and amorphous materials such as glasses is the radiation transmitted with an equal velocity in all directions. Crystals with lower symmetry—optically anisotropic materials such as $CaCO_3$, SiO_2, and H_2O—exhibit the unique property that the incident beam is split into two separate beams upon transmission through the material (see Fig. 15-4). This phenomenon, known as *birefringence*, comes about

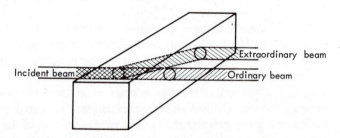

FIG. 15-4   Schematic illustration of double refraction or birefringence

because of the local differences in atomic packing in certain directions and the resultant differences in scattering along these directions. One beam, known as the *ordinary beam*, is the continuation of the incident beam, while the second beam, or *extraordinary beam*, is displaced by an amount related to the thickness of the crystal and the crystallographic differences in *n*. Generally, the difference in refractive indices is small (0.001–0.01) and the displacement between the two beams is also small. However, materials such as $CaCO_3$, with a Δn of 0.17, exhibit rather appreciable separation of the two beams. All optically anisotropic materials have a particular crystallographic direction along which double refraction does not occur. This direction is known as the *optic axis.* The absence of double refraction is related to the fact that atoms are located symmetrically about this axis; consequently, the scattering process is isotropic. Examples of optic axes are the c axes in both hexagonal and tetragonal crystals.

One unique characteristic of optically anisotropic materials is that the ordinary and extraordinary beams are *polarized;* that is, each beam consists of electromagnetic radiation in which the electric field vectors are all parallel. The angle between electric field vectors in the two beams is 90°C. Some materials also have the property of preferentially absorbing radiation that is polarized in a certain direction. In these materials either the ordinary or extraordinary beam will be preferentially absorbed, and the resultant transmitted beam will be strongly polarized. Examples of such materials are Polaroid and tourmaline.

15-3 OPTICAL PROPERTIES AS A FUNCTION OF ATOMIC BONDING

Metallic Bonding

For all wavelengths from the very long radio waves to the middle ultraviolet, metals strongly reflect and/or absorb the incident radiation. Figure 15-5 illustrates this behavior for gold. For shorter wavelengths there is partial transmission of the incident beam. The electronic characteristic of metals that most influences their optical properties is the ready availability of empty energy levels immediately adjacent to the Fermi level. Electrons in the conduction band readily absorb incident photons over a wide wavelength range. Two possibilities exist for the electrons once they absorb energy and move to higher energy levels. First, if the excited electron undergoes a collision with a lattice ion, then the extra energy will be dissipated in the form of lattice vibrations (phonons), and we speak of the incident electromagnetic radiation as being absorbed. For the case where the probability of a collision with an ion is very small, the electron will emit a photon as it drops back to a lower energy level, and the result is a strongly reflected beam.

The characteristic color of some metals is due to the preferential absorption of a portion of the visible spectrum, the details of the absorption process being dependent on the particular electronic structure. For example, in gold there is greater absorption of the green portion of visible spectrum (greater reflection of red and yellow light), giving gold its characteristic appearance. Silver, on the other hand, strongly reflects all portions of the visible spectrum, giving it a white color.

Ionic Bonding

The electronic structure of ionic crystals differs from that of metals in that all the electrons in ionics are tightly bound to atoms and there are no empty energy

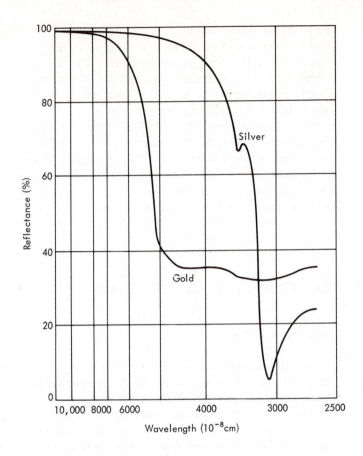

FIG. 15-5   Reflectance spectra for gold and silver

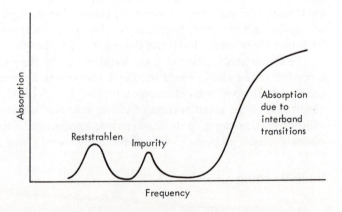

FIG. 15-6   Schematic illustration of absorption spectrum of an ionic crystal

levels immediately adjacent to uppermost filled electron states. Consequently, the electrons cannot absorb photons unless the photon has an energy greater than a certain minimum which is equal to the band gap. Figure 15-6 qualitatively illustrates the absorption properties of an ionic solid, with a few minor exceptions. For all frequencies below that value corresponding to the band-gap energy there is strong transmission of incident radiation. When the photon energy exceeds the band-gap energy, there is a strong probability that the photon will be absorbed, and the incident beam is rapidly attenuated in the material. A typical value of the band-gap energy for ionic crystals is ≈ 6–8 eV (see Table 15-2); consequently, ionics are transparent to the visible spectrum but suffer strong absorption in the ultraviolet.

TABLE 15-2

Position of Absorption Peak Corresponding to Electronic Transition in Pure Alkali Halides

Materials	Absorption Energy (eV)	λ (A)
NaCl	7.8	1580
KCl	7.6	1620
CsCl	7.6	1620
CsBr	6.6	1890
KBr	6.5	1890
KI	5.6	2200
RbI	5.5	2230

The minor absorption peaks that occur at lower frequencies arise from two different interactions. The first, labeled *reststrahlen absorption* in Fig. 15-6, is caused by an interaction of the incident electromagnetic radiation with the ions themselves. This interaction occurs when the frequency of the incident radiation is comparable to the frequency of vibration of the ions. At this point the ions are excited into oscillatory motion by the electric field of the radiation, they suffer collisions with neighboring ions, and the energy of the incident beam is dissipated in the form of thermal energy.

The absorption peak, labeled *impurity absorption* in Fig. 15-6, is caused by impurities or lattice defects in the ionic solid. The presence of such defects locally alters the electronic energy levels. As a simple example, let us consider the case of vacancies in NaCl (Fig. 15-7). A Na vacancy acts effectively as a negative charge (absence of positive charge). Because of this, the outer electrons of the Cl ions adjacent to a Na vacancy are not as strongly bound as normal; that is, it should be easier to excite these electrons to the conduction band because of the effective negative charge of the Na vacancy. In terms of energy levels, these electrons reside at a level somewhere between the valence band and the conduction band (Fig. 15-7). Conversely, a Cl vacancy acts as a positive charge (absence of negative charge) and locally provides unoccupied electronic energy levels that are in the band-gap region and only slightly above the valence band. The presence of these new electronic energy levels gives rise to new absorption levels (often in the visible spectrum) and, consequently, strongly influences the optical properties. Some typical

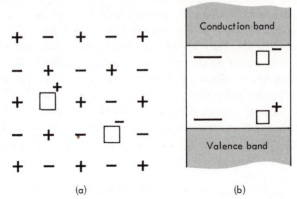

FIG. 15-7   Color centers in ionic crystals: (a) schematic of anion and cation vacancies; (b) electron energy levels arising from these vacancies.

values of absorption energy for electronic states associated with anion vacancies are given in Table 15-3.

Because these absorption levels commonly are in the visible region, they often cause the ionic crystal to exhibit a definite color, the intensity of which depends on the density of defects. Not surprisingly, the defects are often referred to as *color centers*.

In addition to anion and cation vacancies producing localized electronic energy levels, impurity atoms also exhibit characteristic absorption bands in the visible spectrum. Impurity atoms are in fact responsible for the color of many ionic solids. For example, Al_2O_3 is colorless when pure but (1) turns red with small additions of Cr and is known as *ruby*, (2) turns blue with additions of Fe and is known as *sapphire*, (3) turns green when Cl is present and is referred to as *oriental emerald*.

TABLE 15-3

Absorption Energies Associated with Anion Vacancies

Materials	Absorption Energy (eV)
LiF	5.0
NaF	3.6
NaCl	2.7
KF	2.6
KCl	2.2
KBr	2.0
CsCl	2.0

EXAMPLE 15-3   It is observed that initially transparent high-purity KCl turns purple when exposed to high-energy X rays. Also, the purple color gradually fades away and the material becomes transparent if the sample is held at room temperature for a few hours after the X-ray irradiation. The fading process occurs more rapidly if the irradiated KCl is heated above room temperature. Suggest a possible explanation for this behavior.

Solution: The incident X rays have enough energy to produce a fairly high (nonequilibrium) density of anion and cation vacancies. The local electron energy levels associated with these vacancies give rise to absorption and emission spectra which provide the purple color. As the sample is held at room temperature, the nonequilibrium vacancies anneal out and the color fades. If the temperature is raised, the diffusion-controlled annealing process is enhanced and the fading process occurs more readily.

Covalent Bonding

The optical properties of covalent materials are similar to those of ionic solids since both types of bonding result in an energy gap between the filled and empty electronic energy states. Low energy photons cannot excite electrons across this band gap, and, consequently, the covalent materials are transparent to the longer wavelength electromagnetic radiation. Good insulators such as diamond and SiO_2 have large band gaps (≈ 6 eV) and are transparent to radiation of wavelengths down to the ultraviolet region. Semiconductors such as Si, Ge, and GaAs have smaller band gaps (≈ 1 eV) and are transparent only to wavelengths in the infrared region and to longer wavelengths. The metallic luster of these semiconductors comes about by a process similar to that which takes place in metals. Visible radiation excites electrons across the band gap, and the subsequent re-radiation as the electrons drop back to the valence band gives rise to strong reflection.

An example of an absorption-versus-wavelength curve for a semiconductor (silicon) is shown in Fig. 15-8. Strong absorption at a wavelength of about 1.1 μ is due to the excitation of electrons from the valence band to the conduction band. The smaller absorption peaks at longer wavelengths are due to the thin SiO_2 layer on the surface and to impurities (donors and acceptors), which have energy levels somewhere in the forbidden gap region. In covalent materials, impurity absorption

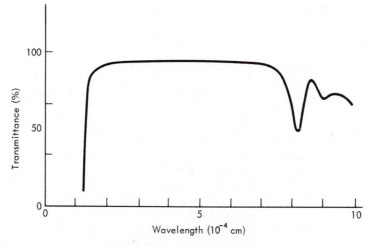

FIG. 15-8   Transmittance of *p*-type silicon at room temperature

in the visible spectrum can be observed only where the band gap is large enough to insure normal transmission of visible wavelengths. Examples of impurity effects here include colored diamonds and the myriad of colored glasses. The most common impurity atoms added to glass to impart color are the transition elements characterized by incomplete d or f electronic shells (usually V, Cr, Mo, Fe, Co, Ni, Cu). Here characteristic absorption occurs either as a result of electronic transitions involving the outer electrons or transitions in the inner incomplete shells. Some typical absorption spectra for different impurities in glass are shown in Fig. 15-9.

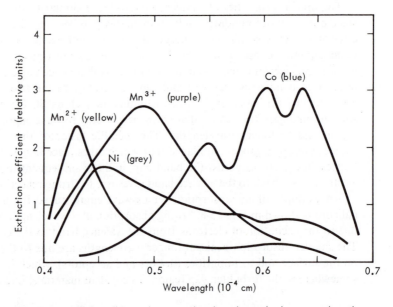

FIG. 15-9 Plot of extinction coefficient (inversely proportional to absorption) vs. wavelength for several ions in glasses

EXAMPLE 15-4   Suppose you have two n-type germanium crystals with different dopant levels. Show by qualitative means how the absorption coefficient varies with wavelength for the low resistivity (high dopant level) and the high-resistivity (low dopant level) crystals.

Solution: The low resistivity crystal has a higher density of conduction electrons and thus a greater absorption coefficient at all wavelengths. At short wavelengths where the incident photon has energy greater than that of the band gap, the density of conduction electron is less important (absorption occurs via interband transitions), and the absorption coefficient is essentially the same for both crystals.

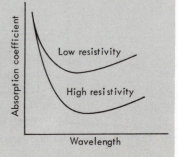

When numerous small, second-phase particles* are present in a glass and the indices of refraction of the particles and matrix are appreciably different, then the glass will usually be opaque. This opacity results from the fact that the incident beam is subject to numerous internal reflections and refractions at the particle-matrix interfaces until it is eventually absorbed. Some typical opacifiers commonly used with silicate glasses are shown in Table 15-4.

TABLE 15-4
Opacifiers Used with Silicate Glasses

Opacifier	$n_{crystal}/n_{glass}$ (n_{glass} = 1.5)
SnO_2	1.33
TiO_2	1.84
Air pores	0.67
CaF_2	0.93
ZrO_2	1.47
$CaTiO_3$	1.57

Van der Waals bonding and Hydrogen Bonding

Materials with Van der Waals bonds, such as Ne, Ar, Kr, CH_4, and Cl_2 are generally insulators with a large band gap. Consequently, these materials are transparent to electromagnetic radiation down to the short ultraviolet, where photons have sufficient energy to cause electronic transitions.

Polymeric materials behave in much the same way as materials with Van der Waals bonds in that the electrons are tightly bound to the atoms. Opacity or color in polymers is usually the result of the presence of a second phase that has been added to the polymer to enhance its properties. One intrinsic aspect of polymers that contributes to opacity is partial crystallinity. The density of amorphous and crystalline phases differs, and hence so does the index of refraction. Thus, radiation incident on a partially crystalline polymer will suffer multiple internal reflections at crystal-amorphous matrix interfaces, and the polymer will have a "milky" appearance. This effect can also be observed if partial crystallinity can be induced by deformation. An initially transparent polymer will become partially opaque as the long-chain molecules become aligned locally due to an applied force.

15-4 LUMINESCENCE

The general process by which an excited electron returns to its ground state by the emission of a photon in the visible or near-visible range is referred to as *luminescence*. We differentiate between different luminescent materials on the basis of how long the electron remains in its excited state before emitting a photon.

---

*"Small particles" means the dimensions of the particles should be of the order of the incident wavelength.

Materials for which this time delay is less than approximately 10^{-8} sec are known as *fluorescent* materials, while materials having longer time delays are called *phosphorescent*. In general, the characteristic emission spectra of luminescent materials are dependent on the presence of impurity atoms, which provide discrete energy levels in the forbidden region between the valance and conduction bands. Each type of impurity atom provides a given emission band, and the presence of more than one type of impurity can extend the emission band over a wide range of wavelengths. Figure 15-10 illustrates the emission spectra of zinc sulfide phosphors with different impurities.

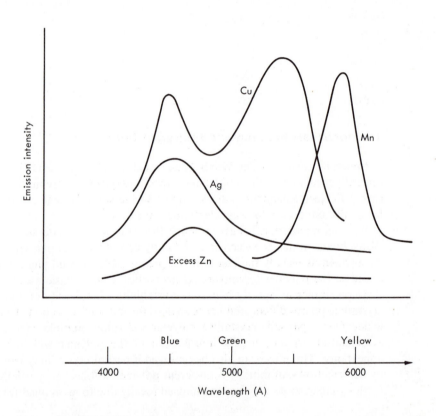

FIG. 15-10 Emission spectra for several zinc-sulfide phosphors

EXAMPLE 15-5

The intensity of luminescence after the excitation radiation has been turned off is called the *afterglow*. If the probability that a conduction electron undergoes a transition back to the ground state is $1/\tau$ per second and the instantaneous density of conduction electrons is $N(t)$, give an expression for the afterglow (intensity) as a function of time.

Solution:

The intensity of luminescence I at any time is equal to the number of photons emitted, or

$$I = -\frac{d N(t)}{dt} = \frac{N(t)}{\tau}$$

Integrating the right-hand side and rearranging, we have

$$N(t) = N_0 \exp\left(-\frac{t}{\tau}\right)$$

where N_0 is the conduction electron density when the excitation light is turned off. The rate of decay of the intensity is then given by

$$I = \frac{N_0}{\tau} \exp\left(-\frac{t}{\tau}\right) = I_0 \exp\left(-\frac{t}{\tau}\right)$$

where I_0 is the initial intensity.

EXAMPLE 15-6

In some phosphors, electrons excited to the conduction band are temporarily trapped in donor levels just below the conduction band before they fall to a lower energy level and emit a photon (see accompanying sketch). If the fractional number of electrons released per second from traps in a phosphor is $10^9 \exp\left(-\epsilon_t/kT\right)$, and phosphorescent intensity is $\frac{1}{2}I_0$ after 0.1 sec at 300°K, what is ϵ_t? What is the time interval required to reach half intensity at 200°K?

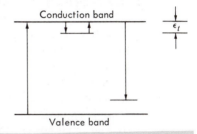

Conduction band

Valence band

Solution:

From Example 15-5 we can write

$$\frac{I}{I_0} = \exp\left(-\frac{t}{\tau}\right)$$

where $1/\tau = 10^9 \exp\left(-\epsilon_t/kT\right)$. Substituting in the appropriate values, we have

$$\frac{1}{2} = \exp\left[-0.1\left\{10^9 \exp\left[-\frac{\epsilon_t}{300}(8.6 \times 10^{-5})\right]\right\}\right]$$

from which $\epsilon_t = 0.48$ eV. To reach half intensity at 200°K, we have

$$\frac{1}{2} = \exp\left[-t\left\{10^9 \exp\left[-\frac{0.48}{200}(8.6 \times 16^{-5})\right]\right\}\right]$$

from which $t = 10^3$ sec.

The different types of luminescence based on the method of electron excitation are summarized in the following paragraphs.

Photoluminescence

In this process the electronic excitation is produced by photon absorption. Typical examples are phosphors used for X-ray detection and in fluorescent lamps. In fluorescent lamps the problem is to convert ultraviolet mercury radiation ($\lambda = 2537$ A) from a glow discharge into visible light. This problem has been solved by coating the inside of the lamp with a complex calcium halophosphate phosphor with antimony and manganese impurities (blue emission and orange emission, respectively). The two different impurities are added to produce a white light.

Cathodoluminescence

Electronic excitation occurs in this case via bombardment by high-energy electrons. Typical phosphors include copper-doped ZnS and (Cd, Zn)S. Applications for this process include cathode-ray oscilloscopes and television tubes.

Electroluminescence

This process occurs in two different situations. First, there is the case in which small particles of a fairly good conductor are distributed throughout a phosphor. Under application of a high alternating electric field (10^4 V/cm), electrons can be liberated from the conductor and can be made to collide with impurities in the phosphor. The ionized impurity will then provide the luminescent action. The second situation in which electroluminescence occurs is where the applied field moves electrons and holes in the vicinity of a *p-n* junction. When the electrons recombine with the holes, radiation is observed. This type of behavior has been observed in numerous semiconductors, including GaAs, GaP, CdS, and SiC. Possible applications for electroluminescence include television screens, light panels, and small power coherent light sources.

15-5 PHOTOCONDUCTIVITY

Photoconductivity is a process by which the electrical conductivity of a semiconductor or insulator is increased by the absorption of electromagnetic radiation. The wavelength of the radiation promoting photoconductivity is often in the visible spectrum, although the same effect is observed with radiation in the infrared, ultraviolet, or X-ray range. In some instances the increase in conductivity can be as much as a factor of 10^{10}, or even larger. The physical basis of photoconductivity is the excitation of electrons from the valence band to the conduction band, creating a pair of free charge carriers (electron plus a hole). The electrical conductivity, therefore, shows a marked increase when the energy of the incident photons is slightly greater than that of the band gap. Some typical examples of photoconductivity are shown in Fig. 15-11. The decrease in photoconductivity with increased photon energy (greater than the band-gap energy) is due to complete absorption of the radiation in a thin surface layer that has an intrinsically lower photosensitivity than the interior.

The important material aspects of photoconductivity are the rate at which electron-hole pairs are produced and the lifetime of these mobile charge carriers. The pair production depends primarily on the wavelength (probability of absorption) and intensity of the incident light. The recombination rate of electron-hole pairs depends on a number of factors that are summarized in Fig. 15-12. One possibility is that direct recombination will occur between an electron in the conduction band and a hole in the valence band. An alternative exists if there are impurities in the material that introduce localized energy levels within the band gap. These localized

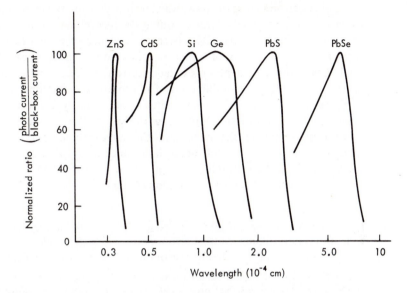

FIG. 15-11 The intrinsic photoconductivity of a number of typical photoconductors as a function of wavelengths

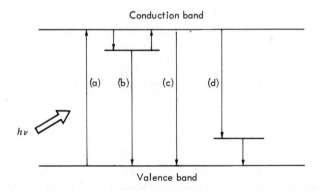

FIG. 15-12 Schematic illustration of electronic processes occurring during photoconductivity: (a) direct electron excitation; (b) trap near conduction band; (c) direct recombination; (d) trap near valence band

levels can act as traps for the photo-excited electrons and holes and affect not only the recombination process but also the instantaneous photo current. Consider briefly two of the cases shown in Fig. 15-12. First, if localized levels exist near the conduction band, these levels may trap electrons from the conduction band, and the time spent in the trap will decrease the effective mobility or velocity of the electron and subsequently reduce the photo current. From the trap the electron either may be excited (by phonons) back to the conduction band or may drop down to the valence band.

If the electron makes the downward transition from the trap to the valence band, then we can think of the trap as aiding the recombination process and refer to it as a *recombination center*. An alternate sequence of events for a recombination center would be capture of a photo-excited hole by a trap near the valence band and then capture of a photo-excited electron, resulting in recombination of the carriers. Generally, recombination of photo-excited electrons and holes occurs in the presence of a recombination center rather than by a direct conduction band-valence band transition. Optimizing the properties of the photoconductor by optimizing the carrier lifetime requires that a recombination center have little probability of capturing an electron (or hole) once it has already captured a hole (or electron). Typical electron lifetimes in CdS range from 10^{-10} to 10^{-2} sec, depending on the nature of the recombination centers.

EXAMPLE 15-7

(a) Suppose you have a germanium single crystal with dimensions $1 \times 1 \times 20$ mm. If the crystal is doped with 10^{16} Sb atoms per cc and the electron mobility is 3600 cm²/V·sec, what is the conductivity of this crystal? Assume all Sb atoms are ionized.

(b) Now suppose the crystal is exposed to a pulse of light such that 10^{14} photons are absorbed. Assuming a quantum efficiency of 1 (that is, each photon absorbed produces one conduction electron), what happens to the conductivity immediately following the irradiation?

Solution:

(a) $\quad$ Conductivity $= \eta e \mu$
$$= (10^{16} \text{ cm}^{-3})(1.6 \times 10^{-19} \text{ C})(3600 \text{ cm}^2/\text{V sec})$$
$$= 5.75 \text{ mho/cm}$$

(b) The volume of the crystal is 0.02 cc. The absorption of 10^{14} photons produces 5×10^{15} electrons/cc, which should increase the conductivity by $5 \times 10^{15}/10^{16}$, or 50%.

Applications of radiation-sensitive materials include radiation detectors (light meters), radiation-controlled electrical switches, and more recently, solar cells. A solar cell is basically a *p-n* junction used to convert light energy to electrical energy. It operates on the principle that incident photons create holes and electrons in the junction region (Fig. 15-13). These holes and electrons are swept to the *p*-type and *n*-type sides, respectively, to lower their potential energy, as discussed in Chapter 13. If there are no external connections to the junction, then the current I_1 produced by the migration of electrons and holes will produce a bias across the junction, which will in turn cause a reverse current flow (I_2) so that there will be no net charge transport; i.e., $I_1 = I_2$. However, if an external load is connected across the junction as shown in Fig. 15-13(b), then part of I_1 flows through this load and the *p-n* junction acts as a solar battery. Silicon has been extensively used for solar batteries because the band gap is large enough so that an extensive portion of the solar spectrum can be used to produce excitations near the junction region. The efficiency of these solar cells in converting solar energy to electrical energy approaches 15%.

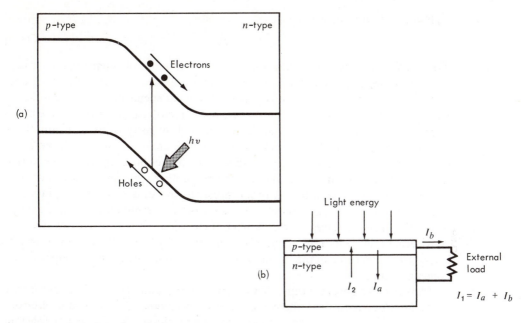

FIG. 15-13 Operation of a solar battery

EXAMPLE 15-8

Derive an expression for relationship between voltage and photo current in a solar cell.

Solution: From Chapter 13 we know the current flow across a *p-n* junction is given by

$$I = I_0 - I_0 \exp\left(\frac{eV}{kT}\right)$$

where I_0 is the dark current (no incident radiation). If light is incident on the junction, a photo current I_1 is produced and the above expression can be written

$$I = I_0 + I_1 - I_0 \exp\left(\frac{eV}{kT}\right)$$

If the external voltage is zero, then we can express the induced photo voltage V in terms of the photo current. Rearranging terms, we have

$$V = \frac{kT}{e} \ln\left(\frac{I_1}{I_0} + 1\right),$$

or a logarithmic relationship between V and I_1 when I_1/I_0 is much greater than 1.

15-6 STIMULATED EMISSION—MASERS AND LASERS

Thus far in our discussions we have listed several optical properties that result from the electronic absorption and emission of photons. We have treated the general case where an electron that changes energy levels by an amount ΔE emits a photon of frequency ν given by

$$\Delta E = h\nu \qquad\qquad (15\text{-}2)$$

The emission process has been assumed to be a random phenomenon. That is, if we have electrons excited to a high energy state, we have assumed that these electrons make the transition to lower energy levels independently of one another in a manner such that (1) there is no definite phase relationship between the emitted photons (noncoherent beam), and (2) the emitted intensity is just the sum of the intensities of the individual photons or is proportional to the sum of the emission centers. In *lasers* and *masers* these conditions no longer hold. Instead, we find that the emitted photons are all in phase with one another (coherent beam), such that the emission intensity (which is proportional to the square of the amplitude) is now proportional to the square of the sum of the amplitudes or, equivalently, proportional to the square of the number of emission centers. Obviously, if we are considering a large number of emitting centers ($>10^{16}$ per cc), then there will be a very great difference in the intensity of a noncoherent beam when compared to a coherent beam.

There are two basic concepts that we must understand before we can appreciate how a laser operates. First, there must be some mechanism that stimulates electrons to make their downward transition in a coherent fashion. This process of coherent emission, or *stimulated emission*, is illustrated in Fig. 15-14. Consider an electron

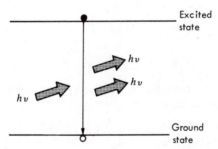

FIG. 15-14 Schematic illustration of the stimulated emission of a photon of energy $h\nu$.

in an excited state occupying an energy level ΔE above the ground state. An incident photon, of frequency corresponding exactly to $\nu = \Delta E/h$, can stimulate the excited electron to make the downward transition and emit a photon of exactly the same energy, frequency, and phase as the incident photon. There is no obvious physical explanation why the incident radiation stimulates the downward transition. The explanation involves a quantum mechanical consideration. Physically, we can imagine that the incident photon helps to push the excited electron down to the lower state. The resultant emission of a photon in phase with the incident photon gives an amplification, since there now are two photons, whereas initially there was only one. This process of amplification is the origin of the words *laser* and *maser* [*l*ight (or *m*icrowave) *a*mplification by the *s*timulated *e*mission of *r*adiation].

The second basic concept is that there must be more electrons in the higher energy nonequilibrium electronic state than in the lower energy state. The reason for this second condition is as follows. For successful operation of the laser we require that some of the emitted radiation be able to escape from the material. This can happen only if there is a greater probability of an atom emitting rather than absorbing a photon, and this will be the case only if a greater fraction of the atoms is in the high energy state.

Let us examine two simple models that qualitatively show how we might "invert" the population so that more electrons will be in the high energy state than the low energy state. Two examples of the *three-level laser* are shown in Fig. 15-15, with

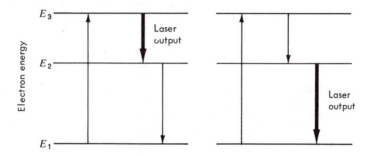

FIG. 15-15   Examples of three-level lasers

the three energy levels involved labeled E_1, E_2, and E_3. In the first example, population inversion can be achieved if the electron lifetime in level E_3 is much longer than that in E_2. That is, the pumping signal (incident radiation) will excite electrons up to level E_3, and since the $E_3 \rightarrow E_2$ transition is much slower than the $E_2 \rightarrow E_1$ transition, we can achieve a much higher electron population in E_3 than in E_2. Subsequent stimulated emission of the electrons in E_3 produces the laser effect. The second three-level model is similar to the first, with the exception that $E_3 \rightarrow E_2$ is much more rapid than $E_2 \rightarrow E_1$, so that a population inversion is built up between levels E_2 and E_1. In this case the laser action comes from the $E_2 \rightarrow E_1$ transition.

The *four-level laser* shown in Fig. 15-16 is qualitatively similar to the three-level laser models. In the four-level model the $E_4 \rightarrow E_3$ and $E_2 \rightarrow E_1$ transitions are rapid, while the $E_3 \rightarrow E_2$ transition is slow. Thus, we build up a population inversion between levels E_3 and E_2, obtaining the laser action from stimulated emission of electrons in the E_3 level.

Thus far we have not considered the origin of the energy levels giving rise to the laser action. In solid state lasers these energy levels are usually associated with impurity atoms such as Cr in Al_2O_3 in the ruby laser. Laser action can also be obtained from gases, where one particular type of gas atom (or molecule) provides the necessary energy levels (e.g., Ne in the He-Ne gas laser). Some typical solid state laser and maser materials are given in Table 15-5.

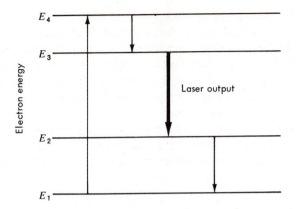

FIG. 15-16 Schematic of the operation of a four-level laser

TABLE 15-5

Laser and Maser Materials with Emitted Wavelengths

Matrix	Impurity Atoms	Wavelength (A)
Al_2O_3	Cr	6,943
CaF_2	U	26,000
CaF_2	Sm	7,100
SrF_2	U	26,000
SrF_2	Nd	10,600
SrF_2	Tm	7,100
$CaWO_4$	Nd	10,600
$CaWO_4$	Pr	10,500
$CaWO_4$	Ho	20,500
Glass	Nd	10,600
Glass	Yb	10,100

To get some feeling for how an actual laser operates, let us examine the ruby laser, which consists of a single crystal of Al_2O_3 with Cr ions as substitutional atoms. The energy band structure associated with the Cr ions is shown in Fig. 15-17. Green and yellow light have sufficient energy to excite electrons to the E_3 level. The $E_3 \rightarrow E_2$ transition is rapid, creating a population inversion between levels E_2 and E_1. The energy difference between E_2 and E_1 corresponds to a wavelength of 6943 A (red light) and accounts for the characteristic red fluorescence of ruby. The actual operation of the ruby laser is shown schematically in Fig. 15-18. The ruby crystal is in the form of a cylinder about 4 cm long and 0.5 cm in diameter. The ends have been carefully polished to create parallel, optically flat faces. One end has been silvered to act as a mirror, while the other end is partially silvered so as to transmit part of an incident beam and reflect part of it. A xenon flash lamp is used to pump the electrons up to level E_3. When a population inversion is achieved, the stimulated emission process is initiated by the random, spontaneous emission of a photon as an electron makes the $E_2 \rightarrow E_1$ transition. As

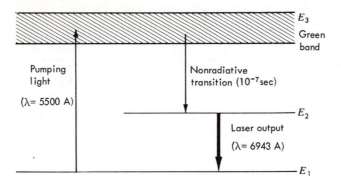

FIG. 15-17 Electron energy level diagram for a ruby laser

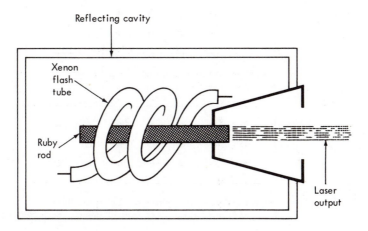

FIG. 15-18 Schematic illustration of a ruby laser in operation

this photon travels along the length of the crystal, it stimulates other $E_2 \longrightarrow E_1$ transitions and produces a coherent beam of light. This beam is completely reflected at one end of the rod and partially reflected at the other, so that a portion of the beam makes many traverses along the length of the crystal, increasing the number of Cr ions stimulated to emit radiation. The laser output comes from the transmitted portion of the beam at the half-silvered end. This process is illustrated in Fig. 15-19.

The emitted beam is not only coherent and of high intensity; it also has very small divergence. The small divergence results from the fact that for a photon to stimulate appreciable intensity, it has to travel back and forth along the length of the crystal many hundreds of times. Only those photons emitted nearly exactly parallel to the cylinder axis can make this many traverses without passing out the side of the crystal. To give some idea of the divergence of the emitted beam, a ruby laser directed at the moon would have a cross section of less than 2 mi by the time it reached the lunar surface.

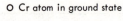

○ Cr atom in ground state

● Cr atom in excited state

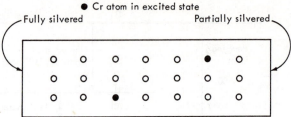

(a) At equilibrium

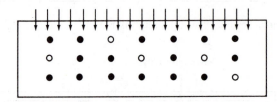

(b) Excitation from flash lamp and nonequilibrium
population inversion

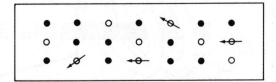

(c) Photons spontaneously emitted parallel to crystal axis
travel back and forth. Other nonaxial photons leave crystal at sides.

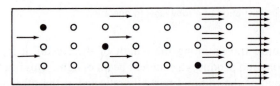

(d) Axial beam is reflected numerous times, stimulates
emission of other photons, increases in intensity, and is
partially transmitted out of crystal at partially-silvered
end.

FIG. 15-19 Illustration of laser action in a ruby crystal

Because the laser beam is monochromatic and highly directional, it can be focused with ordinary glass lenses to a spot size of the order of the wavelength of the light. The power density carried in one of these focused beams can easily exceed 10^8 W/cm^2. This is enough energy to melt and vaporize any known material. However, there are many problems to be solved before a laser can deliver power of this magnitude on a continuous basis. In general, solid state lasers must be operated on a pulse basis to prevent overheating. The pulse lasts for only about 10^{-6} sec. Gas lasers and some solid state lasers can be operated on a continuous basis but at relatively low power levels.

One type of laser that can operate on a continuous basis is the diode laser. If we cause a current to flow through a *p-n* junction, then we can cause a population inversion with a high density of electrons in the conduction band of the *p*-type material and holes in the *n*-type material. Stimulated emission can then accompany the electron transitions to the lower levels. A schematic drawing of a GaAs diode laser is shown in Fig. 15-20. The unique aspect of diode lasers is their efficiency,

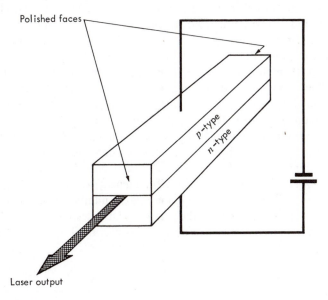

FIG. 15-20 Illustration of a GaAs diode laser

since practically all the electrical energy can be converted to optical radiation. This contrasts with the optically pumped lasers in which most of the incident radiation is lost as heat and only a very small fraction is used to invert the electronic population.

Applications for lasers beams include the following:
1. Use in communications to transmit information in the same way radio waves transmit information. Because of the high frequency and spectral purity of laser radiation, a large number ($\approx 10^8$) of radio or TV channels could be contained in the visible spectrum alone.
2. Use in medical operation or problems involving materials analysis where intense heating is important.
3. Use in photography, especially the new area of holography (information of three dimensions contained on a two-dimensional image).
4. Use in high-precision measurements.

15-7 PHOTOGRAPHIC IMAGES

The ability to record a photographic image depends on the ability of electromagnetic radiation to cause a chemical reaction, such as the decomposition of a silver halide (AgBr or AgI.) The photographic process actually consists of three separate steps. First, the film, which is merely a mixture of small grains of AgBr suspended in gelatin and supported on a suitable substrate (cellulose acetate or glass), is exposed. Incident photons cause the decomposition of AgBr into metallic Ag and Br_2 gas in the following general sequence. The photon produces an electron and a hole in the AgBr crystal. The hole combines with a Br^- ion to form a neutral Br atom, and the electron combines with an interstitial Ag^+ ion to form a neutral Ag atom. Both of these reactions take place at the surface of the AgBr grain. With the absorption of more photons, the process continues with Br_2 molecules escaping from the surface and the formation of a small nucleus of pure Ag. After exposure, AgBr crystals that have absorbed photons contain a small Ag particle, while those AgBr grains that have "seen" no incident photons remain pure AgBr. The second step in the process is to chemically treat the material so that those grains of AgBr with a small Ag nucleus are reduced to pure Ag. This is the *developing* process. The final step consists of *fixing* the film, or dissolving all the remaining AgBr crystals. The final negative, therefore, contains clear areas (no Ag) where the incident photon intensity was zero, and appears opaque or black where the intensity was high and a large number of small Ag crystals had formed. The dark appearance of the small Ag particles results from the fact that any incident radiation is reflected numerous times until it is almost completely absorbed.

PROBLEMS

15-1 The following table lists four insulators and semiconductors, their approximate band gaps, and their color. Explain why each material has its characteristic appearance.

Semiconductor	Energy Gap (eV)	Color
Diamond	5.6	Colorless
Silicon carbide	3.1	Canary yellow
Sulfur	2.4	Yellow
Silicon	1.1	Opaque

15-2 An absorption-versus-wavelength curve for an insulator indicates a strong absorption maximum at a wavelength equal to 1500 A. Calculate the band gap for this material.

15-3 Discuss the influence of a large hydrostatic pressure on the optical properties of a sample of polystyrene that is initially in the rubbery state and has zero birefringence.

15-4 An initially transparent sample of NaCl is heated in the presence of Na vapor and develops a yellow color. Discuss the reason for this color change.

15-5 The most valuable diamonds are those with a slight bluish tint. If the band gap in a diamond is 6 eV and the wavelength of blue light is ~ 4500 A, does it seem likely that the blue tint comes from an intrinsic feature of the diamond lattice or from impurities in the diamond structure?

15-6 The ability of ordinary paints to impart color is due to the reflection at the interface between two transparent media (matrix and dispersion of small particles) of different indices of refraction, giving the paint its reflectivity and color. If the binder or matrix is completely transparent, describe the differences in the small particles (pigment) used to produce white and red paint.

15-7 Suppose the lifetime τ of an electron in the conduction band of GaAs is 10^{-8} sec. If we assume the electron's motion can be approximated to Brownian-like motion with the average mean square displacement $\langle r^2 \rangle$ given by

$$\langle r^2 \rangle = \tfrac{2}{3} l v \tau$$

where l is the distance between collisions (≈ 10 A) and v is the velocity (assume equal to thermal velocity), what is the average displacement of the excited electron during its lifetime?

15-8 An ionic crystal with a high density of color centers can be made colorless by irradiation with the same wavelength light that corresponds to the excited state of the electron-vacancy pair. As the color disappears, it is observed that the electrical conductivity increases. Explain the loss of color and the increase in conductivity.

15-9 Suppose a ruby laser with output intensity of I_1 is used to stimulate the output of a second ruby laser which normally has an output of I_2. Express the intensity of the "double" laser beam in terms of I_1 and I_2.

15-10 Transparent plastic drinking cups are often observed to become "cloudy" when squeezed or deformed elastically. Explain the origin of this loss of transparency.

15-11 Show with the aid of an electron energy level diagram the relative positions of energy levels corresponding to anion and cation vacancies in KCl.

15-12 Suppose you have a piece of green glass. On the same graph plot the fraction of incident intensity that is (a) absorbed and (b) transmitted as a function of incident wavelength.

15-13 Can a *pure* semiconductor look green in white light?

15-14 Suppose you wanted to produce a photographic film that was responsive to all wavelengths in the visible spectrum. How might you go about doing this?

15-15 Imagine that a ruby laser could be designed to operate with a pulsed output of 10^{10} W over a period of 10^{-4} sec. If all this energy were incident on an area of 1 cm^2 of a body, calculate the local stress exerted on the body due to the incident beam.

BIBLIOGRAPHY

L. A. AZAROFF and J. J. BROPHY, *Electronic Processes in Materials*, Ch. 14. New York: McGraw-Hill, 1963.

R. H. BUBE, *Photoconductivity of Solids*. New York: John Wiley, 1960.

A. J. DEKKER, *Solid State Physics*, Ch. 15–16. Englewood Cliffs, N.J.: Prentice-Hall, 1957.

G. DIENES and G. VINEYARD, *Radiation Effects in Solids*. New York: Interscience, 1957.

W. D. KINGERY, *Introduction to Ceramics*, Ch. 15. New York: John Wiley, 1960.

C. KITTEL, *Introduction to Solid State Physics*, Ch. 17–18. New York: John Wiley, 1956.

T. S. MOSS, *Optical Properties of Semiconductors*. London: Butterworth, 1961.

A. L. SCHAWLOW, "Optical Masers," *Scientific American* **204** (1961), p. 52.

A. L. SCHAWLOW, "Advances in Optical Masers," *Scientific American* **209** (1963), p. 34.

W. E. SPICER, "Effect of Electromagnetic Radiation," in *An Atomistic Approach to the Nature and Properties of Materials*, ed. J.A. Pask. New York: John Wiley, 1967.

Appendix

Appendix

Electronic Structure of the Elements

Atomic Number	Element	Electron Configuration			
1	H	$1s^1$			
2	He	$1s^2$			
3	Li	$1s^2$ $2s^1$			
4	Be	$1s^2$ $2s^2$			
5	B	$1s^2$ $2s^2$ $2p^1$			
6	C	$1s^2$ $2s^2$ $2p^2$			
7	N	$1s^2$ $2s^2$ $2p^3$			
8	O	$1s^2$ $2s^2$ $2p^4$			
9	F	$1s^2$ $2s^2$ $2p^5$			
10	Ne	$1s^2$ $2s^2$ $2p^6$			
11	Na				$3s^1$
12	Mg				$3s^2$
13	Al				$3s^2$ $3p^1$
14	Si	Ne core			$3s^2$ $3p^2$
15	P				$3s^2$ $3p^3$
16	S				$3s^2$ $3p^4$
17	Cl				$3s^2$ $3p^5$
18	Ar	$1s^2$ $2s^2$ $2p^6$			$3s^2$ $3p^6$

Electronic Structure of the Elements (Cont'd)

Atomic Number	Element		Electron Configuration
19	K	Ar core	$4s^1$
20	Ca		$4s^2$
21	Sc		$3d^1\ 4s^2$
22	Ti		$3d^2\ 4s^2$
23	V		$3d^3\ 4s^2$
24	Cr		$3d^5\ 4s^1$
25	Mn		$3d^5\ 4s^2$
26	Fe		$3d^6\ 4s^2$
27	Co		$3d^7\ 4s^2$
28	Ni		$3d^8\ 4s^2$
29	Cu		$3d^{10}\ 4s^1$
30	Zn		$3d^{10}\ 4s^2$
31	Ga		$3d^{10}\ 4s^2\ 4p^1$
32	Ge		$3d^{10}\ 4s^2\ 4p^2$
33	As		$3d^{10}\ 4s^2\ 4p^3$
34	Se		$3d^{10}\ 4s^2\ 4p^4$
35	Br		$3d^{10}\ 4s^2\ 4p^5$
36	Kr		$1s^2\ 2s^2\ 2p^6\ 3s^2\ 3p^6\ 3d^{10}\ 4s^2\ 4p^6$
37	Rb	Kr core	$5s^1$
38	Sr		$5s^2$
39	Y		$4d^1\ 5s^2$
40	Zr		$4d^2\ 5s^2$
41	Nb		$4d^4\ 5s^1$
42	Mo		$4d^5\ 5s^1$
43	Tc		$4d^6\ 5s^1$
44	Ru		$4d^7\ 5s^1$
45	Rh		$4d^8\ 5s^1$
46	Pd		$4d^{10}$
47	Ag		$4d^{10}\ 5s^1$
48	Cd		$4d^{10}\ 5s^2$
49	In		$4d^{10}\ 5s^2\ 5p^1$
50	Sn		$4d^{10}\ 5s^2\ 5p^2$
51	Sb		$4d^{10}\ 5s^2\ 5p^3$
52	Te		$4d^{10}\ 5s^2\ 5p^4$
53	I		$4d^{10}\ 5s^2\ 5p^5$
54	Xe		$1s^2\ 2s^2\ 2p^6\ 3s^2\ 3p^6\ 3d^{10}\ 4s^2\ 4p^6\ 4d^{10}\ 5s^2\ 5p^6$
55	Cs	Xe core	$5s^2\ 5p^6\quad 6s^1$
56	Ba		$5s^2\ 5p^6\quad 6s^2$
57	La		$5s^2\ 5p^6\ 5d^1\ 6s^2$
58	Ce		$4f^2\ 5s^2\ 5p^6\quad 6s^2$
59	Pr		$4f^3\ 5s^2\ 5p^6\quad 6s^2$
60	Nd		$4f^4\ 5s^2\ 5p^6\quad 6s^2$
61	Pm		$4f^5\ 5s^2\ 5p^6\quad 6s^2$
62	Sm		$4f^6\ 5s^2\ 5p^6\quad 6s^2$
63	Eu		$4f^7\ 5s^2\ 5p^6\quad 6s^2$
64	Gd		$4f^7\ 5s^2\ 5p^6\ 5d^1\ 6s^2$
65	Tb		$4f^9\ 5s^2\ 5p^6\quad 6s^2$
66	Dy		$4f^{10}\ 5s^2\ 5p^6\quad 6s^2$

Electronic Structure of the Elements (Cont'd)

Atomic Number	Element	Electron Configuration
67	Ho	$4f^{11}\ 5s^2\ 5p^6\qquad\qquad\qquad 6s^2$
68	Er	$4f^{12}\ 5s^2\ 5p^6\qquad\qquad\qquad 6s^2$
69	Tm	$4f^{13}\ 5s^2\ 5p^6\qquad\qquad\qquad 6s^2$
70	Yb	$4f^{14}\ 5s^2\ 5p^6\qquad\qquad\qquad 6s^2$
71	Lu	$4f^{14}\ 5s^2\ 5p^6\ 5d^1\qquad\quad 6s^2$
72	Hf	$4f^{14}\ 5s^2\ 5p^6\ 5d^2\qquad\quad 6s^2$
73	Ta	$4f^{14}\ 5s^2\ 5p^6\ 5d^3\qquad\quad 6s^2$
74	W	$4f^{14}\ 5s^2\ 5p^6\ 5d^4\qquad\quad 6s^2$
75	Re	$4f^{14}\ 5s^2\ 5p^6\ 5d^5\qquad\quad 6s^2$
76	Os	$4f^{14}\ 5s^2\ 5p^6\ 5d^6\qquad\quad 6s^2$
77	Ir	$4f^{14}\ 5s^2\ 5p^6\ 5d^9$
78	Pt	$4f^{14}\ 5s^2\ 5p^6\ 5d^9\qquad\quad 6s^1$
79	Au	$4f^{14}\ 5s^2\ 5p^6\ 5d^{10}\qquad 6s^1$
80	Hg	$4f^{14}\ 5s^2\ 5p^6\ 5d^{10}\qquad 6s^2$
81	Ti	$4f^{14}\ 5s^2\ 5p^6\ 5d^{10}\qquad 6s^2\quad 6p^1$
82	Pb	$4f^{14}\ 5s^2\ 5p^6\ 5d^{10}\qquad 6s^2\quad 6p^2$
83	Bi	$4f^{14}\ 5s^2\ 5p^6\ 5d^{10}\qquad 6s^2\quad 6p^3$
84	Po	$4f^{14}\ 5s^2\ 5p^6\ 5d^{10}\qquad 6s^2\quad 6p^4$
85	At	$4f^{14}\ 5s^2\ 5p^6\ 5d^{10}\qquad 6s^2\quad 6p^5$
86	Rn	$1s^2\ 2s^2\ 2p^6\ 3s^2\ 3p^6\ 3d^{10}\ 4s^2\ 4p^6\ 4d^{10}\ 4f^{14}\ 5s^2\ 5p^6\ 5d^{10}\ 6s^2\ 6p^6$
87	Fr	$7s^1$
88	Ra	$7s^2$
89	Ac	$6d^1\ 7s^2$
90	Th	$6d^2\ 7s^2$
91	Pa	$5f^2\ 6d^1\ 7s^2$
92	U	$5f^3\ 6d^1\ 7s^2$
93	Np	$5f^4\ 6d^1\ 7s^2$
94	Pu	$5f^6\ \ \ \ \ 7s^2$
95	Am	$5f^7\ \ \ \ \ 7s^2$
96	Cm	$5f^7\ 6d^1\ 7s^2$
97	Bk	$5f^9\ \ \ \ \ 7s^2$
98	Cf	$5f^{10}\ \ \ \ \ 7s^2$
99	Es	$5f^{11}\ \ \ \ \ 7s^2$
100	Fm	$5f^{12}\ \ \ \ \ 7s^2$
101	Md	$5f^{13}\ \ \ \ \ 7s^2$
102	No	$5f^{14}\ \ \ \ \ 7s^2$
103	Lr	$5f^{14}\ 6d^1\ 7s^2$

Xe core (elements 67–86)

Rn core (elements 87–103)

Physical Properties of Selected Elements at Room Temperature

Element	Atomic Weight	Crystal Structure	Ionic Radius (10⁻⁸ cm)		Density (gm/cm³)	Melting Temperature (°C)	Coefficient of Thermal Expansion (cm/cm °C) × 10⁶	Thermal Conductivity (cal cm/cm² sec °C)	Specific Heat (cal/gm °C)	Young's Modulus (10⁶ psi)
Aluminum	26.98	FCC	Al^{3+}	0.50	2.69	660	23.6	0.53	0.215	10
Antimony	121.76	Rhombohedral	Sb^{5+}	0.62	6.62	630	≈9	0.045	0.049	11.3
Arsenic	74.91	Rhombohedral	As^{5+}	0.47	5.72	817	4.7	—	0.082	1.8
Barium	137.36	BCC	Ba^{2+}	1.35	3.5	714	—	—	0.068	—
Beryllium	9.01	HCP	Be^{2+}	0.31	1.84	1277	11.6	0.35	0.45	42
Bismuth	209.0	Rhombohedral	Bi^{5+}	0.74	9.80	271	13.3	0.02	0.029	4.6
Boron	10.82	Orthorhombic	B^{3+}	0.20	2.34	≈2030	8.3	—	0.309	50
Bromine	79.92	Orthorhombic	Br^-	1.95	3.12	-7.2	—	—	0.070	—
Cadmium	112.41	HCP	Cd^{2+}	0.97	8.65	321	29.8	0.22	0.055	8
Carbon (graphite)	12.01	Hexagonal	C^{4+}	0.15	2.25	3727	≈2	0.057	0.165	6
Cesium	132.91	FCC	Cs^+	1.69	1.90	29	97	—	0.048	0.2
Chlorine	35.46	Tetragonal	Cl^-	1.81	3.2×10^{-3}	-101	—	0.17×10^{-4}	0.116	—
Chromium	52.01	BCC	Cr^{3+}	0.64	7.19	1875	6.2	0.16	0.11	36
Cobalt	58.94	HCP	Co^{2+}	0.72	8.85	1495	13.8	0.16	0.099	30
Columbium	92.91	BCC	Cb^{5+}	0.70	8.57	2468	7.3	0.12	0.065	16
Copper	63.54	FCC	Cu^+	0.96	8.96	1083	16.5	0.94	0.092	17
Fluorine	19.00		F^-	1.36	1.7×10^{-3}	-220	—	—	0.18	—
Gallium	69.72	Orthorhombic	Ga^{3+}	0.62	5.91	30	18	0.08	0.079	1.4
Germanium	72.60	Diamond Cubic	Ge^{4+}	0.53	5.32	937	5.7	0.14	0.073	23
Gold	197.0	FCC	Au^+	1.37	19.32	1063	14.2	0.71	0.031	11.6
Hafnium	178.58	HCP	Hf^{4+}	0.78	13.09	2222	5.9	0.22	0.035	20.0
Indium	114.82	Face Centered Tetragonal	In^{3+}	0.81	7.32	156	33	0.057	0.057	1.6
Iodine	126.91	Orthorhombic	I^-	2.16	4.94	113	93	10.4×10^{-4}	0.052	—
Iridium	192.2	FCC	Ir^{4+}	0.68	22.5	2454	6.8	0.14	0.031	76
Iron	55.85	BCC	Fe^{2+}	0.75	7.87	1536	11.7	0.18	0.11	29
Lead	207.21	FCC	Pb^{4+}	0.84	11.36	327	29.3	0.083	0.031	2
Lithium	6.94	BCC	Li^+	0.60	0.534	181	56	0.17	0.79	1.6

Element	Atomic Weight	Crystal Structure	Ionic Radius (10^{-8} cm)	Density (gm/cm³)	Melting Temperature (°C)	Coefficient of Thermal Expansion (cm/cm °C $\times 10^6$)	Thermal Conductivity (cal cm/cm² sec °C)	Specific Heat (cal/gm °C)	Young's Modulus (10^6 psi)
Magnesium	24.32	HCP	Mg^{2+} 0.65	1.74	650	27.1	0.37	0.25	6.3
Manganese	54.94	Cubic (complex)	Mn^{2+} 0.80	7.43	1245	22.0	—	0.115	23
Mercury	200.61	Rhombohedral	Hg^{2+} 1.10	13.55	-38.4	—	0.0196	0.033	—
Molybdenum	95.95	BCC	Mo^{+6} 0.62	10.22	2610	4.9	0.34	0.066	47
Nickel	58.71	FCC	Ni^{2+} 0.70	8.90	1453	13.3	0.22	0.105	30
Nitrogen	14.01	Hexagonal	N^{3-} 1.71	1.25×10^{-3}	-209.9	—	6×10^{-5}	0.247	—
Oxygen	16.00	Cubic	O^{2-} 1.40	1.43×10^{-3}	-218.8	—	5.9×10^{-5}	0.218	—
Palladium	106.7	FCC	Pd^{4+} 0.65	12.02	1552	11.7	0.17	0.058	16.3
Phosphorous	30.98	Cubic	P^{3-} 2.12	1.83	44.2	125	—	0.177	—
Platinum	195.09	FCC	Pt^{2+} 0.80	21.45	1769	8.9	0.16	0.031	21.3
Potassium	39.10	BCC	K^{+} 1.33	0.86	63.7	83	0.24	0.177	0.51
Rhenium	186.22	HCP	Re^{4+} 0.72	21.04	3180	6.7	0.17	0.033	66.7
Rhodium	102.91	FCC	Rh^{3+} 0.68	12.44	1966	8.3	0.21	0.059	42.5
Selenium	78.96	Hexagonal	Se^{2-} 1.98	4.79	217	37	12×10^{-4}	0.084	8.4
Silicon	28.09	Diamond Cubic	Si^{4+} 0.41	2.33	1410	≈ 5	0.20	0.162	29
Silver	107.88	FCC	Ag^{+} 1.26	10.49	960.8	19.6	1.0	0.056	11
Sodium	22.99	BCC	Na^{+} 0.95	0.97	97.8	71	0.32	0.295	1.3
Sulfur	32.07	Orthorhombic	S^{2-} 1.82	2.07	119.0	64	6.3×10^{-4}	0.175	—
Tantalum	180.95	BCC	Ta^{5+} 0.68	16.6	2996	6.5	0.13	0.034	27
Thallium	204.39	HCP	Tl^{3+} 0.95	11.85	303	28	0.093	0.031	1.1
Thorium	232.05	FCC	Th^{4+} 1.02	11.66	1750	12.5	0.09	0.034	11.2
Tin	118.70	Tetragonal	Sn^{4+} 0.71	7.29	231.9	23	0.15	0.054	6
Titanium	47.90	HCP	Ti^{4+} 0.68	4.51	1668	8.4	0.22	0.124	16.8
Tungsten	183.86	BCC	W^{4+} 0.70	19.3	3410	4.6	0.39	0.033	56.4
Uranium	238.07	Orthorhombic	U^{4+} 0.97	19.07	1132.3	≈ 10	0.071	0.028	24
Vanadium	50.95	BCC	V^{3+} 0.66	6.1	1900	8.3	0.074	0.119	19
Zinc	65.38	HCP	Zn^{2+} 0.74	7.13	419.5	39.7	0.27	0.092	13.4
Zirconium	91.22	HCP	Zr^{4+} 0.80	6.49	1852	5.8	0.21	0.067	13.7

Index

Index